T0281229

Nanoelectronics
A Molecular View

World Scientific Series in Nanoscience and Nanotechnology*

ISSN: 2301-301X

Series Editor-in-Chief: Frans Spaepen (*Harvard University, USA*)

Published

Vol. 7   Scanning Probe Microscopy for Energy Research:
Materials, Devices, and Applications
*edited by Dawn A. Bonnell (The University of Pennsylvania, USA)
and Sergei V. Kalinin (Oak Ridge National Laboratory, USA)*

Vol. 8   Polyoxometalate Chemistry: Some Recent Trends
*edited by Francis Secheresse (Université de Versailles-St Quentin,
France)*

Vol. 9   Handbook of Biomimetics and Bioinspiration: Biologically-Driven
Engineering of Materials, Processes, Devices, and Systems
*edited by Esmaiel Jabbari (University of South carolina, USA),
Deok-Ho Kim (University of Washington, USA),
Luke P. Lee (University of California, Berkeley, USA),
Amir Ghaemmaghami (University of Nottingham, UK) and
Ali Khademhosseini (Harvard University, USA & Massachusetts
Institute of Technology, USA)*

Vol. 10   Pore Scale Phenomena: Frontiers in Energy and Environment
*edited by John Poate (Colorado School of Mines, USA),
Tissa Illangasekare (Colorado School of Mines, USA),
Hossein Kazemi (Colorado School of Mines, USA) and
Robert Kee (Colorado School of Mines, USA)*

Vol. 11   Molecular Bioelectronics: The 19 Years of Progress (2nd Edition)
by Nicolini Claudio (University of Genoa, Italy)

Vol. 12   Nanomaterials for Photocatalytic Chemistry
by Yugang Sun (Temple University, USA)

Vol. 13   Nanoelectronics: A Molecular View
by Avik Ghosh (University of Virginia, USA)

Volume
13

World Scientific Series in
Nanoscience and Nanotechnology

Nanoelectronics
A Molecular View

Avik Ghosh

University of Virginia, USA

World Scientific

NEW JERSEY · LONDON · SINGAPORE · BEIJING · SHANGHAI · HONG KONG · TAIPEI · CHENNAI · TOKYO

Published by

World Scientific Publishing Co. Pte. Ltd.

5 Toh Tuck Link, Singapore 596224

USA office: 27 Warren Street, Suite 401-402, Hackensack, NJ 07601

UK office: 57 Shelton Street, Covent Garden, London WC2H 9HE

Library of Congress Cataloging-in-Publication Data
Names: Ghosh, Avik, 1970– author.
Title: Nanoelectronics : a molecular view / Avik Ghosh (University of Virginia, USA).
Other titles: World Scientific series in nanoscience and nanotechnology ; v. 13.
Description: Singapore ; Hackensack, NJ : World Scientific Publishing Co. Pte. Ltd., [2016] |
    Series: World Scientific series in nanoscience and nanotechnology,
    ISSN 2301-301X ; volume 13 | Includes bibliographical references.
Identifiers: LCCN 2016029077 (print) | ISBN 9789813144491 (hardcover ; alk. paper) |
    ISBN 9813144491 (hardcover ; alk. paper) | ISBN 9789813146228 (softcover ; alk. paper) |
    ISBN 9813146222 (softcover ; alk. paper)
Subjects: LCSH: Nanoelectronics. | Molecular electronics. | Quantum electronics. |
    Electron transport.
Classification: LCC TK7874.84 .G46 2016 (print) | DDC 621.38101/539--dc23
LC record available at https://lccn.loc.gov/2016029077

British Library Cataloguing-in-Publication Data
A catalogue record for this book is available from the British Library.

Desk Editor: Rhaimie Wahap

Printed in Singapore

To my parents, who taught me to dream, and to my wife and kids who help me keep it real

Preface

A popular movie back in the 60s was "Fantastic Voyage", based on a novel by Harry Kleiner, where a group of scientists were miniaturized and injected into the body of an injured scientist to fix his brain damage. The movie had a more modern update for my generation, "Inner Space", also a fun watch. Beyond the specific plot points, what drove these movies was the venue — imagining the action and adventure swirling around in a miniature unfamiliar world — a universe we carry in us every day but rarely get to observe at the level of detail that the movies deftly portrayed.

The emergent fields of nanoscience and nanotechnology draw natural comparisons to those movies. It is after all, the popular view of 'nano' among much of the public, fueled by novels like Michael Crichton's "Prey". But this book is not about injecting nanobots into the human body. The comparison I draw here is less literal, yet technologically more immediate. This book is also about a fantastic voyage — that of an electron coursing through various materials, negotiating impurities, junctions, surfaces, interfaces. With the advent of advanced spectroscopic tools, growth of emergent materials, the evolution of computational modeling, we can now 'peek under the hood' and participate in this incredible journey directly. The screenplay of this movie is also filled with twists and intrigues — the electrons sometimes behaving like classical particles jostling with each other, sometimes as waves interfering, occasionally shaking up the surrounding lattice of atoms by emitting phonons or absorbing photons, rushing to deplete their charges under the action of a gate, repelling each other through Coulomb interactions, correlating charges, locking spins. It is indeed an action-filled script with a busy cast of characters.

Beyond the intellectual exercise of visualizing and deconstructing electron dynamics, what fascinates is that the emergent fields of nanoscience and technology rely greatly on such a molecular understanding of current flow — in chemical reactions, surface processes, metal insulator transitions or ultrasmall switches. Today's semiconductor chips are getting unmanageably hot from the sheer volume of billions of transistors operating together in our smartphones and tablet computers. As a result, they undergo continuous make-overs, in architecture, material embodiment and even physical principles. Not only logic, but memory, sensing and pattern recognition, solid state lighting, thermoelectrics and solar cells are all seeing their own scaling rules lead them naturally towards exploration and redesign driven by their fundamental molecular structures. A thorough understanding of device conduction down to its molecular detail is often essential to understanding, to its continued miniaturization and sustained performance.

I teach two complementary engineering classes fairly regularly. One is a top-down introduction to solid-state electronic devices based on classical Newtonian physics. A large number of traditional devices such as bipolar junction transistors (BJT) and conventional metal oxide semiconductor field effect transistors (MOSFETs), Schottky diodes, high electron mobility transistors (HEMTs), solar cells and photodetectors can be understood in terms of drift-diffusion equations, often reduced to simple circuit diagrams. The other course I teach, for grads/undergrads, is a bottom-up introduction to nanoelectronics. That course introduces quantum transport physics, and is suitable for emerging materials and devices such as tunnel junctions, graphene, spintronics, silicon nanowires and quantum dots.

Three lessons emerged from these courses, plus my own research over the years as a physicist in an engineering department dabbling with quantum chemistry and materials science.

(1) A proper methodology must be introduced that merges the top down and bottom up viewpoints, connecting established CAD tools with more esoteric transport formalisms used by physicists. This is essential since even conventional devices such as commercial MOSFETs have a large fraction of electrons (currently $\sim 70\%$) acting ballistically, requiring us to study processes beyond top down scattering and diffusion.

(2) These connections are better achieved through hands-on examples and toy models that connect chemistry with bandstructure and many-body

physics. Such a combined viewpoint is not within the purview of isolated physics, chemistry, and engineering courses, and is truly interdisciplinary.

(3) At the same time, students must be able to model real materials hands-on and make quantitative and predictive statements.

This book arose from that realization, and aims for a few specific goals.

1. Combining a bottom-up approach to electronics with classical top-down viewpoints popular in device simulation tools. In fact, going a step further by outlining a mosaic of transport theories by drawing hierarchies and interrelationships connecting them — not only between classical and quantum, but also between collective and single particle behavior.

2. Introduce a comprehensive molecular view that can be used to understand not just molecular electronics, but transport through other 'molecules' such as graphene ribbons, magnetic junctions and quantum dots. A viewpoint that can nonetheless scale up to conventional bulk solids.

3. Emphasize the unique physics that emerges as we shrink down our devices — the orbital symmetries, basis sets, quantum effects like exchange (which device theorists typically include in a limited embodiment of Pauli exclusion) start to become dominant rather than dormant.

4. A unified treatment of scattering ranging from elastic impurity to inelastic and incoherent scattering, going between 'quick and dirty' phenomenological treatments to more rigorous 'molecular' effects such as inelastic tunneling spectroscopy or polaron formation.

5. Handling correlation effects in transport, where electrons find clever ways to lock their dynamics and lower their energy costs. Correlation effects are already relevant for today's technology (semiconductor nanocrystal memory relies on Coulomb blockade), but could be even more relevant in future with Mott switches and other correlation-based devices.

6. Provide a lot of 'hands-on' examples (I refer to those as *case studies*) to illustrate the principles by doing, rather than saying. I believe a few good examples save me several pages of uninspired formalism. In the spirit of the 'molecular view', I frequently illustrate my points with prototypical molecules. I use the hydrogen (H_2) molecule to illustrate bonding,

exact diagonalization, configuration interaction and the Gutzwiller approach, and benzene to illustrate bandstructure, resonance, hybridization, Coulomb blockade, gateability and a scale-up to graphene. And I try to provide words and pictures behind the operational equations, so that they make not just mathematical, but intuitive sense.

7. Connecting these studies with fundamentals of low-power computing, bypassing the Boltzmann limit and exploring some prototypical, unconventional switching devices. Through these examples, I seek to bring physical concepts to the engineers, and bring engineering concepts to the physicists.

This book comes at the heels of a remarkable set of volumes. Volume 1, "Lessons from nanoelectronics — A New Perspective on Transport" by Supriyo Datta provides a wonderful introduction to quantum transport of charges, spins and heat. Volume 2, "Near-equilibrium Transport: Fundamentals and Applications" by Mark Lundstrom focuses on low-bias transport. This volume extends these techniques to bring into its fold chemistry, molecular solids and emerging low-power devices. We start with the motivation — both the intellectual pursuit driving physicists (Chapter 1) and the practical device angle relevant for engineers (Chapter 2). Next, we describe the commonly used quantum mechanical tools in Chapters 3-6, the underlying language for these descriptions. I recap quantum physics, quantum chemistry and band theory in Chapters 3-5. In Chapter 6, I introduce second quantization needed to understand correlation effects. In Chapter 7, I introduce the concept of Green's functions, which lie at the heart of the 'bottom-up' transport theory used all over the book.

It is fair to say that most functional electronic devices, chemical and biological processes involve venturing far from equilibrium, requiring a thorough understanding of transport theory. Chapter 8 gives a bird's eye view of the transport mosaic, from classical to quantum. Chapter 9 starts with classical drift diffusion, connecting it rigorously with our intuitive picture of an overdamped Newtonian particle kicked around by thermal white noise. Chapter 10 discusses semiclassical transport, while Chapter 11 takes us to the Nonequilibrium Green's Function (NEGF) relevant for quantum transport. We follow Datta's derivation of the main NEGF equations from simple one-electron quantum mechanics with open, thermalized boundary conditions. We also describe the more mainstream, Keldysh contour theoretical formulation of NEGF. For weak interactions, the two approaches converge.

For stronger interactions, the Keldysh approach allows us to extend established perturbative techniques from equilibrium many body physics such as Feynman diagrams, albeit on a Keldysh contour. For very strong interactions, we may need to improvise non-perturbative approaches that I discuss in Chapter 27.

Chapters 12-13 deal with simple limits of the NEGF equations. For coherent, ballistic flow, the equations reduce to the Landauer formula for a two-terminal device (and Landauer-Büttiker for multi-terminal). For the opposite end of scattering dominated devices, we recover Ohm's Law. The beauty of NEGF is that it allows us to incorporate partial scattering, so we can interpolate between the two limits, and independently study the role of momentum, phase, spin and energy scattering.

The most voluminous part of this book is on the adventures of the electron in its journey. In contrast to other books or monographs on quantum transport, the classification I choose to pursue is 'behavioral'. My aim here is to start with the flow mechanisms that are simplest to understand and model, and then add complexity in incremental layers from one chapter on to another. Accordingly, I start with quantum tunneling in Chapter 14. Beyond a standard WKB approach, I show how to convert a transmission into an actual current across a thin film and how to extend it to small molecular layers. Chapter 15 extends this to tunneling with an additional symmetry index. I discuss how angular symmetry arising from the orbital chemistry of the tunneling electrons can and already has made an impact in oxide technology, how spin selection rules further control the tunneling probability in tunnel magnetoresistance devices, and finally how the behavior of pseudospins controls Klein tunneling and Berry phase in graphene and topological insulator pn junctions.

From tunneling, we move to regular two-terminal conduction across lower dimensional systems, with emphasis on graphene in Chapter 16 and molecular conduction in Chapter 17. Since these systems are small enough to be near ballistic (i.e., electron does not scatter until it reaches the contacts), the machinery of quantum transport becomes indispensable in explaining their measured transport characteristics even qualitatively.

We spend some time talking of gating and switching in Chapter 18, connecting to the device world and its needs. Beyond conventional electrostatic switching, we talk of unconventional switches such as conformational relays in Chapter 19 (more generally, configurational gating), two terminal switches and resistive memory elements in Chapter 20, connecting to neuromorphic systems and stochastic processes such as random telegraph noise. In Chapter 21, we look at the behavior of weakly interacting electrons in presence of a magnetic field, creating skipping orbits and Landau levels, Hofstader butterflies in a lattice, and quantum Hall plateaus. We discuss how one can use noncollinear magnetization between two magnets to generate a torque in order to write information for scalable non-volatile storage.

Researchers working on NEGF often limit themselves to the simplified Landauer equation valid for non-interacting or weakly interacting systems. A significant difference between Landauer and NEGF manifests itself in presence of scattering (Chapters 22-24) and of many-particle interactions (Chapters 25-27). There are phenomenological ways to introduce scattering such as mean free paths and Büttiker probes — and more rigorous descriptions based on the Keldysh formalism that capture much more intricate microscopic details. A cornucopia of phenomena arise from scattering, from thermalization and current saturation to self-heating and phonon bottlenecks, polaronic shifts and vibrational sidebands. My aim in Chapters 22-24 is to connect and illustrate these approaches — the rigorous and the not-so-rigorous. The Caroli/Meir-Wingreen formula serves as a glue between the Landauer-NEGF approaches. One may wonder why a mainstream device physicist needs to venture beyond Landauer theory to do NEGF, especially as devices head towards ballisticity. However, every serious device theorist should at the least include screened Coulomb interactions in Poisson's equation, and some dephasing to kill spurious quantum interference effects (on occasion we also need inelastic phonon scattering) — whereupon an explicit treatment of the nonequilibrium Green's function G^n becomes mandatory.

Correlated systems provide the ultimate challenge in electronic structure theory. The challenge is significantly compounded for nonequilibrium properties like current, which in general turns out to be much harder to calculate in the face of myriads of excitations out of the ground state. One needs to respect the symmetry/antisymmetry properties of the entire multielectron wavefunction while dealing with its two particle Coulomb interactions. Chapter 25 introduces the many body (Fock) space and Chapter 26 introduces the multielectron transport formalism and the challenges it brings, especially with regard to the proper treatment of coherent broadening of one electron transitions in a many-electron manifold. Chapter 27 introduces some of these broadening effects non-perturbatively through the equation of motion technique. We generalize it to solids to study the Mott transition. Importantly, we argue that we can in principle use conventional NEGF to calculate a correlation current, by including the corresponding non-perturbative self-energy matrix into the Meir-Wingreen formula.

Finally in Chapters 28-30, we bring together our equations and overall understanding to explore examples that are relevant to the fundamental physics of low power, subthermal switching. We speculate on how to beat the Boltzmann tyranny to switch a channel efficiently. Only time will tell if any of them end up being practical. However, they are good test cases to apply our equations and explore the interesting physics that allows them, in principle, to beat the aforementioned limit. More excitingly, as we will argue, almost every example in this category requires the full advanced machinery of quantum and/or correlated transport, well beyond conventional drift-diffusion equation.

A project like this involves many I am indebted to. My family — a constant source of inspiration over the years. My teachers in school and college, humble and yet immensely inspiring, who got me started on the road to scientific exploration. My mentors spanning my research career, each with definitive contributions to quantum transport — John Wilkins, Supriyo Datta and Mark Lundstrom — who taught me to think big and think simple. My scientific collaborators over the years, too many to name individually. Various funding agencies and program managers, from NSF to DoD to SRC that provide the much needed and well appreciated financial resources to pursue the research necessary for the studies. My students and postdocs, often the ones silently carrying out much of the actual gruntwork underlying these studies. A special shout out to Carlos Polanco, and his unflagging commitment to reading and rereading, editing and opining at every stage of this project while balancing his thesis on interfacial thermal conductance. A strong thank you to Jianhua Ma for helping secure permissions and final edits. To the publisher, thank you for providing such an exciting venue to collect my thoughts and cumulative wisdom.

Avik Ghosh
University of Virginia

Contents

Part 2 Equilibrium Tools: Quantum Mechanics and Thermodynamics 31

Part 3   Nonequilibrium Concepts: The Nature of Transport Equations   119

Part 6   A Smattering of Scattering: Resistance, Decoherence and Dissipation   327

PART 1

Nano-Electronics:

Cool Science vs Smart Technology

Chapter 1

The intellectual lure — quantum physics in devices

1.1 The I–V menagerie

We begin with the physical motivation, primarily to explore the dynamics of electrons as they journey along a channel. Along the way, electrons can encounter a rich variety of interactions affecting their dynamics, at times acting as classical point particles ricocheting off each other, alternately as quantum waves interfering and even annihilating each other. Electrons can lock their spins in a magnet or correlate their charges in a condensate, in a delicate dance aimed to reduce their overall energy cost, mediated by quasiparticles such as collective atomic vibrations in a solid.

To do justice to such a wide range of properties especially far from equilibrium, we must start at the bottom. We need to develop the dynamical equations governing electron flow, including its environment at its most fundamental, molecular level. We need to do this in a unified way, spanning organic molecules, solids, magnets, and the continuum of materials in between. The aim of this book is to provide such a 'molecular' view of nanoelectronics, and see how this may help us understand or even design novel nano-electronic devices. This is not just to say that we develop the transport equations 'bottom-up'. We go further by borrowing tools from molecular chemistry, such as orbital basis sets relevant for benzene or Hartree Fock descriptions of exchange, moving up to a large piece of solid, touching systems in between such as two dimensional materials and strongly correlated quantum dots.

What is the merit of such a detailed molecular deconstruction? Many simulators for solid state devices still rely on classical, Newtonian drift diffusion equation, with quantum mechanical effects introduced as parametrized corrections (one may even designate some of them as 'fudge factors'). This worked in the past, even if it required a large proliferation of unphysical parameters in compact models. The situation, however, has changed radically over the last several years. Commercial transistors as of 2014 are already operating at more than 70% of their ballistic limit. This means that the observed current is 0.7 times what it would have been if the electrons went ballistically, i.e., single shot like a bullet with no momentum back scattering. Channels studied in emerging research such as carbon nanotubes or graphene are often measured to be nearly 100% ballistic. The few straggling, diffusing electrons still keep transistors operating fairly classically, but to properly account for ballisticity in emerging devices, we will increasingly need to adopt a 'bottom-up' view of resistance, given that there is seemingly nothing to resist the bullet-like flow of a ballistic electron except the contacts.

There are other quantum effects, notably tunneling — the tail of an electron wave zipping right through a thin barrier — that are mainstream today. Electron tunneling through oxides into the gate drove up the standby power dissipation and was a significant problem in pre high-k insulator technology days. Direct source to drain tunneling still continues to plague ultrashort materials such as organic molecules on metal contacts. At the same time gate controlled tunneling creates new opportunities for switching devices such as resonant quantum dots and subthermal tunnel field effect transistors (TFET). Equally interesting are strong many-body correlations, which are especially challenging to model far from equilibrium, but highly relevant to emerging device concepts such as Mott switches. These effects are hard to describe even qualitatively using bulk descriptions like effective mass and mean-field potentials. All this is not surprising in retrospect, given that one can arrive at classical dynamics as a limiting case for a quantum particle, by adding in dephasing events in bulk materials. However, *there is no honest way to add stuff to classical physics, and hope to emerge naturally with quantum attributes in the end.* The same applies for many body correlations, especially at non-equilibrium; they do not come out naturally as limiting cases of conventional 'mean-field' device models where electrons interact with other electrons through average potentials.

The emphasis in this book is on a molecular view of the flow of current *far from low bias*, where strong nonequilibrium physics persists. Far from equilibrium, a fascinating variety of current voltage (I-V) characteristics can be observed, from linear to exponential to saturating to asymmetric to a sequence of cascaded plateaus (Figs. 1.1-1.3). Much of biology and chemistry involves reaction kinetics and phase transitions that are meaningful away from equilibrium. But even to a device engineer, each such I-V curve bears potential significance in the nonlinear regime — as a tunnel device, rectifier, transistor, modulator or single electron switch for instance. It is the aim of this book to bring such a wide variety of transport phenomena under a common umbrella, especially with strong correlation and incoherent scattering, evolving from a unifying set of principles and equations. We can use these principles to explore new emergent materials (say a topological insulator), or new physical principles (e.g., spintronics). At the same time, they should connect with familiar devices such as a regular CMOS transistor or a pn junction. Thus the aim of this book is not just to study the emerging 'bottom-up' physics, but also the conventional 'top-down', and then see how they meet up in the middle.

1.2 Ohm's law redux

The quantum vs classical distinction manifests itself not just in exotic devices, but in mainstream systems as well. We can trace this by looking at Ohm's law, which is taught in every physics and microelectronics curriculum. The classical resistance of a device is given by the voltage V to current I ratio, $R = V/I = L/\sigma A$, where A is the cross sectional area, L is the length and σ is the conductivity of electron flow. Loosely speaking, the electron flows like fluid in a hose pipe or perhaps grains of sand in a sieve. The fluid resistance is high if we make the bore of the pipe narrow and reduce A, make L very long so the fluid scatters more often from inclusions or pieces of dirt, or increase the 'viscosity' of the fluid by replacing water with tar and thus reduce the conductivity σ. The equation was derived empirically, made intuitive sense and was measured repeatedly over the years.

The new Ohm's law, which has been operational since the 80s, arose when the channel sizes shrunk to the point that the electrons shot across ballistically like a bullet and the average scattering length exceeded the channel length. Since the channel offers little resistance as a result, the resistance of the electron must be mandated by the contacts. Surprisingly however, the ultimate low bias ballistic resistance for a thin, single moded conductor is

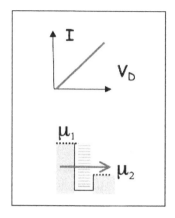

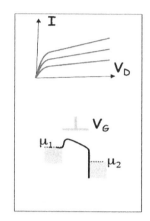

(a) Ohmic current (top) in a metallic channel involving a continuum set of states in green (bottom)

(b) Gate control of charge and current in a Field-effect transistor (FET), where gate fields attract or repel charges in and out of a channel like an electrostatic 'plunger'. See Section 18.1

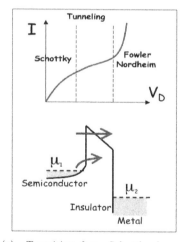

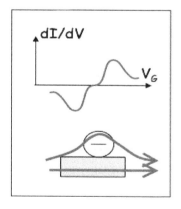

(c) Transition from Schottky barrier to regular tunneling to Fowler-Nordheim Tunneling across a metal-insulator-semiconductor junction (Section 17.4)

(d) Fano interference between a localized state and a continuum (Section 12.5)

Fig. 1.1 *The I-V menagerie showing classical I-Vs (top row) vs quantum systems with tunneling and interference (bottom row). All are described with Landauer theory with conductance set by the quantum transmission probability summed over modes (Eq. (1.1)). Parts 3-5 of the book deal with ballistic quantum transport.*

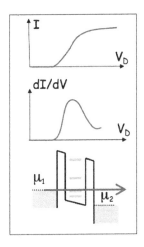

(a) Resonant electron transport through discrete levels (Section 16.3)

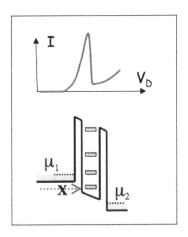

(b) Negative differential resistance in a resonant tunneling diode (RTD) when a level falls out of a band under bias (Section 20.1)

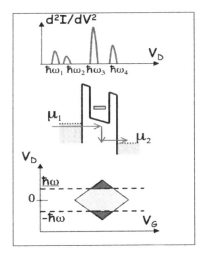

(c) Off resonant Inelastic Electron Tunneling Spectra (IETS) with peaks in d^2I/dV^2, as phonon emission channels open up during electron tunneling (Section 24.4)

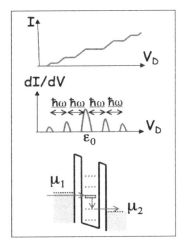

(d) Phonon-assisted tunneling near resonance gives equally separated sidebands (Section 24.5)

Fig. 1.2 *The I-V menagerie, showcasing examples of scattering. The treatment of inelastic scattering (bottom row) requires us to invoke the full machinery of Nonequilibrium Green's functions (NEGF) beyond just the Landauer transmission formalism. Part 6 of the book focuses on the impact of scattering*

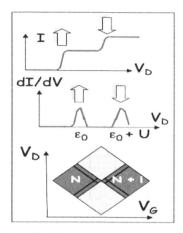

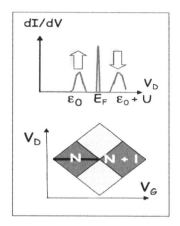

(a) (Top) Coulomb Blockade of a single electron transistor (SET). (Bottom) Diamond diagram showing secondary lines from electronic excitations (Section 26.2.4)

(b) Kondo Resonance in a quantum dot driven by singlet formation with a electrons in the lead, showing a zero bias resonant peak in the odd electron sector (Section 27.6).

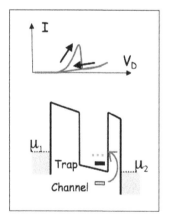

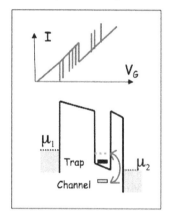

(c) Switching from trap-channel interactions (Section 20.2)

(d) Random Telegraph noise due to charge trapping/detrapping on resonance (Section 20.5)

Fig. 1.3 *The I-V menagerie, showcasing examples of correlated transport in strongly interacting systems that require us to go beyond 'mean field theory'. We can get these I-Vs by working directly with the density matrix in many body configuration space (Part 7 of the book). We could revert to conventional NEGF if we can extract an approximate self energy, typically nonlocal in time and non-perturbative in the interaction parameter.*

independent of the contact properties. In fact, the fundamental resistance ends up as a universal constant $R_0 = h/2q^2 \approx 12.9$ kΩ, where electron charge $q = 1.602 \times 10^{-19}$ Coulombs and $h = 6.602 \times 10^{-34}$ Js. The presence of Planck's constant h suggests that the origin is quantum mechanical.

We will have occasion to discuss this at length in this book, as well as the behavior of the resistance for multimoded conductors far from low bias. The reconciliation of the two limits, classical vs quantum, arises from a restatement of Ohm's law as

$$R = \frac{R_0}{\bar{T}M} \tag{1.1}$$

where M is the number of modes in the channel, and $\bar{T}$ is the average quantum mechanical transmission probability (ratio of transmitted to incident current densities), a number lying between zero and one. We will find that this resistance can be separated out into

$$R = \underbrace{\frac{R_0}{M}}_{\text{quantum}} + \underbrace{\frac{L}{\sigma A}}_{\text{classical}} \tag{1.2}$$

so that the quantum and classical contributions act in series, the former arising at the contacts where there is a significant dilution of modes, and the latter arising inside the channel. While this seems like a simple interpolation between the limits, it does involve a radical change of viewpoint, since we now view a conductor (*even a dirty one!*) as a quantum mechanical waveguide. Under certain conditions we may even see interference effects that cannot be described in a straightforward way as a sum of resistances.

1.3 Fundamentals of quantum transport

The overarching goal of the book is presenting the fundamental physics of current flow in its various incarnations across various systems. This involves developing the equations to capture the physics, discussing the parameter inputs such as bandstructures and scattering rates that go into these equations, and then observing the large variety of emergent phenomena that can come out of those equations, together with their experimental evidence and device relevance. In other words, we will discuss what to calculate, how to calculate them, and what the calculated results tell us about the system and its capabilities.

One of the major points in this book is the novelty of transport and non-equilibrium physics. Non-equilibrium is a very special beast, requiring us to combine dynamics (e.g., Newton's law or Schrodinger equations) with thermodynamics. This is quite different from conventional topics in solid-state physics, where we deal with a fixed number of electrons that are held in equilibrium. In contrast, we are dealing with systems that are open, capable of exchanging charge and energy irreversibly with their surroundings, all the while being held at non-equilibrium — in a state of steady flux maintained by external energy sources. We will not delve into the details of the external world or the energy sources, but rather their effects projected onto the system in question. In that sense, we will inevitably need to deal wih *partial information*, for instance, measurable thermal averages of bilinear products (effectively, a standard deviation or current noise) on which to base our studies on. With the emergence of characterization and simulation tools that allow us to explore these effects right down to their molecular levels, we are now empowered to truly study them 'bottom-up', from atoms to circuits.

Chapter 2

The practical lure — the quest for a new switch

The three terminal switch known as the transistor is one of the great inventions and arguably the greatest economic success story of the past century. In the last several decades, commodity prices typically increased by a factor between 3-10X. The cost of a house in the US soared from a mere $20K in the 70s to an average of $200K around 2008. In contrast, semiconductor unit prices have decreased steadily by *6 orders of magnitude*! This was made possible by continued improvement in transistor design, but more strikingly because of our ability to integrate many such transistors (currently billions) onto a small chip. To our knowledge, such an economic boom is unparalleled in the history of human existence. If the late 18th and early 19th centuries were characterized by the Industrial Revolution, then the mid to late 20th century was definitely characterized by the rise of digital information technology, with the transistor being its main fuel.

The essence of digital logic is the ability to store information in binary, expressed as ones and zeros (as opposed to decimals, which bear significance solely due to our evolutionary inheritance of ten fingers). In charge based digital logic, we can imagine zero referring to the absence of charge at a location, and one signifying its presence. A logic operation requires writing information, a zero or a one, at any location according to a specific algorithm embodying that logic (say adding two numbers). We then need to propagate the information — the sequence of ones and zeros, retrieve and read it as needed, and finally erase it to reset the system for the next logical operation. In short, we need to shuffle around the ones and zeros over time, in the process driving a current. One of the aims of the semiconductor technology roadmap has historically been to identify ways of generating such currents efficiently.

What does efficiency mean for current flow? At the level of a single device, this means the ability to write and read the information bit quickly, accurately, and at a low energy cost (energy being the main currency here). At the level of a circuit, the bit should be able to toggle a device downstream, like a chain of dominos, making sure that the information propagates according to the desired algorithm accurately and without undue latency. This also means that there needs to be a sufficient degree of *gain* to offset any losses along the chain, and that the output of one device can feed efficiently as input to the next device unidirectionally, without in turn being affected by it through any undesirable backpropagation or feedback.

As is now well known, the continued miniaturization of transistors to make them smaller, faster and cheaper drove the celebrated Dennard scaling and ultimately Moore's law for several decades. However, device scaling has been notoriously difficult to sustain of late. In the past, the main challenge was to optimize the **electrostatics**, i.e., to keep the gate capacitance large enough to dominate the channel potential. As we will see in chapter 18, holding the channel bands from slipping under drain bias allows the current to saturate and the device to have appreciable gain. The architecture of the transistor evolved accordingly, from shallow junctions with raised source drain to halo doping to high k dielectrics and recently to 3-D architectures such as FinFETs, primarily to enhance the gate control parameter (i.e., mitigate deleterious 'short channel effects') and provide adequate charge while reducing the 2-D footprint. New high mobility channel materials are also being introduced — such as strained silicon, with a push towards III-V and Ge, to improve the **speed** of the device. What has however spelt the demise of Dennard scaling is its rapidly increasing **thermal budget**. If we continued scaling transistor geometries as before, along with the accompanying scaling rules especially the clock frequency, the chip would attain unsustainable temperatures just from the cost of computing, causing it to melt. (This actually happened with a single core microprocessor that was discontinued since it dissipated more power than an easy bake oven!)

The industrial solution has primarily been to move away from such costly scaling, especially voltage and clock frequency, adopt architectural solutions such as multicore and hyperthreading technologies, growing in the third dimension (tall rather than small) and scaling up the wafer sizes to keep the *economic* aspect of Moore's law (chip performance per dollar) going. In other words, we are moving away from simply improving performance towards reducing the overall footprint. Without delving into such details, let us spend some time trying to understand some of the device metrics and the physical and material parameters that control them.

2.1 Single device level issues: speed and energy

At its heart the transistor is just a gate capacitor driving a semiconducting resistor. The two are coupled electrostatically but isolated physically with a thin insulator separating them. Charging up the capacitor changes the field lines that draw in or push out charge density and thus alter the conductivity of the semiconductor by several orders of magnitude, turning it on or off in the process. A Complementary Metal Oxide Semiconductor (CMOS) inverter consists of two complementary transistors placed in series between a power supply and a ground (Fig. 18.1). The one connected to ground (nMOS) switches on under positive gate bias while the other (pMOS) switches on with negative gate bias. Thus the output terminal drawn from a point between them registers either the ground (a '0') or the power supply (a '1'), opposite to the gate input. We can then connect an odd number of inverters and feedback the last output back into the first input to create a ring oscillator. We can also use a different sequence of inverters to create a NAND or NOR gate and ultimately realize a Universal Boolean logic system (we revisit this later, Fig. 18.12).

The transistor is thus a resistor-capacitor (RC) network. Let us outline some of its performance metrics that engineers care about. Given a power supply voltage V_D, a gate capacitor C_G, charge Q and an ON current I_{on}, the device delay — the time to switch the transistors on or off, is

$$\tau = Q/I_{on} = C_G V_D/I_{on} = RC_G \qquad (2.1)$$

the standard RC time constant to charge or discharge the capacitor.

While the capacitance is fixed by the geometry of the device and the gate insulator, the resistance is set by the material properties of the channel. In particular, we care about the extent to which we can electrostatically alter ('gate') its electrical conductivity, a quantity that varies in the literature over an incredible 20 orders of magnitude! (In comparison, thermal conductivity has only 4 orders to work with). A good modeling platform should be able to predict such a conductivity and its field-dependent modulation. The bulk of this book is on computing the resistance of a nanoscale device.

Tunability however comes at an energy cost. The wider we vary the voltage range, the more dissipative the process is. Consider an RC network where a voltage source set at V_D turns on the capacitive charging, with $V_C = Q/C$ as Q and V_C both ramp up over time. The energy stored in the capacitor at the end of such an operation (i.e., total work done in bringing the charges to the capacitor) is given by

$$E_C = \int_0^{CV_D} V_C dQ = \int_0^{CV_D} \frac{Q}{C} dQ = \frac{CV_D^2}{2} \qquad (2.2)$$

The energy delivered from the battery was of course $QV_D = CV_D^2$, meaning that the rest of the energy, $CV_D^2/2$ was dissipated in the wires. The difference arose because the energy delivered at a constant battery voltage V_D, but the capacitor stored a voltage V_C that ramped up from zero to V_D over time, and the energy stored is thus a fraction of the delivered power. The rest must therefore be dissipated in the wire.

But how can we write down the dissipation in the wire without ever specifying the properties of that wire, notably its resistance? What if we choose a superconducting wire with a small resistance, for instance? If we choose a wire with a very low resistance, we drive a lot of current, which ramps up the power dissipation $P \sim V_D^2/R$. But then the large currents charge up the capacitor very fast in a time $\tau \sim RC$ (Eq. (2.1)), and the energy cost $P \times \tau$ at the end remains independent of the wire resistance. Higher power is compensated by higher speed.

As we argue later, what causes the dissipation is ultimately the inability of the charges to instantly follow the turn on and turn off processes at the battery. One way to mitigate this phase lag and thus cut down the energy dissipation is to make the switching process itself slow.

2.2   Adiabatic switching

We mentioned above that one way to cut down on the energy cost is by doing operations slowly. Our aim in this section is to show why speed is expensive. The gist of the argument is that a circuit has a natural time constant for operating, given by the RC product of the capacitor. When processes run faster than that speed, the electrons cannot keep up with the battery and pick up an additional phase lag. It is these out of phase components that end up dissipating energy and shooting up the energy cost. For the usual process of charging and discharging a capacitor with a DC source, these processes arise from the abrupt turn on of the circuit, which brings in various fast Fourier components that are ultimately dissipative. A cure is to do the operations slower than the natural time scales of the circuit, and live with the consequences of a slower circuit.

Let us look at an RC circuit. Elementary circuit theory starts with Kirchhoff's voltage law between the battery 'B', capacitor 'C' and resistor 'R'

$$V_B(t) = V_C(t) + V_R(t) = \frac{Q}{C} + R\frac{dQ}{dt}. \qquad (2.3)$$

For a voltage $V_B(t) = V_0 \cos \omega t$, we can use standard tricks to solve the first order differential equation. For instance, we can use phasor representation $V_B(t), Q(t) \to V_0 e^{i\omega t}, Q_0 e^{i\omega t}$ at a given frequency, whereby derivatives and integrals of the exponentials give algebraic prefactors. The result

$$V_B(t) = \text{Re}\left[V_0 e^{i\omega t}\right] = \text{Re}\left[Q\left(1/C + i\omega R\right)\right] \quad \text{(from Eq. (2.3))}$$

$$\implies Q(t) = \text{Re}\left[\frac{CV_0 e^{i\omega t}}{1 + i\omega RC}\right]$$

$$\implies I(t) = \frac{dQ}{dt} = \text{Re}\left[\frac{i\omega C V_0 e^{i\omega t}}{1 + i\omega RC}\right] = -\frac{V_0}{R}\sin\left(\omega t - \theta\right)\sin\theta, \qquad (2.4)$$

where the phase angle $\theta = \tan^{-1}(\omega RC)$. We can now find the energy accumulated over a particular time range. It is instructive to work this out over one AC period, $T = 2\pi/\omega$. Each energy is given by

$$E_C(T) = \int_0^T V_C dQ = 0,$$

$$E_R(T) = \int_0^T V_R dQ = \frac{CV_0^2}{2}\left[\frac{2\pi\omega RC}{1 + (\omega RC)^2}\right]$$

$$= \int_0^T V_B dQ = E_B(T). \qquad (2.5)$$

Note that the capacitor does not store any energy over the period. Over the first quarter period (i.e., integrating $E_C(t)$ upto $t = \pi/2\omega$) and the third quarter period we get a stored energy of $(CV_0^2/2)\left[(\omega RC)^2 - 1\right]/\left[(\omega RC)^2 + 1\right]$, while during the second and fourth quarters we restore that energy as the capacitor discharges back to the battery. The dissipation however hurts both ways and the battery loses the corresponding energy irretrievably into the Joule losses in the wires. In fact, if we computed the dissipation and the energy flowing out of the battery over another period, say between $2(n-1)\pi/\omega$ and $2n\pi/\omega$, the dissipative term will keep increasing monotonically by a factor $n^2 - (n-1)^2$ that indicates that the energy actually continually flows out of the battery and into the dissipative losses.

It is easy to see that the energy cost is small when $\omega RC \ll 1$. This is because a slow signal can be easily followed by the electrons moving at their natural RC constant (which happens to be equal to its dielectric relaxation time ϵ/σ, with ϵ: dielectric constant, σ: conductivity, i.e., the rate of charge flow defined by the equation of continuity). The phase lag $\theta = \tan^{-1}(\omega RC)$ vanishes accordingly. Slow signals thereby dissipate less energy by bringing the electrons in phase with the source.

Equation (2.5) suggests that a fast signal with $\omega RC \gg 1$ also dissipates little. That is because we tracked the dissipation only over a single period $2\pi/\omega$ and a fast signal reduces that period. That is simply a sampling artifact. If we tracked it over time continually, a slow signal always dissipates less. We can see this by computing the power dissipated instead

$$P_{\text{diss}}(T) = E_R(T)/T = \frac{V_0^2}{2R}\Big[\frac{(\omega RC)^2}{1+(\omega RC)^2}\Big], \qquad (2.6)$$

which vanishes for a slow signal, but reaches a maximum Joule loss of $V_0^2/2R$ for a fast signal.

2.3   Adding an inductive component

The RC circuit above shuffles energy back and forth at the oscillatory frequency of the forcing system, the battery. We can add an intrinsic oscillatory rate by also including an inductive component. The equation for the driven LCR circuit is easily mapped onto a forced damped oscillator

$$L\frac{d^2Q}{dt^2} + R\frac{dQ}{dt} + \frac{Q}{C} = V \quad \longrightarrow \quad m\frac{d^2x}{dt^2} + \gamma\frac{dx}{dt} + kx = F, \qquad (2.7)$$

showing the resistor R acts as the 'damping' γ, the inductor L as the inertial mass m and the inverse capacitance ('elastance') as the restorative spring constant k. Once again, this can be solved using the phasors. The result is analogous to the RC. Over one period, the reactive components — the capacitor and the inductor, store no energy but simply shuttle it back and forth. At the end of the day the net energy delivered by the battery goes into the dissipative losses in the wires. The dissipation

$$P_R = \frac{1}{T} \int_0^T I^2 R dt = \frac{V_0^2}{2R} \left[\frac{(\omega RC)^2}{(1 - \omega^2/\omega_0^2)^2 + (\omega RC)^2} \right]. \qquad (2.8)$$

Curiously, fastest signals are no longer the most dissipative ($\omega \to \infty$, $P_R \to 0$). Dissipation peaks near the resonant frequency $\omega_0 = 1/\sqrt{LC}$. The previous RC result, where ultrafast signals were dissipative, can now be viewed as a special case where $L = 0$ and $\omega_0 \to \infty$.

2.4   The cost of speed: LCR with sudden turn-on

Let us now assume a sharp turn on for the switch. Assuming $V(t) = V_0\Theta(t)$, where $\Theta(t)$ is the Heaviside step function that is unity for $t \geq 0$ and zero for $t < 0$, we can solve the differential equation. Let us focus on the results.

$$I(t) = \frac{V_0}{\omega_1 L} e^{-Rt/2L} \sin(\omega_1 t), \qquad (2.9)$$

where $\omega_1 = \sqrt{\omega_0^2 - R^2/4L^2}$ and $\phi = \tan^{-1}(R/2\omega_1 L)$. The dissipated energy

$$E_d = \int_0^t I^2 R dt$$

$$= \frac{V_0^2 R}{2\omega_1^2 L^2} \left[\frac{L}{R} \left(1 - e^{-Rt/L} \right) + \frac{\cos(2\omega_1 t + \phi)e^{-Rt/L} - \cos(\phi)}{2\omega_0} \right]. \quad (2.10)$$

We see right away that in the limit $t \to \infty$, $Q = CV_0$ and $E_d = CV_0^2/2$.

So we are now back to the dissipated energy in an RC for all ω_0 values, seemingly regardless of speed! The reason is that unlike the constant or oscillating V previously, the switch here turned on as a sharp step function, bringing in a large mix of fast Fourier components. A segment of these components are faster than the RC time constant that determines the tardiness of the electrons, ending up dissipating the energy. We could of course stop the process precisely at the cyclical charging point when the capacitor voltage first reaches V before it overshoots (Fig. 2.1 middle panel). This happens within the first period of oscillation. However, it requires precise timing information of the LC circuit. As shall see shortly (Section 2.8), gathering information costs time and energy, but could save us on energy dissipation.

2.5   LCR with slow turn-on

We can deliberately craft a turn on voltage that ramps up so that its fastest Fourier component is slower than the RC time constant. Intuition says this should have little dissipation, but the price we pay is that the circuit is as a result ponderously slow. Let us quantify what the trade off is.

We start with a voltage input $V(t) = V_0 \Theta(t)[1 - e^{-\lambda t}]$ to give a gradual turn on. We now have an inhomogeneous equation, whose solution

$$I(t) = I_0 e^{-Rt/2L} \sin(\omega_1 t - \theta) + I_1 e^{-\lambda t}, \qquad (2.11)$$

where the particular integral gives

$$I_1 = \lambda V_0 / [L\lambda^2 - R\lambda + 1/C]. \qquad (2.12)$$

We can fix the amplitude and phase of the first term by requiring that $I(0) = dI/dt(0) = 0$, which gives us

$$\tan \theta = \frac{\omega_1}{\lambda - R/2L}, \quad I_0 = I_1 / \sin \theta. \qquad (2.13)$$

We can now integrate the current to get the charge, and the current squared to get the dissipation. The numerical results are shown in Fig. 2.1. The thing that jumps out is that if we turn on the potential slowly compared to the RC time constant, we get rid of the fast Fourier components so that the charge can follow the potential and stay in phase, thus lowering the dissipation cost. As shown in the figure, the green curve allows the capacitor to be charged to the desired voltage, while the dissipation is much smaller than the $CV_0^2/2$ we were getting used to. One way to understand is to break up the charging process into N steps, so that over each step we incrementally ramp up the voltage by equal steps V_0/N. The energy dissipated then becomes $N \times C(V_0/N)^2/2$, which tends to zero as $N \to \infty$.

To get an estimate for the 'Q-factor', the rate of dissipation each cycle, we assume that $\lambda \ll R/2L < \omega_0$, in other words, still in the under-damped limit but with a very slow turn-on. Specifically, let's assume $\lambda = a(R/2L)$, and $R/2L = b\omega_0$, where a and b are constants less than unity. It is straightforward then to show that

$$\lambda RC = 2ab^2 \ll 1. \qquad (2.14)$$

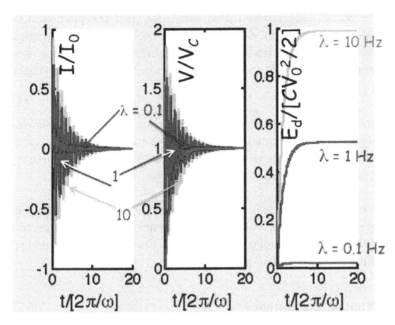

Fig. 2.1 *Charging the LCR with slowly turned on pulses. We use $R = 0.1\Omega$, $V_0 = 1\,V$, $C = 1F$ and $L = 1H$, and plot results in terms of natural limits. Green is for fast turn-on $\lambda = 10Hz$, blue for $\lambda = 1Hz$ and red is for $\lambda = 0.1Hz$. For adiabatic turn-on (red), the voltage reaches the desired value, but the dissipation E_d is low.*

Under these conditions, when the turn-on time is much larger than the RC,

$$E_d/R = \int_0^\infty I_0^2 e^{-Rt/L} \sin^2(\omega_1 t - \theta) + \int_0^\infty I_1^2 e^{-2\lambda t}$$
$$+ 2I_0 I_1 \int_0^\infty e^{-(\lambda + R/2L)t} \sin(\omega_1 t - \theta). \qquad (2.15)$$

Each integral can be calculated analytically, but for small λ, only the second term matters. In other words,

$$E_d \approx \frac{I_1^2 R}{2\lambda}. \qquad (2.16)$$

Also, in the slow turn-on limit, the expression for I_1 simplifies considerably to give us $I_1 \approx \lambda V_0 C$. We thus get

$$E_d \approx \frac{\lambda R C^2 V_0^2}{2} = (\lambda R C) \frac{C V_0^2}{2}. \qquad (2.17)$$

The fraction of the pulse that belonged to the 'fast' components has been removed, reducing the dissipation down to a small 'quality factor' λRC, much less than unity as proved earlier.

While we successfully reduced dissipation, we paid the price of dealing
with a slower circuit, slow enough to allow electrons to keep up with it.
To properly capture this trade off, device engineers often track the 'energy
delay product'. In the example above, the delay is set by $1/\lambda$, giving us

$$E_d\tau \approx \left(\lambda RC\right) C V_0^2 \times \frac{1}{\lambda} = Q^2 R \qquad (2.18)$$

To reduce the energy delay product, we have two choices, (a) either choose
materials with low R, explaining our interest in high mobility materials
such as III-V, Ge, graphene or topological insulators, or (b) reduce the
number of charges Q needed to encode each bit.

In the next sections, we will see what limits the charge and the voltage
in today's devices. At the end of the book, once we have a chance to discuss the thermodynamics of transport, we can discuss how one could bypass
these limitations, at least in principle.

2.6   What limits voltage?

The voltage needed for binary switching is ultimately determined by acceptable error rates in the circuit. In a conventional transistor, the Fermi
energy — the highest occupied energy in the contacts, lies in the bandgap
of its semiconducting channel, creating a barrier E_b to electron or hole current (Fig. 18.2). The transistor turns on with a gate voltage that raises
or depresses the channel potential to enable charge flow. However even in
the off state there is a thermal distribution of contact electrons and holes,
including some at remote energies that can jump over the barrier and support a leakage current (Fig. 2.2(a)). The spontaneous, unplanned jump
contributes to a corresponding error given by Boltzmann statistics

$$\Pi_{\text{err}} = e^{-E_b/k_B T} \qquad (2.19)$$

where T is the temperature in Kelvins and k_B is Boltzmann's constant
(useful to note that at room temperature, $T = 300\text{K}$, and $k_B T \approx 25$ meV).

For voltage gated devices, the voltage needed to switch the device is given
by the probability of the device staying on divided by the probability of it
being off (this ratio is inverse to the error rate). From Eq. (2.19), we get

$$V = \frac{E_b}{q} = \frac{k_B T}{q} \ln\left(\frac{1}{\Pi_{\text{err}}}\right) = \frac{k_B T}{q} \ln\left(\frac{p_{\text{on}}}{p_{\text{off}}}\right). \qquad (2.20)$$

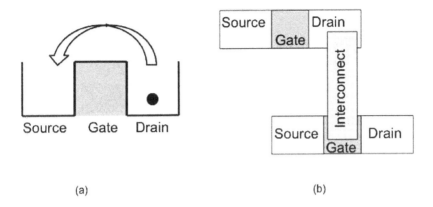

(a)   (b)

Fig. 2.2   *(a) Voltage is set by the tendency for charges to jump back over a barrier due to thermionic emission through the tails of the Fermi Dirac distribution. We want enough of a barrier to minimize such an error not just at the device level, but more aggressively to contain it at a circuit level. (b) Number of charges needed to encode a bit is set by drivability, i.e., the need for charges on a drain to charge up the gate capacitance of a device downstream. In practice, the charges need to be enough to flood the interconnect and ensure adequate density at the gate.*

The equation applies not just to regular CMOS devices, but also to other switches such as voltage-gated ion channels for instance (Fig. 2.3). The roles of 'on' and 'off' are played by 'out' and 'in', the extra vs intra cellular ion concentrations around a lipid bilayer, with different on and off ratios for the various ionic channels among sodium ('Na'), potassium ('K') and chlorine ('Cl'). The resulting Goldman equation looks very similar, except we weight the processes by the ion concentrations (o: out, i: in)

$$V_m = \frac{k_B T}{q} \ln \left(\frac{p_{oK}[K^+]_o + p_{oNa}[Na^+]_o + p_{oCl}[Cl^-]_o}{p_{iK}[K^+]_i + p_{iNa}[Na^+]_i + p_{iCl}[Cl^-]_i} \right). \tag{2.21}$$

The highest error one can tolerate for distinguishability of two states is $\Pi_{err} = 0.5$, so that the fundamental binary switching limit from Eq. (2.20)

$$\boxed{(E_b)_{\min} = k_B T \ln 2 \approx 0.7 k_B T} \tag{2.22}$$

In practice however, we need much lower error rates in a circuit.

What is the acceptable error rate in a circuit? If we consider a circuit as a simplistic array of N_{gate} concatenated gates, then the circuit functions only when every single gate functions. This means the circuit error Π_{ckt} is

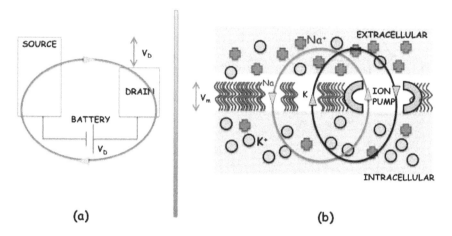

(a)   **(b)**

Fig. 2.3   *(a) Electron flow in a conductor, going towards drain with a positive bias, and then resent back to the source by the battery that works uphill in an attempt to restore equilibrium. (b) Flow of Na and K ions setting up the action potential along an axon. Na flows inside through a Na channel down the density gradient when the channel opens at a critical voltage, while K flows outside at a slight delay at a somewhat slower speed (because of their heavier masses). The ion pump, a Na K ATPase, acts as a battery and moves back the ions uphill to prep for the next action potential pulse.*

related to the device error Π_{err} by

$$\underbrace{1 - \Pi_{\mathrm{ckt}}}_{\text{ckt works}} = \underbrace{\left(1 - \Pi_{\mathrm{err}}\right)^{N_{\mathrm{gate}}}}_{\text{all gates work}} \qquad (2.23)$$

For a circuit to function at $\Pi_{\mathrm{err}} = 0.1\% = 0.001$, assuming 300 million concatenated gates, the device error $\Pi_{\mathrm{err}} \approx 3.3 \times 10^{-12}$. The corresponding voltage extracted from Eq. (2.20) amounts to $V = 26\ k_B T/q \approx 0.6$ V. The actual number is a bit larger once we account for non-ideal electrostatics, barrier fluctuations and delays, and is closer to $\sim 40\ k_B T/q \approx 1$ V.

2.7   What limits charge?

We saw that the dissipative energy $Q\Delta V$ is limited by the switching voltage to ~ 1 V. How about charge? Is there a reason we cannot operate with a single electron? The answer is that we need enough charges to drive the next interconnected device down the line in a circuit (Fig. 2.2(b)). For instance in a ring oscillator, we need the electrons from the first inverter to flip the second and then on, like a chain of dominos. This means we need at least $Q/q = C_G V_D/q$ charges to depress its gate adequately to allow its own charges to flow, C_G being the gate capacitance. For an NMOS with gate capacitance $C_G = 0.03$ fF, we need about 200 electrons to charge up the gate to 1 V — the minimum barrier needed to avoid substantial backflips.

However, the situation is actually worse. Just putting 200 electrons into an interconnect does not guarantee that we have 200 electrons where the gate sits (Fig. 2.2(b)). To ensure adequate gate charging in presence of stochastic electron placement, we need to charge up the entire interconnect capacitance. The capacitance of a local interconnect with a fan-out of 4 is considerably larger (say, 1 fF compared to 0.03 fF for the gate), which means to successfully charge a single gate sitting at the end of the interconnect to 1 V, we need almost 6000 electrons at an energy of $6000eV \approx 240,000 \, k_B T$!

2.8   Reversible computing and the cost of information

It is in principle possible to move electrons with little dissipation if we are clever about it. Going back to our two well one barrier example, the thermal excitation of an electron over a barrier was ultimately dissipation-less. Although energy was blasted out when the electron fell downhill, that energy was borrowed from the environment when it was excited over the barrier. But to make the process unidirectional with no spontaneous backflip, we need to lower the barrier and at the same time introduce a directional asymmetry by making one well lower than the other. In a transistor, the gate voltage lowers the barrier while the drain voltage creates the well asymmetry (Fig. 18.2). This will cause a net dissipation because of the *extra* drop the asymmetry creates. The bleeding of electron energy to the environment happens at the end when friction forces the electron to stick to the bottom of the well once it falls downhill (landing with a thud, rather than rocking around like a skateboarder on a frictionless ramp, or bouncing back like a kid on a trampoline). *Dissipation requires both falling and sticking* — for instance, when friction causes us to abruptly *reset the coordinate* of the electron to the well bottom. Since resetting also erases all information on its past history, dissipation is ultimately tied with the irretrievable loss of information.

In principle however, we could have done the computation reversibly with zero dissipation if we had precise knowledge of the state of the electron. For instance, if we knew in advance which well contained the electron, then we could have raised the other well to make the potential monostable rather than bistable, slowly slid the single well laterally to the location of the other erased well, and then raised a barrier to prevent a return to the original side (Fig. 2.4, top row). We made the potential 'moonwalk', maintaining the electron constantly at its ground state instead of letting it fall downhill, so the energy dissipation was zero.

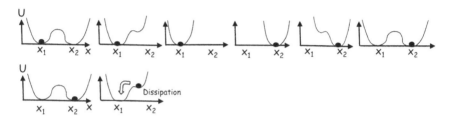

Fig. 2.4   *(Top) From left to right, a sequence of transitions (a 'moonwalk') that morphs a bistable well into a monostable one and then back to bistable, while maintaining the electron at ground state. What makes this possible is prior knowledge of the electron location, in this case the left well, so that we raise the right well without disrupting the electron. (Bottom) In the absence of knowledge of the electron, we will occasionally raise the wrong well, causing the electron to fall downhill and dissipate its energy to the environment when it settles down at the bottom.*

How did we manage to accomplish such a dissipationless transition? To do so, we needed several special conditions. For one, we had to drive the transitions slowly and have some frictional damping so the electron stays stuck at the bottom. For instance, if we did the lateral shift quickly compared to the natural excitation time of the electron (its RC constant), it would escape its ground state, much like kicking rather than slowly sliding a full bucket would cause its water to slosh. At the same time, the process needed to be faster than the equilibration time, else the electron would spontaneously jump over the barrier and spread out evenly between wells. In other words, we need a time window within which to operate. On top of that, we needed some intelligence gathering. Remember that in the conventional switching process, dissipation required a hard reset that erased all information. Instead, we can choose to store the information by making a copy of the initial state, and then use that knowledge to decide which

potential well to raise first. Accurate information gathering is important
— if in fact we did this incorrectly, we would have raised the electron and
forced it to dissipate its energy (Fig. 2.4, bottom row). Instead of playing
a game of Russian roulette on which well is occupied, we can choose to
gather information about the electron state and raise the correct well each
time, thereby avoiding dissipation. We invest some effort instead into the
read and copy operations.

The idea can be extended to data storage in general. Dissipation happens when we throw away information by squeezing the available phase
space — in the above example from a two state system to a single state.
For instance, an AND gate has four possible input states (00, 01, 10 and
11) that collapse to only one output (0 in 3 cases, 1 in 1 case). The outcome
of an AND gate is irreversible. From the output alone we cannot recon-
struct the input in 3 out of 4 cases, and this loss of information inevitably
causes energy dissipation. The loss of information can be quantified from
the change in information entropy that in turn quantifies the squeezing of
phase space. The entropy is a measure of how many given microstates we
can have corresponding to a given macrostate (the latter specified broadly
for instance by total number of particles or total energy). N equally likely
microstates will each have equal probability $p = 1/N$, so that the entropy
(compare Eq. (2.20))

$$S = k_B \Big\langle \ln N \Big\rangle = k_B \Big\langle \ln \Big(\frac{1}{p} \Big) \Big\rangle = -k_B \sum_i p_i \ln p_i, \qquad (2.24)$$

where $\langle \cdots \rangle$ is the thermodynamic average, and p_i is the probability of the
ith microscopic state. We can interpret this expression as a measure of the
information content, as it gives us a non-zero value only when the probabilities are neither zero nor one, i.e., we have uncertainty in the system. We
can therefore view information as a measure of 'surprise'. *A system in a
state of complete equilibrium is utterly predictable, and as such carries no
useful information content.* Dead men tell no tales!

Let us look at the process of writing to a known final state, say, an electron in the right well. If the initial state is unknown (no copy), it has
equal probabilities of being in each well $p_1 = p_2 = 1/2$ and initial entropy
$S = k_B \ln 2$, entropy that is then reduced to zero upon writing to the preset
final with $p_1 = 0$, $p_2 = 1$. The decrease in entropy of the system required
the system to cool down (fewer final microstates means less disorder), which

must then be counter-balanced by a heating of the modes of the environment. A corresponding minimal amount of heat $\Delta Q = T\Delta S = k_B T \ln 2$ per bit has flown into the surrounding environment where the information gets lost among its myriad modes, causing the dissipation. We can view this dissipation as a consequence of information 'erasure' — a state with actual information in it (unknown initial state and thus non-zero entropy) has been squeezed to a known final state with zero information. For the AND gate, we can do an analogous analysis — squeezing the entropy from $S = -k_B \sum_i p_i \ln p_i = k_B \ln 4$ across four equally likely states to a final of $S = -k_B[3/4 \ln (3/4) + 1/4 \ln (1/4)]$ creates a net energy dissipation of $\sim 0.8 k_B T$. Once again, the irreversible erasure of the initial bits forced dissipation.

As we argued above, if we did know the initial state in our bistable well by making a read and copy, then we can do the write at zero dissipation — if the electron was in the first well, we raise the well on the other side and slide the first well before raising a barrier in the middle (Fig. 2.4 top). If it was in the second well already, we just do nothing. The entropies of both the initial and final states are zero since we have in each case one known microstate. The change in entropy is accordingly zero, and there is no squeezing of phase space and thus no need for dissipation; but now the information gathering and the copy processes come with their own energy overheads and delay (recall the cost of speed). An example of such a reversible, erasure-free computation is a controlled NOT gate, which has two inputs and two outputs. The C-NOT gate flips the second bit if and only if the first (control) is unity, thus conserving the information phase space and being completely reversible. Just repeating the C-NOT restores the initial state.

The simple picture above is, just that — simplistic. There are no doubt some nuances to the argument. For instance, we assumed that the potential is modified fast enough that the probabilities p_1 and p_2 do not always equilibrate to $1/2$ through inevitable thermal jumps. In other words, we assumed a metastable initial state, without which we would see no dissipation at all (we would in fact revert to the slow, adiabatic limit where we already saw that we have little dissipation). To capture the impact of the fast evolving potential, we will need to calculate the actual evolution of the probability distribution in the presence of frictional damping. Damping involves a coupling with the environment that allows us to bleed energy unidirectionally into its numerous modes (and in some cases even get energized by it — Section 20.4). Damping is necessary because without damping even falling does not hurt! We can capture the overall physics of the overdamped oscillator in an evolving potential with thermal kicks by numerically solving the Fokker-Planck equation for the electron probability distribution P (Section 9.1)

$$\gamma \frac{\partial P}{\partial t} = \frac{\partial}{\partial x} \left(P \frac{\partial U}{\partial x} \right) + k_B T \frac{\partial^2 P}{\partial x^2}, \tag{2.25}$$

with γ the damping coefficient in kg/s (Eq. (2.7)). We can then calculate the dissipation $E_d = \int_0^\infty dt V(t) I(t)$ in presence of the evolving potential $V(t) = -U/q$, where the current $I(t) = -qdP/dt$. It is also worth remembering that the dissipation occurred when we did a hard reset that could have been avoided by storing all the information or letting the electron continue on a frictionless ramp. We can preserve the information by converting the kinetic energy of the electron into the angular momentum stored by the electron spin, for instance (much like a parkour enthusiast breaking a jump into a roll).

2.9   Circuit level needs: compatibility, isolation and gain

To summarize, the energy and ultimately the voltage are set by the thermodynamic bound of $\sim 0.7 k_B T$ and the rate of error propagation in a concatenated circuit. The number of electrons is limited by drivability, the need to charge up the entire interconnect capacitance to this voltage. The resulting heat generated is considerable. The speed on the other hand is limited by the material properties of the channel. The molecular viewpoint will allow us to explore both the fundamental physics and the materials chemistry needed to address the large thermal budget and the speed constraints. For much of the book, we will calculate material conductances that set the electron speed. At the end of the book, we will speculate on how we could maintain the solidarity and drivability encoded in an array of charges, yet reduce their energy cost, by reducing the switching voltage well below the thermal limit in Eq. (2.20). The final chapters will also be the most speculative and some of the device ideas could well turn out to be impractical. The true value however is to illustrate how we can analyze these unconventional switching mechanisms (as well as new emergent materials and devices), using the modeling framework developed in this book.

While energy, speed and error are the three metrics for a high performance low power device, there are other needs to operate a Boolean circuit built on these devices. The input and output must be **compatible**, so that one can drive the other (a thermal input generating a current output is only useful if we can reconvert the latter into a thermal input for the next device). The input and output must also be **decoupled**, so that the output does not influence the input. Consider a linear chain of dipole coupled nanomagnetic dots where the couplings try to place the spins antiferromagnetically, i.e., along a sequence of alternating ups and downs (Fig. 2.5). If we now want to flip an input from up to down say, the dot next to it ends up in a state of 'frustration', where the neighbor on the newly written input side wants to force it contrary to what the neighbor on the already existing output side wants. In such a system we will need to find a way to temporarily erase the frustrated dot and the output neighbor, let the input neighbor feed unidirectionally, before releasing the two erased dots. In other words, we need a phased clocking scheme. It may seem easier to avoid such a granular clock varying on the scale of individual dots by erasing all spins in the output sector with global clocking, but that prevents us from 'pipelining' signals continuously. Phased clocking allows fast pipelining and reduces latency, but brings in an energy overhead and complexity, once again as an energy-delay trade-off.

Finally, we want the device to have considerable **gain**. An ideal field effect transistor has a capacitive input with zero gate current but non-zero output current in the channel, creating a large current gain (the energy coming from the power supply setting the source to drain bias). We need in addition a voltage gain for analog signal amplification, and signal restoration in digital applications. Later in Fig. 18.1, we draw an inverter curve whose steepness quantifies its gain. The ideal zero and one binary digital bits represent the two ends of the curve. A ring oscillator is created by an odd number of such inverters arranged on a periodic chain (an odd number of signal reversals fed back into the input will keep the circuit oscillating). At each stage of the information transfer between two inverters, the signal can get corrupted by noise, driving the high and the low voltages towards the middle where they get harder to distinguish. A steep gain curve makes the middle an unstable point, so that the signals migrate back to the two ends and restore digital information. As we see later, to get a good gain we will need each transistor to have a well saturated current.

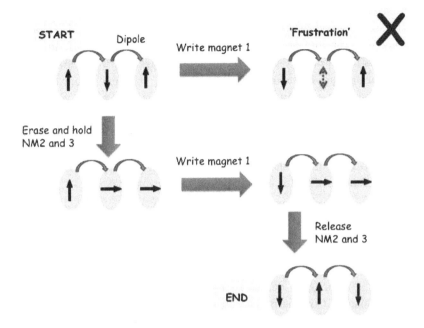

Fig. 2.5 *Input-output isolation needed to switch a set of dipole coupled nanomagnets ('NM') from start to end, while avoiding 'frustration'. A sequence of erase and hold pulses decouple input from output and propagate signals unidirectionally. The erase can involve external fields, spin torque or internal strain, and must be low energy as well.*

2.10 A complex, shrinking phase space

In practice, a circuit craves more than performance — the need for speed. It needs low power — static and dynamic — both in its quiescent off mode and its active switching mode. We can translate these needs in terms of $I - Vs$ for a three terminal device. Device engineers traditionally desire a steep gate transfer characteristic $I - V_G$ for low voltage switching and low dynamic power dissipation. They also prefer a low OFF current for low static power dissipation and a large ON current for high speed operation. The output characteristic $I - V_D$ should have a steep rise followed by a large saturating ON current reached at a similar low voltage (the drain and gate voltages are usually set equal to each other through a common power supply).

The $I - Vs$ require good electrostatics with a large dominant gate capacitance, insulating oxide barriers, channels with adequate band-gaps and high band-offsets across heterojunctions. High mobility operation requires high quality surfaces and interfaces, low phonon scattering rates and a channel with low electron and hole effective masses along the transport direction. These constraints often come with their own baggage. For instance, low mass often appears at the expense of band-gap, graphene being the most egregious offender whose ultralight electrons come tied with a zero band-gap. A low mass tends to increase gate oxide tunneling across ultrathin insulators and elevate the OFF current. Finally and perhaps most significantly, a high ON current requires not just high mobility, but a high carrier density, which paradoxically needs some of the effective masses — specifically, the 'density of states effective mass', to be sizeable.

In short, a well tempered transistor must navigate a complex design space of material and device parameters. Throughout the book, we discuss how numerous technological advances plus our emergent physical understanding have worked and continue to work relentlessly over decades to meet these challenges — from gate design to material engineering to novel switching mechanisms. Continuing the onslaught of device integration towards ever increasing computational power is, in a way, the secondary benefit. The prime benefit is that we have learned an enormous amount on how charge, spin and heat flow at their fundamental dimensions, how we quantify and model them in various embodiments and environments, and how we measure, predict and perhaps even utilize these properties for something potentially useful down the road.

PART 2

Equilibrium Tools:

Quantum Mechanics and Thermodynamics

Chapter 3

First quantization

In this chapter, we will recap quantum mechanics and its classical correspondence, and introduce the compact bra-ket notation. Quantum processes clearly underlie nano-electronics, sometimes implicitly through bandgaps and band effective masses, but often explicitly as in tunneling, energy quantization and interference. First quantization, which occupies much of standard texts in quantum mechanics, deals with the quantization of the electron wavelength through suitable boundary conditions on the electron wavefunction $\psi(\vec{r})$, for instance in a box or an atom. The wavefunction satisfies the nonrelativistic Schrödinger equation

$$i\hbar\frac{\partial\psi(\vec{r},t)}{\partial t} = \left(-\frac{\hbar^2\nabla^2}{2m} + U(\vec{r})\right)\psi(\vec{r},t). \tag{3.1}$$

While the outcome of a single quantum measurement is inherently uncertain, the odds of measuring a particular outcome are given by the probability density $|\psi|^2$ according to the above *deterministic* equation. That identical electrons can act independently, yet somehow build up collectively to an overall probability distribution, opens up a lot of thorny philosophical quandaries that make quantum mechanics so counter-intuitive. What makes it relevant to technology however, is the fact that most instrumental readings average over many single shot measurements, so we need their aggregate outcomes that can be predicted fairly accurately. We can extract the measured spectrum (eigenvalues) as well as expectation values for various observables by operating on ψ with different Hermitian operators, such as the position operator $\hat{x} = \vec{x}$ or the momentum operator $\hat{p} = -i\hbar\vec{\nabla}$, satisfying a commutation relation $[\hat{x}, \hat{p}] = \hat{x}\hat{p} - \hat{p}\hat{x} = i\hbar$. Properties such as the Heisenberg uncertainty principle $\sigma_x\sigma_{p_x} \geq \hbar/2$ follow naturally from the underlying wave mathematics and commutation algebra, as we discuss shortly (around Eq. (3.5)).

3.1 When does a quantum particle act 'classically'?

It is worth asking when a quantum object tends to behave classically. In other words, is it better to interpret an electron as an extended wave or a lumped particle? Since the wave identity of the electron is parametrized by its wavelength, we expect an electron to act classically if *it encounters a potential that varies very slowly over a wavelength*, generating minimal force and a negligible buildup of phase. This is analogous to geometrical optics, where photons follow definite trajectories when their wavelengths are small. The electron de Broglie wavelength $\lambda = h/\sqrt{2mE}$ grows at lower temperatures where $E \sim k_B T$ plummets, so we expect quantum coherence to be prominent at lower temperatures. At higher temperatures, not only does the wavelength shrink, but in addition thermal fluctuations across the environment tend to randomize the phase through multiple scattering events integrated over multiple electron ensembles, making it act classically. We will explore the nature of these decoherence events later in the book.

That the abruptness of the potential determines its wave nature can be established formally by plugging a generic complex wavefunction $\Psi(x) = A(x) \exp[i\phi(x)]$ (A, ϕ real) into the one dimensional Schrödinger equation (Eq. (3.1)), and separating out its real and imaginary parts, yielding

$$\frac{-\hbar^2}{2m}\left[A'' - A(\phi')^2\right] + (U - E)A = 0,$$

$$2A'\phi' + A\phi'' = 0, \tag{3.2}$$

where $' \to d/dx$. The second equation gives us $A \propto 1/\sqrt{\phi'}$. The first can be simplified if we substitute this expression for A and drop the second derivative term A'', which then gives us $\phi(x) = \int_0^x k(y)dy$ where $k(y) = \sqrt{2m[E - U(y)]/\hbar^2}$. The final result then becomes

$$\psi(x) = \frac{1}{\sqrt{k(x)}} \exp\left(i \int_0^x k(y)dy\right). \tag{3.3}$$

What is 'classical' about this equation, and at what point did we invoke a classical approximation? The answer to the second question is that we ignored A'' compared to $2m(E - U)A/\hbar^2 = k^2 A$, whence substituting the two solutions above for ϕ and A, we see that this is equivalent to $(1/2\pi)|d\lambda/dx| \ll 1$, meaning that the change in wavelength over one wavelength must be small and thus the corresponding potential must also vary slowly. The classical aspect enters from the pre-factor $\propto 1/\sqrt{k(x)}$, which simply means the probability $|\psi|^2$ of finding a particle at a given point is inversely related to its velocity. This is consistent with classical physics, where the particle spends most time as it slows down to a small k and lingers around those turning points. Indeed, the exact quantum solutions to various potentials reveal such a peaking near slowdown (Fig. 3.1) only when we go to higher energy eigenstates that enjoy shorter wavelengths compared to the lengthscales of potential variations.

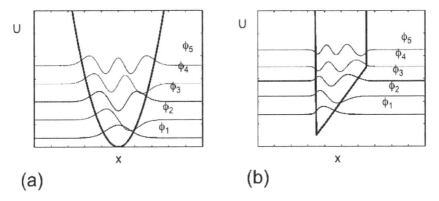

(a) **(b)**

Fig. 3.1 *(a) Eigenfunctions of the harmonic oscillator plotted at heights proportional to the energy eigenvalues. The levels are equidistant, a feature that helps us recognize vibrational modes in a molecular spectrum. The asymptotic solutions at $x \gg a_0 = \sqrt{\hbar/m\omega}$ vanish outside the well as Gaussian functions while those inside the well have a multiplicative factor involving Hermite polynomials to enforce orthogonality among solutions, $\Psi_l(x) \sim H_l(x/a_0) \exp\left[-x^2/2a_0^2\right]$. Relevant to this section however is the observation that the wavefunctions peak towards the edges following classical expectations, except the lowest solution which has no classical analogue. (b) This trend is even more striking for a tilted potential, for instance a quantum well under an electric field $\mathcal{E}_0$, whose eigenstates are Airy functions $A_i(x/b_0)$ where $b_0 = (2mq\mathcal{E}_0/\hbar^2)^{1/3}$. While higher energy solutions peak near the classical turning points away from the well bottom, the lowest quantum state breaks with the crowd by tilting right into it. The charge distribution under an electric field sets the channel capacitance in series with the gate oxide capacitance, and needs to be properly accounted for.*

Despite the inherent uncertainty embodied by the wavefunction, however, it is straightforward to show that *averages of the quantum observables follow classical rules*, as captured in **Ehrenfest's theorem**. Using Hamilton's equations discussed later (Eq. (10.21)) and the algebra of commutation, it is straightforward to show that the averages of position and momentum $\langle x \rangle = \int dx \psi^*(x) \hat{x} \psi(x)$ and $\langle p \rangle = \int dx \psi^*(x) \hat{p} \psi(x)$ satisfy (assuming 1D here)

$$\frac{d\langle x \rangle}{dt} = \frac{\langle p \rangle}{m},$$
$$\frac{d\langle p \rangle}{dt} = -\left\langle \frac{\partial U}{\partial x} \right\rangle. \tag{3.4}$$

These relations are exactly what classical Newtonian physics would predict. Where quantum trajectories differ from classical is in their deviation from these average values. From the commutator relation $[x, p] = i\hbar$, we can then show that the product of the standard deviations from the mean, $\sigma_x = \sqrt{\langle x^2 \rangle - \langle x \rangle^2}$ and $\sigma_p = \sqrt{\langle p^2 \rangle - \langle p \rangle^2}$, has a lower limit

$$\sigma_x \sigma_p \geq \hbar/2, \tag{3.5}$$

meaning that we are fundamentally destined to sacrifice precision on at least one member of the complementary pairs and live with the resulting uncertainty. The inequality follows by requiring that the norm of any linear combination of the deviations δx, δp from their averages is necessarily positive semidefinite, in other words, $\langle |\delta x + i\lambda \delta p|^2 \rangle \geq 0$ for any value of λ.

The uncertainty equation is not surprising from a signal processing point of view. Complimentary Fourier pairs satisfy an 'uncertainty' relation in the sense that signals that are well defined in time t, such as a delta function pulse, decompose into a lot of fast varying Fourier components and are thus spread out (uncertain) in ω space. Conversely, a well localized, monochromatic pulse in ω space corresponds to an extended plane wave in t. In other words, a signal must be either time limited or band limited. A similar uncertainty exists between the complementary Fourier pairs x and k. It is when we interpret k as momentum $p = \hbar k$ that we dive into the counterintuitive nature of quantum mechanics, since without the existence of both position and momentum, we can no longer precisely triangulate a quantum particle (as a GPS would do!), meaning that the quantum particle does not own a precise measurable trajectory *even in principle*.

If pressed, we can perhaps imagine a quantum particle following a blurred out trajectory spread out over a De Broglie wavelength $\sigma_x \sim \lambda$ around a classical average, its momentum equally blurred around a similar classical expectation. A single shot measurement of position or momentum will yield a precise reading lying somewhere on these blurred lines, with a likelihood given by the local intensity of the blur. For objects we encounter in our day to day life, such as a pen or a needle, the uncertainty is too small to matter (thermal jitter alone makes the spread orders of magnitude tighter than nuclear dimensions). For an electron however, this uncertainty is significant enough to be actually measurable.

Let us next simplify the notation to avoid specifying too many indices or stay tied to a particular representation, say x or k.

3.2   Basis sets, bras and kets

At this stage, we introduce two concepts that are helpful. The first is the idea of resolving a wavefunction $\Psi(x)$ into a suitable basis set $\{\psi_n(x)\}$

$$\Psi(x) = \sum_n c_n \psi_n(x). \tag{3.6}$$

Each basis element acts like a coordinate axis for Ψ, much like spatial coordinates describe position in 3D or Fourier components describe a periodic signal. In principle, we can choose any basis set that completely spans the dimension we are interested in. In practice we truncate the sum to a finite set of basis elements, whereupon particular basis sets, say chemical orbitals for molecules, become expedient.

Some of the popular basis sets in quantum chemistry and solid state physics are shown in Fig. 3.2. Plane wave bases are popular for solids, sometimes mixed with 'pseudopotentials' to avoid handling the core regions where Ψ varies a lot (these variations, driven by orthogonality, would require fine numerical meshes while not contributing to bonding chemistry). Periodic solids are often described by mixing plane waves with atomic states to generate extended Bloch states, as we discuss later. We can also study Fourier superpositions of Bloch states to generate Wannier functions that are localized around the atomic cores and look 'molecular'.

For molecules, Slater type orbitals $\sim r^l \exp(-\zeta|r|)$ resemble atomic eigenstates, but are frequently replaced with superpositions of Gaussian type orbitals $\sim r^l \exp(-\zeta r^2)$ for computational simplicity (products of displaced Gaussians are Gaussians and their derivatives are easy to extract). Notations like 6-31G double zeta signify that a core electron was described by 6 Gaussians, while each valence electron was split into one orbital with 3 Gaussians and another with 1 Gaussian. For practicality, we want a basis set that is small enough to be tractable yet flexible enough to capture

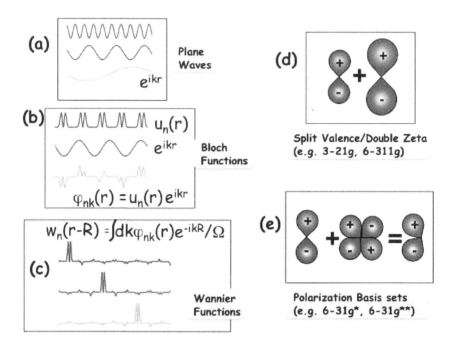

Fig. 3.2   *Popular Basis sets in solid state physics and chemistry. (a) Plane waves with varying wavevector* k, *(b) Bloch eigenstates of periodic solids, consist of a periodic array* $u_n(r)$ *of local atomic orbitals modulated by plane waves. (c) Localized Wannier basis sets in a solid are created by a Fourier superposition of Bloch functions and are mutually orthogonal. (d) Double zeta atomic basis sets consist of superpositions of atomic orbitals with varying radial exponents for flexible sizing. (e) Polarization basis sets mix higher orbitals (e.g.,* $p + d$*) to allow the lower orbitals (e.g.,* p*) to distort in response to an electric field.*

dynamics and mixing, requiring us to frequently go beyond the populated valence electron ground states. We can add higher orbitals to build flexibility into the basis sets, such as split-valence states with variable electron radii (responding for instance to electostatic compression) and polarization bases sets with variable orientation (breaking symmetry along certain directions in response to a field induced Stark effect).

To describe molecular solids, we often mix spherical orbital bases in the core region with plane wave bases outside, generating methods with acronyms such as KKR, APW and OPW. A challenge is to maintain orthogonality among basis sets for simplicity, while resembling true chemical orbitals which do overlap between atoms. There are often trade-offs, for instance localizing the basis sets maximally (e.g., with Wannier states) requires us to go beyond next-nearest neighbor interactions and deal with larger supercells. Choosing a suitable basis set is still much of an art.

The second concept we introduce is that of bras and kets. Once we know our basis set $\{\psi_n(x)\}$, we can express $\Psi(x)$ in Eq. (3.6) simply as the collection of expansion coefficients $\{c_n\}$ and define a *ket*

$$\left| \Psi \right\rangle = \begin{pmatrix} c_1 \\ c_2 \\ \dots \\ c_n \end{pmatrix}. \tag{3.7}$$

By that definition, we can write the basis sets themselves in ket form as

$$\left| \psi_1 \right\rangle = \begin{pmatrix} 1 \\ 0 \\ \dots \\ 0 \end{pmatrix}, \quad \left| \psi_2 \right\rangle = \begin{pmatrix} 0 \\ 1 \\ \dots \\ 0 \end{pmatrix}, \quad \left| \psi_n \right\rangle = \begin{pmatrix} 0 \\ 0 \\ \dots \\ 1 \end{pmatrix}. \tag{3.8}$$

The corresponding conjugate transpose representing a function $\Phi^*(x) = \sum_n b_n^* \psi_n^*(x)$ is called a *bra*

$$\left\langle \Phi \right| = \begin{pmatrix} b_1^* & b_2^* & \dots & b_n^* \end{pmatrix} \tag{3.9}$$

so that

$$\left\langle \psi_1 \right| = \begin{pmatrix} 1 & 0 & \dots & 0 \end{pmatrix}, \quad \left\langle \psi_2 \right| = \begin{pmatrix} 0 & 1 & \dots & 0 \end{pmatrix}, \dots. \tag{3.10}$$

Combining a bra and a ket with matrix multiplication rules, we get a *bra-ket* or *bracket* that represents the overlap of $\Psi(x)$ with $\Phi(x)$,

$$\int \Phi^*(x)\Psi(x)dx = \sum_{mn} b_m^* c_n \underbrace{\int \psi_m^*(x)\psi_n^*(x)dx}_{\delta_{mn}}$$

$$= \sum_n b_n^* c_n = \left\langle \Phi \middle| \Psi \right\rangle. \tag{3.11}$$

We have chosen our basis sets to be orthonormal above to allow the unique representation in Eq. (3.8). It's easy to verify from Eqs. (3.8) and (3.10) that $\left\langle \psi_m \middle| \psi_n \right\rangle = \delta_{mn}$, where δ is the Kronecker delta function.

A complete basis set $\{\psi_n\}$ must also span the entire N-dimensional space, as outlined by the completeness relation

$$\sum_n |\psi_n\rangle\langle\psi_n| = \begin{pmatrix} 1 & 0 & \ldots & \ldots & 0 \\ 0 & 1 & 0 & \ldots & 0 \\ \ldots & & & & \\ 0 & 0 & \ldots & 0 & 1 \end{pmatrix} = I \quad \text{(Identity matrix)}. \tag{3.12}$$

Completeness guarantees that any arbitrary function $|F\rangle$ in N-dimensions is representable in this basis set as a ket with coefficients α_n, since

$$|F\rangle = I|F\rangle = \sum_n |\psi_n\rangle \underbrace{\langle\psi_n|F\rangle}_{\alpha_n} \tag{3.13}$$

The bra ket notation is conceptually and notationally simple in that we can simply refer to it abstractly as a collection of coefficients without explicitly invoking the basis set every time. To get a concrete representation, say in real or momentum space, we just take the overlap of the ket with the suitable bra, $\psi(x) = \langle x|\Psi\rangle$, $\psi_k = \langle k|\Psi\rangle$, etc.

3.3 Representations in quantum mechanics and evolution of quantum states

We now discuss several equivalent ways of approaching the Schrödinger equation. They differ in what variables actually evolve with time, but are consistent in that they maintain the same measurable average.

3.3.1 *Schrödinger representation: wavefunction evolving*

The 'standard' introduction to quantum mechanics starts by assuming that all the time dependence lies in the wavefunction, which satisfies the time-dependent Schrödinger equation

$$i\hbar\frac{\partial}{\partial t}\Psi(t) = \hat{\mathcal{H}}(t)\Psi(t), \tag{3.14}$$

whose formal solution is

$$\boxed{\begin{array}{c} \underline{\text{Schrödinger representation}} \\ |\Psi_S\rangle = \Psi_S(t) = \underbrace{e^{-i\int_{t'}^{t}\hat{\mathcal{H}}(\tau)d\tau/\hbar}}_{U(t,t')}\Psi(t'). \end{array}} \tag{3.15}$$

Thus, the expectation value of an observable, which is ultimately what we care about, can be written in matrix notation as

$$\langle \hat{\mathcal{O}} \rangle = \langle \Psi_S | \hat{\mathcal{O}} | \Psi_S \rangle = \int d\Omega \; \psi^* U^\dagger \hat{\mathcal{O}} U \Psi, \qquad (3.16)$$

where $d\Omega$ is differential volume and U is a unitary operator satisfying $UU^\dagger = U^\dagger U = I$ that enforces time reversal symmetry. Operating by U takes a quantum state forward, and operating thereafter by $U^\dagger$ restores it perfectly. The dagger indicates the combination of conjugation and matrix transpose (turning rows into columns and vice-versa).

3.3.2 *Heisenberg representation: operator evolving*

We could of course preserve this average by shoving the entire time-dependence onto the operators, specifically the combination $U^\dagger \hat{\mathcal{O}} U$ above, by making the latter time-dependent instead of the wave-function

$$\boxed{\begin{array}{c} \underline{\text{Heisenberg representation}} \\[4pt] \hat{\mathcal{O}}_H(t) = U^\dagger \hat{\mathcal{O}} U \end{array}} \qquad (3.17)$$

so that $\langle \hat{\mathcal{O}} \rangle = \int d\Omega \psi^* \hat{\mathcal{O}}_H \psi$.

Let us work out the time evolution equation for the operator. From the definition of U above, we get $dU/dt = -i\hat{\mathcal{H}}U/\hbar$, so that

$$\frac{d\hat{\mathcal{O}}_H(t)}{dt} = U^\dagger \frac{\partial \hat{\mathcal{O}}}{\partial t} U + \underbrace{\frac{\partial U^\dagger}{\partial t}}_{iU^\dagger \hat{\mathcal{H}}/\hbar} \hat{\mathcal{O}} U + U^\dagger \hat{\mathcal{O}} \underbrace{\frac{\partial U}{\partial t}}_{-i\hat{\mathcal{H}}U/\hbar}$$

$$= \left. \frac{\partial \hat{\mathcal{O}}(t)}{\partial t} \right|_H + \frac{i}{\hbar} U^\dagger \left[\hat{\mathcal{H}}\hat{\mathcal{O}} - \hat{\mathcal{O}}\hat{\mathcal{H}} \right] U$$

$$= \left. \frac{\partial \hat{\mathcal{O}}(t)}{\partial t} \right|_H + \frac{1}{i\hbar} \left[\hat{\mathcal{O}}, \hat{\mathcal{H}} \right]_H, \qquad (3.18)$$

giving us the dynamics of the operator that now carries all time evolution information.

3.3.3   *Interaction representation: wavefunction slow, operator fast*

The idea here is to separate the Hamiltonian into a part whose eigenspectrum is known and another 'interacting' part whose response is the main unknown. Under this condition, we can find yet another way to divvy up the time dependence so as to preserve the operator average, one where *the wavefunction only evolves under the (typically small and perturbative) interacting Hamiltonian while the operator evolves under the larger non-interacting Hamiltonian.* In other words, the wavefunction evolves slowly, and the operator evolves fast.

$$\hat{\mathcal{H}} = \hat{\mathcal{H}}_O + \hat{\mathcal{H}}_T(t),$$

$$\frac{d\hat{O}_I(t)}{dt} = \frac{1}{i\hbar}\left[\hat{O}_I, \hat{\mathcal{H}}_O\right],$$

$$i\hbar\frac{\partial\Psi_I}{\partial t} = \hat{\mathcal{H}}_{TI}\Psi_I. \tag{3.19}$$

It is straightforward to show that this can be accomplished by making both the wavefunction and the operator time-dependent, starting with the Schrödinger representation, but instead of using the entire Hamiltonian, separating out the trivial $\hat{\mathcal{H}}_O$ phases associated with the non-interacting part. The interacting part $\mathcal{H}_{TI}$ is hidden implicitly in $\Psi_S(t)$ below, and explicitly shows up in the time evolution of the density matrix (Eq. (3.25)), discussed shortly.

$$\boxed{\begin{array}{c} \underline{\text{interaction representation}} \\[4pt] \hat{O}_I(t) = U_O^\dagger \hat{O}_S U_O \\[4pt] |\Psi_I(t)\rangle = U_O^\dagger |\Psi_S(t)\rangle \\[4pt] U_O(t, t') = \exp\left(-i\int_{t'}^{t}\hat{\mathcal{H}}_O(\tau)d\tau/\hbar\right). \end{array}} \tag{3.20}$$

3.4   Evolution of statistically mixed states

So far, we have been dealing with 'pure' quantum states, such as phase coherent superpositions of eigenstates, $|\Psi\rangle = \sum_n c_n|\psi_n\rangle$. The outcome of a measurement on these states is uncertain and probabilistic, with eigenvalue ϵ_n corresponding to state $|\psi_n\rangle$ appearing with probability $|c_n|^2$. However, the point to emphasize is that the probabilities themselves are completely deterministic, time-reversible, and the underlying superposition state $|\Psi\rangle$ is spelt out with absolute precision. In nature however, a separate kind of randomness and uncertainty arises, such as from thermal fluctuations that create a mixed ensemble out of such pure states. In fact, those are the kinds of states that we will commonly encounter in quantum transport (the channel eigenstates will be populated by a thermalized mixture of electrons injected from the contacts). The thermal average of such a state is then given by a probabilistic mixture

$$\langle A \rangle = \sum_n \underbrace{p_n}_{\text{mix probability}} \times \underbrace{\langle \psi_n | \hat{A} | \psi_n \rangle}_{\text{pure state average}} . \tag{3.21}$$

To formalize this procedure, we define a density matrix

$$\hat{\rho} = \sum_n p_n |\psi_n\rangle\langle\psi_n| = \begin{pmatrix} p_1 & 0 & \dots \dots & 0 \\ 0 & p_2 & 0 & \dots & 0 \\ \dots & & & & \\ 0 & 0 & \dots & 0 & p_N \end{pmatrix}, \tag{3.22}$$

describing the mixing statistics, so that the final *quantum statistical* average is simply obtained by doing a trace

$$\langle A \rangle = \text{Tr}\left(\hat{\rho}\hat{A}\right). \tag{3.23}$$

These equations are valid even for non-orthogonal bras and kets, where the density matrix is non-diagonal (the diagonal elements still designate the occupation probabilities of the states while the off-diagonal elements designate the 'coherence' between the involved states). For a pure state like Eq. (3.6), each entry for p_i is either one or zero, meaning $p_i^2 = p_i$, which in turn implies that the density matrix for a pure state is *idempotent*, i.e., $\hat{\rho}^2 = \hat{\rho}$.

Let us now work out the time evolution of such a mixed state, as a precursor to studying quantum transport. While we will not invoke this equation initially and go with intuitive expressions, we will need to utilize this algebra when approaching a rigorous derivation of quantum flow, as in the last few chapters on correlations. Let us also write this out in the interaction representation. The corresponding density matrix $\hat{\rho}_I(t) = |\Psi_I(t)\rangle\langle\Psi_I(t)|$

satisfies the equation

$$i\hbar \frac{d\hat{\rho}_I(t)}{dt} = \underbrace{i\hbar \frac{d|\Psi_I(t)\rangle}{dt}\langle\Psi_I(t)|}_{\hat{\mathcal{H}}_{TI}|\Psi_I(t)\rangle} + |\Psi_I(t)\rangle \underbrace{\left(i\hbar \frac{d\langle\Psi_I(t)|}{dt}\right)}_{-\langle\hat{\Psi}_I(t)|\hat{\mathcal{H}}_{TI}}$$

$$= \left[\hat{\mathcal{H}}_{TI}(t)\,,\,\hat{\rho}_I(t)\right]. \tag{3.24}$$

In other words,

$$\frac{d\hat{\rho}_I(t)}{dt} = -\frac{i}{\hbar}\left[\hat{\mathcal{H}}_{TI}(t), \hat{\rho}_I(t)\right]. \tag{3.25}$$

We will late encounter a many body Hamiltonian, such as for a quantum dot coupled with contacts (Eq. (6.26)), with a dot Hamiltonian $\hat{\mathcal{H}}_d$, a contact Hamiltonian $\hat{\mathcal{H}}_C$ and a coupling term $\hat{\mathcal{H}}_T$. They will each involve a set of creation/annihilation operators that obey commutation/anticommutation rules in order to enforce symmetry. A distinction we want to exploit is that the contacts are 'simple' and do not evolve from their thermal equilibrium states while the channel along with its own internal interactions evolves in time. In keeping the contact states pinned to equilibrium, they must lose all the phase memory imparted by the channel electrons, and thus rob the channel of its memory — disentangling the states in the process. The formal way to ensure this is as follows: We evolve all operators in the *interaction representation* (described above), separating out the Hamiltonian into the active channel or 'dot' with all the interesting and critical interactions built into it, and a passive 'contact' that does not evolve.

$$\hat{O}_I(t) = e^{i\mathcal{H}_0 t}\hat{O}e^{-i\mathcal{H}_0 t}, \quad \hat{\mathcal{H}}_0 = \hat{\mathcal{H}}_d + \hat{\mathcal{H}}_C. \tag{3.26}$$

Integrating Eq. (3.25) for the many body density matrix gives $\hat{\rho}_I(t) = \hat{\rho}_I(0) - \frac{i}{\hbar}\int_0^t dt'\left[\hat{\mathcal{H}}_{TI}(t'), \hat{\rho}_I(t')\right]$.

Substituting this expression for $\hat{\rho}_I(t)$ into the right side of Eq. (3.25),

$$\boxed{\frac{d\hat{\rho}_I(t)}{dt} = -\frac{i}{\hbar}\left[\hat{\mathcal{H}}_{TI}(t), \hat{\rho}_I(0)\right] - \frac{1}{\hbar^2}\int_0^t dt'\left[\hat{\mathcal{H}}_{TI}(t), \left[\hat{\mathcal{H}}_{TI}(t'), \hat{\rho}_I(t')\right]\right].}$$

$$\tag{3.27}$$

Next, we want to extract the thermodynamic average of the dot over a specific contact state, say the contact designated by α ($\alpha = L, R$), assuming it is in local equilibrium at electrochemical potential μ_α and inverse temperature $\beta_\alpha = 1/k_B T_\alpha$.

$$\langle \hat{\mathcal{O}} \rangle_\alpha = \mathrm{Tr}_{C\alpha} \left[\hat{\rho}_\alpha \hat{\mathcal{O}} \right] = \mathrm{Tr}_{C\alpha} \left[\frac{e^{-\beta_\alpha \left(\hat{\mathcal{H}}_{C\alpha} - \mu_\alpha \hat{N} \right)} \hat{\mathcal{O}}}{\mathcal{Z}_\alpha} \right], \qquad (3.28)$$

where $\mathcal{Z}_\alpha = \mathrm{Tr} \left[e^{-\beta_\alpha \left(\hat{\mathcal{H}}_{C\alpha} - \mu_\alpha \hat{N} \right)} \right]$ is the so-called partition function invoked to enforce conservation of total probability.

Since we are interested primarily in the non-equilibrium properties of the dot, we will trace out both contact states, and work with a 'reduced density matrix'. The next section shows how to do a partial trace.

$$\hat{\mathcal{O}}_d = \mathrm{Tr}_{L,R} \left[\hat{\rho}_L \hat{\rho}_R \hat{\mathcal{O}} \right]. \qquad (3.29)$$

Each matrix $\hat{\rho}_{L,R}, \hat{\mathcal{O}}$ above has the same size spanning all coordinates of the entire left contact-channel-right contact system, but with only relevant coordinate blocks populated (the rest of the blocks amount to identity). The key point is that in the process of reducing our system by voluntarily throwing away detailed contact information and replacing $\hat{\rho}_{L,R}$ with a pre-ordained Boltzmann average (Eq. (3.28)), we introduce an irreversible, *non-unitary* evolution in time that is at the heart of transport. This will require that the evolution $U = e^{-i\hat{\mathcal{H}}t/\hbar}$ incorporate a non-Hermitian component, $\mathcal{H} \to \mathcal{H} + \Sigma$.

To implement the above steps for a many body system, we will discuss creation/annihilation operators in a few chapters. We will pick up the density matrix evolution again later on, when we use it derive the master equation (Section 26.2.3).

3.5 Partial traces and reduced density matrices

Let us work out one example of how we trace out selectively over one variable. Consider a system with a spin (up or down) sitting at each of two sites A and B. The individual basis sets for spin at site A, spin at site B, and jointly occupied spins at sites A and B can be written as

$$\left| \psi^A_{1,2} \right\rangle = \left\{ | \uparrow_A \rangle, | \downarrow_A \rangle \right\}, \quad \left| \psi^B_{1,2} \right\rangle = \left\{ | \uparrow_B \rangle, | \downarrow_B \rangle \right\}$$

$$\left| \psi^{AB}_{11,12,21,22} \right\rangle = \left\{ | \uparrow_A \uparrow_B \rangle, | \uparrow_A \downarrow_B \rangle, | \downarrow_A \uparrow_B \rangle, | \downarrow_A \downarrow_B \rangle \right\}. \qquad (3.30)$$

Note that we can write the joint basis sets as a direct product

$$\psi^{AB}_{ij} = \psi^A_i \otimes \psi^B_j, \qquad (3.31)$$

which simply means we tag the basis sets onto each other combinatorially.

Consider now a superposition state $|\Phi\rangle = (|\uparrow_A\uparrow_B\rangle + 2|\uparrow_A\downarrow_B\rangle + |\downarrow_A\downarrow_B\rangle)/\sqrt{6}$. The corresponding density matrix $\rho = |\Phi\rangle\langle\Phi|$

$$
\rho = \begin{array}{c} \\ \langle\uparrow_A\uparrow_B| \\ \langle\uparrow_A\downarrow_B| \\ \langle\downarrow_A\uparrow_B| \\ \langle\downarrow_A\downarrow_B| \end{array}
\begin{array}{cccc} |\uparrow_A\uparrow_B\rangle & |\uparrow_A\downarrow_B\rangle & |\downarrow_A\uparrow_B\rangle & |\downarrow_A\downarrow_B\rangle \end{array}
\left(\begin{array}{cccc}
1 & 2 & 0 & 1 \\
2 & 4 & 0 & 2 \\
0 & 0 & 0 & 0 \\
1 & 2 & 0 & 1
\end{array}\right)/6.
$$

$$(3.32)$$

We can see it's a pure state because of its idempotency, $\rho^2 = \rho$.

Let us see how to reduce the matrix to basis set ψ^A by partially tracing over states of spin B, or to basis set ψ^B by partially tracing over states of spin A. The ijth element of the reduced matrix $\mathrm{Trace}_B(\hat\rho)$ consists of the value of the spin at A in the ith basis set crossed with the value of the spin at A in the jth basis set, summed over all intermediate spin values for site B. For ρ_A, this amounts to tracing over each 2×2 submatrix. For ρ_B, we skip an element along each diagonal.

$$
\rho_A(1,2) = \left\langle \psi_1^A \middle| \rho_A \middle| \psi_2^A \right\rangle = \sum_i \left\langle \psi_{1i}^{AB} \middle| \rho \middle| \psi_{2i}^{AB} \right\rangle
$$

$$
= \sum_i \left\langle \psi_1^A \otimes \psi_i^B \middle| \rho \middle| \psi_2^A \otimes \psi_i^B \right\rangle = \sum_i \left\langle \uparrow_A \otimes \psi_i^B \middle| \rho \middle| \downarrow_A \otimes \psi_i^B \right\rangle
$$

$$
= \left\langle \uparrow_A\uparrow_B \middle| \rho \middle| \downarrow_A\uparrow_B \right\rangle + \left\langle \uparrow_A\downarrow_B \middle| \rho \middle| \downarrow_A\downarrow_B \right\rangle = (0+2)/6 = 1/3. \quad (3.33)
$$

Continuing this exercise component by component, we find

$$
\rho_A = \mathrm{Trace}_B(\rho) = \frac{1}{6}\begin{pmatrix} 5 & 2 \\ 2 & 1 \end{pmatrix}, \qquad \rho_B = \mathrm{Trace}_A(\rho) = \frac{1}{6}\begin{pmatrix} 1 & 2 \\ 2 & 5 \end{pmatrix}. \quad (3.34)
$$

Quantum transport requires a partial trace over the contact states (Section 26.2.3), a non-unitary operation amounting to throwing away information. We associate the information reset with rapid thermalization in the contacts. The reader can easily verify that unitarity is necessary to preserve the idempotency properties of ρ. Thus the non-unitary operation of a selective trace on the original pure state ends up generating mixed non-pure states. Indeed, it is easy to check from Eq. (3.34) that $\rho_{A,B}^2 \neq \rho_{A,B}$.

Chapter 4

Quantum chemistry: atoms to molecules

4.1 Bonds, bands and resonances

Quantum mechanics stipulates that the dynamics of an electron is governed by a wave equation that sets its probability amplitude. Imposing boundary conditions on the wave limits their allowed wavelengths, just like the cut off modes in a waveguide or the strings of a guitar. In quantum physics, we relate the momentum of the electron, a particulate property, with its wavelength, a wave property (it is this correspondence that often makes the topic counter intuitive). This means that the energy levels of an electron in a confined space, such as a potential box or an orbiting ring, are also discretely quantized.

The simplest 1D quantum problem we learn to solve is a particle in a box, whose potential is zero inside a box and infinite outside. Classically, the particle would have a uniform distribution over the length of the well. Quantum mechanically however, each allowed wavefunction ϕ_n must vanish at the edges. The electron has zero probability of being outside, and wavefunction continuity means that they must vanish at the ends. At the same time, the eigenfunctions must satisfy three constraints:

- they must each try to minimize their curvature when sorted by energy (since the curvature $\sim \nabla^2$ is proportional to the kinetic energy, the potential energy being zero here).
- They must develop nodes (i.e., zeros) in space in order to be orthogonal to each other. The lowest energy solution has the least nodes to minimize curvature, and the number grows up from there. The mathematical origin of orthogonality is the Hermiticity of the Hamiltonian. Physically, we expect each 'mode' to be independent to the others with zero overlap.

- Finally, we set each eigenfunction to be normalized so that the total probability is unity. In other words, $\int dx \phi_m(x)\phi_n(x) = \delta_{mn}$.

The solutions are superpositions of free electron plane waves satisfying the correct boundary conditions, in this case, sinusoids $\phi_n(x) = \sqrt{(2/L)} \sin(n\pi x/L)$, $0 \leq x \leq L$ with integral n. We next show how they lead to the emergence of orbital symmetries, bonds and bands in multiple dimensions.

4.2 The hydrogen atom

Perhaps the only atomic system for which the Schrödinger equation can be solved analytically is the hydrogen atom. Hydrogen has a single electron and thus lacks any complication with electron correlation. The total Hamiltonian can be separated conveniently into radial and angular parts using spherical coordinates (r, θ, ϕ)

$$\hat{\mathcal{H}}\psi = \left(\frac{-\hbar^2 \nabla^2}{2m} - \frac{q^2}{4\pi\epsilon_0 r} \right)\psi$$

$$= \frac{-\hbar^2}{2mr}\frac{\partial^2(r\psi)}{\partial r^2} + \left(\frac{\hat{L}^2(\theta,\phi)}{2mr^2} - \frac{q^2}{4\pi\epsilon_0 r} \right)\psi, \tag{4.1}$$

where the angular part of ∇^2 can be recast entirely in terms of the angular momentum squared $\hat{L}^2$ divided by the moment of inertia mr^2. The wavefunction is accordingly separable as well,

$$\psi = C\frac{R_{nl}(r)}{r}Y_{lm}(\theta,\phi) \tag{4.2}$$

with C being an overall normalization constant in 3D. The hydrogen atom solution is worked out in standard quantum mechanics textbooks:

Hydrogen Atom Eigenvalues and modes

$$E_n = \underbrace{\left(\frac{-q^2}{8\pi\epsilon_0 a_0} \right)}_{-13.6\ eV} \times \frac{1}{n^2}, \quad (n = 1, 2, 3, \ldots), \qquad a_0 = \underbrace{\frac{4\pi\epsilon_0 \hbar^2}{mq^2}}_{0.529\ \mathring{A}}$$

$$Y_{lm}(\theta,\phi) \propto P_{lm}(\cos\theta)e^{im\phi}$$

$$R_{nl}(r) \propto \rho^{l+1}e^{-\rho}L_{n-l-1}^{2l+1}(\rho), \quad \rho = r/na_0,$$

$$\tag{4.3}$$

where L_{n-l-1}^{2l+1} is an associated Laguerre polynomial and P_{lm} is an associated Legendre polynomial. The angular part of the wavefunction is a spherical harmonic $Y_{lm}(\theta, \phi)$, which is the simultaneous eigenstate of the commuting angular momentum operators $\hat{L}^2$ and $\hat{L}_z = p_\phi = -i\hbar\partial/\partial\phi$, supporting eigenvalues $l(l+1)\hbar^2$ and $m\hbar$, with $m = -l, -l+1, -l+2, \cdots, l-2, l-1, l$.

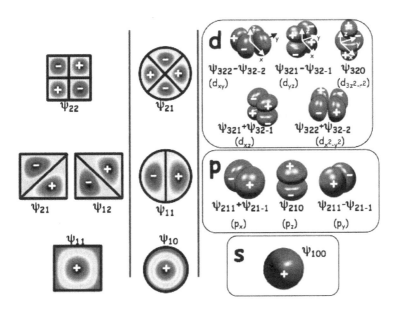

Fig. 4.1 *(Left) Angular modes in a square well, obtained by pairing the individual x and y directed sinusoidal 'particle in a box' solutions. (Middle) Evolution of angular modes into a circular well, separating the radial nodes with 2D angular orbitals. (Right) Evolution onto a sphere, generating spherical harmonics that describe s, p and d orbitals (the individual orbitals are actually the real and imaginary parts of the spherical harmonics, labeled below each plot). The number of nodal lines in 2D and nodal planes in 3D increases with angular momentum quantum number l, making the modes more 'lobey' and allowing them to interlace and maintain orthogonality with each other*

Let us try to understand the nature of the solutions. Since the Coulomb potential is spherically symmetric (independent of angle), the Hamiltonian does not vary with θ, ϕ and the corresponding angular momenta p_θ, p_ϕ are conserved according to Hamilton's equations (Eq. (10.15)). The angular part $Y_{lm}(\theta, \phi)$, operates effectively as a free particle in angular space, since the Coulomb potential does not vary along θ, ϕ and the corresponding forces are zero. We expect the solution of a free electron to be a plane wave, and indeed this is what we see for the ϕ part, $\sim e^{im\phi}$. The ϕ

rotation takes an electron along a line of latitude with angle designated by m and axial wavevector $\hat{z}$. Since $\hat{z}$ is fixed in space, the rotation gives us the expected plane wave solution. In contrast, the θ rotation proceeds along a line of longitude. The axial vector perpendicular to the longitudinal plane changes from point to point, making even the free electron solution along θ complicated. However, there are some patterns one can establish with the θ solutions. For instance, as the angular momentum quantum number l increases so does the kinetic energy and the curvature of the modes. As a result, the θ solutions grow increasingly 'lobey', picking up more and more nodal planes where they vanish, starting from zero nodes for $l = 0$, one node for $l = 1$, two for $l = 2$ and so on.

The angular nodes all interlace neatly among themselves to enforce spatial orthogonality in the θ, ϕ plane (for instance, orthogonality among 2s and 2p), and are called 'orbitals'. The orbitals represent the equivalent of fundamental modes that one sees in an excited drum membrane (Fig. 4.1 middle), except the membrane here is stretched out over a sphere (Fig. 4.1 right). The convoluted shapes also enforce completeness, so that the sum of orbital probability densities within a given l equals a constant.

Substituting the spherical harmonics in Eqs. (4.1) and (4.2), we find that the radial part of the wavefunction lives in an effective one dimensional potential that is constructed out of the attractive Coulomb potential plus a repulsive centrifugal potential. The centrifugal force $F_C = mv^2/r = L^2/mr^3$, leading to the centrifugal potential $V_C(r) = -\int_\infty^r F_C(r')dr' = L^2/2mr^2$ classically, and $\hat{L}^2/2mr^2$ quantum mechanically, with eigenvalues $l(l+1)\hbar^2$ for $\hat{L}^2$

$$\left(\frac{-\hbar^2}{2m}\frac{d^2}{dr^2} + \underbrace{\frac{l(l+1)\hbar^2}{2mr^2}}_{\text{Centrifugal}} - \underbrace{\frac{q^2}{4\pi\epsilon_0 r}}_{\text{Coulomb}} \right) R_{nl} = E_n R_{nl}. \tag{4.4}$$

The resulting 1d potential well determines the quantized levels of the hydrogen atom.

The radial solution $R_{nl}(r)/r$ extracted from Eq. (4.4) has three parts, as seen in Eq. (4.3). At large distances, the wavefunction decays as a WKB solution $\sim \exp(-\kappa r)$ with κ related to the total binding energy $|E|$ (the relations spelt out just before Eq. (14.1)). Higher energy solutions decay faster. At low distances near $r = 0$, the radial function follows a power law profile $\sim r^l$ which makes the electron wavefunction vanish at the nucleus, kept away by centrifugal repulsion. The exceptions are s electrons ($l = 0$) that have zero angular momentum, no centrifugal barrier and display a prominent cusp near $r = 0$. Note that we pulled out a $1/r$ factor at the start in Eq. (4.2) so that $R_{nl}(r)$ in Eq. (4.4) truly sees a 1-D environment, normalized over the one dimensional radial length dr rather that the three dimensional radial volume $r^2 dr$. At intermediate distances, the radial function has $n - l - 1$ nodes where the wavefunction vanishes. These nodes arising from the Laguerre polynomials enforce orthogonality among the radial modes, for instance, between the 1s and 2s states, and define the limits, $l = 0, 1, 2, \ldots, n - 1$.

4.3 Beyond hydrogen: electron-electron interactions

When we go down the periodic table from hydrogen, the main complication is the emergence of electron-electron Coulomb repulsion. This complicates the physics, because of the simultaneous need to maintain an exchange term (Pauli exclusion) between parallel spins, as well as higher order correlations involving the admixture of all possible excited states. The accurate way to deal with these elements would be to solve their spectra using full configuration interaction. We will discuss this in the chapter on correlation. However, for elements with lower atomic number, we can simplify the physics considerably, pretending that the electron electron repulsion acts as an average field that retains spherical symmetry, (or that associated with the underlying lattice periodicity, as we typically do in Schrödinger-Poisson solvers for device band structures). Under this condition, we can still mandate a separation of variables $\psi = C R_{nlm}(r) Y_{lm}(\theta, \phi)/r$ as in Eq. (4.2), and keep the angular part $Y_{lm}(\theta, \phi)$ and the underlying orbital structure intact. We assume each electron swims in an average, spherically symmetric field created and felt equally by all the other $Z - 1$ electrons (Z is the atomic number), which is the essence of the Hartree Fock approximation (the precise nature of the 'averaging' is discussed later in Eqs. (27.1) and (27.2)). The radial equation modifies to include the Coulomb corrections

$$\left(\frac{-\hbar^2}{2m} \frac{d^2}{dr^2} + \frac{l(l+1)\hbar^2}{2mr^2} - \frac{q^2}{4\pi\epsilon_0 r} + \underbrace{V_H(r)}_{\text{Hartree}} \right) R_{nlm}(r)$$

$$+ \underbrace{\int dr' V_F(r, r') R_{nlm}(r')}_{\text{Fock}} = E_{nlm} R_{nlm}(r) \qquad (4.5)$$

where we sum over occupied states 'occ'

$$V_H(r) = \frac{q^2}{4\pi\epsilon_0} \frac{Z-1}{Z} \int \sum_{abc}^{occ} \frac{|R_{abc}(r')|^2}{|\vec{r} - \vec{r'}|} dr'$$

$$V_F(r, r') = -\frac{q^2}{4\pi\epsilon_0} \frac{Z-1}{Z} \sum_{abc}^{occ} \frac{R_{abc}^*(r') R_{abc}(r)}{|\vec{r} - \vec{r'}|}.$$ (4.6)

The kernel can be simplified using standard identities, $1/|\vec{r} - \vec{r'}| = \sum_{l=0}^{\infty} r_<^l P_l(\cos\theta)/r_>^{l+1}$, where $r_< = \min(r, r')$, $r_> = \max(r, r')$ and θ is the angle between $\vec{r}, \vec{r'}$. For spherically symmetric distributions like we assume above in our mean-field approximation, we only take $l = 0$ and keep track of $r_{<,>}$ when doing the radial integral over all space.

We will derive the Hartree-Fock equations shortly, when we discuss many electron wavefunctions (Eqs. (6.25)). We can however try to interpret the terms physically. The Hartree term is the usual repulsion from all other occupied states that Poisson's equation takes into account. The Coulomb repulsion arises from the charge density of all occupied states $\sim |R_{abc}|^2$, scaled by $(Z-1)/Z$ to exclude self interaction (a crude way to account for the fact that an electron should not feel a potential due to itself). The equation must be evaluated self consistently since the potential determines R (Eq. (4.5)) and is in turn determined by it (Eq. (4.6)). While the full potential outlined later in Eq. (6.25) includes electronic interactions precisely, by extracting an average one electron potential above by summing over occupied states, invoking a set of approximations that we spell out later in detail (Eq. (27.1)), we have in fact ignored correlations in the system.

We will see later how the nonlocal Fock term enforces Pauli exclusion and arises from electron wavefunction antisymmetry. Device models often ignore the Fock term explicitly, but account for it *ex post facto* when filling up the electronic levels with a maximum of two opposite spins per level.

4.4   Role of direct Coulomb repulsion: screening

4.4.0.1   *Screening in molecules*

What is the main role of Coulomb interaction? We start with the Hartree part. The repulsion of outer valence electrons by inner core electrons leads to an effective 'screening' of the nuclear potential, so that the outer electrons see less of an attraction, $Z \to Z_{\text{eff}}$. We can test this with Helium. Inspired by the lowest energy 1s wavefunction for hydrogenic electrons, we generalize the solution in Eq. (4.3) to 1 electron with Z protons,

$$\psi_{H_{1s}}(r) = C R_{00}(r) Y_{00}(\theta, \phi)/r = \sqrt{\frac{Z^3}{\pi a_0^3}} e^{-Zr/a_0}. \tag{4.7}$$

We can conjecture a similar two electron wavefunction for Helium, however one that perceives a screened nuclear charge density $Z \neq 2$

$$\psi_{He}(\vec{r}_1, \vec{r}_2) = \psi_{H_{1s}}(\vec{r}_1)\psi_{H_{1s}}(\vec{r}_2) = \frac{Z^3}{\pi a_0^3} e^{-Z(r_1 + r_2)/a_0}. \tag{4.8}$$

We can then calculate the expected energy of the atom by writing down the Hamiltonian for the two Helium electrons

$$\hat{\mathcal{H}} = -\frac{\hbar^2 \nabla_1^2}{2m} - \frac{\hbar^2 \nabla_2^2}{2m} - \frac{2q^2}{4\pi\epsilon_0 r_1} - \frac{2q^2}{4\pi\epsilon_0 r_2} + \frac{q^2}{4\pi\epsilon_0 r_{12}}. \tag{4.9}$$

Each of the integrals involved in computing the average of $\hat{\mathcal{H}}$ in our guessed wavefunction $\psi_{He}(r_1, r_2)$ can be calculated analytically in elliptical coordinates. This is a useful mathematical exercise, but let's jump to the final answer. The expected value of the energy is given by

$$E = \left\langle \hat{\mathcal{H}} \right\rangle = \int \psi_{He}^* \hat{\mathcal{H}} \psi_{He} d^3 r_1 d^3 r_2 = \frac{q^2}{4\pi\epsilon_0 a_0} \left(Z^2 - 4Z + 5Z/8 \right). \tag{4.10}$$

If we treat Z as a variational parameter for our trial wavefunction, then the lowest energy solution is obtained by minimizing the expected energy above with respect to Z. The result is quite illuminating

$$Z_{\min} = 27/16 \approx 1.7$$
$$E = -77.5 \text{ eV}. \tag{4.11}$$

The Helium atom sees almost 15% screening on average that reduces its perceived nuclear charge. The total energy of the atom is given by adding the energy to take out the first electron in presence of the full screening influence of the second (reducing the hydrogen binding energy $2^2 \times 13.6 = 54.4$ eV to only about 25 eV; keep in mind that $Z = 1.7$ is the average screening over both electrons), plus the energy to take out the second electron in absence of any screening (54.4 eV), giving us a total energy of 79.4 eV, quite comparable with the number we worked out above.

As we move down the periodic table to higher electron numbers, the screening creates a pattern in the way electrons fill the atomic levels, generating the so called $n + l$ rule (also called the Madelung-Klechkowski rule).

Simply put, the energy is lowest for electronic states with lowest number of nodes, corresponding to minimizing the curvature (expectation value of the kinetic energy) of the wavefunction. Since the (n, l)th mode has $n - l - 1$ radial nodes from $R_{nl}(r)$ and $2l + 1$ angular nodes from $Y_{lm}(\theta, \phi)$, the total number of states lying energetically below this mode is given by $(n - l - 1) + (2l + 1) = n + l$. This in turn sets the overall nuclear charge and thus the depth of the Coulomb well accommodating these modes, the precise relation being $(6Z)^{1/3} = n + l$. The relation can be derived rigorously by counting the number of electrons with a given l and then dividing by $2(2l + 1)$ to get the number of shells. The result of this exercise is the **Madelung** $n + l$ **rule** that dictates that electrons fill the shells along lines of constant $n + l$. To this we also add **Hund's rules** which also seek to reduce the overall Coulomb cost. The first requires us to maximize the spin in order to maximize the exchange interaction. The second rule maximizes the orbital angular momentum in order to reduce the angular overlap of wavefunctions. Finally we maximize (minimize) the total angular momentum J for greater than (less than) half-filled shells, to cut down on the energy due to *spin-orbit coupling* (Section 5.7).

To get a crude estimate for the Coulomb repulsion, we use the long-wavelength, Thomas-Fermi approximation. For an atom with a small nuclear charge Z, the screened electron nuclear term works out to be

$$V_{\text{el-nucl}}(r) \approx \frac{-Zq}{4\pi\epsilon_0 r(1 + r/\lambda)^2}, \qquad \lambda = \frac{0.55 a_0}{Z^{1/3}}. \tag{4.12}$$

This expression gives a weakened attraction as we venture far from the nucleus or else increase the atomic charge, which in turn increases the number of intervening electrons that screen the nuclear potential. Substituting the expression for Z from above, we get

$$\lambda = \frac{a_0}{n + l}, \quad Z_{\text{eff}}(r) = \frac{Z}{\left[1 + r(n + l)/a_0\right]^2}. \tag{4.13}$$

4.4.0.2 *Screening in solids*

We can contrast this screening in a molecule above with that in a solid. The Hartree term is the solution to Poisson's equation,

$$\nabla^2 \delta U = -\frac{q^2 \delta n}{\epsilon}. \tag{4.14}$$

For a degenerate semiconductor (Fermi energy near band) or a metal (Fermi energy inside band) we can write this in terms of the density of states D

at the Fermi energy (the linearization works for $\delta U \ll E_F$, in other words when the potential varies slower than λ_F)

$$\delta n = \underbrace{\frac{\partial n}{\partial E}}_{D} \delta U. \tag{4.15}$$

Substituting in Eq. (4.14) gives us the long wavelength Thomas Fermi equation $(\nabla^2 + 1/\lambda^2)\delta U = 0$ with $\lambda = \sqrt{\epsilon/q^2 D}$. Its solution, matching up with the Coulomb potential at a point charge, is the screened electron nuclear potential in a degenerate semiconductor or metal

$$V_{el-nucl}(r) \approx \frac{-Zq}{4\pi\epsilon_0 r} e^{-r/\lambda}, \qquad \lambda = \sqrt{\frac{\epsilon}{q^2 D}} \tag{4.16}$$

in contrast with Eq. (4.12) for a molecule, and Debye-Hückel screening in a nondegenerate semiconductor where we replace $D \to 1/k_B T$.

Higher order corrections to the screening term arise at shorter wavelengths for fast varying charge distributions. Near the Fermi energy in a metal, there is a sharp cutoff in the Fermi distribution and the linearized derivative expression cannot be used any more. The term $\partial n/\partial E$ gets replaced by a frequency dependent susceptibility $\chi_q(\omega)$. It is convenient to work this out in a plane wave basis and derive the susceptibility using first order time dependent perturbation theory ($\phi_k \sim e^{i\vec{k}\cdot\vec{r}}$ unperturbed, ψ_k perturbed).

$$|\psi_k\rangle \approx |\phi_k\rangle + \sum_q \frac{\overbrace{\langle\phi_{k+q}|H|\phi_k\rangle}^{V_q}}{E_k - E_{k+q} + \hbar\omega}|\phi_{k+q}\rangle \quad \text{(1st order perturbation)}$$

from which we calculate the perturbed electron density

$$\delta n(r) = \sum_k f_k \left[\langle\psi_k|\psi_k\rangle - \langle\phi_k|\phi_k\rangle\right],$$

f_k is the electron occupancy given by the Fermi-Dirac distribution that looks like a step function around the Fermi energy at low temperature. We then get the k space susceptibility

$$\chi_q(\omega) = \lim_{V_q \to 0} \frac{\delta n_q}{V_q}, \qquad n_q = \int d^3 r \delta n(r) e^{iqr} \to \text{simplify}$$

$$= \sum_k \frac{f_{k-q} - f_k}{E_{k-q} - E_k + \hbar\omega} \quad \text{(Lindhard Equation)} \tag{4.17}$$

to describe dynamic screening. The dynamic nature, captured by the frequency dependence of χ, arises as the screening charges cannot follow the perturbing fields instantly. Charges have a natural tardiness because of their finite mass. Signals varying faster than the charge's RC time constant $\sim \Delta E / \hbar$ are inadequately screened, and this is what the frequency dependent susceptibility captures. Near $q = 2k_F$, the static ($\omega = 0$) susceptibility picks up a logarithmic singularity as the denominator in Eq. (4.17) vanishes due to Fermi surface nesting, i.e., $E_{k-2k_F} = E_k$. An example can be seen with the square Fermi surface for a 2D simple cubic lattice (Fig. 5.12, contour plot on the floor of the middle figure). The singularity gives a branch cut in energy that creates an additional slowly decaying oscillatory term in the screening charge, $\sim \cos{(2k_F r)}/r^3$ known as Friedel oscillations. The oscillations are similar to the 'Gibbs phenomenon' for Fourier transforms at a jump discontinuity, or 'ringing' in a low pass filter.

What is the underlying physics of these oscillations? The oscillations happen because the system wants to minimize energy by placing screening charges as close to an external charge as possible while maintaining overall charge neutrality. Instead of pushing all screening charges inward and shoving all the neutralizing counter charges to the other end, it is energetically favorable to interlace them as closely as possible. However the smallest possible wavelength corresponds to the largest scattering wavevector allowed on a sharp Fermi sphere, $q = 2k_F$, determining the wave vector of the charge density wave oscillations.

4.5   Atom to molecule

So far, we worked out the energetics for electrons in a single atom. The simplest approach is to solve a 1-electron Schrödinger equation with a self-consistent ('Hartree') potential due to the other electrons. A more sophisticated approach accounts for the exchange interaction arising from the antisymmetry of the wave-functions. A more accurate approach also accounts for relaxation of the orbitals due to mixing of virtual unoccupied state configurations, leading to correlation effects (we discuss this configuration interaction in detail in the chapter on correlation).

When we go to a multinuclear molecule, we will need a completely different approach. For starters, we cannot pretend to have a spherically symmetric with a 'central' (angle independent) potential due to the presence of multiple nuclear centers. This makes the numerical problem on a multireference grid intractable, and calls for simpler approaches. Fortunately, we can indeed simplify the problem greatly using chemical intuition. For instance, it is reasonable to assume that the ground state of the hydrogen molecule will be a mixture of the ground state wavefunctions of the individual hydrogen atoms, so that we can treat each molecular orbital as a linear combination of atomic orbitals (it will turn out that even this assumption needs to be exercised with caution! We discuss later in the chapter on correlation). The reason this simplifies the problem greatly is because we have already invested a lot of effort in extracting properties of the individual hydrogen atom wavefunctions, so all we need to do now is to estimate the coefficients of the mixture — a much easier computational problem!

We thus define a set of atomic orbital *basis sets*, which will be in effect, our chemical 'coordinates' in terms of which we will decompose the description of the molecular states. For instance, we can write

$$\Psi_{MO}(\vec{r}) = \sum_{\beta=1}^{\infty} c_\beta \phi_\beta(\vec{r}), \tag{4.18}$$

$\{\phi_\beta(\vec{r})\}$ is a complete atomic orbital basis set. The Schrödinger equation

$$\hat{H}\Psi_{MO} = E\Psi_{MO}. \tag{4.19}$$

Left multiplying by $\phi_\alpha^*(\vec{r})$ and integrating over 3D volume, we get

$$\sum_\beta H_{\alpha\beta} c_\beta = E \sum_\beta S_{\alpha\beta} c_\beta, \tag{4.20}$$

where the Hamiltonian and overlap matrix elements

$$H_{\alpha\beta} = \int d^3\vec{r}\, \phi_\alpha^*(\vec{r}) \hat{H} \phi_\beta(\vec{r}), \qquad S_{\alpha\beta} = \int d^3\vec{r}\, \phi_\alpha^*(\vec{r}) \phi_\beta(\vec{r}). \tag{4.21}$$

The eigenvalue problem in Eq. (4.20) can thus be written in matrix form as

$$\begin{pmatrix} H_{11} & H_{12} & \cdots & H_{1,N-1} & H_{1N} \\ H_{12} & H_{22} & H_{23} & \cdots & H_{2N} \\ \cdots & \cdots & \cdots & \cdots & \cdots \\ \cdots & \cdots & \cdots & \cdots & \cdots \\ H_{N1} & H_{N2} & \cdots & H_{N,N-1} & H_{NN} \end{pmatrix} \begin{pmatrix} c_1 \\ c_2 \\ \cdots \\ \cdots \\ c_N \end{pmatrix} = E \begin{pmatrix} S_{11} & S_{12} & \cdots & S_{1,N-1} & S_{1N} \\ S_{12} & S_{22} & S_{23} & \cdots & S_{2N} \\ \cdots & \cdots & \cdots & \cdots & \cdots \\ \cdots & \cdots & \cdots & \cdots & \cdots \\ S_{N1} & S_{N2} & \cdots & S_{N,N-1} & S_{NN} \end{pmatrix} \begin{pmatrix} c_1 \\ c_2 \\ \cdots \\ \cdots \\ c_N \end{pmatrix} \tag{4.22}$$

and the eigenenergies are obtained from the matrix $S^{-1}H$, or equivalently but recast in a more symmetric form, the matrix $S^{-1/2}HS^{-1/2}$. A phenomenological example is Extended Huckel Theory, where we use $H_{\alpha\beta} = KS_{\alpha\beta}(U_\alpha + U_\beta)$ with K a constant and U_α the on-site Hubbard energy (Section 6.4).

4.6 Bonding in H_2^+ molecular ion

Let us start with two 1s orbitals spanning the atoms in an H_2^+ molecule (a very similar 2×2 model works for ethylene). The eigenvalue equation is

$$\begin{pmatrix} H_{AA} & H_{AB} \\ H_{BA} & H_{BB} \end{pmatrix} \begin{pmatrix} c_A \\ c_B \end{pmatrix} = E \begin{pmatrix} S_{AA} & S_{AB} \\ S_{BA} & S_{BB} \end{pmatrix} \begin{pmatrix} c_A \\ c_B \end{pmatrix}. \qquad (4.23)$$

If we turn off the couplings between the atoms, then $H_{AA} = H_{BB} = \epsilon$ boils down to the onsite energy of each 1s H orbital (at energy -13.6 eV). The other matrix elements $H_{AB} = H_{BA} = -t_0$ represent the hopping term between two neighboring atoms driven by wavefunction overlap. $S_{AA} = S_{BB} = 1$, assuming the 1s orbitals are normalized, while $S_{AB} = S_{BA} = s$ is the overlap of the neighboring wavefunctions. We thus get

$$\begin{pmatrix} \epsilon & -t_0 \\ -t_0 & \epsilon \end{pmatrix} \begin{pmatrix} c_A \\ c_B \end{pmatrix} = E \begin{pmatrix} 1 & s \\ s & 1 \end{pmatrix} \begin{pmatrix} c_A \\ c_B \end{pmatrix} \qquad (4.24)$$

or in other words,

$$\underbrace{\begin{pmatrix} E - \epsilon & Es + t_0 \\ Es + t_0 & E - \epsilon \end{pmatrix}}_{M} \begin{pmatrix} c_A \\ c_B \end{pmatrix} = 0. \qquad (4.25)$$

The product of the two matrices is zero but the coefficients $c_{A,B}$ cannot vanish (in fact, we want the sum of their squares to equal unity), meaning the determinant of matrix M must vanish. This gives us two solutions

$$E_{\pm} = \frac{\epsilon \pm t_0}{1 \mp s}. \qquad (4.26)$$

Plugging these back, into Eq. (4.25) we then get the coefficient ratio

$$c_B = \mp c_A, \qquad (4.27)$$

which coupled with the normalization condition

$$1 = \int d^3\vec{r} |\Psi_{H_2}|^2 = c_A^2 \underbrace{\int d^3\vec{r} \phi_A^* \phi_A}_{1} + c_B^2 \underbrace{\int d^3\vec{r} \phi_B^* \phi_B}_{1} + 2 c_A c_B \underbrace{\int d^3\vec{r} \phi_A^* \phi_B}_{s}$$

$$= c_A^2 + c_B^2 + 2 c_A c_B s \qquad (4.28)$$

and substituting into Ψ in Eq. (4.18) gives us

$$\Psi_{H_2^+} = \frac{\phi_A \mp \phi_B}{\sqrt{2(1 \mp s)}}. \qquad (4.29)$$

We see immediately that the low energy sector is the 'bonding' configuration that is symmetric in the basis sets,

$$\Psi_- = \frac{\phi_A + \phi_B}{\sqrt{2(1+s)}}, \quad E_- = \frac{\epsilon - t_0}{1+s}. \tag{4.30}$$

The electrons can be seen to be delocalized between the atoms. How does this reduce the energy? *By reducing the curvature of the wavefunction, which as you recall reduces the kinetic energy.* The reduction in energy is driven by the strength of the hopping term t_0, which depends on orbital orientation. Orbitals pointing towards each other, such as $p_x - p_x$ orbitals, will have lower kinetic energy compared to orbitals pointing away such as $p_z - p_z$ orbitals (assuming the bonds are in the x-y plane).

In contrast, the high energy sector is the 'antibonding' configuration where the electrons stay as far away from each other as possible, reducing the overlap and increasing the curvature (kinetic energy) of the wavefunctions.

$$\Psi_+ = \frac{\phi_A - \phi_B}{\sqrt{2(1-s)}}, \quad E_+ = \frac{\epsilon + t_0}{1-s}. \tag{4.31}$$

There could also be orientations that give zero overlap, for instance, between an s orbital and a neighboring p_z orbital, generating *non-bonding* states. Note that so far the only terms we have considered in our Hamiltonian are the electron-nuclear potentials around each of the two possible nuclei (already included in the basis set of each H-atom), and the kinetic energy term — partly already included in the H-atom basis sets, but the rest arising due to hopping and captured by $-t_0$. What we haven't included is the electron-electron Coulomb cost, in the absence of which the bonding state ended up being energetically more favorable. In the opposite limit of large charging energy, the situation reverses and the antibonding state becomes preferable as it avoids double occupancy that is promoted by hopping.

4.6.0.1 *Variational treatment of H_2^+*

We can do a variational analysis of the H_2^+ molecule much like we did for Helium, by choosing hydrogenic orbitals as before (Eq. (4.7)) with screening Z, centered around their own respective nuclei, the nuclear separation being another parameter R. The Hamiltonian is a bit different from Eq. (4.32),

$$\hat{\mathcal{H}} = -\frac{\hbar^2 \nabla^2}{2m} - \frac{q^2}{4\pi\epsilon_0 r_1} - \frac{q^2}{4\pi\epsilon_0 r_2} + \frac{q^2}{4\pi\epsilon_0 r_{12}}. \tag{4.32}$$

This is again solvable analytically in the hydrogenic basis sets, using elliptical coordinates $(r_1 \pm r_2)/2$. We get $s = e^{-ZR}[1 + ZR + Z^2R^2/3]$, $\epsilon = Z^2/2 - (Z+1/R)(1-e^{-2ZR})$, and $t_0 = Z^2 s/2 - Z(Z-2)(1+ZR)e^{-ZR}$. For each nuclear separation R we can calculate the minimizing Z and plot the ground state energy. We find that for the unoptimized $Z = 1$, the minimum energy $E = -1.76$ eV occurs at $R = 1.32$ Å. If we allow Z to be a variational parameter for each R, then the optimal bond length $R = 1.07$ Å and bonding energy $E = -2.35$ eV. The experimental numbers are $R = 1.06$ Å and bonding energy $E = -2.79$ eV.

Loosely speaking, the kinetic energy (bonding) term promotes double occupancy, which by Pauli exclusion implies that the electron spins must be antiparallel (antiferromagnetic), while the Coulomb term deters double occupancy, and Pauli exclusion helps in this regard by making the spins parallel (ferromagnetic). The actual state that an electron ends up in depends on a delicate balance between these two considerations primarily, which is complicated in real materials immensely by screening, the localization of the orbitals (e.g., d vs s), crystal structure, strain and so on.

4.7   Bond mixing and hybridization

Methane (CH_4) has a tetrahedral structure, with the carbon atom bonding with four hydrogens, as a group IV element tends to do. Electron–electron repulsion makes their bonds stay as far away from each other as possible (this is called the Valence Shell Electron Pair Repulsion or VSEPR model). From 3D geometry, one can show that the four bonds are oriented at an angle of $109°28'$. Since the bonds are not at $90°$, we cannot describe them as simple p_x, p_y, p_z orbitals. It is thus natural to wonder what describes the bonds that point towards the atoms, and how they are formed.

We can take combinations of the carbon s and p orbitals to construct a new set of basis orbitals that point along the tetrahedral directions. Those hybridized basis sets overlap strongly with their partners and thus enjoy strong bonding antibonding splittings, resulting in structurally stable σ bonds with energies sitting far from the Fermi energy. While the bonds are not oriented perpendicular to each other, we can still choose admixtures and signs of multiple orbitals to keep their wavefunctions mutually orthogonal. The way this works is as follows. The valence electron configuration

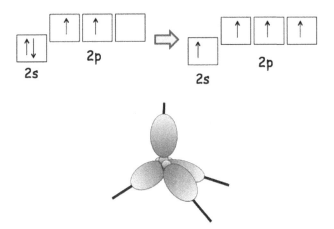

of carbon is $2s^2 2p^2$, which argues for the involvement of only 2 p orbitals. However, it is energetically favorable for the molecule to promote one of the $2s$ electrons to the third p orbital and then mix with all three symmetrically to create four equivalent sp^3 bonds, recovering the lost excitation energy through the bond formation energy in the process. The coefficients of the s and p orbitals are chosen to enforce wavefunction orthogonality.

$$\phi_{sp^3}^A = \frac{1}{2}\left[\phi_{2s} + \phi_{2p_x} + \phi_{2p_y} + \phi_{2p_z}\right]$$

$$\phi_{sp^3}^B = \frac{1}{2}\left[\phi_{2s} - \phi_{2p_x} - \phi_{2p_y} + \phi_{2p_z}\right]$$

$$\phi_{sp^3}^C = \frac{1}{2}\left[\phi_{2s} - \phi_{2p_x} + \phi_{2p_y} - \phi_{2p_z}\right]$$

$$\phi_{sp^3}^D = \frac{1}{2}\left[\phi_{2s} + \phi_{2p_x} - \phi_{2p_y} - \phi_{2p_z}\right] \tag{4.33}$$

It is easy to verify that the p components are oriented tetrahedrally at 109°28', and that the presence of the spherical s orbitals provides an extra 'knob' to make these orbitals orthogonal.

Carbon is perhaps the only element in the periodic table with comparable sp^2 and sp^3 bond energies, equally likely to form 3D tetrahedral as well as 2D planar structures (explaining why much of biology, from flesh and bones to skin, involves carbon, and why carbon has so many allotropes from diamond to graphene to nanotubes to buckyballs). We can construct a planar sp^2 series from two of the p orbitals and one s orbital, with the third p_z orbital sticking out of the plane, as in benzene and graphene. The three orthogonal 120° separated bonding orbitals are

$$\phi_{sp^2}^A = \sqrt{\frac{1}{3}}\phi_{2s} + \sqrt{\frac{2}{3}}\phi_{2p_x},$$

$$\phi_{sp^2}^B = \sqrt{\frac{1}{3}}\phi_{2s} + \sqrt{\frac{2}{3}}\left[-\frac{1}{2}\phi_{2p_x} + \frac{\sqrt{3}}{2}\phi_{2p_y}\right],$$

$$\phi_{sp^2}^C = \sqrt{\frac{1}{3}}\phi_{2s} + \sqrt{\frac{2}{3}}\left[-\frac{1}{2}\phi_{2p_x} - \frac{\sqrt{3}}{2}\phi_{2p_y}\right]. \qquad (4.34)$$

The three hybridized orbitals generate bonding-antibonding splittings and stable σ and σ^* bonds. Transport is dominated by electrons hopping between the overlapping π bonds constructed from the residual p_z orbitals.

Finally, we mix just one s and one p orbital to construct linear sp bonds,

$$\phi_{sp}^A = \sqrt{\frac{1}{2}}\left[\phi_{2s} + \phi_{2p_x}\right],$$

$$\phi_{sp}^B = \sqrt{\frac{1}{2}}\left[\phi_{2s} - \phi_{2p_x}\right]. \qquad (4.35)$$

The overlap of directed orbitals in molecules creates energy-stabilized covalent bonds separated from higher energy antibonds. We next discuss how extending such bonding-antibonding split across a crystal leads to the emergence of a population of allowed bands and forbidden band-gaps. Our ability to engineer these bands and populate them selectively allows us to realize heterojunction electronic and optical devices of intricate complexity.

Chapter 5

Bonds to bands: molecules to solids

As we extend a set of bonds to form a periodic network, a different symmetry emerges among the eigenstates. The periodicity of the atomic cores in a solid naturally introduces a Fourier space description with an associated wavevector $\vec{k}$. The allowed eigenstates can now be labeled in terms of $\vec{k}$, as well as a separate band index at each $\vec{k}$ representing the orbital symmetries of the underlying atomic valence electrons. Near the high symmetry points these $E - k$s typically look parabolic, and can be interpreted as 'free' electrons, albeit with an effective mass m^* (typically a tensor representing anisotropy). Generally these masses are smaller than the free electron mass, because a valence electron gets pulled in by the ionic cores and slingshots quickly from one atomic well to another in the process.

We can extract a lot of technologically useful information from these bands and their underlying orbital symmetries. For instance, the direct bandgap of GaAs allows it to emit light efficiently compared to Silicon. Since the speed $v = \Delta E/\hbar \Delta k$ of a photon is much larger than an electron, photons that are energy matched to an electronic band-gap ΔE cannot absorb much momentum Δk from the electrons and allow only vertical transitions along the $E - k$. Another technologically useful quantity is the transport effective mass that determines the electron mobility (GaAs and Ge are better than Si in that way), but also the role of tunneling (OFF currents in n-Ge(111) are higher than n-Ge(100)). The ON current is determined by the carrier density which relies on the density of states effective mass, making ballistic n-Ge(100) a higher current carrying device than n-Ge(111) (we discuss this in section 18.2).

5.1   Benzene: a quantized 1D bandstructure

Let's use the π basis to work out the low energy bandstructure of benzene. The six p_z orbitals sit symmetrically on a ring. In as much as it's equivalent to a 1D chain of atoms with a periodicity of six, it gives us an insight into how bands are formed. The time independent Schrödinger equation for the nth atom leads to a 1-D tight binding model (Eq. (4.22) n^{th} row, with $H_{nn} = \epsilon_0$, $S_{nn} = 0$, $H_{n,n\pm1} = -t$)

$$E\psi_n = \epsilon_0\psi_n - t\psi_{n+1} - t\psi_{n-1}, \quad n = 1, \ldots, 6. \quad (5.1)$$

For periodic boundary conditions, we conjecture a plane wave solution $\psi_n = \psi_0 e^{ikna}$, substituting which in Eq. (5.1) we get

$$E = \epsilon_0 - 2t\cos ka, \quad (5.2)$$

which we plot in Fig. 5.1. Periodicity imposes a quantization condition on the wavevectors so that $\psi_6 = \psi_0$ and $e^{6ika} = 1$, giving us the quantized k's (marked out in circle in Fig. 5.1), the corresponding energy levels (degenerate pairs are artificially separated for visual clarity, by introducing a small asymmetry), and also the six corresponding normalized eigenstates in

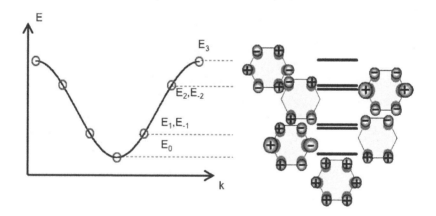

Fig. 5.1   *(Left) 1D tight binding band, with quantized k states for benzene marked out in circles. (Right) Corresponding eigenenergies and eigenfunctions for benzene (red positive, blue negative).*

Fig. 5.1. Just like the orbitals for hydrogen, we plot the real and imaginary parts in Fig. 5.1 by taking linear combinations of equal energy (degenerate) solutions. As expected, within the six pz orbital basis set, the lowest energy

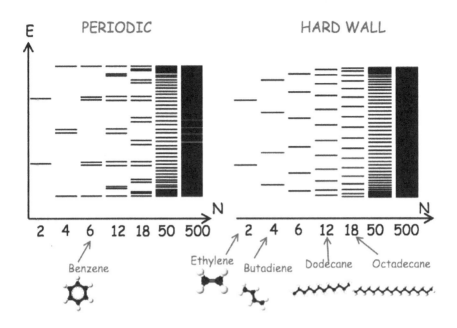

Fig. 5.2 *(Left) Periodic vs (Right) hard wall boundary conditions describing a ring vs a linear alkane chain of N atoms, showing the evolution of discrete levels into a metallic band. All energies are normalized to the same overall bandwidth $4t_0$.*

solution is the one with a homogeneous electron cloud (lowest curvature), while the highest is the one with the fastest varying cloud with shortest wavelength. While benzene has a pronounced bandgap, this is due to the discreteness of the spectrum enforced by quantization of k. If we now expand to a larger and larger ring, the k states become more and more dense till we will cover the entire $E - k$ shown in Fig. 5.1. The evolution of the electronic states is shown in Fig. 5.2, with both the ring structure (periodic boundary condition) and a linear structure where the wavefunctions vanish at the ends (hard wall boundary condition). The finite size solutions are obtained either by taking doubly degenerate plane waves with $k = \pm 2n\pi/L$ as for a ring, or by superposing plane wave solutions to get sinusoids with $k = n\pi/L$ as for a chain. In each case, we get a 1D metallic band, assuming of course that the underlying atomic structures are structurally stable. What is critical to this metallicity is the fact that the hopping elements between the p_z orbitals is repeated. Although the valency of carbon mandates alternating single and double bonds for benzene, the two ways to lay out these alternating bonds 'resonate' in energy, giving on average 1.5 bonds running around the ring.

5.2   Alternating bonds: semiconductors and solitons

When we cut open the benzene ring to create a linear chain, the bandstructure depends critically on the dimerization pattern. For alkanes where each carbon is connected to two hydrogens, the resulting single bonded structure has an underlying metallic band (Fig. 5.2, right). However if we kept the single hydrogen atoms attached to carbon to create a chain of alkenes, then valency satisfaction requires us to alternate single and double bonds and thus create a dimerization pattern. Such a dimerization opens a bandgap, as we see in Fig. 5.3 with trans polyacetylene.

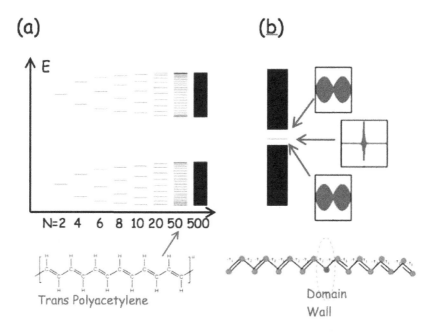

Fig. 5.3   *(Left) A chain of dimerized units evolves into a bandgapped semiconductor. (b) Switching the dimerization pattern in the middle creates a defect state (a soliton or a domain wall), with localized wavefunction shown in blue (the neighboring energy states have delocalized sinusoidal states). These solitons can hop under an electric field, making the conjugated polymers conducting upon doping.*

If the hopping terms oscillate between t_1 and t_2 with an onsite term ϵ_0, then it is straightforward to write down by inspection the equation for the nth dimer, consisting of a single and a double bond.

$$E\{\psi\}_n = [\alpha]\{\psi\}_n - [\beta]\{\psi\}_{n+1} - [\beta]^\dagger\{\psi\}_{n-1}$$

$$[\alpha] = \begin{bmatrix} \epsilon_0 & -t_1 \\ -t_1 & \epsilon_0 \end{bmatrix}, \qquad [\beta] = \begin{bmatrix} 0 & 0 \\ t_2 & 0 \end{bmatrix}. \tag{5.3}$$

If we can again assume periodic boundary conditions and invoke a plane wave solution (Bloch's theorem) $\propto e^{ikna}$, we get the energies

$$E\{\psi\}_n = \underbrace{\left([\alpha] - [\beta]e^{ika} - [\beta]^\dagger e^{-ika}\right)}_{H_k}\{\psi\}_n$$

$$E_\pm = \epsilon_0 \pm \sqrt{t_1^2 + t_2^2 - 2t_1t_1\cos ka} \qquad \text{(eigenvalues of } H_k\text{)}, \quad (5.4)$$

where a is the dimer to dimer distance. For a continuous set of k points (infinite chain), this structure evolves into a semiconducting band with bandgap $2|t_1 - t_2|$ and equal conduction and valence band widths of $2|t_2|$. If we further allow each t to be replaced by a matrix corresponding to many orbitals instead of just a single p_z, then the bands have considerable more complexity. An example is in Fig. 5.4, where a single thiophene ring has been extended into a polymer of polythiophene. It is notable how extending the chain lowers the bandgap (consistent with the particle in a box and the uncertainty principle). Further modification of the bands happens due to lateral cross linkings between polymer chains, often in a disordered fashion to give a relatively flat bandstructure with very little dispersion.

One may wonder how conjugated polymers with alternating single and double bonds can conduct electricity, as they are in fact known to do. After all, we got a sizeable bandgap simply due to the dimerization. In low dimensions, such a dimerization tends to be the routine rather than the exception (in other words, the linear alkenes are energetically stabler than the alkynes in a hydrogen rich environment). The dimerization is driven by the Peierls transition, whereby the lowered total energy due to the gap opening compensates for any increase in nuclear potential due to the bond squeezing (the energy gain from the gap goes as $\sim \ln\delta$ while the loss from bond

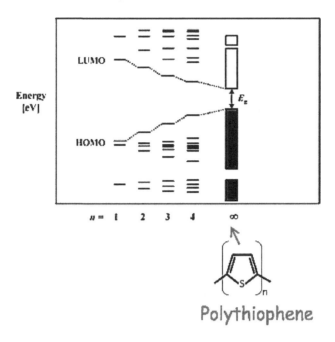

Fig. 5.4 *Evolution of bandgap with increasing chain length of polythiophene (Ref. A. Ajayaghosh, 'Donoracceptor type low band gap polymers: polysquaraines and related systems', Chem. Soc. Rev. 32, 181 (2003)).*

distortion is $\sim \delta^2$, with δ being the fractional difference in bond length across a dimer). How can such a bandgapped polymer conduct electricity?

Conducting polymers have a domain wall, where an alternating single-double bond pattern starting at the left conflicts with a complementary pattern starting at the right, resulting in a defect in the middle (Fig. 5.3, right). Such a defect introduces a localized state in the middle of the bandgap, whose wavefunction decays exponentially on either side. Unlike a donor or acceptor in a semiconductor, however, such a defect stays inactive and neutral even at high temperatures as the polymer bandgaps are pretty large. However, we can dope this defect with an extra electron or hole and mobilize it with an electric field (i.e., the defect site itself hops from site to site as double and single bonds swap places). The hopping of this doped defect state ultimately renders the polymer conductive.

5.3   Repeating benzene: aromatic hydrocarbons to graphene

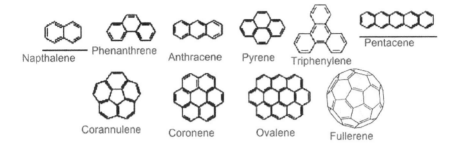

Napthalene   Phenanthrene   Anthracene   Pyrene   Triphenylene   Pentacene

Corannulene   Coronene   Ovalene   Fullerene

Note that a lot of the molecular chemistry is hidden in the underlying matrices (e.g., α, β) for the chains. But treating each matrix as a single unit, we can readily use our 1D bandstructure idea to extend this to chains of such units, and even extend this to 2D and 3D. We can thus evolve from benzene to an entire class of aromatic hydrocarbons as above (ring compounds like coronene or chains like anthracene or pentacene). All of them will have small bandgaps similar to benzene, because of the delocalization of the π electron cloud across the conjugated rings (the small gap often lies in the visible spectrum). To these hexagonal subunits, we can add pentagonal rings to allow the hexagons to turn around corners, giving us more complex structures such as corannulene and ultimately buckyballs. Some of these ring compounds could be used to cage metallic elements, for instance magnesium in chlorophyll and iron in hemoglobine, endowing those compounds with a rich metallic color. Finally, extending the hexagons indefinitely on a plane leads ultimately to graphene, where the constancy of the hopping elements and the symmetry of the dimer pz orbitals leads to the elimination of the bandgap altogether.

Bandgaps thus require oscillating band parameters — either the hopping terms as in polymers, or the onsite energies as in superlattices. The underlying molecular chemistry is always critical to understanding solid state bandstructures. For instance, silicon has a bandgap, which suggests a dimerization pattern in all directions. The dimerization is obvious along the (111) directions, the cubic diagonals. However along the (100) directions or the cubic axes themselves, the projected atomic coordinates do not show any dimerization (Fig. 5.5) and would normally predict a metallic state. An indirect bandgap opens up because of the presence of higher order p and d orbitals, whose directional bonds do not project symmetrically along the (100) surfaces and generate the required dimerization.

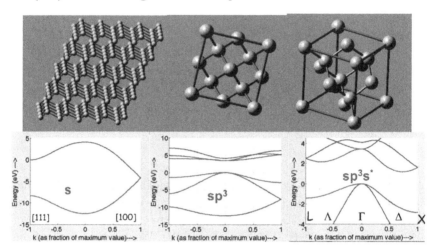

Fig. 5.5   *Top L to R: Si crystal, and unit cell showing symmetric projections on (100) plane and dimerization along (111) direction. Bottom L to R: tight binding $E - k$ with just s orbitals gives metallic bands along (100) because of the lack of dimerization on the projected surfaces. An sp^3 basis gives a bandgap (albeit direct) as the directional p bonds do not project symmetrically along (100). A full $sp^3 s^*$ basis gives the desired indirect band. Other details (e.g., transverse effective mass, and dimer bonding along the surfaces) require more variables, such as d orbitals, next-nearest neighbor interactions, non-orthogonality, and increasingly, tightly bound Wannier-like basis sets (Fig. 3.2).*

5.4   The eigenmodes of monolayer graphene

We start with a tight-binding model for graphene in a single p_z orbital per atom basis (Fig. 5.6). The honeycomb lattice needs a dimer basis set (shown in dashed boxes) to account for the missing atoms at the hexagon centers. The lattice vectors are $\vec{R}_{1,2} = a\hat{x} \pm b\hat{y}$, where $a = 3a_0/2$ and $b = a_0\sqrt{3}/2$, a_0 being the C-C bond-length in graphene. The dimer wavefunctions can be written as a linear combination of the atomic p_z orbitals A and B.

$$\Psi = c_A \psi_A + c_B \psi_B \tag{5.5}$$

where $\psi_{A,B}$ are the p_z orbitals of the dimer atoms A and B of the two interpenetrating triangular sublattices. We collect the coefficients $c_{A,B}$ into a 2×1 vector, so that the corresponding Hamiltonian is 2×2 in this minimal

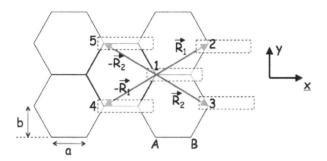

Fig. 5.6 *Dimer unit cells and lattice vectors in graphene, closest dimers labeled 1 to 5*

basis set. Taking just nearest neighbor interactions with hopping strength $t_0 \approx 2.4\text{eV}$, starting with the onsite dimer and then proceeding clockwise from the top right interaction vector, the 2×2 k-space Hamiltonian in the dimer basis set is given by summing over nearest neighbor interacting dimers as we did in Eq. (5.4)

$$H_k = \sum_{mn} H_{mn} e^{i\vec{\kappa}\cdot(\vec{R}_m - \vec{R}_n)} = \underbrace{\begin{bmatrix} 0 & -t_0 \\ -t_0 & 0 \end{bmatrix}}_{H_{11}}$$

$$+ \underbrace{\begin{bmatrix} 0 & 0 \\ -t_0 e^{i\vec{\kappa}\cdot\vec{R}_1} & 0 \end{bmatrix}}_{H_{12}} + \underbrace{\begin{bmatrix} 0 & 0 \\ -t_0 e^{i\vec{\kappa}\cdot\vec{R}_2} & 0 \end{bmatrix}}_{H_{13}} + \underbrace{\begin{bmatrix} 0 & -t_0 e^{-i\vec{\kappa}\cdot\vec{R}_1} \\ 0 & 0 \end{bmatrix}}_{H_{14}} + \underbrace{\begin{bmatrix} 0 & -t_0 e^{-i\vec{\kappa}\cdot\vec{R}_2} \\ 0 & 0 \end{bmatrix}}_{H_{15}}$$

$$= \begin{bmatrix} 0 & -t_0\left(1 + e^{-i\vec{\kappa}\cdot\vec{R}_1} + e^{-i\vec{\kappa}\cdot\vec{R}_2}\right) \\ -t_0\left(1 + e^{i\vec{\kappa}\cdot\vec{R}_1} + e^{i\vec{\kappa}\cdot\vec{R}_2}\right) & 0 \end{bmatrix} \qquad (5.6)$$

The two sets of eigenvalues describing the conduction and valence band, are given by solving $\det(EI - H_k) = 0$. Using $\vec{R}_{1,2} = a\hat{x} \pm b\hat{y}$, we get

$$E_{\pm} = \pm t_0 |1 + e^{i\vec{\kappa}\cdot\vec{R}_1} + e^{i\vec{\kappa}\cdot\vec{R}_2}| = \pm t_0 |1 + 2e^{i\kappa_x a}\cos\kappa_y b| \qquad (5.7)$$

which merge when $\kappa_x a = 0$, $\kappa_y b = 2\pi/3$. This point happens to be exactly the corner of the hexagonal Brillouin zone, so that graphene acts as a gapless material. Furthermore, since each carbon atom donates one electron, the band is half-filled, so that the Fermi energy for undoped graphene passes exactly through this special point, called 'Dirac point' (and five other equivalent points over the Brillouin zone). The six points reduce to two distinct 'valleys', the rest of the points relatable through the k-lattice vectors $K_{1,2} = \dfrac{2\pi}{a}\hat{x} \pm \dfrac{2\pi}{b}\hat{y}$ and are thus equivalent. Taylor expanding around the

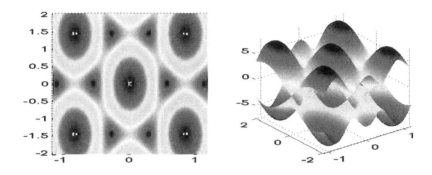

Fig. 5.7   *(Left) colorplot and (Right) sideview of graphene Ek, showing semimetallic gapless dispersion at the six corners of the hexagonal Brillouin zone*

Dirac points, keeping in mind that $a = 3a_0/2$ and $b = \sqrt{3}a_0/2$, we get

$$e^{i\kappa_x a} \approx 1 + i\kappa_x a = 1 + 3ik_x a_0/2 \qquad (5.8)$$

where $k_x = \kappa_x$ and

$$\cos\kappa_y b = \cos\left[2\pi/3 + (\kappa_y b - 2\pi/3)\right]$$
$$= -1/2\cos\left(\kappa_y b - 2\pi/3\right) - \sqrt{3}/2\sin\left(\kappa_y b - 2\pi/3\right)$$
$$\approx -1/2 - \sqrt{3}/2(\kappa_y b - 2\pi/3) \qquad (for\ \kappa_y \approx 2\pi/3b)$$
$$= -1/2 - 3k_y a_0/2 \qquad (5.9)$$

with $k_y = \kappa_y - 2\pi/3b$. $\{k_x, k_y\}$ are the small deviations from the Dirac points. This gives us

$$1 + 2e^{i\kappa_x a}\cos\kappa_y b \approx 3ia_0(k_x + ik_y) \qquad (5.10)$$

so that the 2×2 Hamiltonian near the Dirac point can be written as

$$H = \begin{bmatrix} 0 & \hbar v_F(k_x - ik_y) \\ \hbar v_F(k_x + ik_y) & 0 \end{bmatrix} = \hbar v_F \vec{\sigma} \cdot \vec{k} \qquad (5.11)$$

where $v_F = 3a_0 t_0/\hbar \approx 10^8 \text{cm/s}$ is the Fermi velocity, $\vec{k} = \{k_x, k_y\}$ is the 2-D quasimomentum vector and $\vec{\sigma}$ is the 2-D Pauli matrix, with components

$$\sigma_x = \begin{pmatrix} 0 & 1 \\ 1 & 0 \end{pmatrix}, \quad \sigma_y = \begin{pmatrix} 0 & -i \\ i & 0 \end{pmatrix} \qquad (5.12)$$

The two eigenvalues of this equation are obtained by solving $\det(EI_{2\times 2} - H) = 0$, giving us the two eigenvalues

$$E_{\pm} = \pm \hbar v_F \sqrt{k_x^2 + k_y^2} = \pm \hbar v_F |\vec{k}| \qquad (5.13)$$

with eigenvectors controlled by the chiral angle $\theta = tan^{-1}(k_y/k_x)$

$$\Psi_{\pm} = \begin{pmatrix} 1 \\ \pm e^{i\theta} \end{pmatrix} e^{i(k_x x + k_y y)} \qquad (5.14)$$

(+ for conduction band, − for valence band, with signs flipping for the other valley). Note that the states at $\theta = 0$ and π are orthogonal, which gives a very simple explanation for the semi-metallicity of graphene. Level repulsion and band 'anticrossing' arises from the off diagonal Hamiltonian elements, as in the 1-D semiconducting dimer chain (Section 5.2). Since this term vanishes between orthogonal eigenstates above, the resulting states must cross, - a manifestation of a more general group theoretical principle called the *Von Neumann-Wigner theorem*. The chiral nature of these eigenvalues (where the phase is connected to momentum) will come in handy later on when we discuss chiral tunneling in graphene. We will label the two states corresponding to the bonding and antibonding combinations of the p_z orbitals as 'up' and 'down' pseudospins (Fig. 5.8)

5.4.1 *Electron mass in graphene*

What is striking about this linear bandstructure is its impact on the electron mobility. Scientists often extract a rest mass for graphene by fitting a relativistic equation $E = \pm\sqrt{p^2c^2 + m^2c^4}$ to the E-k. That mass of course vanishes everywhere along the linear graphene E-k, so it is common to interpret graphitic electrons as 'massless Dirac fermions'. However, a separate definition of band effective mass enters the conventional definitions of mobility $\mu = q\tau/m^*$ and cyclotron level spacing qB/m^*, usually by fitting a parabola locally. It is the mass we used for silicon, diamond, benzene, coronene and all intervening structures leading up to graphene, for all of which we actually solved the *non-relativistic* Schrodinger equation in tight binding. The band effective mass obtained by fitting a parabolic free electron band to an $E - k$ can be easily shown to vary inversely proportional to the band curvature, $m^* = 1/\hbar^2[\partial^2 E/\partial k^2] = \partial p/\partial v$. This may sound counter-intuitive — how can a more confining band with fast varying $E - k$ actually see lighter electrons? The answer lies in the inverse relation between Fourier transform pairs — a steeper E-k actually corresponds to a shallower, less confining set of coupled quantum wells or atoms, i.e., smaller 't', in real space. Extending this m^* formula to graphene, however, is not obvious. A linear E-k may suggest zero curvature and infinite effective mass. Conversely, if we think about a small band-gap between the upper and lower Dirac cones, then near the apex of each cone the E-k varies very rapidly, and the effective mass suddenly seems to vanish. What is the 'proper' way to interpret the band effective mass?

We must remember that the effective mass-curvature relation only works by fitting a parabola to the bottom of a smoothly varying band-edge, whereby the coefficient of the quadratic Taylor expansion term pulls out the curvature. However for a strongly non-parabolic band as in graphene, we need a different, more general interpretation of m^*. We will use instead a more general definition, $m^* = p/v$ instead of $\partial p/\partial v$. This definition is consistent with the cyclotron effective mass, which determines the period of an electronic orbit around the Fermi surface in presence of a magnetic field.

$$m^* = \frac{\hbar^2}{2\pi} \frac{\partial A}{\partial E} \qquad (5.15)$$

where A is the Fermi surface area enclosed by the orbit, in this case, a simple circle. In other words,

$$m^* = \frac{\hbar^2}{2\pi} \frac{\partial(\pi k^2)}{\partial E} = \hbar^2 k \frac{dk}{dE} = \frac{\hbar k}{dE/\hbar dk} = \frac{p}{v} \qquad (5.16)$$

Near the bottom of a parabolic band, we can use L'Hospital's rule to give

$$\lim_{k \to 0} m^* = \lim_{k \to 0} \frac{\hbar k}{dE/\hbar dk} = \frac{\hbar^2}{d^2E/dk^2} = \frac{\partial p}{\partial v} \qquad (5.17)$$

which is the usual definition of effective mass inversely proportional to the band-curvature. For a linear band however, using the more general expression $m^* = p/v$, we see that approaching the Dirac point the numerator $p = \hbar k \to 0$, but the denominator v given by the slope of the E-k *stays constant*, so that m^* is *strongly energy-dependent* and more importantly $m^* \to 0$, and the mobility $q\tau/m^*$ shoots up.

The reader may struggle to interpret the action of an electric field on an electron endowed with such a photon like bandstructure. After all, how can we impart an electrostatic force on an electron if its speed (slope of the E-k) stays unchanged? What an applied force does is change the overall momentum of the electron. For free electrons, this involves an actual speed-up, while for graphitic electrons the momentum change manifests as an increase in band effective mass. Instead of speeding up, the electron appears to get heavier as an electric field operates on it. The classical analogue is a rocket that gains thrust in the absence of an external force by constantly jettisoning its fuel mass.

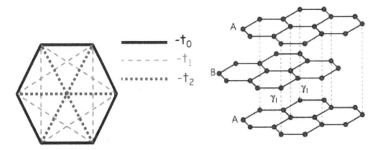

Fig. 5.8 *(a) Beyond nearest neighbor interactions stay masked in graphene and yield the same band dispersion as nearest neighbor models; however, breaking the symmetry with a nanoribbon causes these interactions to open a bandgap, in addition to the contribution of strains at the edges. (b) Interactions across Bernally stacked bilayer graphene, ultimately responsible for two parabolic bands with zero bandgap.*

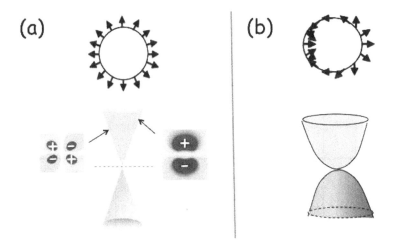

Fig. 5.9 *(a) Eigenstates of graphene can be labeled with a pseudospin index denoted by an arrow. This pseudospin represents the phase angle θ that determines the phase mixing of the two dimer basis sets underlying the wavefunctions. The forward and reverse components are the bonding and antibonding (+ and -) combinations of the orbitals (Eq. (5.14)). Notably, they are orthogonal. (b)The winding angle is twice as fast for bilayer graphene (Eq. (5.22)), so the forward and reverse modes precisely match up.*

5.5 Other variants: nanoribbons and bilayer graphene

The absence of a bandgap makes graphene hard to gate. The expected ON to OFF ratio is much lower than silicon CMOS. We will work out the current through graphene later (Eq. (16.4)). There are however ways to open a bandgap in graphene (Fig. 5.8). One way is to fashion a graphene nanoribbon (GNR), where a bandgap is opened by quantization. The second is to use Bernally stacked bilayer graphene (BLG), where we can open a bandgap by forcing an asymmetry between the top and bottom layers, such as by using a vertical electric field.

The bandgap of a GNR arises partly from the edge atoms straining towards dimerization like benzene, but more prominently because of residual next nearest neighbor interactions. These longer ranged interactions are usually camouflaged in monolayer graphene. Consider a variant of the tight-binding model we just described for graphene, but with nearest neighbor interactions $-t_0$, next nearest neighbor interactions $-t_1$ (such as between atomic sites 1 and 2 in Fig. 5.6) and next to next nearest neighbor interactions $-t_2$, running diametrically across the hexagon from each site (Fig. 5.8). Going through the same tight binding exercise as before,

$$
H_k = \underbrace{\begin{bmatrix} 0 & -t_0 \\ -t_0 & 0 \end{bmatrix}}_{H_{11}} + \underbrace{\begin{bmatrix} -t_1 & -t_2 \\ -t_0 & -t_1 \end{bmatrix} e^{i\vec{\kappa}\cdot\vec{R}_1}}_{H_{12}} + \underbrace{\begin{bmatrix} -t_1 & -t_2 \\ -t_0 & -t_1 \end{bmatrix} e^{i\vec{\kappa}\cdot\vec{R}_2}}_{H_{13}}
$$

$$
+ \underbrace{\begin{bmatrix} -t_1 & -t_0 \\ -t_2 & -t_1 \end{bmatrix} e^{-i\vec{\kappa}\cdot\vec{R}_1}}_{H_{14}} + \underbrace{\begin{bmatrix} -t_1 & -t_0 \\ -t_2 & -t_1 \end{bmatrix} e^{-i\vec{\kappa}\cdot\vec{R}_2}}_{H_{15}} + \underbrace{\begin{bmatrix} -t_1 & -t_2 \\ -t_2 & -t_1 \end{bmatrix} e^{i\vec{\kappa}\cdot(\vec{R}_1-\vec{R}_2)}}_{H_{16}}
$$

$$
+ \underbrace{\begin{bmatrix} -t_1 & -t_2 \\ -t_2 & -t_1 \end{bmatrix} e^{-i\vec{\kappa}\cdot(\vec{R}_1-\vec{R}_2)}}_{H_{17}} + \underbrace{\begin{bmatrix} 0 & 0 \\ -t_2 & 0 \end{bmatrix} e^{i\vec{\kappa}\cdot(\vec{R}_1+\vec{R}_2)}}_{H_{18}} + \underbrace{\begin{bmatrix} 0 & -t_2 \\ 0 & 0 \end{bmatrix} e^{-i\vec{\kappa}\cdot(\vec{R}_1+\vec{R}_2)}}_{H_{19}}
$$

$$(5.18)$$

It is straightforward to verify that at the two inequivalent Brillouin zone points where $\vec{\kappa} = \pm(2\pi/3b)\hat{y}$, the Hamiltonian simplifies

$$
H_k = \begin{bmatrix} 3t_1 & 0 \\ 0 & 3t_1 \end{bmatrix}
$$

$$(5.19)$$

which gives once again zero band-gap. In fact, expanding around the two inequivalent Brillouin zone points where $\vec{\kappa} = \pm(2\pi/3b)\hat{y}$, the Hamiltonian simplifies and the eigenvalues can be shown to be

$$E_\pm \approx 3t_1 \pm \frac{\sqrt{3}}{2} t_{\text{eff}} a k \qquad (5.20)$$

where $t_{\text{eff}} = t_0 - 2t_2$. The longer ranged interactions simply renormalize the electron velocity but do not alter the graphene bandstructure, which stays linear. However, truncating the interactions as in a nanoribbon quantizes the k states and breaks the angular symmetry of the k-sum, opening a bandgap that oscillates with chirality (nanoribbon width). Often though the oscillations are washed out by edge roughness.

In Bernally stacked Bilayer Graphene (BLG) two graphene sheets are oriented at sixty degrees relative to each other, so that the the A and B sublattice atoms get coupled across the planes (Fig. 5.9(b)). This extra coupling turns the linear dispersion into a quadratic one. We can treat this as simply two copies of the original monolayer graphene (MLG), but tied together by only the $A_2 - B_1$ terms. The other off-diagonal terms can be ignored for now. The bilayer Hamiltonian then becomes

$$
\begin{array}{cccc}
\mathbf{A_1} & \mathbf{B_1} & \mathbf{A_2} & \mathbf{B_2}
\end{array}
$$

$$
H = \begin{bmatrix}
0 & \hbar v_F(k_x - ik_y) & 0 & 0 \\
\hbar v_F(k_x + ik_y) & 0 & \gamma_1 & 0 \\
0 & \gamma_1 & 0 & \hbar v_F(k_x - ik_y) \\
0 & 0 & \hbar v_F(k_x + ik_y) & 0
\end{bmatrix}
\qquad (5.21)
$$

with an interlayer coupling $\gamma_1 \approx 0.4$ eV. The eigenvalues are $\approx \pm\gamma_1/2(\pm 1 + 1 + 2v_F^2 p^2/\gamma_1^2)$ for low energies compared to γ_1. The two lowest energy solutions are parabolic. They arise from the $A_1 - B_2$ sector, and can be rewritten as $\pm p^2/2m^*$, where $m^* = \gamma_1/2v_F^2 \approx 0.033m_0$, a very low effective mass about half that of GaAs. The corresponding two lowest energy eigenvectors are given by

$$\Psi_\pm = \begin{pmatrix} 1 \\ \pm e^{2i\theta} \end{pmatrix} e^{i(k_x x + k_y y)} \qquad (5.22)$$

that gives twice the winding number for the phase angle compared to MLG (Eq. (5.14)). While graphene will be seen to promote Klein tunneling at a

PN junction (no reflection for zero angle modes, section 15.3), the factor of two in the phase angle of BLG gives anti-Klein tunneling (no transmission at zero angle).

5.6   Berry phase and Chern numbers

Note that our designation of the angle θ, in fact the up or downness of the pseudospins in Fig. 5.8 is arbitrary. We can course choose a different convention which would make the pseudospin arrows look different. But their rate of winding around the Fermi circle will stay unchanged. Accordingly, we expect a 'topological invariant' that will give us this rate of winding, independent of how we label the pseudospins. As we go around the Fermi circle by slowly varying $\vec{k}(t)$, we can expect the electron to stay tied to its instantaneous eigenstate of $\mathcal{H}(\vec{k}(t))$, except for a phase factor γ. For the nth eigenstate,

$$\left|\Psi_n(\vec{k}(t))\right\rangle = \underbrace{e^{i\gamma_n(t)}}_{\text{Berry term}} e^{-i/\hbar \int_0^t dt'\,\epsilon_n(\vec{k}(t'))}\left|n(\vec{k}(t))\right\rangle \qquad (5.23)$$

If we plug this into the time-dependent Schrödinger equation $i\hbar\partial|\Psi_n\rangle/\partial t = \mathcal{H}|\Psi_n\rangle$, and use $\mathcal{H}|n(k(t))\rangle = \epsilon_n(k(t))|n(k,t)\rangle$, we get

$$\gamma_n = i\int_0^t dt'\left\langle n(\vec{k}(t'))\left|\frac{d}{dt'}\right|n(\vec{k}(t'))\right\rangle = i\int_{\vec{k}(0)}^{\vec{k}(t)} d\vec{k}\cdot\underbrace{\left\langle n(\vec{k})\left|\vec{\nabla}_k\right|n(\vec{k})\right\rangle}_{\vec{\mathcal{A}}_n} \quad (5.24)$$

The Berry term is reminiscent of the phase picked up by a free particle in a magnetic field, $\sim e^{i(p-qA)x/\hbar}$ (see Eq. (21.8)), except this is picked up in k-space (or any other convenient parameter space). The equivalent of the magnetic vector potential is the Berry connection, $\vec{\mathcal{A}}_n$, the corresponding magnetic field $\vec{\Omega}$ is the Berry curvature, and the phase picked up on completing a circuit is proportional to the flux

$$\vec{\mathcal{A}}_n = i\left\langle n(\vec{k})\left|\vec{\nabla}_k\right|n(\vec{k})\right\rangle$$

$$\vec{\Omega} = \vec{\nabla}_k \times \vec{\mathcal{A}}_n = i\left\langle\frac{\partial\phi_n}{\partial\vec{k}}\left|\times\right|\frac{\partial\phi_n}{\partial\vec{k}}\right\rangle$$

$$\gamma_n = \oint d\vec{k}.\vec{\mathcal{A}}_n = \int d\vec{S}_k.\vec{\Omega} \quad \text{(Using Gauss' Theorem)} \qquad (5.25)$$

where $\vec{S}_k$ is the k-space area element. The Chern theorem says that the surface integral of the Berry curvature is a 'topological invariant' and equal to $2\pi n$ where n is the 'charge', so that the Berry flux is quantized. This is analogous to the Gauss-Bonet theorem in differential geometry, where we relate the integral of the curvature of a solid, such as a donut or a sphere, to its genus — the number of holes that the solid contains (a topological invariant, unchanging as we continuously deform the donut or the sphere).

Let us verify this invariance with graphene. Using Eq. (5.14) for the eigenstates $|\Psi_\pm\rangle$ of graphene, we get the Berry connection

$$\vec{\mathcal{A}}_n(k) = i\langle\Psi_\pm|\vec{\partial}_k|\Psi_\pm\rangle = \vec{\partial}_k\theta/2 = \frac{-\hat{x}k_y + \hat{y}k_x}{2k^2} = \frac{-\hat{x}\sin\theta + \hat{y}\cos\theta}{2k} \tag{5.26}$$

We also write

$$d\vec{k} = k(-\hat{x}\sin\theta + \hat{y}\cos\theta)d\theta \tag{5.27}$$

so that the Berry phase

$$\gamma_n = \frac{1}{2}\oint d\theta = \pi \tag{5.28}$$

corresponding to a Chern number of $1/2$. For bilayer graphene, everything remains the same except the angle in the wavefunction is 2θ (Eq. (5.22)), so that the Berry phase $\gamma_n = 2\pi$ and the Chern number is 1.

One can visualize the Berry angle as the angle picked up as we 'parallel transport' the vector along the closed loop, maintaining a constant angle between the vector and the local tangent to the loop. Figure 5.9 suggests that the angles ought to be the double of what we get, 2π for monolayer and 4π for bilayer graphene. The reason we get half-angles is because we are dealing with *spinors* rather than *vectors*. One way to 'see' this half-angle is to note that regular vectors are orthogonal at 90^0 (e.g., the x and y axes), while spins or pseudospins are only orthogonal when they run antiparallel to each other at 180^0.

5.7   Spin-orbit coupling and topological insulators

Topological insulators are an emerging class of insulating materials whose surface bands resemble that of graphene, except they are spin (rather than pseudospin) polarized. Much like graphene, the integral of their Berry curvature is again, a topological invariant. To understand how such spin polarized metallic surface states arise, we need to first understand how spins couple with electron motion — a concept called spin-orbit coupling (SOC).

Spin-orbit couplings arise for heavier atoms from an effect that can be considered 'relativistic'. From the perspective of an electron zipping around its orbit, there is an equivalent ionic core orbiting around it, carrying a charge Zq. The motion of this charge creates an electric current, which in turn generates a magnetic field by the usual Biot-Savart law of electromagnetism. Spin-orbit coupling arises when this internal magnetic field tries to align the electronic spin with itself (Fig. 5.10). By controlling the orbital motion of an electron externally, such as with a gate field, we can then control its spin orientation as well.

The magnetic field created by the nuclear motion around the electron can be written as $\vec{B} = (\vec{v} \times \vec{\mathcal{E}})/2$. This equation follows from relativistic Lorentz transformations connecting two inertial frames moving at a constant velocity relative to each other (the factor of 2 is a correction called 'Thomas precession' that accounts for the fact that the frame is actually rotating, i.e., accelerating). The coupling of this nuclear motion-generated magnetic field with the electronic spin is then given by the Zeeman formula

$$\mathcal{H}_{so} = -\frac{g\mu_B \vec{S} \cdot \vec{B}}{\hbar} = -\frac{g\mu_B \vec{S} \cdot \left(\vec{v} \times \vec{\mathcal{E}}\right)}{2\hbar} \tag{5.29}$$

where in MKS units the constant connecting the energy with field (Joules with Tesla) is the Bohr magneton $\mu_B = q\hbar/2m \approx 9 \times 10^{-26} J/T$, while $g \approx 2$ is the Lande g-factor for electrons.

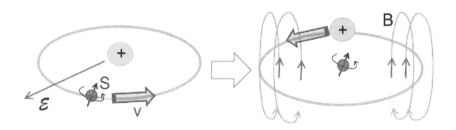

Fig. 5.10   *(Left) Spinning electron orbiting around a nucleus, in presence of a radial Coulomb field. (Right) From the perspective of the electron, the charged nucleus revolves around it, creating a current loop and thus a magnetic field which aligns the electron spin. The resulting potential is called spin-orbit coupling, and can be used to control the electron's spin by controlling its momentum with an external gate.*

For a single atom, the nuclear electric field is oriented radially following Coulomb's law while the velocity $\vec{v} = \vec{p}/m$ relates to the linear momentum, in which case

$$\mathcal{H}_{so} \propto -\vec{S} \cdot \left(\vec{p} \times \vec{r} \right) = \xi \vec{L} \cdot \vec{S} \tag{5.30}$$

where $\vec{L} = \vec{r} \times \vec{p}$ is the angular momentum. The spin-orbit coupling exploits the internal magnetic field to trade off orbital angular momentum against spin. In fact, it accounts for part of Hund's rules of maximum multiplicity that we pay attention to when filling up atomic shells with electrons.

For a solid where the electric field preferably acts along one of the crystallographic axes, the spin-orbit coupling can take other forms. For instance for a top gated structure with a vertical electric field along the z-axis, $\vec{\mathcal{E}} \parallel \hat{z}$, the Hamiltonian simplifies to give us a Rashba coupling

$$\mathcal{H}_{\text{Rashba}} \propto \vec{S} \cdot \left(\vec{k} \times \hat{z} \right) = \alpha_R \left(S_x k_y - S_y k_x \right) \tag{5.31}$$

The Rashba term allows us to control the rate of precession of spins with an external gate field. It lies at the heart of a 'gedanken' device loosely known as the Datta-Das *spin transistor* — in actuality, a *modulator* whose current oscillates with gate bias. Varying the gate field controls the rate of precession of the spins injected into a semiconducting channel from a magnetic contact (assumed to be perfectly polarized). Depending on whether the rotated spin at the end of the channel aligns or misaligns with a magnetized drain contact, the current swings between maximum and minimum.

In heavy materials like HgTe, the spin-orbit coupling is sizeable and can in fact mess up the band orders significantly. For normal 3-D band insulators, the bulk conduction bands are derived from s orbitals and are spherically symmetric, while the valence bands are derived from atomic p orbitals, are 3-fold degenerate (6 if we include spin) and significantly non-parabolic. The parity symmetry of the orbitals is notably different between conduction and valence band. Spin-orbit coupling affects the p orbitals which have non-zero angular momentum, usually splitting off one of the tridegenerate levels from the other two. When the spin-orbit coupling is significant, the p-band can actually move far enough to venture above the original s-type conduction band, leading to a band-inversion where now the conduction band is p-type and valence band is s-type.

One way to realize a topological insulator is to put a normal and an inverted band insulator next to each other, such as a layer of HgTe sandwiched

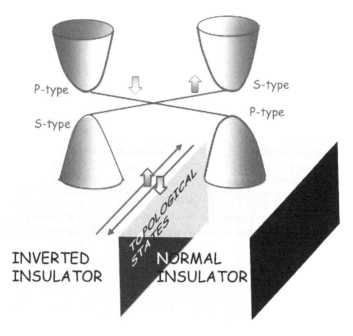

Fig. 5.11 *Normal insulator on right (s-type conduction band, p-type valence band), ad-joining an inverted insulator with p-type CB and s-type VB, the inversion created by strong spin-orbit coupling. Bands cross at the interfaces to create Dirac like metallic states that are in addition spin-polarized. The oppositely directed spin currents have or-thogonal symmetry and create high mobility 1-D topologically protected quantum states.*

between thin layers of CdTe, so that the HgTe quantum wells support a bulk band-gap between inverted sub-bands (Fig. 5.11). Symmetry rules require that the electron states evolve continuously between the parabolic bands on either side so that bands with the same orbital symmetry connect with each other (we will again encounter such symmetry constraints when con-sidering complex bands that connect bands across MgO in Section 15.2). For the 1-D interface in films of CdTe/HgTe/CdTe, we need accordingly that p-type conduction states in HgTe connect with p-type valence states in CdTe, and similarly for s-type states. The result is a linear crossing of the different symmetry bands at the interface between HgTe/CdTe, once again invoking the Von Neumann-Wigner theorem. These surface states are now Dirac like because the band-crossing closes the bandgap at the interface. In other words, they are metallic 1-D edge states residing only at the HgTe/CdTe interfaces (the bulk states are still insulating). Further-more, these surface metallic states are also labeled by their spin states to allow the adiabatic connection between band edges. In other words, the forward moving band is spin 'up' while the backward band is spin 'down'. In fact, it is the spintronic analogue of the pseudospin separated states for graphene (Eq. (5.14)). In simple terms, we have an internal magnetic field that sets up counter-propagating spin currents along the HgTe/CdTe interface. Since opposing spin states cannot overlap, they are symmetry protected from back-scattering. If we can create an imbalance under bias to generate a net nonequilibrium spin current, its mobility is arguably sizeable.

The argument can in fact be extended to 3-D solids like Bi_2Se_3 which also has a strong band inversion, between the frontier p_z orbitals that antibond in Se, bond in Bi and then switch energy in presence of SOC compared to the atomic limit. Accordingly, the top and bottom surfaces of a slab adjoin a normal insulator (in this case, vacuum), so that each surface supports 2-D Dirac cones spin polarized very similar pseudospins in graphene. Assuming the surfaces are in the $x - y$ plane, the corresponding low energy Dirac Hamiltonian for the surface states can be written in a form similar to the Rashba Hamiltonian,

$$\mathcal{H}_{3DTI} \propto v_F \hat{z} \cdot \left(\vec{S} \times \vec{k} \right) \tag{5.32}$$

5.8 Bands to Density of States

For quick estimates of transport characteristics such as a conductance, a useful quantity is the density of states $D(E)$ — the number of distinct quantum states within a given energy range (Fig. 5.12). From the band-structure $E - k$, we can directly count the number of states in each energy range (each $\vec{k}$ represents two spin states) and get the density of states

$$D(E) = 2 \sum_{\vec{k}} \delta \left(E - E_{\vec{k}} \right) \tag{5.33}$$

which can be simplified for simple materials and bandstructures. Since D has units of /eV/volume and k is /m, it is clear that for single subbands in d-dimensions, $D(E) \sim k^d / |E|$. This gives $D \sim |E|^{d/2-1}$ for parabolic bands and $D \sim |E|$ for graphene which has a linear dispersion in 2D (Eq. (16.3)). A numerical plot of the density of states, obtained by straightforward binning of the states vs their energies, is shown in Fig. 5.12 for a simple cubic lattice in multiple dimensions.

The integral of density of states upto the Fermi energy gives us the static (DC) capacitance-voltage characteristic at low temperature. More relevant to quantum transport are the number of propagating modes $M(E)$, proportional to the density of states times the projected velocity components along the transport direction, whose energy integral over a voltage bias window yields the ballistic currrent and the conductance-voltage curve underlying charge transport. We discuss these in detail in chapters 12 and 13 in the context of the Landauer formula.

The purpose of this chapter was to illustrate how molecular wavefunctions in solids overlap to form bands. For much of device physics, all we need is

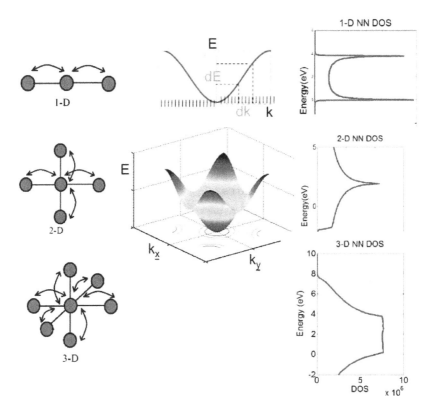

Fig. 5.12   *(Left) Simple cubic crystal, (Middle) computed band, and (Right) numerical Density of States computed by 'binning' - compiling a histogram of k-points corresponding to each energy bin. At the bottom of the bands where a continuum free-electron model suffices, the DOS goes as $D \sim E^{d/2-1}$ in d-dimensions, as argued in the text.*

the curvature of the bands around the high symmetry points, captured by a set of effective mass tensors. These masses can be extracted from the $E - k$s that enumerate the overall eigenvalues of the solid. As we will see down the road, a lot more richness lurks in the geometry of the Bloch eigenstates, such as their spin or pseudospin textures, whose symmetries derive directly from those of the underlying orbital and spin states. In a later chapter (Chapter 15), we will see how such symmetries in turn influence quantum transport.

Chapter 6

Second quantization and field theory

Let us now discuss how to go beyond single electron quantum mechanics to that of multiple electrons. We will discuss how to account for Pauli exclusion principle properly in the bandstructure, and not just in the occupation statistics. To explore many particle rules systematically, we will need to understand second quantization. Readers may choose to skip this chapter and come back to it later when we discuss interacting systems, particularly, scattering and correlation effects. We will introduce creation and annihilation operators as a way to enforce wavefunction symmetry under particle exchange, and how it leads to the second quantized Hamiltonian. A simple limit is the Hartree Fock theory that marries Coulomb interactions with Pauli exclusion. In this context, we discuss Feynman diagrams as a way to visually describe the underlying processes. We also discuss symmetry of phonon operators, the use of Green's functions to solve their responses to external perturbations, and the concept of self energies as a way to capture complicated interactions not directly included in the electronic potential.

6.1 Symmetry and identical particles

Second quantization, which leads to the quantization of the particle number, is related ultimately to the fact that electrons are **indistinguishable** particles. Let us start by looking at objects that are distinguishable, such as a pair of twins A and B that can still somehow be distinguished (say a mole on one's cheek or a different hair density). The two configurations corresponding to the twins swapping positions are in fact, distinct and demonstrably distinguishable, so that their joint probabilities of occupancy $P(\vec{r}_1, \vec{r}_2)$ and $P(\vec{r}_2, \vec{r}_1)$ are unequal and not simply related (for instance, $\vec{r}_1$ maybe a location A likes to frequent more often than B). Indistinguishable

particles like electrons and phonons are, on the other hand, identical under
swap, so that $P(\vec{r}_1, \vec{r}_2) = P(\vec{r}_2, \vec{r}_1)$. Since $P \propto |\Psi|^2$ in quantum mechanics,
this gives us the following possibilities

$$
\begin{aligned}
\Psi(\vec{r}_1, \vec{r}_2) &= -\Psi(\vec{r}_2, \vec{r}_1) \quad \text{Fermions (e.g., electrons)} \\
\Psi(\vec{r}_1, \vec{r}_2) &= +\Psi(\vec{r}_2, \vec{r}_1) \quad \text{Bosons (e.g., photons and phonons)}
\end{aligned}
\tag{6.1}
$$

In principle we have other possibilities, such as $\Psi(\vec{r}_1, \vec{r}_2) = e^{i\theta}\Psi(\vec{r}_2, \vec{r}_1)$, $0 < \theta < \pi$, or $\Psi(\vec{r}_1, \vec{r}_2) = \Psi^*(\vec{r}_2, \vec{r}_1)$, but we will ignore
them here, and assume Ψs are single-valued functions.

The difference between fermions and bosons stems from just the above
sign difference. That difference goes on to explain why electrons need to
obey the Pauli exclusion principle, while at the same time photons can Bose
condense and lead to lasing. Furthermore, symmetry arguments allow us
to derive the spin-statistics theorem under exchange — particles with angular momenta that are half integral multiples of $\hbar$ are Fermions, while
those with integer multiples are Bosons. Finally, the Boltzmann principle
describing the occupation probability $P_N \propto \exp\left(-E_N + E_F N\right)/k_B T$ of a
many body state of N electrons and total energy E_N leads ultimately to the
Fermi-Dirac and Bose-Einstein equilibrium thermal distributions. In short,
all of quantum statistical physics hails from the above sign discrepancy.

Let us discuss how to construct many-body Fermion wavefunctions that
satisfy the above sign rule.

6.1.1 *Slater determinants: towards Pauli exclusion*

One may be tempted to construct a many-body wavefunction as a simple
product of the wavefunctions for each occupied one-electron state $\psi_{\alpha_i}(\vec{r}_\beta)$
(α_i denotes for the ith orbital, β denotes a particular spatial coordinate
labeling the βth electron)

$$
\Psi(\vec{r}_1, \vec{r}_2, \ldots, \vec{r}_N) = \psi_{\alpha_1}(\vec{r}_1)\psi_{\alpha_2}(\vec{r}_2)\ldots\psi_{\alpha_N}(\vec{r}_N)
\tag{6.2}
$$

where each one-electron wavefunction, assumed ortho-normal (i.e.,
$\int \psi_{\alpha_i}^* \psi_{\alpha_j} = \delta_{\alpha_i \alpha_j}$), follows the one-electron Schrödinger equation

$$
\hat{h}_\beta \psi_{\alpha_i}(\vec{r}_\beta) = \epsilon_\alpha \psi_{\alpha_i}(\vec{r}_\beta)
\tag{6.3}
$$

Under this condition, the energy of the many electron system is a simple sum, $E = \sum_\alpha \epsilon_\alpha$ over the occupied states. Standard device models implicitly assume such a scenario where we solve for the one electron and compose the many body system trivially out of them to the point that we don't even worry about symmetry. But Eq. (6.2) ignores the quantum mechanical fact just asserted, namely, that *many-electron wavefunctions are antisymmetric under exchange*. In other words, if we swap any two electrons within this many-body wavefunction, the latter must pick up a negative sign. The sign clearly matters. As we will now see, Pauli exclusion arises out of it.

The way to enforce the antisymmetry property is to create the many-body wavefunction not out of simple products, but out of superpositions of products. A collection of N complete 1-electron basis sets will necessarily span the N-electron space, but we will need to carefully choose the superposition coefficients to enforce the required antisymmetry. Thus for instance, the two-electron superposition state

$$\Psi(\vec{r}_1, \vec{r}_2) = \frac{1}{\sqrt{2}} \left[\psi_1(\vec{r}_1)\psi_2(\vec{r}_2) - \psi_2(\vec{r}_1)\psi_1(\vec{r}_2) \right] \tag{6.4}$$

is properly antisymmetrized, and in this case, normalized.

We can generalize this principle to generate the prototypical antisymmetrized N-electron wavefunction that leads to a Slater determinant

$$\Psi(\vec{r}_1, \vec{r}_2, \ldots, \vec{r}_N) = \frac{1}{\sqrt{N!}} \begin{vmatrix} \psi_{\alpha_1}(\vec{r}_1) & \psi_{\alpha_2}(\vec{r}_1) & \cdots\cdots & \psi_{\alpha_N}(\vec{r}_1) \\ \psi_{\alpha_1}(\vec{r}_2) & \psi_{\alpha_2}(\vec{r}_2) & \cdots\cdots & \psi_{\alpha_N}(\vec{r}_2) \\ \cdots & \cdots & \cdots\cdots & \cdots \\ \cdots & \cdots & \cdots\cdots & \cdots \\ \psi_{\alpha_1}(\vec{r}_N) & \psi_{\alpha_2}(\vec{r}_N) & \cdots\cdots & \psi_{\alpha_N}(\vec{r}_N) \end{vmatrix} \tag{6.5}$$

It is straightforward to see that

- swapping any two electron coordinates amounts to swapping two rows of the matrix, which pulls out a negative sign and makes the wavefunction antisymmetric.
- A trivial consequence is that the probability of any two electrons sitting on top of each other is zero (i.e., Pauli exclusion). This follows because putting them at the same point makes the corresponding two rows identical, whereupon the determinant, and thus the wavefunction and joint occupation probability vanishes.

6.1.2   *Electron creation and annihilation operators*

The need to deal with Slater determinants poses a severe burden. However, it is a necessary burden. In this way, we capture accurately the effect of exchange on the energy levels, as opposed to semiclassical device models that introduce Pauli exclusion 'after the fact' only when filling up the levels. A many-body state can be designated as $|\alpha_1, \alpha_2, \ldots, \alpha_N\rangle$, a sorted list of singly occupied orbitals, which in position representation and bra-ket notation means

$$\left\langle \vec{r}_1, \vec{r}_2, \ldots, \vec{r}_\mu, \ldots, \vec{r}_N \, \middle| \, \alpha_1, \alpha_2, \ldots, \alpha_\mu, \ldots, \alpha_N \right\rangle$$

$$= \frac{1}{N!} \sum_P (-1)^P \psi_{\alpha P_1}(\vec{r}_1) \psi_{\alpha P_2}(\vec{r}_2) \ldots \psi_{\alpha P_N}(\vec{r}_N) \tag{6.6}$$

where each $\psi_i(\vec{r}_j)$ is a single particle state satisfying a one electron Hamiltonian, and P is the number of permutations from a reference configuration. The sign $(-1)^P$ leads to the aforementioned Slater determinant. The antisymmetry property can be restated as

$$\left| \alpha_1, \alpha_2, \ldots, \underline{\alpha_\mu}, \ldots, \underline{\alpha_\nu}, \ldots, \alpha_N \right\rangle = -\left| \alpha_1, \alpha_2, \ldots, \underline{\alpha_\nu}, \ldots, \underline{\alpha_\mu}, \ldots, \alpha_N \right\rangle \tag{6.7}$$

We need a systematic way to handle this sign change, which is critical to quantum chemistry (after all it is what leads to Pauli exclusion, which is what ultimately prevents me from sinking into the floor!) Slater determinants take care of the *order* of the electron states, but notationally they are cumbersome. Second quantization moves the burden of antisymmetry to a set of evolving time-dependent Fermionic creation $c^\dagger$ and annihilation c operators, that respectively create or remove an electron out of a given reference state (see Heisenberg representation, Section 3.3.2). For instance, we can define an $N+1$ particle state by introducing a creation operator acting on an N particle state

$$c^\dagger_{\alpha_\mu} \underbrace{\left| \alpha_1, \alpha_2, \ldots, \alpha_N \right\rangle}_{\text{N particle state}} = \underbrace{\left| \underline{\alpha_\mu}, \alpha_1, \alpha_2, \ldots, \alpha_N \right\rangle}_{\text{N particle state}} \tag{6.8}$$

where the creation operator creates an additional electron at the leftmost end (the 'leftness' is arbitrary, and we could choose to count states to the right instead, as long as we stay consistent with our choice, because the order matters!). Each entry of the many-body state designates the one-electron state α_μ of an existing single electron. The electron addition could represent a physical process such as the photoexcitation of electron-hole pairs, or perhaps current flow involving addition and removal of electrons taking us between an N and an $N \pm 1$ particle state. Thus, we can write the many-body state itself as starting with vacuum $|0\rangle$, and then adding in the electrons at its left-most end, one at a time

$$\left|\alpha_1, \alpha_2, \ldots, \alpha_N\right\rangle = c^\dagger_{\alpha_1} c^\dagger_{\alpha_2} \ldots c^\dagger_{\alpha_N} |0\rangle \tag{6.9}$$

If we now want to add the state α_ν not at the left end, but somewhere in the middle of the many-particle state, say, at position $\vec{r}_\nu$, then we just create the electron at the left most end and ripple that state in through $\nu - 1$ swaps. Each swap gives a negative sign because of the overall wavefunction antisymmetry (Eq. (6.1)), so that

$$c^\dagger_{\alpha_\nu} \left|\alpha_1, \alpha_2, \ldots, \alpha_N\right\rangle = (-1)^{\nu-1} \left|\underbrace{\alpha_1, \alpha_2, \ldots,}_{\nu-1 \text{ states}} \underline{\alpha_\nu}, \ldots, \alpha_N\right\rangle \tag{6.10}$$

where the sign ends up being determined by *the number of filled states $\nu - 1$ to its left*. Similarly, the annihilation operator c_ν destroys the state by rippling through $\nu - 1$ filled states to access and destroy the νth state.

$$c_{\alpha_\nu} \left|\underbrace{\alpha_1, \alpha_2, \ldots,}_{\nu-1 \text{ states}} \underline{\alpha_\nu}, \ldots, \alpha_N\right\rangle = (-1)^{\nu-1} \left|\alpha_1, \alpha_2, \ldots, \alpha_N\right\rangle \tag{6.11}$$

It is straightforward to show then that the creation and annihilation operators satisfy the anticommutation rules

$$\begin{aligned} \{c^\dagger_\mu, c^\dagger_\nu\} &= c^\dagger_\mu c^\dagger_\nu + c^\dagger_\nu c^\dagger_\mu = 0 \\ \{c_\mu, c_\nu\} &= c_\mu c_\nu + c_\nu c_\mu = 0 \\ \{c_\mu, c^\dagger_\nu\} &= c_\mu c^\dagger_\nu + c^\dagger_\nu c_\mu = \delta_{\mu,\nu} \end{aligned} \tag{6.12}$$

The equations are easy to interpret. For instance, the operators $c^\dagger_\mu c^\dagger_\nu$ and $c^\dagger_\nu c^\dagger_\mu$ create almost the same state except in the first case μ sits to the left of ν, and to the right in the second case. This flip in order introduces an extra negative sign, leading to the first equation.

As long as we can express our Hamiltonian and key variables in this second-quantized notation, antisymmetry is guaranteed by the anticommutation properties of these operators, and we do not need to worry about Slater determinants any more.

6.1.3 *Field operators*

We have now established how to write down the many body wavefunction $\Psi(\vec{r}_1, \vec{r}_2, \ldots)$ in terms of individual one-electron wavefunctions $\psi_{\alpha_i}(\vec{r}_j)$, as well as the creation and annihilation operators $c^\dagger$ and c that create or destroy a one electron state. We can now define a field operator $\Psi^\dagger(\vec{r})$ which simply generates the electron wavefunction including its spatial dependence out of vacuum. We can then create the entire many body wavefunction one at a time through repeated operation. In Dirac bra-ket notation, this means

$$\Psi^\dagger(\vec{r})|0\rangle = |\vec{r}\rangle$$
$$\Psi^\dagger(\vec{r}_2)\Psi^\dagger(\vec{r}_1)|0\rangle = |\vec{r}_2, \vec{r}_1\rangle$$
$$\ldots \tag{6.13}$$

Expanding $|\vec{r}\rangle$ in a complete single particle basis set, inserting the completeness relation (Eq. (3.12)) $\sum_\alpha |\alpha\rangle\langle\alpha| = I$, we get from the definitions of Eqs. (6.7) and (6.6)

$$|\vec{r}\rangle = \sum_\alpha \underbrace{|\alpha\rangle}_{c_\alpha^\dagger|0\rangle} \underbrace{\langle\alpha|\vec{r}\rangle}_{\psi_\alpha^*(\vec{r})} = \sum_\alpha c_\alpha^\dagger \psi_\alpha^*(\vec{r})|0\rangle \tag{6.14}$$

From Eqs. (6.13) and (6.14), we can then infer a straightforward relation between the field operator and the one-electron version, as well as its Hermitian conjugate

$$\boxed{\begin{aligned} \Psi^\dagger(\vec{r}) &= \sum_\alpha c_\alpha^\dagger \psi_\alpha^*(\vec{r}) \\ \Psi(\vec{r}) &= \sum_\alpha c_\alpha \psi_\alpha(\vec{r}) \end{aligned}} \tag{6.15}$$

The anticommutation of the $c, c^\dagger$ operators and the orthonormality of the ψs translate to the anticommutation property of the field operators

$$\{\Psi^\dagger(\vec{r}), \Psi^\dagger(\vec{r}')\} = \{\Psi(\vec{r}), \Psi(\vec{r}')\} = 0$$
$$\{\Psi(\vec{r}), \Psi^\dagger(\vec{r}')\} = \delta(\vec{r} - \vec{r}') \tag{6.16}$$

We now have the wavefunctions in second quantized language. What about the operators and the Hamiltonian itself?

6.2 The second quantized Hamiltonian

The non relativistic Hamiltonian of a system of electrons and nuclei can be written as the sum of nuclear kinetic energy, electron kinetic energy, electron-nuclear attractive interaction, electron-electron repulsion and nuclear-nuclear repulsion, shown here in sequence.

$$
\hat{\mathcal{H}} = \underbrace{\sum_{\alpha} -\frac{\hbar^2 \nabla_\alpha^2}{2M_\alpha}}_{T_{\text{nucl}}} + \underbrace{\sum_{i} -\frac{\hbar^2 \nabla_i^2}{2m}}_{T_{\text{el}}} - \underbrace{\sum_{i\alpha} \frac{Zq^2}{4\pi\epsilon_0 |\vec{r}_i - \vec{R}_\alpha|}}_{U_{\text{el-nucl}}}
$$

$$
+ \underbrace{\sum_{ij} \frac{q^2}{4\pi\epsilon_0 |\vec{r}_i - \vec{r}_j|}}_{U_{\text{el-el}}} + \underbrace{\sum_{\alpha\beta} \frac{Z^2 q^2}{4\pi\epsilon_0 |\vec{R}_\alpha - \vec{R}_\beta|}}_{U_{\text{nucl-nucl}}} \qquad (6.17)
$$

In general, this is a really complicated dynamical problem to solve for the electrons and atoms. To proceed, one makes the following assumptions:

- **Fast electrons zipping past slow atoms.** The atoms, each a thousand times heavier than the electrons and thus incomparably sluggish, are assumed to stay frozen when we work out the electron dynamics (this is the *Born Oppenheimer approximation*). The static locations of those atoms, $\{\vec{R}_0\}$, will later be calculated separately. The approximation means that the first and last terms in the above Hamiltonian are constants as far as the electron is concerned, and we only need to solve the rest of the Hamiltonian for the electron wavefunction with the atomic coordinates treated as constant parameters, giving us the conditional wavefunction $\Psi(\{\vec{r}\}, \{\vec{R}\}) \to \Psi(\{\vec{r}\} | \{\vec{R}_0\})$ and electronic energy $E_{\text{el}}(\{\vec{R}_0\})$,

$$
\left(\hat{T}_{\text{el}} + \hat{U}_{\text{el-nucl}} + \hat{U}_{\text{el-el}} \right) \Psi\left(\{\vec{r}\} \big| \{\vec{R}_0\}\right) = E_{\text{el}}\left(\{\vec{R}_0\}\right) \Psi\left(\{\vec{r}\} \big| \{\vec{R}_0\}\right) \quad (6.18)
$$

This is what we will focus on in this section, i.e., solving the middle three Hamiltonian terms. Out of these, the first two terms lead to conventional band theory. The third is usually approximated with Poisson's equation, but is in general really complicated!

- **Slow atoms moving amidst relaxed electrons.** To get the atomic coordinates $\{\vec{R}_0\}$, we use the solved electronic energies $E_{\text{el}}(\{\vec{R}\})$ from

Eq. (6.18) for each frozen atomic configuration and calculate the nuclear potential, obtained by adding back the nuclear nuclear repulsion term to the calculated electronic energy $E_{el}(\{\vec{R}\})$. Once again, the separate speeds of the fast electrons vs slow atoms play a role, in that we assume the electrons instantly relax to their lowest energy ground state as we vary the atomic coordinates. The resulting nuclear potential looks like an interatomic Morse potential, and determines both the equilibrium atomic coordinates $\{\vec{R}_0\}$ and the vibrational dynamics around them. The quantized vibrational modes inside the nuclear potential give us a set of normal modes called *phonons*.

- **Deviations from atomic equilibrium sites**. The electronic energies were calculated with the atoms frozen at their equilibrium sites. If we want to include the effect of slow oscillations around those sites, we can expand the electron nuclear potential around those points to give us the electron phonon coupling

$$U_{\text{el-nucl}}\left(|\vec{r}_i - \vec{R}_\alpha|\right) \approx U_{\text{el-nucl}}\left(|\vec{r}_i - \vec{R}_{0\alpha}|\right)$$
$$+ \underbrace{\delta\vec{R}_\alpha \cdot \vec{\nabla}_\alpha U_{\text{el-nucl}}\Big|_{\vec{R}_{0\alpha}}}_{\text{el-phonon}} + O\left(\delta\vec{R}_\alpha \delta\vec{R}_\beta\right) \quad (6.19)$$

6.2.1 *Hartree-Fock: Poisson meets Pauli*

Let us first try to write down the electronic Hamiltonian. We will do so in the second quantized notation, keeping the creation/annihilation operators as the principal time-evolving entities, and moving the one-particle wavefunctions $\psi_\alpha(\vec{r})$ inside spatial integrals used to define some matrix components. We can then look at 0-electron, 1-electron and 2-electron potentials, the number designated by how many electron coordinates constitute the potential operator. We expand $\hat{\mathcal{H}}$ in 0, 1 or 2-electron basis sets and use the definition of the field operator (Eq. (6.15)) and the orthonormality of ψ

$$\int \underbrace{d\Omega}_{\text{volume}} \Psi^\dagger(\vec{r})\Psi(\vec{r}) = \sum_{\alpha,\beta} \underbrace{\int d\Omega\, \psi_\alpha^*(\vec{r})\psi_\beta(\vec{r})}_{=\delta_{\alpha,\beta}} c_\alpha^\dagger c_\beta$$

$$= \sum_\alpha \underbrace{c_\alpha^\dagger c_\alpha}_{\hat{n}_\alpha} \quad \text{(electron density)} \quad (6.20)$$

Next, we write down the one electron term

$$U^{(1)} = \int d\Omega \Psi^\dagger(\vec{r}) \left(T_{el} + U_{el\text{-}nucl}(\vec{r}) \right) \Psi(\vec{r}) \qquad \text{(1-electron potentials)}$$

$$= \underbrace{\sum_\alpha \epsilon_\alpha \hat{n}_\alpha}_{\text{onsite energy}} + \underbrace{\sum_{\alpha \neq \beta} t_{\alpha\beta} c_\alpha^\dagger c_\beta}_{\text{hopping}} \qquad (6.21)$$

where

$$\epsilon_\alpha = \int d\Omega \psi_\alpha^*(\vec{r}) \left(T_{el} + U_{el\text{-}nucl}(\vec{r}) \right) \psi_\alpha(\vec{r})$$

$$t_{\alpha\beta} = \int d\Omega \psi_\alpha^*(\vec{r}) \left(T_{el} + U_{el\text{-}nucl}(\vec{r}) \right) \psi_\beta(\vec{r}) \qquad (6.22)$$

The one electron term is at the heart of tight binding theory, extended by chemists to nonorthogonal Hückel and Extended Hückel theories in minimal basis sets with much better transferability of parameters between diverse environments (see Section 6.4).

Finally, the electron electron interaction term $U_{el\text{-}el}$ in Eq. (6.17)

$$U^{(12)} = \int d\Omega_1 d\Omega_2 n(\vec{r}_1) \frac{q^2}{4\pi\epsilon_0 |\vec{r}_1 - \vec{r}_2|} n(\vec{r}_2) \quad \text{(2-electron potentials)}$$

$$= \int d\Omega_1 d\Omega_2 \Psi^\dagger(\vec{r}_1) \Psi(\vec{r}_1) \frac{q^2}{4\pi\epsilon_0 |\vec{r}_1 - \vec{r}_2|} \Psi^\dagger(\vec{r}_2) \Psi(\vec{r}_2) \quad \begin{array}{l} \text{(Use Eq. (6.15)} \\ \text{for } \Psi, \Psi^\dagger) \end{array}$$

$$= \sum_{\alpha\beta\gamma\delta} U_{\alpha\beta\gamma\delta} c_\alpha^\dagger c_\beta c_\gamma^\dagger c_\delta \quad \text{(ripple } c_\beta \text{ to the right using Eq. (6.12))}$$

$$= \sum_{\alpha\beta\delta} U_{\alpha\beta\beta\delta} c_\alpha^\dagger c_\delta + \sum_{\alpha\beta\gamma\delta} U_{\alpha\beta\gamma\delta} c_\alpha^\dagger c_\gamma^\dagger c_\delta c_\beta \qquad (6.23)$$

The first term can be absorbed into the hopping term in Eq. (6.21), while the second involves the integral

$$U_{\alpha\beta\gamma\delta} = \int d\Omega_1 d\Omega_2 \psi_\alpha^*(\vec{r}_1) \psi_\beta(\vec{r}_1) \frac{q^2}{4\pi\epsilon_0 r_{12}} \psi_\gamma^*(\vec{r}_2) \psi_\delta(\vec{r}_2) \qquad (6.24)$$

For tightly bound electrons, the overlap of wavefunctions is zero unless their orbital indices match. Pairing up the operators in this case, we will get two terms, one for $\alpha = \beta$, $\gamma = \delta$ and one for $\alpha = \delta$, $\beta = \gamma$. The result

$$U_{\alpha\alpha\beta\beta} = \int d\Omega_1 d\Omega_2 n_\alpha(\vec{r}_1) \frac{q^2}{4\pi\epsilon_0 r_{12}} n_\beta(\vec{r}_2) \quad \text{(Direct or Hartree)}$$

$$U_{\alpha\beta\beta\alpha} = \int d\Omega_1 d\Omega_2 n^*_{\alpha\beta}(\vec{r}_1, \vec{r}_2) \frac{q^2}{4\pi\epsilon_0 r_{12}} n_{\alpha\beta}(\vec{r}_1, \vec{r}_2) \quad \text{(Exchange or Fock)}$$

$$U^{(12)} = \sum_{\alpha\neq\beta} \left[U_{\alpha\alpha\beta\beta} - U_{\alpha\beta\beta\alpha} \right] \hat{n}_\alpha \hat{n}_\beta + \text{hopping terms} \quad (6.25)$$

where $n_\alpha(\vec{r}) = |\psi_\alpha(\vec{r})|^2$, $n_{\alpha\beta}(\vec{r}_1, \vec{r}_2) = \psi^*_\alpha(\vec{r}_1)\psi_\beta(\vec{r}_2)$ and $\hat{n}_\alpha = c^\dagger_\alpha c_\alpha$. The first term describes the usual Coulomb repulsion between two electrons which we commonly introduce through Poisson's equation in device models. The second describes the cost of exchange, in effect enforcing Pauli exclusion. We see clearly that beyond just the statistics, exchange can also influence the energetics, opening up ways to lower the energy by correlating spins, a fact that ferromagnets make good use of.

Assuming the basis sets are localized enough that only onsite terms survive and nonlocal terms do not, the Hamiltonian simplifies further. Onsite Hartree terms with *parallel spins* get precisely cancelled by the corresponding Fock terms. Introducing non-interacting contacts with dispersion $\epsilon_{\alpha k\sigma}$ and contact-dot couplings $t_{\alpha k\sigma}$, we get the *Single Impurity Anderson Hamiltonian* (SIAM), indices 'd' and 'k' relating to the dot and the momentum in the leads, α the lead index (left, right) and spin index $\sigma = \uparrow, \downarrow$.

$$
\begin{aligned}
\hat{\mathcal{H}}_{el} = &\underbrace{\sum_\sigma \epsilon_{d\sigma} \hat{n}_{d\sigma} + U_H \hat{n}_{d\uparrow} \hat{n}_{d\downarrow}}_{\hat{\mathcal{H}}_d} \quad \textbf{(Dot Term)} \\[2mm]
+ &\underbrace{\sum_{\substack{k\sigma \\ \alpha=L,R}} \epsilon_{\alpha k\sigma} \hat{n}_{\alpha k\sigma}}_{\hat{\mathcal{H}}_C} \quad \textbf{(Contact Term)} \\[2mm]
+ &\underbrace{\sum_{k\sigma\alpha} t_{\alpha k\sigma} c^\dagger_{\alpha k\sigma} c_{d\sigma} + hc}_{\hat{\mathcal{H}}_I} \quad \textbf{(Hybridization)}
\end{aligned}
$$

$$(6.26)$$

the notation 'hc' referring to *Hermitian conjugate* of the term it adjoins. If we allow the dot electrons to hop around itinerantly,

$$\hat{\mathcal{H}}_d \rightarrow \sum_i \epsilon_i \hat{n}_i + \sum_{ij} t_{ij} c_i^\dagger c_j + hc + \sum_i U_i \hat{n}_{i\uparrow} \hat{n}_{i\downarrow} \tag{6.27}$$

as in a chain of coupled dots, we get the *Hubbard model*.

The advantage of writing the many body Hamiltonian in this second quantized notation is that we no longer need to worry about the antisymmetry of the many-body wavefunctions. The anticommutation relations of the $c, c^\dagger$ operators in Eq. (6.12) take care of that automatically.

6.2.2 How far does exclusion extend?

We have already seen the effect of Pauli exclusion earlier, when we encountered an oscillatory behavior of the screening charges in solids (Section 4.4). Those oscillatory terms actually arose from the Fock part of the potential that enforced Pauli exclusion and ultimately the sharp cut off at the Fermi energy. It is natural to ask at this stage how far the influence of Pauli exclusion extends, because we clearly expect two electrons sitting far away with minimal overlap to act fairly independently in every way. The answer is easier to see in a solid, where the molecular (Bloch) part of the wavefunction arising from the underlying atomic orbitals get modulated by an overall plane wave, associated with the periodicity of the unit cells. The exchange density $n_{\alpha\beta}$ in Eq. (6.25) gets replaced with a different set of indices, wavevectors instead of the orbitals. The Hartree Fock exchange density (Eq. (6.25)) for parallel spins can be written, ignoring the Bloch part of ψ and working just with the plane wave component

$$n_{ex}(\vec{r}_1, \vec{r}_2) = \frac{1}{\Omega} \sum_{\vec{k}_1, \vec{k}_2} \Psi^*_{\vec{k}_1}(\vec{r}_1) \Psi^*_{\vec{k}_2}(\vec{r}_2) \Psi_{\vec{k}_1}(\vec{r}_2) \Psi_{\vec{k}_2}(\vec{r}_1)$$

$$\approx \frac{1}{\Omega} \sum_{\vec{k}_1, \vec{k}_2} e^{i(\vec{k}_1 - \vec{k}_2) \cdot (\vec{r}_2 - \vec{r}_1)} = \frac{1}{\Omega} \int \frac{d^3\vec{k}_1 d^3\vec{k}_2}{(2\pi)^6 / \Omega^2} e^{i\Delta\vec{k} \cdot \Delta\vec{r}}$$

$$\tag{6.28}$$

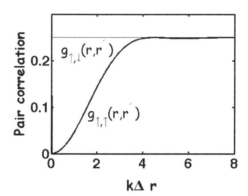

Fig. 6.1   *Pair correlation function $g(r, r')$ represents joint two electron occupation probability. In the absence of strong Coulomb correlation (e.g., in a metal), this term is flat for opposite spins, but it kicks in for parallel spins over a distance related to the 'size' of an electron wavefunction, preventing dual occupancy and enforcing Pauli exclusion. Note that Pauli exclusion has a sphere of influence — two electrons in two atoms sitting far away do not exclude each other.*

We can change variables from $\{\vec{k}_1, \vec{k}_2\}$ to $\vec{K} = (\vec{k}_1 + \vec{k}_2)/2$ and $\Delta\vec{k} = \vec{k}_1 - \vec{k}_2$, which preserves the volume $d^3\vec{k}_1 d^3\vec{k}_2 = d^3\vec{K} d^3\Delta\vec{k}$, to give us

$$n_{ex}(\vec{r}_1, \vec{r}_2) = \frac{\Omega}{(2\pi)^6} \underbrace{\int d^3\vec{K}}_{(2\pi)^3/\Omega} \int d^3\Delta\vec{k} \; e^{i\Delta\vec{k}\cdot\Delta\vec{r}}$$

$$= \frac{1}{(2\pi)^3} \cdot 2\pi \int_0^{k_F} (\Delta k)^2 d\Delta k \int_{-1}^{1} d(\cos\theta) e^{i\Delta k \Delta r \cos\theta}$$

$$= \frac{3n}{2} \left[\frac{\sin(k_F \Delta r) - k_F \Delta r \cos(k_F \Delta r)}{(k_F \Delta r)^3} \right] = \frac{3n}{2} \left[\frac{j_1(k_F \Delta r)}{k_F \Delta r} \right] \quad (6.29)$$

(using $k_F^3 = 3\pi^2 n$, Eq. (13.9), for free electrons in 3D) and the definition of the spherical Bessel function of the first kind, $j_1(x) = [\sin x - x \cos x]/x^2$. We can then calculate the pair correlation function, $g(\vec{r}_1, \vec{r}_2) = [n(\vec{r}_1)/2]^2 - |n_{ex}(\vec{r}_1, \vec{r}_2)|^2$, which represents the probability of joint occupancy and shows a clear exchange-correlation 'hole' (Fig. 6.1) gouged out of the background electron density by Pauli exclusion. The repulsive potential responsible for creating that hole is given by the Fock term (Eq. (6.25))

$$U_F = \frac{1}{4} \iint \frac{|n_{ex}(r_1, r_2)|^2}{r_{12}} d^3\vec{r}_1 d^3\vec{r}_2 \quad \text{switch variables to } \vec{R} = (\vec{r}_1 + \vec{r}_2)/2,$$
$$\vec{\Delta r} = \vec{r}_1 - \vec{r}_2$$

$$= \frac{1}{4} \iint \frac{|n_{ex}(R, \Delta r)|^2}{\Delta r} d^3 R d^3 \Delta r$$

$$= \frac{9}{16} \int n^2(R) d^3 R \int \left[\frac{\sin(k_F \Delta r) - k_F \Delta r \cos(k_F \Delta r)}{(k_F \Delta r)^3} \right]^2 \underbrace{\frac{d^3 \Delta r}{\Delta r}}_{4\pi \Delta r d \Delta r}$$

$$= \frac{9\pi}{4} \int \frac{n^2(R)}{k_F^2} d^3 R \underbrace{\int_0^\infty \frac{(\sin t - t \cos t)^2}{t^5} dt}_{1/4} \quad (\text{Use Eq. (13.9), } k_F^3 = 3\pi^2 n)$$

$$= C_X \int [n(R)]^{4/3} d^3 R, \qquad C_X = \frac{3}{4} \left(\frac{3}{\pi} \right)^{1/3} = 0.7386 \tag{6.30}$$

In other words, $U_F \propto \int n d^3 R / r_0$, which we can interpret as the energy cost from an exchange-correlation 'hole' (Fig. 6.1) of size r_0 equal to an electron it expelled through Pauli exclusion, $(4\pi r_0^3/3)n = 1$. The expulsion of the parallel spin state allows the electrons to lower the Coulomb cost — leading ultimately to Hund's rule coupling and ferromagnetism. In our simplified analysis, only opposite spins are anticorrelated (they stay away), but like spins do not see such a hole despite Coulomb repulsion. This is a shortcoming of the mean field approach which sees no Coulomb correlations, and calls for better electronic structure theory.

6.3 Capturing exchange-correlation in practical bandstructure theories

An accurate electronic structure theory requires us to explore various many body configurations combinatorially, as discussed in the chapter on correlation. The challenge is to account for effects like self-interaction, since an electron must not feel a potential due to itself. The process simplified with the emergence of Density Functional Theory (DFT), which asserts that the *ground state* energy of the system is a functional $E[n(R)]$ of the scalar electron density alone, meaning we can stop tracking the many body configurations. Furthermore the Kohn Sham approach that reduces the many body problem to an equivalent one particle problem, of a quasiparticle moving in an exchange correlation potential $V_{xc} = \delta E_{xc}[n]/\delta n$, where we have separated out from E the kinetic energy, Hartree and the response $-F\delta n$ to an external field F. Unfortunately, this functional $E[n(R)]$ is not known except for very simple systems (materials with near uniform electron density for instance), although its properties, e.g., the need for an orbital-dependent potential and an associated 'derivative discontinuity' (Section 17.1), are known.

The lack of precise knowledge of the exchange-correlation function has led to a 'Jacob's ladder' of approximations of increasing complexity, starting with the Local Density Approximation (LDA) that makes $V_{xc}(R)$ a local function of $n(R)$. We expect this to work for systems with nearly homogeneous electron distributions, for instance, metals with adequate screening. Corrections due to spatial variations in charge density can be incorporated through terms like $\nabla n(R)$ in the Generalized Gradient Approximation (GGA), or $\nabla^2 n(R)$ in meta-GGA. Model expressions for these potentials exist in the literature and are improved upon constantly. Significant improvements have been achieved by mixing model potentials with exact exchange using Hartree Fock, to generate 'hybrid functionals' with acronyms such as B3LYP, B3PW91, HSE, all designed to capture self interaction correction more and more accurately. We will discuss electronic structure again in the chapter on molecular resonance (Section 17.1).

6.4 Phenomenological treatments of band structure: EHT and PPP models

In molecular modeling, it is convenient to use localized basis sets as in tight-binding, but include chemical attributes like non-orthogonality to resemble atomic basis sets. Atomic states (e.g. s-orbitals) are non-orthogonal between neighboring atoms, while Wannier superpositions of atomic Bloch states in a solid are orthogonal. A semi-empirical bandstructure theory popular among chemists is Extended Huckel Theory (EHT), a variant of non-orthogonal tight-binding, except with explicit basis sets and simple rules to compute the matrix elements, even in presence of strain or bonding. The non-orthogonality often allows band parameters to be transferable among different environments, among bulk, surfaces and interfaces. The presence of explicit basis sets enables recomputing the matrix elements among those environments, and also matching/best-fitting interfacial Green's functions between different bandstructure theories.

In semi-empirical EHT, the off diagonal couplings are given by averaging the on-site Hubbard energies, so that the main chemical ingredient of wavefunction overlap S_{ij} is explicitly accounted for in the bonding energies. The Coulomb kernel is obtained using the Pariser-Parr-Pople (PPP) model within the Mataga-Nishimoto approximation.

$$H_{ij} = K S_{ij}(U_{ii} + U_{jj}), \quad U_{ij} = \frac{q^2}{4\pi\epsilon_0 \left(|\vec{r}_i - \vec{r}_j| + 2q^2/(U_{ii} + U_{jj}) \right)} \quad (6.31)$$

The EHT approach does not directly include any electron-electron interactions. To incorporate interactions, several semi-empirical approximations have been adopted by quantum chemists over the years, such as CNDO, MNDO etc (with the "NDO" signifying various degrees of neglect of orbital overlap).

A case study is shown for bulk silicon (references at the end of the book for this chapter). Figure 6.2 shows the bandstructure of bulk silicon in EHT, constructed to fit 18 experimental parameters. Using the same exact basis set, EHT was then extended to two different reconstructed silicon surfaces, with the only adjustment being the recomputation of the overlap matrix elements S_{ij} and thus the Hamiltonian using Eq. (6.31). The reconstructions in silicon create new surface states whose bands are shown to the right and compared with experiments/GW calculations. In other words, we have a practical bandstructure tool that combines the flexibility of tight-binding (e.g. the ability to fit bulk bandstructure parameters) with the predictive transferability and chemical relevance of DFT.

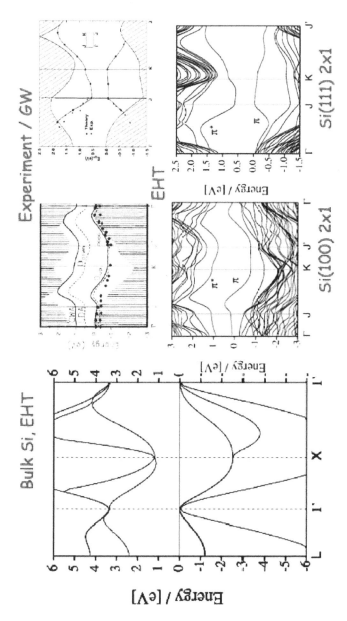

Fig. 6.2   *(Left) EHT bulk bandstructure of Si calibrated against 18 experimental parameters. (Right) Surface bandstructures for reconstructed Si surfaces computed with the same EHT parameters, compared against accurate but computationally expensive GW approximation and experimental data. See paper by Kienle et al. in references to this chapter.*

Chapter 7

Interactions and Green's functions

7.1 How interactions mess things up

A natural question that may arise at this stage is why all this mathematical jugglery is necessary, given that textbooks on solid-state electronics as well as compact models in industry describe operational devices without the need for second quantization or many-body wavefunction descriptions. It turns out that for systems with no interactions (or weak interactions) among its variables, whether electrons or phonons, the physics simplifies considerably. This is because in the absence of interactions the Hamiltonian is quadratic (e.g., in c, $c^\dagger$), and can be easily diagonalized with just a basis transfer within the one-electron subset (it is a bit trickier than simply completing squares, as we need to preserve the commutation/anticommutation rules for the diagonalizing combination of operators, but it is still quite straightforward). This means we can stick to the diagonalized Hamiltonian and its independent one-electron eigenstates, leading to conventional band theory. Strong interactions however give higher order (e.g., cubic, quartic) terms in the many-body Hamiltonian that cannot be diagonalized in the one-electron basis set. They need a systematic perturbation theory while obeying their symmetry rules, and this is best accomplished with the employment of creation and annihilation operators.

7.1.1 *Quadratic electron Hamiltonian: tight-binding*

In absence of interactions, the Hamiltonian for a chain of quantum dots or atoms (Eq. (6.21)) is

$$\hat{\mathcal{H}}_{\text{el}} = \sum_i \epsilon_i \hat{n}_i + \sum_{i,j} t_{ij} c_i^\dagger c_j \tag{7.1}$$

where we are now using a lattice grid point basis instead of an orbital basis (i.e., i,j are nearest neighbor spatial coordinates — the nearest neighbor approximation valid if the basis sets are tightly bound around their respective nuclear lattice points). In the simple case where we have a 1-D chain of N equidistant atoms (separation a) with $\epsilon_i = \epsilon_0$ and $t_{ij} = -t_0$, the algebra becomes particularly simple (Eq. (7.1)). Basis transforming from the lattice basis sets ψ localized around the grid points $x_n = na$ to a Bloch wavevector set

$$c_n = \frac{1}{\sqrt{N}} \sum_k e^{ikx_n} c_k, \qquad c_k = \frac{1}{\sqrt{N}} \sum_n e^{-ikx_n} c_n \qquad (7.2)$$

the Hamiltonian can be recast easily as

$$\hat{\mathcal{H}}_{\text{el}} = \sum_k \epsilon_k c_k^\dagger c_k = \sum_k \epsilon_k \hat{n}_k \qquad (7.3)$$

giving us a band

$$\epsilon_k = \epsilon_0 - 2t \cos ka \qquad (7.4)$$

in 1D, and a much more intricate one in 3D involving multiple orbitals and k points in a complicated Brillouin zone, obtained by diagonalizing

$$[H]_{\vec{k}} = \sum_{mn} [H]_{mn} e^{i\vec{k} \cdot (\vec{R}_m - \vec{R}_n)} \qquad (7.5)$$

Since there are no cross-terms here but just a sum over independent particle contributions in k-space, we can work simply with the electronic bands ϵ_k, as customary in solid state device textbooks, and forget about second quantization altogether.

7.1.1.1 *Visualizing Coulomb contributions*

The Coulomb interaction $\sum_{\alpha\beta} U_{\alpha\beta} \hat{n}_\alpha \hat{n}_\beta$ produces higher order quartic corrections to this Hamiltonian that require special treatment. In many solids, it is screened by other electrons and is small compared to the kinetic energy, whereupon it can be treated as a small perturbation and dealt with using conventional Poisson's equation. However for systems with poor screening, such as lower dimensional molecules, wires or quantum dots, or even bulk insulators or materials with localized partially filled d or f electron shells, the Coulomb term is really hard to simplify. It has a strong nontrivial effect and can be responsible for a rich variety of exotic phenomena ranging from superconductivity to Mott transitions.

For a limited number of orbitals, we may be able to solve the entire many-body Hamiltonian (size $2^N \times 2^N$ for N spinless orbitals) numerically, as we do in the chapter on correlation. For an infinite chain, this becomes really hard to solve, allowing only a few exact solutions (e.g., Bethe ansatz). In other cases such as ferromagnets or a superconducting Cooper pair, we rely on inspired guesses. A common technique is to postulate an average 'field' (a scalar) to stand in for part of the operator product so that the rest of the Hamiltonian turns quadratic, then evaluating the spectra under self-consistency conditions for those non vanishing averages (Section 7.1.3).

The language of Feynman diagrams can be quite useful for dealing with these additional terms in the Hamiltonian in a systematic, perturbative way. We saw earlier that the Coulomb term $\sum_{\alpha\beta} U_{\alpha\beta}\hat{n}_\alpha\hat{n}_\beta$, when expanded in creation/annihilation language, leads to a Hartree term $U_{\alpha\beta\alpha\beta}$ describing direct Coulomb repulsion and a Fock term $U_{\alpha\beta\beta\alpha}$ describing quantum mechanical exchange. We can expect the same terms in a solid, except we will replace the orbital index with a momentum index corresponding to Bloch like plane wave basis sets. The Feynman rules (appendix) allow us to visualize these processes schematically as a set of electrons moving forward in time (upward in Fig. 7.1), positrons moving backward in time, and a Coulomb interaction represented by a wiggly line connecting them. The direct and exchange terms, for instance, can be interpreted as a self energy inserted within the otherwise noninteracting propagator lines. The self energies, following Feynman rules (Section 7.3), amount to

$$\Sigma_H(k,\omega) = \left[iG_0(k,\omega)\right]^2 \sum_q \int \frac{d\epsilon}{2\pi} \left[-iU_{kkqq}\right] \left[iG_0(q,\epsilon)\right] e^{i\epsilon 0^+} \underbrace{(-1)}_{\text{Fermion loop}}$$

$$\Sigma_F(k,\omega) = \left[iG_0(k,\omega)\right]^2 \sum_q \int \frac{d\epsilon}{2\pi} \left[-iU_{qkkq}\right] \left[iG_0(q,\epsilon)\right] e^{i\epsilon 0^+}$$

$$(7.6)$$

I realize some of these terms will not makes sense yet (I haven't even defined Green's functions yet!). But you can see the direct and exchange terms U_{kkqq} and U_{kqkq} (Eq. (6.25)) sitting in Σ and the diagram gives us a way to visualize the corresponding interactions.

The real advantage of Feynman diagrams is that they provide a convenient way to write down many of the self energies Σ by inspection. Their use in quantum electrodynamics is beyond doubt, but there are many examples

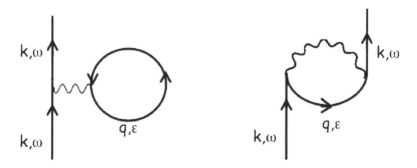

Fig. 7.1 *Feynman diagrams showing various Coulomb self energies in momentum space. Time runs forward along the vertical axis. The left figure shows a Coulomb scattering (wiggly line) between free electrons (solid arrows) with a Coulomb term defined in Eq. (23.13) that lists from left (r_1) to right (r_2) the sequence of creation, destruction, creation and destruction operators, giving us $-iU_{kkqq}$, which comparing with Eq. (6.25) is the Hartree term. On the right, we get $-iU_{qkkq}$ which is the Fock term in Eq. (6.25).*

in condensed matter physics where they have been invoked successfully, for instance to calculate the scattering amplitudes in superconductors and Kondo insulators, or the measured reduction ('renormalization') of the optical band gap in a molecule or a quantum dot by screening processes from polarizing charges in a metal.

There are conditions underlying the applicability of diagrammatic perturbation theory, namely for small perturbing potentials around a noninteracting (quadratic) Hamiltonian. They can be readily extended to finite temperature, by extending the Green's functions along an *imaginary time* axis (the Boltzmann distribution $\hat{\rho} \propto e^{-\beta E}$ can be written as the Hamiltonian evolution over imaginary time, $e^{iE\tau/\hbar}$ with $\tau = i\beta\hbar$). The process gets trickier for nonequilibrium (next chapter), where we must bring in complex time for a different reason, to define a 'Keldysh contour' that starts and ends at prespecified initial eigenstates.

Once again, the main selling point is the simplicity of these diagrams and their ability to help us visualize various underlying processes. We will describe a concrete example in the next chapter, the self energy for three phonon scattering. It is noteworthy that for a lot of practical problems, many body perturbation theory could turn out to be an overkill, while for many others, perturbation may not work at all, in the absence of a suitable small parameter.

7.1.2 *Quadratic lattice Hamiltonian: phonons*

Phonons are quantized lattice vibrations, obtained from the parabolic part of the periodic nuclear potentials (Eq. (6.19)) that can be represented by a set of coupled masses and springs. The phonons are bosons and their creation and annihilation operators satisfy *commutation* rather than anti-commutation rules for electrons (Eq. (6.12)), since they retain the same sign upon particle exchange. We will shortly see that their creation and annihilation operators satisfy

$$
\begin{aligned}
[a_\mu^\dagger, a_\nu^\dagger] &= a_\mu^\dagger a_\nu^\dagger - a_\nu^\dagger a_\mu^\dagger = 0 \\
[a_\mu, a_\nu] &= a_\mu a_\nu - a_\nu a_\mu = 0 \\
[a_\mu, a_\nu^\dagger] &= a_\mu a_\nu^\dagger - a_\nu^\dagger a_\mu = \delta_{\mu,\nu}
\end{aligned}
\tag{7.7}
$$

Thus, swapping operators at distinct locations picks up the same sign. This means that unlike electrons, where we can at most add one electron to a one particle state, we can keep adding to a phonon state and the number of phonons has no upper bounds.

Let us see how these commutation relations come about. The quadratic lattice Hamiltonian from the nuclear potentials can be written in 1-d as a chain of springs connecting masses

$$
\hat{\mathcal{H}}_{\text{ph}} = \sum_n \frac{p_n^2}{2M_n} + \sum_n K_n \frac{(x_n - x_{n+1})^2}{2}
\tag{7.8}
$$

For equal masses and spring constants we again take advantage of periodicity by going to Fourier space, $x_k = \sum_n x_n e^{ikna}/\sqrt{N}$, whereupon the quadratic potential becomes

$$
\sum_n (x_n - x_{n+1})^2 = \frac{1}{N} \sum_{kk'} x_k x_{k'} \underbrace{\sum_n e^{i(k+k')x} \left(1 - e^{ika}\right) \left(1 - e^{ik'a}\right)}_{N\delta(k+k')}
$$

$$
= \sum_k x_k x_{-k} \underbrace{\left(2\left[1 - \cos ka\right]\right)}_{=M\omega_k^2/K}
\tag{7.9}
$$

where $\omega_k = 2\sqrt{K/M}\,|\sin{(ka/2)}|$ in 1D (and once again, much more involved in 3D with in general, longer ranged interactions). The Fourier transform also simplifies the kinetic energy term to give $\sum_k p_k p_{-k}/2M$. The Hamiltonian thus becomes

$$\hat{\mathcal{H}}_{\mathrm{ph}} = \sum_k \frac{p_k p_{-k}}{2M} + \sum_k \frac{M\omega_k^2}{2} x_k x_{-k} \tag{7.10}$$

This structure of $\hat{\mathcal{H}}$ would have represented independent particles (i.e., normal modes) were it not for the fact that the states actually mix $\pm k$ states together. To diagonalize the Hamiltonian, we take a basis transformation that mixes x and p operators while obeying commutation $[x, p] = i\hbar$, which underlies quantization and the uncertainty principle. We choose the constants to make sure the final mixing operators a, $a^\dagger$ are dimensionless. The transformation generates the phonon creation and annihilation operators mentioned earlier

$$a_k = \sqrt{\frac{NM\omega_k}{2\hbar}}\left[x_k + i\frac{p_{-k}}{M\omega_k}\right], \quad x_k = \sqrt{\frac{\hbar}{2NM\omega_k}}\left[a_k + a^\dagger_{-k}\right]$$

$$a_k^\dagger = \sqrt{\frac{NM\omega_k}{2\hbar}}\left[x_{-k} - i\frac{p_k}{M\omega_k}\right], \quad p_{-k} = -i\sqrt{\frac{\hbar NM\omega_k}{2}}\left[a_k - a^\dagger_{-k}\right] \tag{7.11}$$

Substituting, we see now that the Hamiltonian is diagonalized as the desired sum over unmixed k-states, corresponding to independent normal modes.

$$\hat{\mathcal{H}}_{\mathrm{ph}} = \sum_k (\underbrace{a_k^\dagger a_k}_{\hat{n}_k} + 1/2)\hbar\omega_k \tag{7.12}$$

where the phonon creation and annihilation operators satisfy commutation relations readily derived from $[x, p] = i\hbar$, amounting to (compare with Eqs. (6.10)–(6.12))

$$
\begin{aligned}
a_k^\dagger|n_1, n_2, \ldots, n_k, \ldots, n_N\rangle &= \sqrt{n_k + 1}\,|n_1, n_2, \ldots, n_k + 1, \ldots, n_N\rangle \\
a_k|n_1, n_2, \ldots, n_k, \ldots, n_N\rangle &= \sqrt{n_k}\,|n_1, n_2, \ldots, n_k - 1, \ldots, n_N\rangle \\
[a_k^\dagger, a_{k'}^\dagger] &= [a_k, a_{k'}] = 0 \\
[a_k, a_{k'}^\dagger] &= \delta_{kk'}
\end{aligned}
$$

$$\tag{7.13}$$

There is no restriction on the number n_k of particles in a given one-phonon state (repeated $a_k^\dagger$ keeps adding particles!), so that the generic many-body wavefunction is designated by applying the creation operator sequentially on vacuum $|0\rangle$, just like fermions, except we can repeat each creation term as many times as desired (n_i: arbitrary integer)

$$\Psi = \left| n_1, n_2, \ldots, n_k, \ldots, n_N \right\rangle = \frac{(a_1^\dagger)^{n_1}(a_2^\dagger)^{n_2}\ldots(a_N^\dagger)^{n_N}}{\sqrt{n_1! n_2! \ldots n_N!}} |0\rangle \qquad (7.14)$$

The creation operator for the νth phonon state then looks like

$$a_\nu^\dagger = \begin{bmatrix} 0 & \sqrt{1} & 0 & \ldots\ldots\ldots & 0 \\ 0 & 0 & \sqrt{2} & \ldots\ldots\ldots & 0 \\ \ldots\ldots & \ldots\ldots\ldots\ldots & \ldots \\ \ldots\ldots & \ldots\ldots\ldots\ldots & \ldots \\ \ldots\ldots & \ldots\ldots\ldots\ldots & \sqrt{n-1} \\ 0 & 0 & 0 & 0 & 0 & 0 & 0 \end{bmatrix} \quad \text{(obeying Eq. (7.13) line 1)} \quad (7.15)$$

Once again, the quadratic Hamiltonian can be efficiently decoupled in Fourier space with phonon bandstructures ω_k. Higher order terms, however, will need the full machinery of second quantization. A simple example is if there is any anharmonicity that gives a cubic term in the Hooke's law part, $\sim (x_i - x_{i+1})^3$, or other more complex interactions. These higher order terms need the machinery of second quantization, and can be interpreted as phonon–phonon interaction terms responsible for their scattering. Much like the Coulomb term, we can use the machinery of Feynman diagrams to write down various perturbative orders of phonon phonon interactions.

7.1.2.1 *Electron phonon coupling*

The electron–phonon coupling term (Eq. (6.19)) can also be written in a second quantized notation. The term looks like $\hat{n}\hat{\delta x}$ (the electron density arises from the sum $\sum_i$ over electron coordinates in Eq. (6.17)), which can then be written using Eqs. (6.20) and (7.11)

$$\hat{\mathcal{H}}_{\text{el-ph}} = \sum_{i\alpha} M_{i\alpha} \underbrace{c_i^\dagger c_i}_{\hat{n}_i} \underbrace{(a_\alpha + a_\alpha^\dagger)}_{\sim \hat{\delta x}} \qquad (7.16)$$

7.1.3 *From many body to quadratic: BCS superconductivity*

In many cases, a quartic model can also be reduced to a quadratic, non-interacting particle model by treating the rest of the interactions on average, described by a *scalar* 'order parameter' rather than an *operator*. For instance, we can start with the the Hubbard like Coulomb term $\sum_{k1,k2} U_{k1,k2} c_{k1,\uparrow}^\dagger c_{-k1,\downarrow}^\dagger c_{-k2,\downarrow} c_{k2,\uparrow}$, simplified here by assuming overall zero spin zero momentum. We can reduce this to an effective quadratic Hamiltonian by assuming that certain operator combinations, e.g., a net spin zero momentum zero pair, can be replaced by a scalar average,

$$c_{-k\downarrow}c_{k\uparrow} = \underbrace{\left\langle c_{-k\downarrow}c_{k\uparrow} \right\rangle}_{b_k \ (\text{scalar})} + \underbrace{\left[c_{-k\downarrow}c_{k\uparrow} - \left\langle c_{-k\downarrow}c_{k\uparrow} \right\rangle \right]}_{\delta_k \ll 1 \ (\text{operator})} \quad (7.17)$$

Expanding the Hubbard $\mathcal{H}$ to lowest order in the fluctuation δ_k around the scalar average b_k, we get upto a constant $b_{k1}^\dagger b_{k2}$

$$\mathcal{H} = \sum_k \epsilon_k c_k^\dagger c_k + \sum_{k_1,k_2} U_{k1,k2} \underbrace{\left(b_{k_1}^\dagger + \delta_{k_1}^\dagger \right)\left(b_{k_2}^\dagger + \delta_{k_2} \right)}_{\text{quartic}}$$

$$= \sum_k \epsilon_k c_k^\dagger c_k + \sum_{k_1,k_2} U_{k1,k2} \left[\underbrace{b_{k_1}^\dagger b_{k_2}}_{\text{const}} + b_{k_1}^\dagger \underbrace{\delta_{k_2}}_{c_{-k2,\downarrow}c_{k2,\uparrow} - b_{k2}} + \underbrace{\delta_{k_1}^\dagger}_{c_{k1,\uparrow}^\dagger c_{-k1,\downarrow}^\dagger - b_{k1}^\dagger} b_{k2} + \underbrace{\delta_{k_1}^\dagger \delta_{k_2}}_{\text{drop}} \right]$$

$$\approx \underbrace{\sum_k \epsilon_k c_k^\dagger c_k - \sum_k \left[\Delta_k^* c_{-k,\downarrow}c_{k,\uparrow} + \Delta_k c_{k,\uparrow}^\dagger c_{-k,\downarrow}^\dagger \right]}_{\text{quadratic!}}, \quad \text{where} \quad \boxed{\Delta_k = -\sum_{k'} U_{k,k'} b_{k'}}$$

$$(7.18)$$

We can now diagonalize the quadratic Hamiltonian in the basis set $\{c_{k\uparrow}^\dagger, c_{-k\downarrow}\}$. The resulting 2×2 Bogoliubov-de Gennes Hamiltonian, with energies around $E_F = 0$ (electrons have energy ϵ_k, holes $-\epsilon_k$)

$$\mathcal{H} = \begin{matrix} & c_{k\uparrow} \quad c_{-k\downarrow}^\dagger \\ \begin{matrix} c_{k\uparrow}^\dagger \\ c_{-k\downarrow} \end{matrix} & \begin{pmatrix} \epsilon_k & -\Delta_k \\ -\Delta_k^* & -\epsilon_k \end{pmatrix} \end{matrix} \quad (7.19)$$

has eigenvalues $E_\pm = \pm\sqrt{\epsilon_k^2 + |\Delta_k|^2}$ and eigenvectors set by the ratios of matrix elements plus anticommutation rules $\{\gamma_{k\sigma}, \gamma_{k'\sigma'}^\dagger\} = \delta_{kk'}\delta_{\sigma\sigma'}$, so that

$$\gamma_{k\uparrow} = u_k c_{k\uparrow} - v_k c_{-k\downarrow}^\dagger, \qquad \gamma_{-k\downarrow}^\dagger = v_k^* c_{k\uparrow} + u_k^* c_{-k\downarrow}^\dagger$$

$$v_k = \mp\sqrt{(E_\pm - \epsilon_k)/2E_\pm}, \quad u_k = \sqrt{(E_\pm + \epsilon_k)/2E_\pm} \quad (7.20)$$

The eigenvalue equation leads us to view Δ_k as a gap parameter.

But what prevents us from taking any operator combination and replacing with an average to diagonalize the Hamiltonian? The exercise makes sense only if fluctuations are small and the order parameter large, an assumption that needs to be *vetted by thermodynamic self-consistency.* Using Eq. (7.20)

$$\Delta_k = -\sum_{k'} U_{kk'} \langle c_{-k\downarrow} c_{k\uparrow} \rangle = -\sum_{k'} U_{kk'} \underbrace{u_{k'}^* v_{k'}}_{\Delta_{k'}/2E_{k'}} \left\langle 1 - \gamma_{k'\uparrow}^\dagger \gamma_{k'\uparrow} - \gamma_{-k'\uparrow}^\dagger \gamma_{-k'\downarrow} \right\rangle$$

(7.21)

For thermodynamic stability, the pair occupancies $\langle \gamma^\dagger \gamma \rangle$ must follow equilibrium (Fermi-Dirac) statistics, so $\langle \cdots \rangle \to 1 - 2f(E_{k'}) = \tanh(\beta E_{k'}/2)$.

If U is a repulsive potential, then the only solution to the above equation $\Delta \propto -U\Delta \tanh(\beta E/2)/2E$ is $\Delta = 0$. BCS superconductivity arises when we make U attractive, which can happen within a small Debye window $\hbar\omega_D$ around the Fermi energy where slow phonon mediated interactions create a net *attractive* potential between electrons. In short, an imprint of an electron's passage lingers on the surrounding lattice (sort of like a memory foam mattress), pulling in the next electron long after the Coulomb repulsion from the first electron has disappeared. Using $U_{kk'} = -V$ over that energy range, the self-consistency Eq. (7.21) simplifies. With a few lines of algebra, it is easy to write down a non-zero gap parameter for $T < T_c$

$$\Delta(T) = \Delta(0)\sqrt{1 - T/T_c}, \quad \Delta(0) = 1.77k_B T_c \approx 2\hbar\omega_D e^{-1/D(E_F)V} \quad (7.22)$$

The finite gap for $T < T_c$ separates ground state from all excitations, leading to dissipationless flow of the superconducting condensate and a corresponding vanishing of DC resistance.

7.2 Green's functions: why we need them

Solving the many body Hamiltonian in second quantized notation beyond 'mean field' is already a tall order. We will however switch gears and add another complexity to it, a *non-equilibrium* perturbation to the system to drive electron transport. Such a process of interrogating the system is at the heart of transport, which requires us to shake up the system in order to inject and remove charges. Dealing with these external sources requires solving an inhomogeneous equation, which is where the Green's functions come in handy.

Typical measurements deal with the response to an external perturbation — current from an applied voltage, thermal conduction from a temperature gradient, magnetization from an applied field. They all require us to calculate the change in wavefunction because of an external 'source'. *A Green's function is the response to a delta function source, i.e., a spike of unit strength.* The linearity of our equations and the additivity of our solutions means that we can build any source out of a combination of such spikes, and thereby our desired response function out of an equivalent combination of Green's functions. In mathematical terms, in order to solve the operator equation ($\mathcal{L}$ a linear operator, say a time derivative) that is inhomogeneous (i.e., with an arbitrary source that can be decomposed into spikes)

$$\mathcal{L}V(t) = \underbrace{S(t)}_{\text{source}} = \underbrace{\int S(t')\delta(t - t')dt'}_{\text{weighted sum of spikes}} \tag{7.23}$$

we first solve for the case with a unit spike source

$$\mathcal{L}G(t,t') = \underbrace{\delta(t - t')}_{\text{spike}} \tag{7.24}$$

and then construct the desired output as a homogeneous solution (solving Eq. (7.23) in absence of S) and a 'particular integral' obtained by adding the responses together with the same weighting functions $S(t')$

$$V(t) = V_0 + \int S(t')G(t,t')dt' \tag{7.25}$$

It is straightforward to verify this equation by simple substitution into Eq. (7.23).

The merit of calculating Green's functions is that we can do so just from the structure of the main evolution operator $\mathcal{L}$, without worrying about $S(t)$. In fact in quantum transport, we will be deliberately coy about spelling out S, but volunteer instead thermodynamic bilinear averages $\langle SS^\dagger \rangle$. Such coyness is in keeping with the contacts viewed as thermal reservoirs, held in equilibrium by unspecified external forces.

One can define several kinds of Green's functions, some physically relevant, some purely for mathematical expediency. In quantum transport, we will deal with the open boundary Schrödinger equation (Eq. (11.9)). A quantity we care about is the time ordered or causal Green's function or *propagator*, describing the propagating response at position x, time t to a spike at position x' at time t'. All measureable quantities can be extracted from this propagator. Using the bra ket notation

$$G^T(x,t,x',t') = \frac{-i}{\hbar} \frac{\left\langle \Psi_0 \middle| T\psi(x,t)\psi^\dagger(x',t') \middle| \Psi_0 \right\rangle}{\left\langle \Psi_0 \middle| \Psi_0 \right\rangle} \qquad (7.26)$$

Remember that the ψs are described in the interaction representation (Section 3.3), and come with their own evolution operators. The time ordering operator T simply *stacks later events to the left*, keeping track of sign changes upon particle exchange (fermions like electrons give a negative sign, bosons like phonons stay positive). T gives us a $\Theta(t-t')$ when t is to the left, describing a propagating electron created at x',t' with $\psi^\dagger$ and removed later at x,t with ψ. It gives us $-\Theta(t-t')$ when we switch orders and t is to the right, describing a propagating hole going from x,t to x',t'. In other words, we start earlier from the right and end at the left, and include a sign switch to account for fermion antisymmetry.

The brackets $\langle\ldots\rangle$ imply a quantum statistical average that includes the quantum average over the wavefunctions and a thermodynamic average over the Boltzmann weight $e^{-\mathcal{H}/k_B T}|\Psi_0\rangle\langle\Psi_0|$. The catch is that the above Green's function depends on the exact eigenstate $\left|\Psi_0\right\rangle$ of an interacting Hamiltonian $\hat{\mathcal{H}} = \hat{\mathcal{H}}_0 + \hat{V}$, but we do not know the solutions to the Hamiltonian *a priori* (in fact, we are trying to solve it!). We must therefore relate this to the unperturbed eigenstate $\left|\Phi_0\right\rangle$ corresponding to a simpler, dominant Hamiltonian that is quadratic. We can do that using the S matrix and the Gellmann Low theorem. Simply put, we identify the perturbation potential $\hat{V}(t)$, which we assume is turned on and then turned off *slowly* so that at times $t = \pm\infty$ it does not exist and the exact solution

$$\lim_{t\to\pm\infty} \left|\Psi_0\right\rangle = \left|\Phi_0\right\rangle$$

Under this condition, we can write the unitary evolution adiabatically from the initial condition at $t = -\infty$ to $t = 0$ when the interaction $\hat{V}$ has fully turned on (perhaps worth emphasizing once more that we are using the 'interaction representation',where the wavefunctions evolve according to the perturbation Hamiltonian $\hat{V}$). Interpreting the time dependent Schrödinger equation as a first order equation to be solved using an integrating factor, we can formally write down an equation that connects the unperturbed solution at $t = -\infty$ with the exact solution at $t = 0$.

$$\left|\Psi_0\right\rangle = \underbrace{T e^{-i \int_{-\infty}^{0} dt_1 \hat{V}(t_1)/\hbar}}_{S(0,-\infty)} \left|\Phi_0\right\rangle \qquad \textbf{Gellmann Low Theorem} \quad (7.27)$$

The time order T sets the sequence of operators in the expanded integrals.

We can similarly write $\left\langle\Psi_0\right| = \left\langle\Phi_0\right| S(\infty, 0)$ describing the turn off of the potential from $t = 0$ to $t = \infty$ adiabatically. Since we already have a time ordering operator T, we can freely move around the S matrices, combine them into a product $S(\infty, -\infty) = S(\infty, 0)S(0, -\infty)$, and write

$$G^T(x, t, x', t') = \frac{-i}{\hbar} \frac{\left\langle\Phi_0\middle|S(\infty,-\infty)T\psi(x,t)\psi^\dagger(x',t')\middle|\Phi_0\right\rangle}{\left\langle\Phi_0\middle|S(\infty,-\infty)\middle|\Phi_0\right\rangle} \qquad (7.28)$$

involving the *unperturbed* eigenstates Φ_0.

The main thing to do now is to Taylor expand the S matrix, $S(0, -\infty) = T \exp\left[-i \int_{-\infty}^{0} dt_1 \hat{V}(t_1)/\hbar\right]$ for weak perturbations $\hat{V}$. Each term includes multiple ψ and $\psi^\dagger$ terms (e.g., $\hat{V}$ has it from the Coulomb charges, while the $\left\langle\ldots\right\rangle$ introduces a few more such as in the Boltzmann factor $e^{-\beta\mathcal{H}}$). We can then factorize the single average of products into products of averages involving all possible combinatorial pairings of an even number of ψ and $\psi^\dagger$ operators, and represent each average with a Green's function. Such a paired factorization is accorded by *Wick's theorem*, which applies when the unperturbed Hamiltonian representing the eigenstates $\left|\Phi_0\right\rangle$ is *quadratic* and thereby allows Gaussian integrals. It is well known that the various moments of a Gaussian distribution are determined by its second moments with combinatorial coefficients, for instance, $\langle x^4\rangle = 3\langle x^2\rangle^2$. In general, one can prove that $\langle A_1 A_2 \ldots A_{2N}\rangle = \sum_P (-1)^P \langle A_{i1}A_{i2}\rangle \ldots \langle A_{i_{2N-1}}A_{i_{2N}}\rangle$ for all possible permutations for pairing the individual terms, with a corresponding sign flip for electrons. The resulting pairwise 'contractions' are easier to evaluate, and can be described visually with a corresponding Feynman diagram. While we get a lot of diagrams brute force, it turns out that the only ones that matter are topologically distinct, disconnected diagrams.

In the next section, we will focus on the language of the many body operators at equilibrium. In the next chapter, we will discuss nonequilibrium, whereupon the methodology of this chapter breaks down. Under nonequilibrium conditions, we can no longer assume that after we turn off the interaction the system goes back to the unperturbed ground state, i.e., $\lim_{t \to \infty} |\Psi_0\rangle \neq |\Phi_0\rangle$. Under this condition, the fix is to deform the time axis to go back to $t = -\infty$ to get a reference state again ($\langle \Phi_0 |$), but now distinguishing between forward and reverse branches of the contour.

While we can now write down equations for the current, *what makes a many body problem insoluble in general is the fact that the unperturbed Hamiltonian could still be interacting and thus beyond quadratic, whereby the Wick decomposition and Feynman diagrammatic expansion of many body perturbation theory does not apply.*

7.2.1 *Adding interactions: self-energy and Dyson equation*

In addition to the time ordered Green's function describing propagating electrons and holes, we will also encounter the retarded Green's function for the *causal* propagation of electrons, and advanced Green's function for holes. The nice feature of these two Green's functions is that they have poles localized exclusively in one half of the complex energy plane (lower for electrons, upper for holes). We will use the retarded and advanced electron Green's functions often (we will later designate them simply $G \equiv G^R$ and $G^\dagger \equiv G^A$), and define again explicitly in Eq. (11.27)

$$G^R(x,t,x',t') = \frac{-i\Theta(t-t')}{\hbar} \underbrace{\left\langle \psi^\dagger(x,t)\psi(x',t') + \psi(x',t')\psi^\dagger(x,t) \right\rangle}_{= \{\psi(x',t'),\psi^\dagger(x,t)\}}$$

$$G^A(x,t,x',t') = \frac{i\Theta(t'-t)}{\hbar} \underbrace{\left\langle \psi^\dagger(x,t)\psi(x',t') + \psi(x',t')\psi^\dagger(x,t) \right\rangle}_{= \{\psi(x',t'),\psi^\dagger(x,t)\}} \quad (7.29)$$

so that the time ordered Green's function is simply the sum $G^T = G^R + G^A$.

We mentioned earlier that all Green's functions solve differential equations with a unit delta function perturbation. In the next chapter, we will see that the retarded Green's function above is the solution to the Schrödinger equation with a delta function perturbation.

We can derive a nice relation between the Green's functions with and without a particular scattering mechanism, G and G_0, say the interaction with contacts or with Coulomb forces. The interaction is described by a self energy matrix Σ, which is complex in general. As we shall see in the next chapter, the real parts of the self energy matrix provide local onsite potentials shifting the energy levels of the unperturbed Hamiltonian H, and the imaginary parts playing the role of incoherent broadening of those levels. The real and imaginary parts are in general energy dependent, accounting for effects that are nonlocal in time (and equivalently, reflecting the internal energy structure of the scattering centers). Furthermore, these functions are related by a mathematical Hilbert transform, which implies that the self energy can change the shape of the density of states but must conserve its total integrated area and associated sum rule. Now, in the time independent limit, the two Green's functions satisfy (see Eqs. (7.23), (11.11))

$$\left[E - H\right]G_0(E) = 1, \quad (\delta(t - t') \to 1, \ i\hbar\partial/\partial t \to E \text{ upon Fourier transforming})$$

$$\left[E - H - \Sigma(E)\right]G(E) = 1 \quad (\text{time convolution} \to \text{energy product on Fourier transforming})$$

$$(7.30)$$

We can then write the relation between G and G_0 as the Dyson equation

$$\boxed{\begin{array}{c} \underline{\text{Dyson equation}} \\ G^{-1} = G_0^{-1} - \Sigma \end{array}} \quad (7.31)$$

that relates the unperturbed and the interacting Green's functions in terms of the missing self energy. This equation can be equivalently written by left multiplying Eq. (7.31) by G_0 and right multiplying by G

$$G = G_0 + G_0 \Sigma G$$
$$= G_0 + G_0 \Sigma G_0 + G_0 \Sigma G_0 \Sigma G_0 + \dots \quad (7.32)$$

described visually in Fig. 7.2 (note $G_0 \Sigma G$ represents matrix products). The second line schematically underscores the role of various scattering processes, especially when the matrix multiplications are expressed in terms of their explicit components including the intermediate summed indices. While the Dyson equation, as we shall see, captures the equilibrium properties of the system such as the density of states, we will need a separate Green's function (the 'nonequilibrium' Green's function), G^n, to describe how the states are filled at non-equilibrium. For a noninteracting system, G and G^n are simply related but for an interacting nonequilibrium system, discussed in the next chapter, they need to be treated separately.

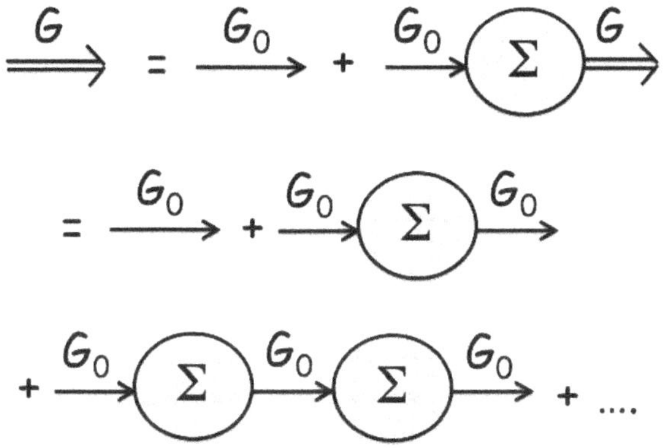

Fig. 7.2   *The Dyson equation relates an interacting G with a noninteracting G_0 related through the self energy Σ. Symbols are explained at the end of the chapter*

7.3   Appendix: Feynman rules

The Feynman rules allow us to calculate the contribution of various 'diagrams' that each represent a physical process, and list them at each perturbative level of the interaction terms. For instance, the diagrams in Fig. 7.1 represent the Hartree term and the Fock term arising out of Coulomb interactions. The rules can be summarized readily.

- Each solid arrow shows an electron with Green's function $i\hbar G_0$ at the momentum and frequency designated on the line, propagating forward in time (a positron is a backward line). We discuss the free electron Green's function in the next chapter (Eq. (11.9)). An arrowed double line describes an interacting Green's function $i\hbar G$.

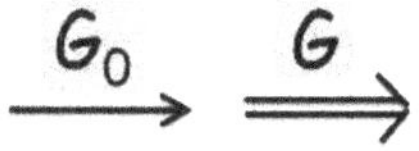

- The Coulomb interaction described by the wiggly line joining $\vec{r}_1$ and $\vec{r}_2$ is $-iU_{\alpha\beta\gamma\delta}/\hbar$ with the indices ordered according to Eq. (23.13). The first two describe respectively arrows leaving and entering point $\vec{r}_1$ while the next two describe arrows leaving and entering point $\vec{r}_2$.

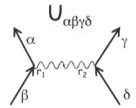

- We conserve momentum and energy at each vertex and sum over all the intermediate momenta and energy.
- A factor (-1) accompanies each Fermion loop, e.g., the Hartree diagram.
- A free phonon is described by a looped line representing the phonon Green's function $i\hbar\Pi_0(\vec{q},\omega)$, introduced in Section 11.4.

- At the junction between two electron propagators and one phonon propagator, we need a vertex $-iM_q/\hbar$ representing the electron phonon coupling term (Eq. (7.16)) $\sum_{i\alpha} M_{i\alpha}c_i^\dagger c_i(a_i^\dagger + a_i)$ in real space and $\sum_{kq} M_q c_k^\dagger c_{k+q}(a_q^\dagger + a_q)$ in momentum space. We will discuss the impact of this term later in the chapter on scattering.

We discuss electron and phonon Green's functions in the next chapter. The diagrammatic techniques conveniently allow us to keep track of perturbative corrections to the wavefunctions. For nonequilibrium, we can write them down using two modifications: applying them to a Keldysh contour and using Langreth's rules to separate out the various products.

PART 3

Nonequilibrium Concepts:

The Nature of Transport Equations

Chapter 8

The transport menagerie

The aim of this and the following chapters is to discuss some of the core physics behind transport equations that we will use through the rest of the book. We discuss how irreversibility and damping arise out of the evolution equations, and how entropic forces mix with dynamics to generate classical drift diffusion and the Boltzmann equation. Drift diffusion with recombination-generation leads to the minority carrier diffusion equation traditionally used to discuss classical switches and transistors. We then introduce ballistic quantum transport theory, where the entropic terms are conveniently hidden inside the contacts. We derive the nonequilibrium Green's function (NEGF) formalism using simple one electron theory, but present later the many electron counterpart that is especially relevant for strong interactions.

Typical equations of evolution for a closed system have the form

$$\left[\mathcal{L}(x,t)\right]V(x,t) = 0 \tag{8.1}$$

where $\mathcal{L}$ is a linear operator and V is the evolving variable — the displacement of an oscillator or the wavefunction of an electron for example. To create a steady transport process, we need a way to take stuff (energy, particles etc.) out of the system and siphon them back in continuously (Fig. 2.3). This means we must deal with a subsystem that is a part of a larger closed system (Fig. 8.1). The subsystem must have open boundaries capable of exchanging particles and/or energy with the rest of the closed system as it evolves. Without plunging into detailed examples rightaway, it suffices to say that for the subsystem, the evolution equation must pick up multiple outflow and inflow terms, with each outflow term depending on what is inside the subsystem, i.e., V, while each inflow term driven by the world immediately outside, the environment, conveniently assumed to

121

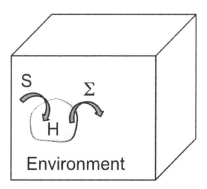

Fig. 8.1 *An open subsystem enclosed in a larger environment evolves according to its own Hamiltonian H, while also seeing an outflow of particles described by a complex, energy dependent matrix $\left[\Sigma(E)\right]$ and an inflow from a source S.*

be much larger and therefore in local equilibrium. The equation then looks like

$$\underbrace{\left[\mathcal{L}(x,t)\right]V}_{\text{evolution}} - \underbrace{\left(\Sigma_1 + \Sigma_2 + \ldots\right)V}_{\text{outflow}} = \underbrace{S_1 + S_2 + \ldots}_{\text{inflow}} \qquad (8.2)$$

where in general the multiplication $\Sigma_i V$ is a *matrix multiplication* in space time coordinates, so that spelt out explicitly, it becomes $\int dt' \int dx' \Sigma_i(x,t;x',t')V(x',t')$, with an integration kernel Σ representing nonlocal contributions from surrounding points in space-time. If the outside consisted of a single whole, then our subsystem would reach equilibrium with it eventually. But if the outside consists of several large 'reservoirs' separated by voltage or thermal sources (for instance, source and drain contacts bias separated by a battery), then our subsystem tries simultaneously to reach all the separate equilibria maintained in the reservoirs, setting off a frenzy of particle flow and as a result gets driven away from equilibrium.

So far, we have merely dealt with the purely dynamic evolution of the system. However, we now need to add statistics to the dynamics, arising from thermal interactions. Such a thermal average will clearly scramble the pristine evolution trajectories, leading to a time average evolution that a recording device can track and measure. Such a thermal averaging gives us the drift diffusion equation for Newtonian physics, and the Quantum Boltzmann equation for Schrödinger physics.

An enormous simplification has happened over the last decade as we moved from the *micro* to the *meso* and now to the *nano* domain, namely, that the channel itself has become smaller than typical mean free paths for scattering. This in fact is one of the simplifying features of the 'molecular' point of view, namely, that for a small molecular sized conductor, any scattering occurs outside the molecular sized channel and inside the macroscopic contacts, *directly influencing just the source terms $S_1 + S_2 + \ldots$* The channel variable V gets randomized only in response to the scrambling of S, but not independently. This is a qualitative difference from macroscopic systems, where the contacts play a minor role and the variables get thermally smeared through scattering processes intrinsic to the channel itself. While S and Σ describe the inflow-outflow terms for the quantum dynamics, we will need separate knowledge of their thermodynamics (i.e., the 'scrambling' forces), to determine the energy distribution of the electrons under nonequilibrium. The thermodynamic requirement involves an entirely different set of terms describing the escape and entry of not just particles, but

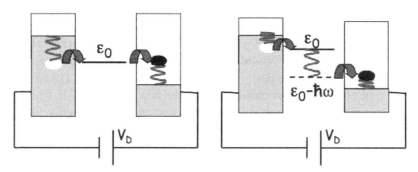

Dissipation mainly in contacts   Dissipation mainly in channel

Fig. 8.2 *(Left) How can a nanosized conductor support a large electric field (volts across nanometers) without burning up? Most of their electrons flow elastically through a single molecular level ϵ_0, conserving energy until they encounter the contacts, where they turn into a 'hot' electron in the drain and a cold 'hole' in the source. These need to sink down or float up to their respective quasiFermi energies through rampant inelastic processes in the contacts, quickly restoring the contacts to their local equilibrium Fermi Dirac distributions. In other words, the reason the channel does not burn up is because the dissipation is in the contact. The extra electron flows back into the battery and is redirected to fill up the hole in the source, maintaining the flow of current through the conducting window between the two Fermi energies. (Right) Inelastic scattering in the channel can also create local dissipation, provided the electrons can be slowed down to sense the atomic motions (equivalently, the electronic levels sharp enough to be influenced by phonon sidebands (Fig. 24.6).*

of entropic forces. The entropic terms enter as recombination-generation $R(x,t), G(x,t)$ in the classical drift-diffusion equation, the scattering matrix $(df/dt)_{sc}(x,x',t)$ in semiclassical Boltzmann Transport, and the in and out-scattering functions $\Sigma^{<,>}(x,x',t,t')$ in the Keldysh treatment of quantum transport that we will discuss shortly.

8.1 The transport mosaic

A snapshot of the hierarchy of transport equations, proceeding horizontally from noninteracting to strongly interacting, and vertically from classical to quantum, is showcased in Fig. 8.3. For a single particle, classical physics starts with Newton's Law along with environmental damping and external forcing, amounting to the Lángevin equation (Eq. (9.3)) in the overdamped limit. A quantum particle on the other hand, sees an open boundary Schrödinger equation with a similar Σ and S — outflow and inflow (Eq. (11.9)).

If we now extend Newton's Law to a weakly interacting gas of particles, we get classical drift diffusion (DD) for the particle density $n(x,t)$ (Eq. (9.16)), while the corresponding Hamilton's equation leads to Boltzmann Transport Equation (BTE) for the distribution function $f(x,k,t)$ (Eq. (10.18)). A smattering of quantum mechanics enters the semiclassical BTE when we use quantum band theory to calculate the effective masses, band velocities and scattering rates for the otherwise classical particles.

We can also work out the collective dynamics derived from Schrödinger equation, and this will lead to the Keldysh-Kadanoff-Baym nonequilibrium Green's function (NEGF) method that forms the core of this book (Eqs. (11.16), (11.17)). The nonequilibrium Green's function $G^<$ has four indices x,x',t,t'. A slightly different analysis in Fourier space $x + x' \to x$, $x - x' \to k$ and with one time index $(t = t')$ leads to Wigner functions $W(x,k,t)$. Although W is reminiscent of f in BTE, the interpretation is somewhat different. Since the quantum mechanical uncertainty principle disallows specification of x and k simultaneously, W cannot be interpreted as a probability density and is not always positive definite.

Finally, we can turn up many body interactions that make the electrons look more particulate. Classically, we have the emergence of strong non-

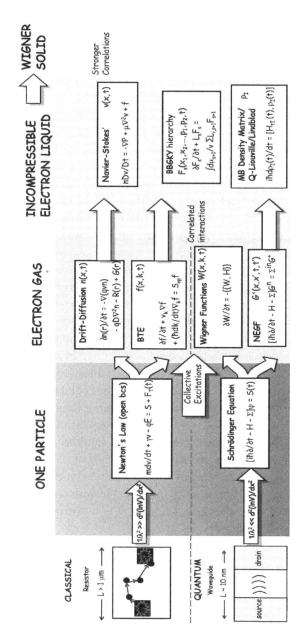

Fig. 8.3 *The evolution of flow equations from classical to quantum, with increasing interaction that drives it from an electron gas to incompressible liquid to solid.*

linearity and turbulence through the inclusion of the flow term $(\vec{v} \cdot \vec{\nabla})\vec{v}$ when going from a single particle to a collective evolution. For plasmas, this leads to a Bogoliubov Born Green Kirkwood Yvon (BBGKY) hierarchy. For fully quantum particles, the many body Hamiltonian enters the quantum Liouville equation for the density matrix. In certain cases typically involving weak coupling with the environment, the formalism can be simplified. Assuming the contact states disentangle with the channel (the *Born approximation*) and also rob the channel of its phase memory (the *Bloch-Redfield approximation*), we can then derive a multielectron master equation (Eq. (26.12)) in the many body (Fock) space of the electrons, which we can also extend nonperturbatively using an equation of motion technique (Section 27.4). Much like BBGKY, we will close the loop of higher order Green's functions using an approximate decoupling scheme.

What thermodynamics introduces into the system are dissipation and recovery. Transport is fundamentally a nonequilibrium process, and it is worth pondering for a second where irreversibility crept into our equations (we will discuss this at length shortly and also in the chapters on scattering). Regarded as a closed system, the channel and its environment evolve according to Eq. (8.1), which is perfectly reversible in time. In other words, the set of channel and environment phase space variables $\{x(t), p(t), x_j(t), p_j(t)\}$ are completely time reversible as long as they are entangled (switching sign on all phase variables *simultaneously* preserves the Hamiltonian). However, we choose to track just the subsystem and postulate that the part we do not track, i.e., the environment, maintains an equilibrium completely independent of the channel dynamics. This ignorance of the environmental dynamics (formally, a trace over just the environment variables x_j, Eq. (3.29) involving their equilibrium density matrices) introduces irreversibility into our system, since the subsystem $\{x(t), p(t)\}$ is not expected on its own to obey time reversal symmetry.

8.2   Quantum transport: an overview

Let me give a broad overview of quantum transport — the reader is encouraged to return to this section after going through the equations later in this chapter. As we will shortly see, the gist of the inflow and outflow in quantum transport are captured by the two quantities Σ and $\Sigma^{<,>}$ (we will alternately refer to the latter as $\Sigma^{\text{in,out}}$, as shown in Fig. 8.4). The NEGF transport equations are easier to get to using the one-electron Schrödinger equation, following the elegant derivations by Datta (Section 11.2). For completeness' sake, we will also outline the 'conventional' derivation of the Keldysh equation (Section 11.5) using Many Body Perturbation Theory (MBPT). Ultimately though the transport equations are neither limited to nor bound by their historical MBPT origins. The latter are relevant in as much as they may allow us to calculate the ingredients for our equations.

This is a subtle point worth emphasizing. The **transport formalisms themselves DO NOT specify what** Σ **and** $\Sigma^{<,>}$ **to use**, any more than Schrödinger's equation instructs us on how to calculate the Hamiltonian! **NEGF simply shows us how to use the self energies**, using the Keldysh and Meir-Wingreen equations (Fig. 8.4, Eq. (11.17)). We do not know how to calculate the self-energies for a general, nonequilibrium strongly interacting system (we do not always know them even at equilibrium). We can invoke reasonable assumptions such as Fermi's Golden Rule for metallurgical junctions (Eqs. (11.10), (11.14)), Hartree Fock for

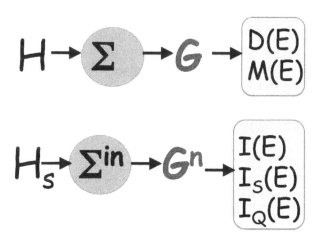

Fig. 8.4 *A system Hamiltonian H allows us to extract the retarded self energy* Σ, *retarded Green's function G and ultimately equilibrium properties such as density of states D and mode distribution M. Adding in a set of sources maintained at different potentials leads to the nonequilibrium in-scattering self energy* Σ^{in} *and the correlation Green's function* G^n *for electrons, which allows us to obtain transport properties such as charge, spin and heat currents I, I_S and I_Q. The hierarchy of quantum transport equations, from incoherent/interacting NEGF to coherent, non-interacting NEGF to Landauer theory, is shown later (Fig. 11.4)*

electronic interactions (Eq. (27.15)), or self consistent Born for linear electron phonon coupling (Eq. (24.16)). The formalism police might advocate elaborate diagrammatic MBPT methods to compute them, which indeed may work in certain cases. But how accurate a result we get out of the transport formalism depends on how clever we are in building interactions into these self energies (the 'garbage in garbage out' principle). After all, not all successful methods will be perturbative in the interaction parameter. In fact, we will see later how a *non-perturbative* self energy extracted from the equation of motion technique can give us considerable nonequilibrium correlation physics out of traditional quantum transport theory (Section 27.5).

Before going into specifics, let us discuss a few generic features of these inflow and outflow terms. The inflow term depends on what is sitting 'outside', i.e., in the contacts (in principle, there should be several inflow terms added together, each carrying an index corresponding to the contact it belongs to). We will see that each S expression contains a hybridization term binding the contact to the channel, times a term representing the intrinsic properties of the channel unperturbed by the contact (Eq. (11.10)). This decoupling, of course, is an approximation — sometimes called the *fundamental embedding assumption*, and creates an instant distinction between the contact and the channel, even if the two are fashioned out of the same material. For transport, we will often replace S by its bilinear thermodynamic average $\langle SS^\dagger \rangle$, dictating an equilibrium distribution on them. This helps us avoid solving for the contact explicitly — after all, why partition the problem into channel and contacts if the contacts aren't somehow 'simple'!

The self-energy Σ term is very tricky. It has an imaginary part, and represents damping and irreversibility (i.e., the particle escapes irretrievably into the contacts). Because damping tends to cause the particle to lag in phase, this self-energy enters as a nonlocal time kernel inside an integral, i.e., it pays attention to the history of the particle at earlier times t'. *Note that this is the precise point where there is a disconnect between Hamiltonian physics and transport theory.* All Hamiltonian evolutions are unitary, e.g., the solution to $i\hbar \partial \Psi / \partial t = H\Psi$ is of the form $\Psi(t) = \exp[-iHt/\hbar]\Psi(0)$. But this is completely reversible; going back in time implies acting on $\Psi(t)$ with its adjoint, $\exp[iH^\dagger t/\hbar]$, and since $H = H^\dagger$, the product of the two evolution operators gives us the identity matrix and the particle is fully recovered.

To get a non-unitary evolution corresponding to irreversibility, we will necessarily need a non-Hermitian matrix, and that is precisely what Σ in the outflow term must be. Upon operating back and forth in time, we are left with an exponential $\exp\left[-i(\Sigma - \Sigma^\dagger)t\right]$, and since Σ is non-Hermitian, we get a reduction in signal $\propto \exp\left[-\Gamma t\right]$, where Γ is twice the antiHermitian part of Σ, i.e., $\Gamma = i(\Sigma - \Sigma^\dagger)$. The exponential term does its job in describing dissipation. The particle decays exponentially with time and is lost forever in the contacts.

How a matrix Σ with an imaginary component crops out of a fully real Hamiltonian is subtle (in truth, it cannot and should not). In anticipation of transport, we will need to somehow 'cheat' by adding in that imaginary part by hand, as we will see shortly. We justify it in various ways by saying that this is the Poincaré recurrence time for a particle to recover, or that we need this to make our exponentials well defined with the complex time domain. We will elaborate on these in the next section. But suffice it to say that we cannot get transport and irreversibility out of a Hamiltonian alone.

8.3 Proceeding from a closed to an open system

To get into transport, we need to introduce two concepts into a Hamiltonian evolution, namely dissipation and recovery, both from contacts with the surrounding. How would we get such a dissipative term from the contacts? We start with a simple model consisting of a single oscillator (the classical channel) coupled linearly to a chain of oscillators (the environment). Our aim is to see how the elimination of the environment variables translates to the introduction of damping and external forcing (outflow and inflow) in the channel subsystem. The classical Hamiltonian is

$$\hat{\mathcal{H}} = \underbrace{\frac{m\dot{x}^2}{2} + \frac{m\omega_0^2 x^2}{2}}_{\text{channel}} + \underbrace{\sum_n \left(\frac{m_n \dot{x}_n^2}{2} + \frac{m_n \omega_n^2 x_n^2}{2} \right)}_{\text{environment}} + \underbrace{\sum_n \lambda_n x x_n}_{\text{coupling}} \qquad (8.3)$$

The terms with $\dot{x}, \dot{x}_n$ contribute to the kinetic energy (strictly speaking, we should write the Hamiltonian in terms of the canonical momentum p),

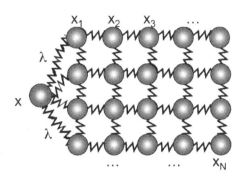

while all the other terms can be thought of as a combined potential energy $U(x, x_n)$ where both the channel and the environment variables reside. Let us calculate the classical evolution of this system.

Newton's law for classical mechanics give us

$$m\ddot{x} = -\partial U/\partial x = -m\omega_0^2 x - \sum_n \lambda_n x_n$$

$$m_n \ddot{x}_n = -\partial U/\partial x_n = -m_n \omega_n^2 x_n - \lambda_n x \qquad (8.4)$$

Solving the second equation for the environment variables gives us

$$x_n(t) = x_n(0)e^{i\omega_n t} - (\lambda_n/m_n) \int dt' x(t') \sin \omega_n(t - t') \quad \text{Integrate by parts}$$

$$= x_n(0)e^{i\omega_n t} + \frac{\lambda_n}{2m_n\omega_n} \int dt' \dot{x}(t') \cos \omega_n(t - t') \qquad (8.5)$$

Plugging the environment solution back into the channel, we get an equation for the channel variable that now has a damping term and looks like a forced damped oscillator

$$m\ddot{x} + m\omega_0^2 x + \int dt' \dot{x}(t') \underbrace{\sum_n \frac{\lambda_n^2}{2m_n\omega_n} \cos \omega_n(t - t')}_{\text{Damping } \Gamma(t-t')} = \underbrace{-\sum_n \lambda_n x_n(0)e^{i\omega_n t}}_{\text{Source S}}$$

$$(8.6)$$

of the same exact form as described in Eq. (8.2). Assuming the environment has very little 'memory', $\Gamma(t - t') = \gamma\delta(t - t')$, we get the celebrated equation for a forced damped oscillator (Eq. (2.7)).

How exactly did irreversible damping creep into our model? Note that the cosine term and thus the Γ are each an even function of $t - t'$, meaning that it sums over times t' that are both earlier than t and later than t. As it stands this doesn't really represent damping, where the response must be *delayed* compared to the input (what we have is some sort of unphysical, 'reversible' damping). Our intuition tells us that $x(t)$ cannot be influenced by events that occur after t, so that a causal damping term must have a time asymmetric $\Gamma(t-t') \propto \Theta(t-t')$ where Θ is the step function equal to unity when the argument is positive and zero when negative. This causality was implicit in the boundary term we dropped when we did the integration by parts a few steps ago, but *must be included by hand*. Damping does not arise in a straightforward way from quantum mechanics. After all, unlike thermodynamics, the dynamical equations contained in Newton's equation (or for that matter Schrödinger's equation) do not share our prejudice towards causality.

Let us revisit the environment variable evolution equation in time domain. We find that the environment Green's function (i.e., response to a unit source) is given by

$$m_n \ddot{g}_n(t,t') = -m_n \omega_n^2 g_n(t,t') + \delta(t-t') \qquad (8.7)$$

So far, everything is real and reversible. But we have to do some mathematical jugglery to actually pull off its Fourier transform. The Fourier transform $g(\omega) = \int_{-\infty}^{\infty} d(t-t') g_n(t-t') e^{i\omega(t-t')}$ can be converted into a contour integral over the real time axis and a semicircular arc closing its ends. To make this work using the residue theorem, the integral must vanish on the arc of this contour. The g_n matrix must pick up an upfront factor of $\Theta(t-t')$, i.e., the response must come after the delta function impulse satisfying causality. For $t > t'$, the way to make the integral above converge is to replace $\omega \to \omega + i\eta$ where η is a very small but positive quantity (how small depends on the radius of the arc), so that $Re\left(e^{i(\omega_n + i\eta)(t-t')}\right) = \cos \omega_n (t-t') e^{-\eta(t-t')}$ sitting inside the integral. But this means that the environment Green's function and the self-energy above should actually involve $\omega + i\eta$. From Eqs. (8.6)–(8.7),

$$m_n g_n(\omega) = \frac{1}{(\omega+i\eta)^2 - \omega_n^2}, \qquad \Sigma(\omega) = \sum_n \frac{\lambda_n^2/m_n}{(\omega+i\eta)^2 - \omega_k^2} \qquad (8.8)$$

which immediately imparts an imaginary part to g_k and Σ. We can write

$$g_n(\omega) = \frac{1}{2m_n\omega_n}\left[\frac{1}{(\omega+i\eta) - \omega_n} - \frac{1}{(\omega+i\eta) + \omega_n}\right] \qquad (8.9)$$

We can easily see that the imaginary part of this will look like a Lorentzian with a small η broadening approaching a delta function. Indeed, this follows from the readily verifiable Pleimjl formula

$$\lim_{\eta \to 0}\left[\frac{1}{A + i\eta}\right] = P\left(\frac{1}{A}\right) - i\pi\delta\,(A) \tag{8.10}$$

where P represents the Cauchy principal part. The last term shows that even a seemingly innocuous, vanishingly small imaginary number, is not really that innocuous after all. In fact, it bears a lasting, integrable effect when the real part vanishes faster than the imaginary part. g_n can now be seen to have an imaginary part

$$a_n = i(g_n - g_n^\dagger) = \frac{\pi}{2m_n\omega_n}\left[\delta(\omega - \omega_n) - \delta(\omega + \omega_n)\right] = \pi\delta(m_n\omega^2 - m_n\omega_n^2)$$
$$\tag{8.11}$$

The function spikes everytime the frequency matches one of the eigenfrequencies ω_n. Indeed, it is easy to show that summing over the a_ns, we recover the phonon density of states (to within a factor of 2π). The imaginary part $\Gamma = i(\Sigma - \Sigma^\dagger)$ of the self-energy is

$$\Gamma(\omega) = \sum_n \lambda_n^{\,2}\pi\delta(m_n\omega^2 - m_n\omega_n^2) \tag{8.12}$$

which Fourier transformed back gives us $\Gamma(t - t') = \Theta(t - t')\sum_n \frac{\lambda_n^2}{2m_n\omega_n}\cos\omega_n(t - t')$. If λ_n is independent of k, we can pull it out of the sum, and then are left with a term $\propto \sum_n \delta(m_n\omega^2 - m_n\omega_n^2)$ in $\Gamma(\omega)$. Astute readers will recognize this as proportional to the phonon density of states $D_{ph}(\omega)$, and the corresponding simplified expression for Γ as Fermi's Golden Rule for phonons, once we identify the scattering rate $1/\tau = \Gamma/\hbar = \pi\lambda^2 D_{ph}(\omega)/\hbar$

But where did the imaginary frequency $i\eta$ physically come from? We introduced it seemingly for mathematical convenience alone! What η represents is the rate of particle extraction from a large but finite system irretrievably into the battery, and the associated broadening of the energy levels due to such a finite lifetime. Any finite system with a finite phase space will show a Poincaré recurrence time during its phase space evolution that is fundamentally at odds with irreversibility. For a large but finite metallic contact with N atoms and energy bandwidth W, we have a near continuum of energy levels with average level separation $\delta\epsilon \sim W/N$. The Poincaré recurrence time is driven by the slowest period in the problem, the atomic Bohr frequency $\omega_B = \delta\epsilon/\hbar = W/N\hbar$. As long as N is finite, so is the recurrence time $2\pi/\omega_B$, and the system is reversible. This is a problem when we are looking for steady-state DC transport that represents $t \to \infty$ and must capture the correct physics beyond this slowest recurrence time arising purely from a finite size effect. To bypass this inherent reversibility for a finite system, we will need to artificially introduce an irreversible escape time τ by hand (representing if you will the final emancipation of the charge straight into the battery), and that is precisely what the $i\eta$ accomplishes. In order to be effective, this τ must operate faster than the recurrence time, meaning $\tau < 2\pi/\omega_B = hN/W$. Such an escape rate will translate into a corresponding broadening $\eta = \hbar/\tau$ as per the energy time uncertainty principle (η is the width of a Lorenzian in frequency domain, obtained upon Fourier transforming a decaying exponent in time). The above inequality implies $\eta > \delta\epsilon$, meaning that our artificial escape rate introduces a broadening that removes the finite size effects by converting the discrete energy spectrum into a continuum density of states.

Note that the final equation we now have has the structure of a forced-damped oscillator, with the damping hidden in the imaginary part of Σ and the forcing coming from the source S. We now need to make some assumptions about the source.

8.4   The special status of contacts

If the sources were explicitly specified, we'd be back to a deterministic evolution. What we want to do instead, is to accord the contacts with a special status, i.e., assume that they are always thermalized and in local equilibrium. This is a thermodynamic constraint and not a material imposition. Atomistically, the channel and contacts could be identical, but conceptually we need to endow each contact with a thermalization property that distinguishes it from the channel. In practice, this is often enforced by a structural difference between the channel and contacts, for instance, a strong doping variation or a dilution of modes with numerous modes within the contact that slow the electron down when injecting into the fewer modes in the channel (but not the other way round — picture a four lane highway merging into a two lane highway and the resulting bottleneck when traffic proceeds from the former to the latter). Electrons in the numerous contact modes scatter profusely among themselves, and aided by just a hint of in-elastic processes that we implicitly assume but care not to model or dwell upon, reach local equilibrium (Fig. 8.5).

Fig. 8.5   *Ways to enforce a distinction between contacts and channel. From left to right, we have a variation in doping, variation in cross-section and variation in scattering/impurity concentration to allow each contact to self-equilibrate under a source to drain bias. Eliminating that distinction by making the contacts resemble the channel will reduce the interfacial resistance, but will also extend the voltage drop across the contact region once it looks just like the channel, creating thereby a large undesirable series resistance.*

Thermalization removes all phase information within the contact states and makes them stochastic. A thermalized gas of electrons generates a white noise, so that upon averaging over the ensemble of possible states, we get a memory free process $\langle S(t) \rangle = 0$, $\langle S(t)S(t') \rangle \propto \delta(t - t')$. Under this condition, Eq. (8.6) simplifies to yield the Lángevin equation, which with a simplified self-energy kernel (specifically, instantaneous damping), reads

$$m\ddot{x} + \gamma \dot{x} = -V'(x) + S \qquad (8.13)$$

with the above stochastic property imposed on S. We'll see shortly that this thermalized Lángevin can be recast in the form of a Fokker-Planck equation that presages classical drift-diffusion, and the Boltzmann distribution comes out as a special case in equilibrium.

What we are about to accomplish is quite incredible. We started with fully time reversible Hamiltonian equations, and just through the almost imperceptible insertion of a miniscule $i\eta$ term at some point, turned it into an irreversible, diffusive, entropy driven equation! While we did this for classical particles, the derivation of quantum transport equations will exactly mirror this procedure, i.e., we will eliminate the contacts with the inclusion of an imaginary term, and then postulate that the contacts are in thermal equilibrium. The noise in the contacts will have a little bit of memory associated with the presence of a Fermi level (Eq. (11.19)) that acts like a low pass filter in frequency domain.

Chapter 9

Classical transport

9.1 Classical physics and drift-diffusion

Let us first discuss the scrambling that happens in bulk even without specifying an outside source. We can ignore the external source when the contacts are large enough and conductive enough to be irrelevant as a result. What we will show now is precisely how classical Newtonian physics and statistical thermodynamics entangle to result in the popular drift diffusion equation. This equation can be 'derived' much more succinctly by just writing down the Boltzmann equation. However, the more laborious process below helps to see the various assumptions that are implicit in these derivations, such as the nature of the noise statistics and the duration of noise 'memory', for instance.

We start with Newton's equation in presence of damping (recall that we discussed the origin of damping earlier)

$$\underbrace{m\ddot{x}}_{\text{inertial}} + \underbrace{\gamma\dot{x}}_{\text{damping}} = \underbrace{-\frac{\partial U}{\partial x}}_{\text{external force}} \tag{9.1}$$

For instance, the Hooke's Law term $m\omega_0^2 x$ in Eq. (8.5) arose from an oscillatory potential $U = m\omega_0^2 x^2/2$.

In presence of thermal scattering, we will now (a) add a thermal noise term, (b) drop the inertial term $m\ddot{x}$ assuming high damping, and (c) average over a Gaussian distribution of the thermal noise, and invoke a memory free 'Markovian' process, in order to extract the final thermally averaged probability distribution $P(x,t)$. This distribution will be shown to satisfy the Fokker-Planck equation, just a step away from the very intuitive drift-diffusion equation that is the starting point for analyzing classical semiconductor devices such as p-n junction diodes.

Newton's law in presence of a stochastic thermal noise force $F(t)$ is

$$m\ddot{x}_F + \gamma\dot{x}_F = -\partial U/\partial x_F + F(t) \qquad (9.2)$$

where x_F evidently depends on F. We derived this equation from the elimination of contact variables in a coupled system two sections ago (Eq. (8.6)). In the overdamped limit with large γ, we drop the inertial term and end up with the **Lángevin equation**:

$$\boxed{\dot{x}_F = W(x_F) + F(t)/\gamma} \qquad (9.3)$$

where $W = -\partial U/\gamma\partial x_F$. The acceleration drops out as the system reaches steady flow, much like the terminal velocity on a parachutist, whereupon the damping, thermal and external forces cancel each other. For a fixed realization F, we can solve the above equation to get a corresponding cleanly defined Newtonian trajectory $x_F(t)$. This is the usual way stochastic Monte Carlo works, i.e., we throw a numerical die that specifies an F using a random number generator, and then average over many tosses of that 'die' to extract a histogram and obtain some statistics.

Since we anticipate F to vary widely during thermalization, it is more convenient to look at the ensemble of possibilities and evaluate the probability distribution P for the electrons, thereby extracting the evolution of all its averages (and possibly even higher moments, and fluctuations around those moments). We will now do a few pages of algebra to demonstrate that the averaging procedure gives an evolution equation for the electron density n that satisfies the **drift-diffusion** equation (**DD**), written out here only for a non-degenerate semiconductor

$$\boxed{\frac{\partial n}{\partial t} = \underbrace{\frac{\partial(\sigma\mathcal{E}/q)}{\partial x}}_{\text{drift}} + \underbrace{\mathcal{D}\frac{\partial^2 n}{\partial x^2}}_{\text{diffusion}}} \qquad (9.4)$$

We introduce the electric field $\mathcal{E}$, conductivity σ and diffusion constant $\mathcal{D}$ shortly.

9.2 Deriving DD: time-average to ensemble average

Instead of following $x_F(t)$, i.e., a restricted set of allowed trajectories stipulated by the Lángevin equation, we now ask for the probability density of accessing *any* coordinate x at a given time. In other words, we will calculate $P(y, t)$, where y is no longer a function of time but *is a free independent variable*. The resulting Fokker-Planck equation, a precursor to drift-diffusion equation, must only involve P but not F, the latter being integrated over its own distribution function.

Evidently for a fixed F, this $P(y, t)$ flocks precisely to the corresponding clean trajectory $x_F(t)$, becoming in fact a delta function $\delta(y - x_F(t))$. In general, F can have a wide range of values with its own probability distribution $\Pi(F)$, which is why P will eventually be a continuum distribution on y. The probability distribution is obtained once we take the fixed delta-function trajectories, and allow the noisy thermal forces driving the stochastic dynamics to vary over their thermalized distribution $\Pi(F)$.

$$P(y, t) = \int_{-\infty}^{\infty} dF \Pi(F) \delta(y - x_F(t)) \tag{9.5}$$

where $\Pi(F)$ is the probability of realizing a particular value of F.

Let us now find how P varies with time. We take partial derivative of P relative to t at fixed y. The only explicitly t-dependent term is x_F in Eq. (9.5), so this gives us

$$\frac{\partial P(y, t)}{\partial t} = \int_{-\infty}^{\infty} dF \Pi(F) \frac{\partial \delta(y - x_F(t))}{\partial x_F} \cdot \left[\frac{dx_F(t)}{dt} \right] \quad \text{(Chain Rule)}$$

$$= \int_{-\infty}^{\infty} dF \Pi(F) \left[-\frac{\partial \delta(y - x_F(t))}{\partial y} \right] \cdot \left[\frac{dx_F(t)}{dt} \right]$$

$$= \int_{-\infty}^{\infty} dF \Pi(F) \left[-\frac{\partial \delta(y - x_F(t))}{\partial y} \right] \cdot \left[W(x_F) + F(t)/\gamma \right] \quad \begin{array}{l} \text{(Using Langevin,} \\ \text{Eq. (9.3))} \end{array}$$

$$= -\frac{\partial}{\partial y} \int_{-\infty}^{\infty} dF \Pi(F) \delta(y - x_F(t)) \cdot \left[W(x_F) + F(t)/\gamma \right]$$

$$= \underbrace{-\frac{\partial}{\partial y} \int_{-\infty}^{\infty} dF \Pi(F) \delta(y - x_F(t)) W(x_F)}_{T_1} \underbrace{- \frac{\partial}{\partial y} \int_{-\infty}^{\infty} dF \Pi(F) F(t) \delta(y - x_F(t))/\gamma}_{T_2}$$

$$\tag{9.6}$$

where we have replaced $\dot{x}_F$ with the Lángevin solution from Eq. (9.3) in line 3 above, then pulled the $\partial/\partial y$ outside the bracket, noting that no term except the delta function has an explicit dependence on y.

The first term of the equation, T_1, can be simplified by using the delta function to replace $W(x_F)$ with $W(y)$, pulling it out of the integral, and then using the definition of Eq. (9.5), to give us

$$T_1 = -\partial\Big[P(y,t)W(y)\Big]/\partial y \tag{9.7}$$

Let us focus now on the second term, T_2.

9.3 Invoking thermal scattering and white noise

The second term has the product $F\Pi(F)$. Here's where we need some information about $\Pi(F)$. We will assume it is a thermalized Gaussian white noise. The thermal force F/γ has standard deviation $\mathcal{D}$ (diffusion constant), so that

$$\Pi(F) = \frac{1}{\sqrt{2\pi\gamma^2\mathcal{D}}}e^{-F^2(t)/2\gamma^2\mathcal{D}} \tag{9.8}$$

which gives us the thermal mean and correlation function, $\langle F(t)\rangle = 0$ and $\langle F(t)F(t')\rangle = \gamma^2\mathcal{D}\delta(t-t')$, meaning that the noise has zero time average and is also delta-correlated in time, retaining no 'memory'. The Gaussian distribution can be used then to write the sought after product

$$F\Pi(F) = -\gamma^2\mathcal{D}\frac{\partial\Pi(F)}{\partial F} \tag{9.9}$$

Thus, the term inside the integral in T_2 can be simplified to

$$-\gamma\mathcal{D}\int_{-\infty}^{\infty} dF\delta(y-x_F(t))\frac{\partial\Pi(F)}{\partial F}$$

$$= \gamma\mathcal{D}\int_{-\infty}^{\infty} dF\Pi(F)\frac{\partial\delta(y-x_F(t))}{\partial F} \quad \text{(integrating by parts, noting } \Pi(\pm\infty)=0)$$

$$= \gamma\mathcal{D}\int_{-\infty}^{\infty} dF\Pi(F)\frac{\partial\delta(y-x_F(t))}{\partial x_F}\cdot\left[\frac{dx_F}{dF}\right] \quad \text{(chain rule)}$$

$$= -\gamma\mathcal{D}\int_{-\infty}^{\infty} dF\Pi(F)\frac{\partial\delta(y-x_F(t))}{\partial y}\cdot\left[\frac{dx_F}{dF}\right] \tag{9.10}$$

using integration by parts at step 2, along with the boundary condition $\Pi(F) = 0$ for $F = \pm\infty$ satisfied by the Gaussian noise. We once again go back to our Lángevin equation to evaluate the last term, which is driven by a Markov (memory free) approximation.

$$\frac{d}{dt}\left[\frac{dx_F}{dF(t_1)}\right] = \frac{d}{dF(t_1)}\left[\frac{dx_F}{dt}\right]$$

$$= \frac{d}{dF(t_1)}\left[W + F(t)/\gamma\right] = \delta(t-t_1)/\gamma \qquad (9.11)$$

and upon integration, the last line gives $dx_F/dF(t) = 1/\gamma$. This means that the integral in T_2 becomes

$$-\mathcal{D}\int_{-\infty}^{\infty}dF\Pi(F)\frac{\partial\delta(y-x_F(t))}{\partial y} = -\mathcal{D}\frac{\partial}{\partial y}\int_{-\infty}^{\infty}dF\Pi(F)\delta(y-x_F(t)) \quad (9.12)$$

and thus

$$T_2 = \mathcal{D}\frac{\partial^2}{\partial y^2}\int_{-\infty}^{\infty}dF\Pi(F)\delta(y-x_F(t))$$

$$= \mathcal{D}\frac{\partial^2 P(y,t)}{\partial y^2} \qquad (9.13)$$

Putting together terms $T_{1,2}$, arrived at the **Fokker-Planck equation**

$$\boxed{\frac{\partial P(y,t)}{\partial t} = -\frac{\partial\left[P(y,t)W(y)\right]}{\partial y} + \mathcal{D}\frac{\partial^2 P(y,t)}{\partial y^2}} \qquad (9.14)$$

where $W(y) = -\partial U(y)/\gamma\partial y$.

9.4 Fokker-Planck to drift-diffusion

The FPE can be recast rather trivially as a **drift-diffusion equation**, once we apply it to the average electron charge density, $\rho_n = -qP$ (q is positive and equal to 1.602×10^{-19} Coulombs) and the electron particle density $n = P$. We get a charge conserving equation of continuity

$$\frac{\partial n}{\partial t} = \frac{1}{q}\frac{\partial J_n}{\partial y} \qquad (9.15)$$

where the electron current density is given by

$$J_n = \underbrace{\frac{nq^2}{\gamma}\frac{\partial(U/q)}{\partial y}}_{\text{drift}} + \underbrace{q\mathcal{D}\frac{\partial n}{\partial y}}_{\text{diffusion}} \qquad (9.16)$$

Since $-U/q$ is the voltage and its negative derivative is the electric field $\mathcal{E}$, we can simplify the first term to the right and cast it in the familiar form

$$J_{\text{drift}} = nqv_{\text{drift}}, \quad v_{\text{drift}} = \mu\mathcal{E}, \quad \mu = q/\gamma \quad (9.17)$$

to identify the mobility as $\mu = q/\gamma$. We can solve the original Newton's equation, Eq. (9.1), in absence of any external or thermal forces, to get $x = x_0 e^{-t/\tau}$, with $\tau = m/\gamma$ the scattering time associated with damping. This gives us the familiar equation for mobility,

$$\mu = q\tau/m \quad (9.18)$$

What was the point of such a laborious derivation of the very intuitive drift diffusion? The exercise was useful to illustrate how entropic forces mixed with Newtonian mechanics generates the diffusive term. It also emphasized the assumptions that went into our derivation, so we know what the fix is once we violate those assumptions. We assumed a free electron gas, which is valid only for 'non-degenerate' semiconductors where the Fermi energy sits several $k_B T$ away from band-edges. The conductivity $\sigma = J_{\text{drift}}/\mathcal{E}$ amounted to $nq^2\tau/m$, where n must be the number of 'free' electrons. A more generic expression for σ (next section) makes this explicitly a Fermi-surface property and removes any ambiguity about what 'free' means.

We also assumed thermal white noise with zero memory, which helped terminate our drift-diffusion equation at the second derivative. Noise sources are often colored (e.g., $1/f$ or random telegraph noise), and in presence of any 'memory' in the noise, the truncation to second derivative becomes problematic (device engineers often build complex noise statistics in time-domain Monte Carlo, but if we want to study the ensemble average as in drift diffusion, we will need to be careful!). For an arbitrary noise source, we will need to invoke a generalized drift-diffusion consisting of an infinite series called the Kramers-Moyal expansion $\sum_m \mathcal{D}_m \partial^m n/\partial x^m$. There is indeed a rich history of generalized diffusion equations for experiments on colored noise and biophysical phenomena, such as the ratcheted motion of kinesin and myosin along cytoskeletal tracks, or the pumping of ions across Na-K ATPase in axons. The underlying commonality in these systems is non-equilibrium, non-white noise which all translate to an unconventional drift-diffusion equation. On occasion, the higher derivatives can be subsumed into an effective position-dependent diffusion constant. The upshot of this process is that diffusion and the Boltzmann distribution can now be rationalized microscopically as the outcome of a Gaussian distribution of thermal kicks with no memory.

Chapter 10

From classical to semiclassical transport

10.1 Deriving Boltzmann

The solution to the Fokker-Planck equation is instructive at equilibrium, where we can drop the left term in Eq. (9.14). We can then solve for $P_{eq}(y)$, with boundary conditions $P_{eq}(\pm\infty) = 0$, to give us

$$P_{eq}(y) = P_0 e^{\int_0^y dx W(x)/\mathcal{D}} = P_0 e^{-U(y)/\gamma\mathcal{D}} \tag{10.1}$$

where $U(0)$ is absorbed in the prefactor constant P_0. If we now use the Einstein relation between diffusion constant and mobility. related to damping γ through $\gamma = q/\mu$, ie, $\mathcal{D} = \mu k_B T/q = k_B T/\gamma$ (proved later), we get

$$\boxed{P_{eq}(y) = P_0 e^{-U(y)/k_B T}} \tag{10.2}$$

which we recognize as **Boltzmann's equation**. *The Boltzmann distribution thus arises in the trivial equilibrium limit of Fokker-Planck, as expected.*

But the real merit of the FPE is the generalization to non-equilibrium, for instance, when U changes with time. A trivial example is a gate bias that is ramped up at a given scan rate (see Eq. (2.25)) — if the ramp rate is 'fast' compared to the relaxation rate given by γ^{-1}, the output should deviate significantly from Boltzmann. In fact, it can give rise to hysteretic loops.

10.2 Quasi-Fermi level vs $\mathcal{E}$-field: what drives a current?

The current equation (Eq. (9.16)) be written in terms of the variation of a single position dependent quantity,

$$\boxed{J_n = \frac{\sigma}{q} \frac{\partial \mu_n(y)}{\partial y}} \tag{10.3}$$

Device physicists call μ_n the quasi-Fermi level while chemists call it the electrochemical potential. σ is the conductivity of the sample. If the conductivity σ ends up varying with position, as it well might, it would need to sit inside the derivative as $\partial(\sigma\mu_n)/q\partial y$.

At equilibrium, the Fermi energy of a system (i.e., the largest energy upto which the energy levels are occupied at zero temperature) must be invariant across the sample in order to maintain zero current. Even at non-equilibrium we can mentally partition the sample into small cells around each grid point y and width Δy, serially probed with non-invasive voltmeters to gauge their local environments. The segment should come in equilibrium with the voltage probes (unaffected by the other segments and cross-talk if we assume that the segments are each larger than the inelastic scattering length needed to relax energies within a segment, $\Delta y > \lambda_E$). The locally equilibrated Fermi energy is $\mu_n(y)$. This is, of course, a concept harder to impose if the entire sample acts ballistically, whereupon $\mu_n(y)$ depends sensitively on the voltage probes.

There is often some confusion about the origin of current, which is frequently mis-attributed to the electric field alone as the driving force, $J = \sigma\mathcal{E}$. This is only true for the drift current, and would be sufficient if the sample had no boundaries or if we dealt with AC fields that establish local currents through displacement fields $J_d \propto \partial\mathcal{E}/\partial t$ without the need to encounter any boundaries. But there is also a diffusion current created simply by the pile-up of electrons that then scatter away toward sparser regions (the diffusion current of atoms is what allows us to smell a perfume, or for a cup to cool down). An ammeter measures the total current, drift plus diffusion, and it is this total current we must be after. If we imagine a finite semiconducting sample disconnected with any leads but subjected to an external DC field (e.g., with remote capacitor plates), we do not expect to get any steady-state current even though an unscreened field clearly persists. The same applies to a pn junction just sitting at equilibrium, facing a sizeable local electric field across the depletion junction but clearly seeing no current in absence of a battery! In all these systems, the electric fields migrate the charges towards the boundaries, where the latter tend to build up and create a counter-propagating diffusion current $\propto \partial n/\partial y$, which ultimately must cancel the drift current and reestablish equilibrium. When driven by a battery, the built-up charges are siphoned out through the adjoining leads, but keep getting replenished by freshly injected charges from other leads, and now the *difference* between the drift and diffusion terms is what shows up as a net current.

It is thus worth emphasizing is that *what drives net current is NOT a varia-tion of electrostatic potential $-U/q$ (i.e., an electric field $\mathcal{E} = \partial U/q\partial y$), but a variation of electrochemical potential μ_n as captured in Eq. (10.3).* The charge density depends on the *chemical potential* — the difference between electrochemical and electrostatic potentials (picture the energy gap between the Fermi level and the bottom of the conduction band, μ_n determining the former and U setting the latter). We can thus write

$$n = F\left(\underbrace{[\mu_n - U]}_{\text{chemical potential}} \right), \tag{10.4}$$

with a function F that is, for instance, a Boltzmann distribution for a classical electron gas or a non-degenerate semiconductor, and a Fermi-Dirac distribution for a degenerate solid. This means that $\partial n/\partial y = (\partial \mu_n/\partial y - \partial U/\partial y)F'$ where $F' = dF/d\mu_n$. Substituting in Eq. (10.3) and adding and subtracting a term, we get

$$J_n = \frac{\sigma}{q}\partial \mu_n/\partial y = \frac{\sigma}{q}\left(\partial \mu_n/\partial y - \partial U/\partial y\right) + \sigma\frac{\partial(U/q)}{\partial y}$$

$$= \underbrace{\frac{\sigma}{qF'}\frac{\partial n}{\partial y}}_{\text{diffusion}} + \underbrace{\sigma\mathcal{E}}_{\text{drift}} \tag{10.5}$$

Comparing with the expression for diffusion current in Eq. (9.16) we get the generalized Einstein relation in terms of the density of states $D = F' = dn/d\mu_n$

$$\sigma = q^2 \mathcal{D}D \tag{10.6}$$

What about thermally driven currents? We can generalize our earlier state-ment to state that in general, a current is set up by a variation of an *in-tensive variable* (one that is independent of size), such as electrochemical potential $\Delta\mu$, temperature gradient ΔT, a gradient in magnetic field ΔH, or a pressure gradient ΔP, the changes all inter-related by the **Gibbs-Duhem relation**

$$S\Delta T = \Omega\Delta P - M\Delta H - N\Delta\mu \tag{10.7}$$

The first law of thermodynamics connects *extensive variables*, - the particle count N, internal energy U, entropy S, magnetization M and volume Ω

$$T\Delta S = \Delta U + P\Delta\Omega - \mu\Delta N - H\Delta M \tag{10.8}$$

10.3   Recombination-generation: restoring balance

For a closed system at steady-state, current $J = 0$ (it clearly vanishes at
the impervious boundaries, and since n becomes independent of t and con-
sequently J becomes independent of y at steady-state, the current density
inside the boundaries must vanish as well). To get a current going, either
we need a non steady-state situation, for instance using an AC field, or
boundaries to the region that allow us to take out and reinject electrons.
We shall see later that this explicitly creates outflow and inflow terms like
we claim earlier. The Lángevin equation gets modified as

$$m\ddot{x}_F + \gamma\dot{x}_F - \underbrace{\Sigma x_F}_{\text{outflow}} = -\partial U/\partial x_F + F(t) + \underbrace{S}_{\text{inflow}} \qquad (10.9)$$

with Σx again generalized to a convolution $\int dx' \Sigma(x_F, x')x'$. In classical
device physics, however, these outflow and inflow terms are simply added
in by hand, giving us drift-diffusion with recombination-generation

$$\boxed{\frac{\partial n}{\partial t} = \frac{1}{q}\frac{\partial J_n}{\partial y} - \underbrace{R_n}_{\text{recombination}} + \underbrace{G_n}_{\text{generation}}} \qquad (10.10)$$

In silicon device models, recombination usually involves trap capture and
emission. Direct band-to-band recombination is disallowed for indirect
bandgap semiconductors because of the large mismatch between photon
and electron velocities, which makes it impossible to conserve both energy and momentum involving light alone (recall that the ratio between
energy and momentum is the band velocity). The trap recombination rate
depends on the size (capture cross-section) and density of traps and the
electron speed. The recombination term alone takes care of the outflow
and inflow processes discussed earlier, depending on its sign (in that sense,
a negative recombination is in effect a generation process). But what we
include explicitly in G_n is generation from external sources such as incident
photons in a solar cell.

The Fokker-Planck equation or the drift-diffusion equation is quite interesting in that it started with a purely dynamical equation that was completely
time-symmetric, and ended up with a statistical equation for a probability
distribution that clearly violates this time-symmetry. That subtlety crept
in by asserting that the thermal force obeyed a stochastic equation specified by a zero mean and a standard-deviation that is delta correlated, i.e.,

an infinitely wide-band white noise. The resulting Markov approximation (Eq. (9.11)) led to the erasure of any 'memory' effects.

10.4 Minority carrier diffusion equation

Many of the classical electronic devices such as bipolar junction transistors (BJTs) can be understood from the simple drift-diffusion equation with RG, under simple assumptions. In addition to the electron current equation, there is a hole equation, the two further coupled by Gauss' law. In 1-D

$$
\left.\begin{aligned}
\frac{\partial n}{\partial t} &= \frac{1}{q}\frac{\partial J_n}{\partial y} + G_n - R_n \\[2mm]
\frac{\partial p}{\partial t} &= -\frac{1}{q}\frac{\partial J_p}{\partial y} + G_p - R_p
\end{aligned}\right\} \quad \text{(equation of continuity)}
$$

$$
\left.\begin{aligned}
J_n &= qn\mu_n\mathcal{E} + q\mathcal{D}_\mathcal{N}\frac{\partial n}{\partial y} \\[2mm]
J_p &= qp\mu_p\mathcal{E} - q\mathcal{D}_\mathcal{P}\frac{\partial p}{\partial y}
\end{aligned}\right\} \quad \text{(drift diffusion)}
$$

$$
\frac{\partial(\epsilon\mathcal{E})}{\partial y} = q\left(p - n + N_D^+ - N_A^-\right) \quad \text{(Gauss' law)}
$$

$$
\tag{10.11}
$$

with N_D^+ and N_A^- the ionized donor and acceptor densities. For typical classical devices, we assume no generation (except for solar cells), $G_{n,p} = 0$, no E-field ($\mathcal{E} \approx 0$), low-level injection (i.e., $n(y,t) = n_{eq}(y) + \Delta n(y,t)$ with $\Delta n \ll n_{eq}$, and similarly for holes). For recombination, we invoke the so-called relaxation time approximation for low level injection that posits that recombination is rate-limited by minority carrier density

$$
R_n \approx -\Delta n/\tau_n, \quad R_p \approx -\Delta p/\tau_p \tag{10.12}
$$

with τ being the minority carrier lifetime. Thus we arrive at the **minority carrier diffusion equation (MCDE)**, that deals with minority carriers (i.e., electron density Δn_p in the p-heavy regions and hole density Δp_n in the n-heavy regions)

$$
\begin{aligned}
\frac{\partial \Delta n_p}{\partial t} &= \mathcal{D}_N \frac{\partial^2 \Delta n_p}{\partial y^2} - \frac{\Delta n_p}{\tau_n} \\
\frac{\partial \Delta p_n}{\partial t} &= \mathcal{D}_P \frac{\partial^2 \Delta p_n}{\partial y^2} - \frac{\Delta p_n}{\tau_p}
\end{aligned}
\tag{10.13}
$$

Classical device physics derives from these equations. Nonequilibrium enters through quasi-Fermi levels in the p and n regions separated by an applied bias, the effect ultimately appearing as a boundary condition, known as the 'Law of the Junction'. Based on this, we can derive the Shockley equation for an ideal diode or the gain and Ebers-Moll circuit model in a bipolar junction transistor for instance. In conventional device modeling, a lot of the intricate scattering physics is woven parametrically into the field, doping and temperature dependent mobilities $\mu_{n,p}$. In quantum transport, we accomplish the same with the self-energies Σ and Σ^{in}. In both cases, we combine different scattering processes by adding their scattering times $\propto 1/\mu, 1/\Gamma$.

10.5   Semiclassical physics: towards Schrödinger

A major development in classical mechanics happened in the evolution from Newton's equations to Hamilton's equations. To start with, the variables changed from $(x, \dot{x}, t)$, to (x, p, t), where the velocity $\dot{x}$ was replaced by the canonical momentum p as an independent variable. This wasn't just a trivial substitution — it was a major change in viewpoint. Instead of the evolution of a particle set by its local metrics like immediate position and speed, the motion is set instead by evaluating entire sets of trajectories with fixed end points in phase space — i.e., a global constraint. In this phase space representation, we can track the motion of the occupation probability $f(x, p, t)$ using the chain rule

$$
\frac{df(x, p, t)}{dt} = \frac{\partial f}{\partial t} + \frac{\partial f}{\partial x} \underbrace{\dot{x}}_{v} + \frac{\partial f}{\partial p} \underbrace{\dot{p}}_{F}
\tag{10.14}
$$

The term that governs the evolution is the Hamiltonian, $H(x, p) = T(p) + U(x)$ where T is the kinetic energy expressed in terms of the generalized momentum p, and U is the potential energy. Hamilton's equations are

$$\dot{x} = \frac{\partial H}{\partial p}, \quad \dot{p} = -\frac{\partial H}{\partial x} \tag{10.15}$$

The first equation is familiar to students of solid-state devices, where the electron velocity is given by the slope of the band, $v = \partial E / \partial p$. The second is simply the familiar Newton's Law, $\dot{p} = F = -\partial U / \partial x$.

More generally, we can look at the time evolution of ANY operator in Hamiltonian mechanics in terms of the so-called Poisson bracket,

$$\frac{dO}{dt} = \frac{\partial O}{\partial t} + \left\{ O, H \right\} \tag{10.16}$$

where the Poisson bracked is defined as

$$\left\{ A, B \right\} = \frac{\partial A}{\partial x} \frac{\partial B}{\partial p} - \frac{\partial A}{\partial p} \frac{\partial B}{\partial x} \tag{10.17}$$

Setting $A = x$ and $B = p$ for the two independent variables x and p, we get Hamilton's equations (Eq. (10.15)) as a special case.

If we now focus on a specific function, the phase space density of the electrons, then the starting point is Liouville's theorem which says that in absence of scattering forces, the probability density $f(x, p, t)$ does not evolve, so that $df/dt = 0$. In the presence of scattering, however, this derivative changes only as much as the scattering imposes, giving us the Boltzmann Transport Equation (BTE)

$$\frac{\partial f}{\partial t} + v \frac{\partial f}{\partial x} + F \frac{\partial f}{\partial p} = \left. \frac{\partial f}{\partial t} \right|_{sc} \tag{10.18}$$

The scattering term is often simplified as the relaxation time approximation relative to the equilibrium distribution function $f_0(x, p)$

$$\left. \frac{\partial f}{\partial t} \right|_{sc} \approx -\frac{f - f_0}{\tau} \tag{10.19}$$

Note that so far, our equations are not even semi-classical, but purely classical (after all, Boltzmann predated the birth of quantum mechanics). With the advent of quantum mechanics, however, bits and pieces of quantum physics percolated into the BTE equation ($\hbar$ enters the equation). These inclusions happened primarily in the way we calculate the quantities v, F,

f_0 and τ using a quantum mechanical band $E(k)$. From Hamilton's equations, we can now write down $v = \partial E / \hbar \partial k$ and $F = \hbar dk/dt$ extracted from the potential derivative $-\partial U / \partial x$.

The generalized Poisson bracketed form of the classical evolution equations shows the relation with quantum mechanics. When moving to QM, there are primarily two changes — for starters, the quantities A and B in Eq. (10.21) are elevated from scalar functions to operators, typically representable as matrices in a given basis set. This means the order of A and B must be preserved in the above equation. Secondly, the variables x and p satisfy the commutation relation instead of the Poisson bracket relation

$$\underbrace{\left\{ x, p \right\}}_{\text{classical}} = 1 \rightarrow \underbrace{\left[\hat{x}, \hat{p} \right] = \hat{x}\hat{p} - \hat{p}\hat{x} = i\hbar}_{\text{quantum}} \tag{10.20}$$

Accordingly, the corresponding Poisson bracket containing the differential product $\Delta x \Delta p$ in the denominator of the derivative gets replaced by the commutator divided by $i\hbar$

$$\frac{dO}{dt} = \frac{\partial O}{\partial t} + \frac{1}{i\hbar} \left[O, H \right] \tag{10.21}$$

which amounts to the Heisenberg representation of quantum mechanics (Eq. (3.18)).

Chapter 11

Quantum transport: the Non-Equilibrium Green's Function formalism

In this section, we will look at the quantum flow of electrons, following a procedure similar to how we derived the classical drift diffusion equation. The latter came from the open boundary Newton's equation, ensemble averaged to describe collective flow. A very similar pathway will now lead to the Keldysh-Kadanoff-Baym Non Equilibrium Green's Function (NEGF) formalism. For simple noninteracting systems, this will simplify significantly, yielding the two terminal Landauer or the multiterminal Landauer Büttiker formalism.

We start with the one-electron Schrödinger equation and tag thermodynamic open boundary conditions onto them to extract their collective flow, leading to a non-equilibrium expression for current. We will mentally separate a 'channel', with well specified dynamical properties stemming from a Hamiltonian, and a 'contact' whose properties will only be known on average. In other words, there is enough scattering in the contacts that any charge escaping into them reach back to local thermal equilibrium (these are the unspecified inelastic processes marked by red wiggly lines in Fig. 8.2). The advantage of this distinction is that we can mentally separate the quantum dynamics in the channel from the thermodynamics in the contact.

In the interest of simplicity, let us first discuss what current flow through a small weakly interacting quantum system looks like. We will then generalize it to derive the full NEGF equations, using a process very similar to the derivation of Drift Diffusion. In this one-electron derivation, pioneered by Datta, self-energies arise as bilinear thermal averages of sources. For completeness' sake, we will eventually outline the many body derivation of NEGF at the end of this chapter, although as we point out earlier

(Section 8.2), *it is critical to dissociate the equations themselves from both their historical underpinnings and the way to calculate its ingredients.*

11.1 'Toy' model for current flow

We start with a simple toy model for current flow for a set of energy levels described by an overall density of states $D(E)$ (Eq. (9.16)). The levels are coupled to a source and a drain with coupling strengths $\gamma_{1,2}$ whose inverses $\hbar/\gamma_{1,2}$ give escape times into the contacts. What makes the system nonequilibrium is that a battery separates their equilibrium Fermi functions $f_{1,2}$ by the drain voltage, and the channel tries simultaneously to reach equilibrium with both system, as shown in Fig. 8.2 (left). As a result, a current is set up in the energy window between the Fermi energies $\sim qV$, and the channel reaches steady state with a nonequilibrium charge density $G^n(E)/2\pi$ (the 2π is introduced for convenience as we will see later). Since each contact will keep injecting as long as $G^n/2\pi \neq f_{1,2}D$, the current at the two contacts can be written down readily as

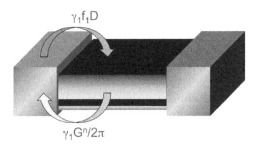

$$I_{1,2} = \int dE \frac{q\gamma_{1,2}}{\hbar}\left(f_{1,2}D - G^n/2\pi\right) \tag{11.1}$$

At steady state, the total current $I_1 + I_2$ is zero, which gives

$$G^n/2\pi = \frac{\gamma_1 f_1 + \gamma_2 f_2}{\gamma_1 + \gamma_2}D, \tag{11.2}$$

as a weighted average of $f_{1,2}$, with weighting factors given by the couplings $\gamma_{1,2}$. Substituting G^n back into Eq. (11.1), we get the corresponding steady state current (per spin)

$$I_1 = -I_2 = \frac{q}{\hbar}\int dE \underbrace{\frac{\gamma_1\gamma_2}{\gamma_1 + \gamma_2}}_{\gamma_{\text{eff}}} D(f_1 - f_2). \tag{11.3}$$

The equation indicates that current can only flow in a slim energy window $\sim |\mu_1 - \mu_2| = qV$ over which the fs differ. Morover, the current is inversely proportional to an escape time $\hbar/\gamma_{\text{eff}} = \hbar/\gamma_1 + \hbar/\gamma_2$, i.e., equal to the sum of the escape and injection times. The transit time within the channel is zero, since the channel is treated here as a single point, with its entire resistance arising from the contacts. We can augment this by adding drift and diffusion terms describing delays within the channel, so that

$$\frac{\hbar}{\gamma_{\text{eff}}} \approx \underbrace{\frac{\hbar}{\gamma_1}}_{\text{injection}} + \underbrace{\frac{\hbar}{\gamma_2}}_{\text{removal}} + \underbrace{\frac{L}{v}}_{\text{ballistic}} + \underbrace{\frac{L^2}{2\mathcal{D}}}_{\text{diffusive}} \tag{11.4}$$

where $\mathcal{D}$ is the diffusion constant and L is the channel length.

Note that the density of states needs to account for any local potential U set up by the contacts (which could include gates or substrates), as well as any screening or Coulomb interactions. In general, U is obtained from Poisson's equation, which for a single point in space can be simplified to $U = U_L + U_0(N - N_0)$, with N_0 the number of electrons in the channel at equilibrium. The first term $U_L = -q(\alpha_S V_S + \alpha_D V_D + \alpha_G V_G)$ arises from the Laplace equation corresponding to a neutral channel ($N = N_0$) with boundary conditions established by the contacts set at their respective bias voltages. The second is a Poisson contribution obtained from the channel charge N in the presence of grounded contacts. $U_0 = q^2/C_E$ is the single electron charging energy set by the source, drain and gate capacitances in parallel, $C_E = C_S + C_D + C_G$, also setting in the process the capacitive transfer factors $\alpha_{S,D,G} = C_{S,D,G}/C_E$ that set the individual electrode contributions to the potential.

The transport equations can now be gathered together and written as

Minimal Transport Equations

$$N = \int dE G^n/2\pi = \int dE D(E-U)\frac{\gamma_1 f_1 + \gamma_2 f_2}{\gamma_1 + \gamma_2}$$

$$I_1 = -I_2 = \frac{q}{h}\int dE\, 2\pi \underbrace{\frac{\gamma_1 \gamma_2}{\gamma_1 + \gamma_2} D(E-U)(f_1 - f_2)}_{MT}$$

$$U = U_L + U_0(N - N_0)$$

$$\tag{11.5}$$

with the potential U assumed to shift the density of states D rigidly without altering its shape.

What we have above is the essence of the Landauer equation, which suggests that *the current is simply given by the number of modes $M(E)$ (more generally, quantum mechanical transmission $\bar{T}(E)$ summed over modes) lying within a bias window around the contact Fermi energy.*

We have simplified the Poisson's equation for a point channel. In general, we need to write $U_i = U_{Li} + \sum_{j \neq i} U_{ij}(N_j - N_{j0})$, where U_{ij} is the screened Coulomb interaction between two point charges in the presence of the surrounding dielectrics, with the various terminals grounded to zero, including as relevant exchange terms. *Note that the equations must be self-consistent, since U depends on N which in turn depends on U.*

For a *ballistic, well contacted* material, we expect the contact couplings to be very strong, $\gamma_{1,2} \gg \hbar v / L$, in which case we can write Eq. (11.4) as

$$\boxed{\left. \gamma_{\text{eff}} \right|_{\text{ballistic, good contacts}} \approx \frac{\hbar v}{L}} \tag{11.6}$$

where in general the velocity is the angle averaged, positive z-directed component from source to drain, $v = \langle |v_z| \rangle$.

The above basic equations often suffice to describe a wide variety of experimental I-Vs quantitatively, requiring lumped parameters such as the density of states D, contact broadening functions $\gamma_{1,2}$ (simple estimate in Eq. (11.14)), and the capacitances $C_{S,D,G}$. The equations are however a lumped circuit distribution and do not capture nuances of spatial properties, not to mention more involved quantum interference effects. *To do so, we will need to replace these parameters with matrices, so that the various diagonal components carry spatially resolved information and off diagonal parts carry phase differences relevant for interference.*

11.2   General quantum transport equations

We will now derive the NEGF quantum transport equations for a general interacting system. The process is very similar to our derivation of drift diffusion from Lángevin, i.e., writing out the open boundary Schrödinger equation, and then imposing local thermodynamic distributions on bilinear averages of the source term. We will later see under what conditions these equations simplify to the toy equations we just wrote down.

A. *The quantum problem*

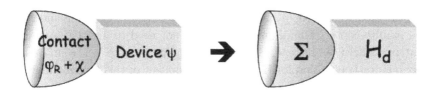

Fig. 11.1   *Toy model for device-contact assembly, showing the wavefunctions in the different regions, and the eventual replacement of the contacts with a self-energy.*

The time independent Schrödinger equation for the entire device-contact system is given for a single contact by

$$E\Psi = [H]\Psi \tag{11.7}$$

Separating out the device ('d') and contact ('c') subspace basis sets explicitly, we have

$$[H] = \begin{pmatrix} H_d & \tau \\ \tau^\dagger & H_c - i0^+ \end{pmatrix}$$

$$\Psi = \begin{pmatrix} \psi \\ \phi_R + \chi \end{pmatrix} \tag{11.8}$$

where ϕ_R is the wavefunction in the contact when it is isolated, ψ is the transmitted wavefunction into the device when ϕ_R is incident on it, and χ is the part reflected back into the contact. As we discussed earlier, $i0^+$ is an artificial escape rate to remove electrons irreversibly to the battery (in fact, it broadens the spectrum so that every energy E that ψ corresponds to also includes the eigenspectrum of ϕ_R). We solve for χ from the second line of the matrix equation (11.7), keeping in mind that ϕ_R satisfies $E\phi_R = (H_c - i0^+)\phi_R$ for the isolated contact. Substituting this expression for χ in the first line of Eq. (11.8), we then get the resulting *open boundary Schrödinger equation* for the device, very reminiscent in structure of the Lángevin equation discussed earlier.

$$(EI - H_d)\psi - \underbrace{\Sigma\psi}_{\text{outflow}} = \underbrace{S_R}_{\text{inflow}} \tag{11.9}$$

where

$$\Sigma = \tau g_c \tau^\dagger$$
$$g_c = [(E + i0^+)I - H_c]^{-1}$$
$$S_R = \tau \phi_R \tag{11.10}$$

I is the identity matrix, g_c is called the contact Green's function and H_c is the infinite contact Hamiltonian (one does not need to invert this infinite matrix explicitly — instead, we can use periodicity to derive a recursive matrix quadratic equation for the *surface* component of this Green's function g that couples with the channel through nearest neighbor block interactions τ).

The solution to the above Schrödinger equation (11.9) in the time-independent limit depends on ϕ_R, which is the (as yet unspecified) contact wavefunction in the absence of coupling with the channel.

$$\psi = GS_R$$
$$G = [EI - H_d - \Sigma]^{-1} \tag{11.11}$$

where the retarded Green's function G represents the delayed (causal) channel response to a unit source, and is the matrix analogue of G introduced later (Eq. (11.23)).

The Green's function helps us obtain static properties of the channel, such as the spatially resolved local density of states (LDOS). For an isolated system with Hamiltonian H, eigenkets $\{|\Phi_n\rangle\}$ and eigenvalues $\{\epsilon_n\}$, we can use the completeness relation (Eq. (3.12)) to expand $I = \sum_n |\Phi_n\rangle\langle\Phi_n|$ and insert into the above expression for the Green's function for the density of states, as we did earlier in Eq. (8.11) for classical oscillators.

$$G_0 = \sum_n \frac{|\Phi_n\rangle\langle\Phi_n|}{E - \epsilon_n + i0^+}$$
$$\underbrace{A_0}_{\text{spectral function}} = i(G_0 - G_0^\dagger) = 2\pi \underbrace{\sum_n |\Phi_n\rangle\langle\Phi_n|\delta(E - \epsilon_n)}_{\text{local density of states}} \tag{11.12}$$

with the local density of states $D(x, E) = \sum_n \langle x|\Phi_n\rangle\langle\Phi_n|x\rangle\delta(E - \epsilon_n) = \sum_n |\Phi_n(x)|^2\delta(E - \epsilon_n)$. In deriving the above, we used the mathematical Pleimjl formula once again (Eq. (8.10)).

In presence of a self energy Σ, for instance from the presence of contacts, it is easy to see that the levels get broadened by the interactions. For an energy independent Σ, we can diagonalize $H + \Sigma$ to extract its complex eigenvalues $\bar{\epsilon}_n = \epsilon_n^* - i\gamma_n/2$, eigenstates $|\bar{\Phi}_n\rangle$ and their 'dual' bra $\langle\tilde{\phi}_n|$, so that the imaginary part gives us a Lorentzian broadening weighted spatially by the probability distribution. The broadened density of states

$$D(x, E) = \frac{A(x, x)}{2\pi} = \sum_n i\left\langle\tilde{\phi}_n(x)\Big|\bar{\Phi}_n(x)\right\rangle\left(\frac{1}{E-\epsilon_n^*+i\gamma_n/2} - \frac{1}{E-\epsilon_n^*-i\gamma_n/2}\right)$$

$$= \sum_n \underbrace{\left\langle\tilde{\phi}_n(x)\Big|\bar{\Phi}_n(x)\right\rangle}_{\text{spatial weight}}\underbrace{\left[\frac{\gamma_n}{(E-\epsilon_n^*)^2+(\gamma_n/2)^2}\right]}_{\text{Lorentzian broadening}} \tag{11.13}$$

The broadening matrix $\Gamma = i(\Sigma - \Sigma^\dagger)$ in turn can be expanded using Eq. (11.10) as $\tau a_c \tau^\dagger$, the contact spectral function $a_c = i(g_c - g_c^\dagger)$ proportional to the surface density of states $D_c = a_c/2\pi$ of an isolated contact.

In a one-level version we now have a very convenient estimate for the broadening in terms of physical parameters of the system

$$\boxed{\gamma = 2\pi|\tau|^2 D_S} \quad \text{Fermi's Golden Rule} \tag{11.14}$$

where D_S is the surface term of the contact density of states D_C, and τ is the chemical bond energy or coupling between the contact surface atom and the channel linker atom.

The multi level version of the broadening equation gets complicated because H and $\Sigma(E)$ in general do not commute and are not simultaneously diagonalizable. One could still work with the eigenstates of the Fock matrix $H + \Sigma(E)$, but even these need to be calculated carefully because (a) the matrix is nonHermitian and its eigenstates are not guaranteed to be orthogonal, and (b) the energy dependent Σ means that the eigenstates, which differ from the states of H as they are 'dressed' by interactions, must be calculated self consistently (i.e., guess an energy ϵ_n and test to see if it forms an eigenstate of $H + \Sigma(\epsilon_n)$).

While Eq. (11.11) gives a formally exact solution of the Schrödinger equation for the device wavefunction ψ, one still needs information about the source term S_R.

B. The thermodynamic problem

We deal with thermodynamics imposed by the contacts when instead of explicitly specifying S_R (i.e., solving the entire infinite problem), we merely specify their bilinear averages $\langle S_R S_R^\dagger \rangle$. Such an averaging procedure implies that off-diagonal coherences in the contact are ignored, while for the diagonal terms we assume thermal white noise with no memory, just as we did in deriving Fokker Planck/drift diffusion. We then get the Keldysh equation for the electron correlation function $G^n(E)$ that gives the energy-resolved level occupancies and coherences. The correlation function is often referred to as the nonequilibrium Green's function, and depends explicitly on the thermodynamics of the source. Using Eq. (11.11)

$$G^n(E) = \langle \psi \psi^\dagger \rangle = G \underbrace{\langle S_R S_R^\dagger \rangle}_{\Sigma^{in}} G^\dagger \qquad (11.15)$$

Note that the Keldysh equation is quite general, in that it can capture higher order correlated interactions as long as we can describe them as an effective 'source' in terms of a matrix equation like Eq. (11.8). We can

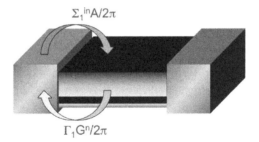

Fig. 11.2 *Replacing the scalar model for transport in a non-interacting system with a matrix model for transport in a general interacting system*

extend Eq. (11.8) to two contacts and rederive Schrödinger's equation with two open boundary conditions. The current arises as the derivative of the number operator. Starting with the definitions (dropping the index 'd' for convenience)

$$G = [EI - H - U - \Sigma_1 - \Sigma_2]^{-1}$$
$$A = i(G - G^\dagger) \qquad \text{(spectral function)}$$
$$\Sigma^{in}_{1,2} = \langle S_{1,2} S_{1,2}^\dagger \rangle \qquad \text{(inscattering function)}$$
$$\Sigma^{in} = \Sigma^{in}_1 + \Sigma^{in}_2 + \Sigma^{in}_s \qquad \text{(including scattering)}$$

$$G^n = G\Sigma^{\text{in}}G^\dagger \qquad \text{(correlation function)}$$
$$\Gamma_{1,2} = i[\Sigma_{1,2} - \Sigma_{1,2}^\dagger] \qquad \text{(broadening)} \qquad (11.16)$$

we get the generalized version of Eq. (11.5) (Fig. 11.2)

> ### General Interacting NEGF Equations
>
> $$\rho = \int dE \frac{G^n(E)}{2\pi}$$
> $$G^n = G\Sigma^{\text{in}}G^\dagger \qquad\qquad \text{(Keldysh equation)}$$
> $$I_1 = \frac{q}{h} \int dE Tr[\Sigma_1^{\text{in}} A - \Gamma_1 G^n] \quad \text{(Meir-Wingreen)}$$

$$(11.17)$$

To derive Meir-Wingreen, we used the time-dependent version of the open boundary Schrödinger equation (Eq. (11.9)) to separate out the ψ and $\psi^\dagger$ derivatives, and then rearranged the terms. It is worth emphasizing that to get the *terminal current*, we must only use the evolving terms corresponding to that one terminal instead of the entire steady-state equation (the latter would give zero total current, consistent with Kirchhoff's laws and can be separated out into a difference between terminal currents)

$$I_1 = q\frac{d}{dt}\text{Tr}\langle\psi^\dagger\psi\rangle = \frac{q}{i\hbar}\text{Tr}\langle\psi^\dagger\left(H\psi + \Sigma_1\psi + S_1\right) - \left(\psi^\dagger H + \psi^\dagger\Sigma_1^\dagger + S_1^\dagger\right)\psi\rangle$$

$$= \frac{q}{i\hbar}\text{Tr}\langle\psi\psi^\dagger\underbrace{\left(\Sigma_1 - \Sigma_1^\dagger\right)}_{-i\Gamma_1} + S_1\underbrace{\psi^\dagger}_{S_1^\dagger G^\dagger} - \underbrace{\psi}_{GS_1} S^\dagger\rangle \qquad (11.18)$$

where we used $\psi = GS_1$ for terminal 1 (Eq. (11.11)) and $\Gamma_{1,2} = i(\Sigma_{1,2} - \Sigma_{1,2}^\dagger)$ (Eq. (11.16)). Using in addition $G^n = \langle\psi\psi^\dagger\rangle$ (Eq. (11.15)), $\Sigma_{1,2}^{\text{in}} = \langle S_{1,2}S_{1,2}^\dagger\rangle$, and $A = i(G - G^\dagger)$ (Eq. (11.16)), we get the NEGF current in Eq. (11.17). The equations can be readily generalized to time-dependent problems by expanding the indices of the matrices to include two times t, t' related to the energy-dependent versions through Fourier transforms.

The real strength of these equations is that they can capture *any degree and kind of scattering* all the way from ballistic to diffusive, not just by metallurgical contacts, but also by virtual dephasing contacts such as phonons, photons and spins. Given a microscopic Hamiltonian describing these scattering events, we can write down expressions for Σ_s^{in} by just identifying the 'source' term acting on the channel (this is usually easy to do perturbatively in the interaction parameter with that source). Furthermore, one can calculate other kinds of currents, such as spin current $\vec{I}_S = d/dt \operatorname{Tr}(\langle \psi^\dagger \vec{\sigma} \psi \rangle)$ and heat current $I_Q = d/dt \operatorname{Tr}(\langle \psi^\dagger E \psi \rangle)$ where $\vec{\sigma}$ is the Pauli spin matrix and E is the energy. Readers are directed to the book by Datta on how NEGF can be used to calculate a rich variety of transport coefficients such as thermoelectric figures of merit, spin Hall and Seebeck coefficients.

For a simple noninteracting system where the only sources are the source and drain contacts maintained at their own private equilibria, we can use Eqs. (11.16), (11.10) to get

$$\Sigma^{in}_{1,2} = \langle S_{1,2} S^\dagger_{1,2} \rangle = \tau \underbrace{\langle \phi_{R1,R2} \phi^\dagger_{R1,R2} \rangle}_{a_{c1,c2} f_{1,2}} \tau^\dagger = \Gamma_{1,2} f_{1,2} \qquad (11.19)$$

where we assumed that the isolated contact wavefunctions are kept at equilibrium and their occupancies are set by their density of states and local Fermi Dirac functions. In the last step, we used the expression for Γ discussed immediately after Eq. (11.13). Substituting this simplification into Eqs. (11.16), (11.17) and simplifying the algebra, we get the Caroli formula and the Fisher Lee equations for coherent flow

---

Coherent NEGF Equations

$$\rho = \int dE \frac{G(\Gamma_1 f_1 + \Gamma_2 f_2) G^\dagger}{2\pi} \qquad (11.20)$$

$$I_1 = \frac{q}{h} \int dE \underbrace{\operatorname{Tr}(\Gamma_1 G \Gamma_2 G^\dagger)}_{MT} (f_1 - f_2) \qquad \text{(Fisher-Lee)}$$

The current depends on the total transmission by the various modes $\sum_i T_i = MT$, determined by the device Hamiltonian H_d including its self-consistent potential, and the contact self energies $\Sigma_{1,2}$ which set the broadenings $\Gamma_{1,2}$. The self-energy matrices have the same size as the Hamiltonian, as they describe the projection of the semi-infinite leads onto the channel.

How do we calculate the self energies in practice? Consider a one dimensional semi-infinite wire with its atoms arranged periodically, Fig. 11.3(a), supporting an onsite Hamiltonian element H_{on} and an offsite nearest neighbor block H_{off}, analogous to ϵ and t in tight binding (Eq. (7.1)). We can express the Green's function g_c of the contact surface atom in terms of the surface atom of the remaining contact (Eqs. (11.16) and (11.10)). Since these atoms are identical for a periodic chain, we get a recursive equation set

$$g_1 = (\alpha - \beta^\dagger g_1 \beta)^{-1}, \qquad g_2 = (\alpha - \beta g_2 \beta^\dagger)^{-1} \qquad (11.21)$$

where $\alpha = EI - H_{\text{on}}$ and $\beta = -H_{\text{off}}$. Solving each matrix quadratic equation numerically, we can then extract the self energies $\Sigma_{1,2} = \tau_{1,2} g_{1,2} \tau_{1,2}^\dagger$, where τ is the Hamiltonian matrix block coupling the channel with the contact. For a 1D contact with scalar entries for H_{on}, H_{off}, it is straightforward to show that the contact self energy terms are each equal to

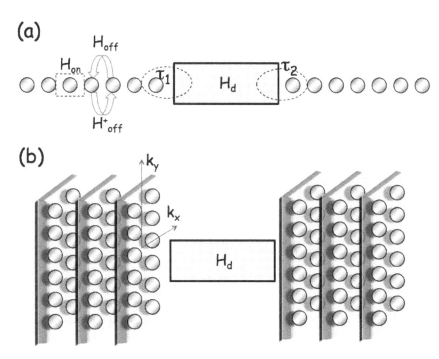

Fig. 11.3 *(a) 1D chain with Hamiltonian blocks specified. (b) For a 3D-1D-3D structure, it is convenient to Fourier transform to (k_x, k_y), whereupon each $\vec{k}$ point acts as a 1D chain above, albeit with $\vec{k}$-dependent H_{on} and H_{off} matrices.*

$$\Sigma_{1D}(E) = H_{\text{off}} e^{ika} \qquad (11.22)$$

where k is obtained by inverting the 1D tight binding equation (Eq. (7.1)), $E = H_{\text{on}} + 2H_{\text{off}} \cos ka$.

The process is easily extended to a 3D contact, Fig. 11.3(b), by using their in-plane periodicity to Fourier transform the matrix elements (Fig. 11.3). Each $\vec{k}_\perp$ component acts as an independent one dimensional wire, so that we can extract the $\vec{k}_\perp$-space surface Green's function and then inverse transform to extract any real space block of the surface Green's function.

How can we relate these equations to the toy equations we derived earlier (Eq. (11.5))? To do so, we will need to draw a connection between the charge occupancy G^n, the density of states D with simple Lorentzian broadening, and the Green's function G (outlined in Eq. (11.11))

$$G(E) = \frac{1}{E - \epsilon_0 + i(\frac{\gamma_1 + \gamma_2}{2})} \tag{11.23}$$

A Lorentzian density of states can then be expressed as the anti-Hermitian part of G

$$D(E) = \frac{1}{2\pi}\left[\frac{\gamma_1 + \gamma_2}{(E - \epsilon_0)^2 + (\frac{\gamma_1 + \gamma_2}{2})^2}\right] = i\left(\frac{G - G^*}{2\pi}\right) = \frac{|G|^2(\gamma_1 + \gamma_2)}{2\pi} \tag{11.24}$$

The transport equation set (11.5) can then be expressed in terms of G as well

$$G(E) = \frac{1}{E - \epsilon_0 - U + i(\frac{\gamma_1 + \gamma_2}{2})}$$

$$N = \int dE \frac{|G(E)|^2(\gamma_1 f_1 + \gamma_2 f_2)}{2\pi}$$

$$I_1 = -I_2 = \frac{q}{h}\int dE \underbrace{|G(E)|^2\gamma_1\gamma_2}_{MT}(f_1 - f_2) \tag{11.25}$$

which we can immediately identify as the scalar version of the coherent NEGF equations above. The latter can be thought of as a matrix generalization of our toy model for coherent transport. In a real space basis set, the diagonal entries can be used to describe spatially varying properties, while the off diagonal components describe spatial coherences that can lead to quantum interference.

$$\epsilon \rightarrow [H] \qquad \text{(device Hamiltonian)}$$
$$\gamma_{1,2} \rightarrow [\Sigma_{1,2}] \qquad \text{(self-energy matrices)}$$
$$[\Gamma_{1,2}] \qquad \text{(broadening matrices)}$$

$$D \to [A]/2\pi \qquad \text{(spectral function)}$$
$$N \to [\rho] \qquad \text{(density matrix)}$$
$$[G^n]/2\pi \qquad \text{(electron Correlation function)}$$
$$M \to [\Gamma A] \qquad \text{(mode count)}$$
$$U \to [U]$$
$$G \to [G] \qquad\qquad\qquad\qquad\qquad (11.26)$$

The size of each matrix is given by that of the basis set spanning the device (in a continuum approximation, it is the number of real space grid points; in a chemical model, the number of orbitals). The diagonal entries allow us to track the variation of the properties along the orbital grid, for instance, spatial variations. The off diagonal entries describe 'coherences' that capture interference effects between electron states occupying two different orbitals corresponding to their row and column indices.

11.3   What is 'nonequilibrium' about NEGF?

To summarize, for any transport problem, the two key quantities are G and G^n above, defined formally in Eq. (11.27) shortly. *The non-equilibrium Green's function (NEGF) refers to G^n*, and explicitly describes the nonequilibrium nature of the transport through the various in-scattering functions Σ^{in}, from both real and 'virtual' contacts kept at separate equilibria. G, the retarded Green's function, represents the static properties of the channel such as its density of states and transmission. For a noninteracting system, the two are simply related, and as a consequence the G we evaluate is ostensibly in equilibrium (Fig. 11.4). However even standard device models usually need to include a Poisson contribution proportional to $N = \int dE G^n/2\pi$ inside the self consistent potential U. For such interacting systems, G and G^n are not proportional and the full machinery of the NEGF equations must be summoned. In particular in the presence of incoherent interactions such as from electron–phonon or electron–electron scattering, separate Hamiltonians and microscopic equations are needed for Σ^{in}. The accuracy of our NEGF current will invariably depend on how precisely we can construct the input Σ^{in} matrices. For very complicated systems with strong nonequilibrium correlations, this can only be done approximately, as we shall see later.

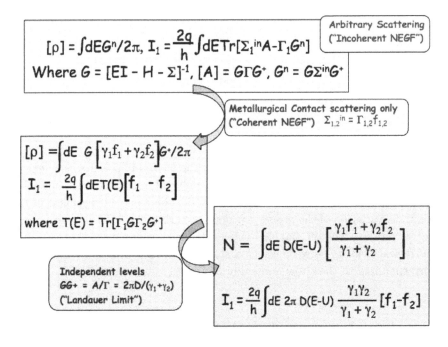

Fig. 11.4   *NEGF Quantum transport equations. In the most general case, we get the Meir-Wingreen form on top, which explicitly involves a correlation Green's function G^n and an in-scattering function Σ^{in}. This equation can handle incoherence, and with a proper self-energy, many-body correlations as well (we provide an explicit nonperturbative example in the chapter on correlation). If the only scattering comes from metallurgical contacts that are held in local equilibrium, the Σ^{in}s simplify and we get the coherent NEGF equations with the Fisher-Lee form for the transmission, shown in the middle. Finally for a system of independent levels with no coherence (all matrices simultaneously diagonalized), the equation reduces to a scalar version of Landauer's equation shown at the bottom.*

11.4   Formal definition of one-particle Green's functions

We defined the retarded Green's function $G_{\alpha,\beta}(t,t')$ as the response at time t and 'location' α (position or orbital or whatever α represents) to a delta function perturbation at time t' and location β (Eq. (7.29)). For this to be retarded, we need $t > t'$, which is consistent with our expectations of causality. The correlation Green's function $G^n_{\alpha,\beta}(t,t')$ is the generalization of energy-resolved density matrix $\hat{\rho}$ for the electrons, with the equal time diagonal entry denoting the electron density $G^n_{\alpha,\alpha}(t,t) = n_\alpha(t)$, and the off diagonal terms with unequal indices or arguments representing the coherent superposition of the corresponding electronic states. Using bra ket notation and creation annihilation operators, we write

$$G_{\alpha\beta}(t,t') = \frac{-i\Theta(t-t')}{\hbar}\left\langle\{c_\beta^\dagger(t'), c_\alpha(t)\}\right\rangle$$

$$G_{\alpha\beta}^n(t,t') = \left\langle c_\beta^\dagger(t')c_\alpha(t)\right\rangle \tag{11.27}$$

where $\langle\ldots\rangle$ denotes a thermodynamic average that amounts to tracing over a product with the *nonequilibrium* density matrix operator $\hat{\rho}$. A similar 'formal' definition of the phonon Green's functions are

$$\Pi_{ij}(t_1, t_2) = -i\frac{\Theta(t_1-t_2)}{\hbar}\left\langle\left[x_i(t_1), x_j(t_2)\right]\right\rangle$$

$$\Pi_{ij}^n(t_1, t_2) = \left\langle x_i(t_1)x_j(t_2)\right\rangle \tag{11.28}$$

where x_i represents the displacement from equilibrium at site i. It is easy to interpret the diagonal elements of the equal time correlation GF, Π_{ii}^n as the fluctuations $\langle x_i^2\rangle$ in the oscillator coordinate about the mean.

It is instructive to work these quantities out for a free electron and a free phonon, since we defined them more intuitively earlier in energy domain through the relations $\Psi = GS_R$ and $G^n = \langle\psi\psi^\dagger\rangle$ (Eqs. (11.11), (11.15)). Let us work this out explicitly for phonons and leave the electronic case as an exercise.

From Eq. (11.28) we get the retarded phonon Green's function. To simplify, we use the quantized description (Eq. (7.11)), inverse transformed to get the real-space coordinates in the Heisenberg representation

$$x_i(t) = \sum_k \sqrt{\frac{\hbar}{2NM\omega_k}}\left[a_k e^{i(kx_i-\omega_k t)} + a_k^\dagger e^{-i(kx_i-\omega_k t)}\right] \tag{11.29}$$

Substituting and expanding, we will get four commutator terms, where we use the commutation rules (Eq. (7.13)) and get

$$\Pi_{ij}\underbrace{(t_1-t_2)}_{\tau} = -i\sum_k \frac{\Theta(t_1-t_2)}{2NM\omega_k}\left[e^{ik(x_i-x_j)-i\omega_k(t_1-t_2)} - h.c.\right] \tag{11.30}$$

The angular averages $\langle\ldots\rangle$ do not play a role here, as the commutators of the a and $a^\dagger$ operators give us simple algebraic constants.

We can now Fourier transform the time difference τ into frequency and thence energy. To get the Fourier transform and frequency-dependent phonon Green's functions, we need to put a complex correction to the frequency, i.e., replace $\omega \to \omega + i\eta$, where η is a vanishingly small positive number. This makes the exponents $e^{i\omega\tau}$ in each term within the square brackets above pick up an extra factor $e^{-\eta\tau}$ which vanishes at the upper limit of $\tau = \infty$ and makes the Green's function well behaved. The result

$$\Pi_{ij}(\omega) = \int_{-\infty}^{\infty} d\tau G_{ij}(\tau) e^{i(\omega+i\eta)\tau}$$

$$= -i \sum_k \frac{1}{2NM\omega_k} \left[\frac{-e^{ik(x_i-x_j)}}{i(\omega+i\eta-\omega_k)} - \frac{-e^{-ik(x_i-x_j)}}{i(\omega+i\eta+\omega_k)} \right] \quad (11.31)$$

The sum over k converts to an integral over the phonon density of states $D_{ph}(\omega_k)$.

The phonon correlation Green's function (Eq. (11.28)) can also be analogously calculated. The angular averages are not always straightforward, as they depend on the density matrix over which we do the thermodynamic averaging and depend on the thermodynamic environment. However, if we assume the phonons are *incoherent*, then

$$\left\langle a_k a_l \right\rangle = \left\langle a_k^\dagger a_l^\dagger \right\rangle = 0 \quad (11.32)$$

Furthermore, if we assume the phonons are *thermalized*, i.e., in equilibrium and non-interacting, then the density matrix $\hat{\rho} = \exp\left[-\sum_i \hbar\omega_i a_i^\dagger a_i / k_B T\right]/\mathcal{Z}$, and the angular average just gives us the equilibrium phonon emission and absorption rates

$$\left\langle a_k^\dagger a_l \right\rangle = \delta_{kl} N_k, \quad \left\langle a_k a_l^\dagger \right\rangle = \delta_{kl}(N_k+1), \quad N_k = \left[e^{\hbar\omega_k/k_B T} - 1\right]^{-1}$$

$$(11.33)$$

so that

$$\Pi_{ij}^n(\tau) = \sum_k \frac{1}{2NM\omega_k} \left[(N_k+1) e^{i[k(x_i-x_j)-\omega_k\tau]} + N_k e^{-i[k(x_i-x_j)-\omega_k\tau]} \right]$$

$$(11.34)$$

The equivalent noninteracting electronic Green's functions at equilibrium are (work this out!)

$$G_{\alpha\beta}(E) = \frac{1}{E - \epsilon_\alpha + i\eta} \delta_{\alpha\beta}, \quad G_{\alpha\beta}^n(E) = f_0(E)\delta_{\alpha\beta}\delta(E - \epsilon_\alpha) \quad (11.35)$$

11.5   The nonequilibrium Green's function ($-iG^<$ or G^n) as a contour ordered formalism

It is worth looking at the formal structure of the transport equations and how they come about in general. The self energies, especially the incoming one Σ^{in}, depend on the nature of the 'source'. There is no simple prescription for writing it down under nonequilibrium conditions in a correlated system, because the equilibrium part can still be interacting and non quadratic, and Wick's decomposition can no longer be justified. We can however write down the equation for the electron occupancy and the current flow in terms of the self energies, culminating in the Keldysh equation for G^n and the Meir Wingreen equation for current I (Eq. (11.17)). Our purpose in this section is to show how these equations, derived earlier from the one electron Schrödinger equation, can be derived from the full many body Hamiltonian. The results are the same, but the formal derivation is nonetheless instructive.

In the chapter on equilibrium quantum physics, we discussed the time ordered causal Green's function that can be expanded perturbatively in the interaction strength $V(t)$ through a series of Feynman diagrams. The starting point was the Gellman Low theorem that allowed us to connect the interacting states with the known noninteracting eigenstates using the S matrix. More specifically, we started with $\left|\Psi_0\right\rangle = S(0, -\infty)\left|\Phi_0\right\rangle$ and simultaneously, $\left\langle\Psi_0\right| = \left\langle\Phi_0\right|S(\infty, 0)$. The first was justified by the act of turning on the interaction to drive the state from $\left|\Phi_0\right\rangle$ to $\left|\Psi_0\right\rangle$, while the second came from bringing it back by turning off the interaction. Note that the first equation simply guarantees that $\left\langle\Psi_0\right| = \left\langle\Phi_0\right|S^\dagger(0, -\infty)$, and it is only when we assume a return to equilibrium after the interaction is turned off, that we can justify the substitution $S^\dagger(0, -\infty) \to S(\infty, 0)$ upto a trivial cancellable phase factor. Clearly this is hard to justify for irreversible processes. Electrons escape into the contact irretrievably during current flow, phonons emitted equilibrate with a thermal bath and conformational changes could shift the channel to a long lived metastable state that does not represent such a return to equilibrium. For a nonequilibrium system, the $t = \infty$ state is unknown and we cannot use that as a reference.

The Keldysh approach bypasses that unspecified time instant by folding the time axis backwards to revisit the $t = -\infty$ solution $\left|\Phi_0\right\rangle$, but keeping tabs

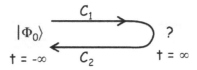

Fig. 11.5

on which contour we sit on. In other words, we now have a contour ordered
Green's function with contour indices τ, τ' (compare with Eq. (7.28))

$$G_C(x,\tau,x',\tau') = \frac{-i}{\hbar} \frac{\left\langle \Phi_0 \left| S_C(-\infty,-\infty) T_C \psi(x,\tau)\psi^\dagger(x',\tau') \right| \Phi_0 \right\rangle}{\left\langle \Phi_0 \left| S_C(-\infty,-\infty) \right| \Phi_0 \right\rangle} \quad (11.36)$$

At finite temperature, we need to extend this contour further onto the
imaginary time axis, as we argued earlier.

There now are four possible ways to place the space time pairs on the
two branches of the contour C_1 and C_2, leading to four Green's functions
(only two of which are independent). These are in turn the hole correlation
function G^p, electron correlation function G^n (also called the nonequilibrium Green's function), time ordered Green's function G_T and antitime
ordered Green's function $\bar{G}_T$.

$$G_C(x,\tau,x',\tau')$$
$$= \begin{cases} \frac{-i}{\hbar}\left\langle \psi(\tau)\psi^\dagger(\tau') \right\rangle = \frac{-i}{\hbar}G^p(x,t,x',t') = G^>(x,t,x',t') & \text{if } \tau \in C_2, \tau' \in C_1 \\ \frac{i}{\hbar}\left\langle \psi^\dagger(\tau')\psi(\tau) \right\rangle = \frac{i}{\hbar}G^n(x,t,x',t') = G^<(x,t,x',t') & \text{if } \tau \in C_1, \tau' \in C_2 \\ G_T(x,t,x',t') & \text{if } \tau,\tau' \in C_1 \\ \bar{G}_T(x,t,x',t') & \text{if } \tau,\tau' \in C_2 \end{cases}$$
$$(11.37)$$

Note that $G_T = \theta(t-t')G^> + \theta(t'-t)G^<$, $\bar{G}_T = \theta(t'-t)G^> + \theta(t-t')G^<$,
and $G^R = \theta(t-t')\left(G^> - G^<\right)$, $G^A = \theta(t'-t)\left(G^< - G^>\right)$. This means
$G^R - G^A = G^< - G^>$.

The key simplification at this stage is the existence of a set of Langreth
rules for simplifying products of Green's functions on the contour. Given
a product AB that we wish to contour order, we can look at all possible

locations of their arguments on the two contours and rewrite them as above. The rules are quite simple

<u>Langreth Rules</u>

$$
\begin{aligned}
(AB)^R &= A^R B^R, \quad (AB)^A = A^A B^A \\
(AB)^< &= A^< B^A + A^R B^<, \quad (AB)^> = A^> B^A + A^R B^> \\
(A \otimes B)^< &= A^< B^<, \quad (A \otimes B)^> = A^> B^> \\
(A \otimes B)^R &= A^< B^R + A^R B^< + A^R B^R
\end{aligned}
$$

(11.38)

where a product implies a matrix product along the time axis, $\left[AB\right](t,t') = \int d\tau A(t,\tau)B(\tau,t')$, while a direct product implies an element by element multiplication along time, $\left[A \otimes B\right](t,t') = A(t,t')B(t,t')$.

As an exercise, let us derive the second equation, keeping track of all possible locations of the intermediate time coordinate. Since we are looking at a 'lesser' function, the first time index must belong to the forward contour C_1 and the second on the returning contour C_2.

$$
\begin{aligned}
(AB)^<(t_1,t_2) &= \int d\tau_3 A(t_1,\tau_3)B(\tau_3,t_2) \qquad \text{(Matrix product)} \\
&= \int_{\tau_3 \in C_1} d\tau_3 A(t_1,\tau_3)B^<(\tau_3,t_2) + \int_{\tau_3 \in C_2} d\tau_3 A^<(t_1,\tau_3)B(\tau_3,t_2)
\end{aligned}
$$

(11.39)

We now vary the placement of τ_3, deforming the contour it sits on so it is separated from t_1 or t_2, whereupon the unindexed terms are forced to become either a lesser or a greater function. For instance in the first term, if τ_3 is to the left of t_1, we have a 'greater' A function, while if it is to the right, we turn the contour back towards $-\infty$ from a time instant between t_1 and t_3 so it is now a 'lesser' function, giving us

$$
\begin{aligned}
\int d\tau_3 A(t_1,\tau_3)B^<(\tau_3,t_2) &= \int_{t_3=-\infty}^{t_1} dt_3 A^>(t_1,t_3)B^<(t_3,t_2) \\
&\quad + \int_{t_3=t_1}^{-\infty} dt_3 A^<(t_1,t_3)B^<(t_3,t_2) \\
&= \int_{-\infty}^{\infty} dt_3 A^R(t_1,t_3)B^<(t_3,t_2) = A^R B^< \quad (11.40)
\end{aligned}
$$

using one of the identities spelt out after Eq. (11.37). The second integral $\int_{\tau_3 \in C_2} d\tau_3 A^<(t_1, \tau_3) B(\tau_3, t_2)$ becomes $A^< B^A$, so that we get the Langreth rule on the second line of Eq. (11.38).

After this laborious exercise, culminating in the Langreth rule, let us look at two quick applications.

11.6   Case study: rederiving Keldysh equation

A straightforward repeated application of these rules to the Dyson equation (Eq. (7.32)) gives us the Keldysh equation.

$$G^< = G_0^< + \underbrace{(G_0 \Sigma G)^<}_{\text{Use Eq. (11.38)}}$$

$$= G_0^< + G_0^< \Sigma^A G^A + G_0^R \Sigma^< G^A + G_0^R \Sigma^R G^< \quad \text{(keep using Dyson equation)}$$

$$= \dots$$

$$= (1 + G^R \Sigma^R) G_0^< (1 + \Sigma^A G^A) + G^R \Sigma^< G^A \tag{11.41}$$

We can pull out G^R from the first factor on the right as $G^R\big[(G^R)^{-1} + \Sigma^R\big] = G^R\big[EI - H_0 - \Sigma^R + \Sigma^R\big] = G^R(G_0)^{-1}$ sitting before $G_0^<$. If we assume that the system was noninteracting in the past, then we can use H_0 and apply it to $(i\hbar\partial/\partial\tau - H_0)G_0(\tau, \tau') = \delta(\tau - \tau')$, implying $(G_0)^{-1}G_0^< = 0$ (because the lesser has different time indices that do not satisfy the delta function). Thereupon the entire first term to the right of Eq. (11.41) (quantifying initial condition) drops out, and we are left with the familiar Keldysh equation, (Eq. (11.17) second line), but written slightly differently in our new notation

$$\boxed{\begin{array}{c} \underline{\text{Keldysh Equation}} \\ G^< = G^R \Sigma^< G^A \end{array}} \tag{11.42}$$

or in our old notation, $G^n = G\Sigma^{\text{in}}G^\dagger$.

11.7   Case study: anharmonic phonon self-energy

Let us now work out the self-energy for phonon–phonon scattering, such as due to anharmonic processes, and see how the Langreth rules emerge once again. A proper treatment of thermal current in NEGF requires such 3 and 4 phonon processes to keep the current from blowing up. The Hamiltonian can be written as

$$H = \frac{1}{2}\sum_i M_i \dot{x}_i^2 + \sum_{ij} K_{ij} x_i x_j + \sum_{ijk} V_{ijk} x_i x_j x_k \qquad (11.43)$$

which gives us the following equation of motion

$$M_i \ddot{x}_i + \sum_j K_{ij} x_j = S_i(t) \qquad (11.44)$$

where the 'source' term due to phonon–phonon scattering is

$$S_i = \sum_{jk} V_{ijk} x_j x_k \qquad (11.45)$$

From this source, we can easily calculate the in scattering function (Eq. (11.16))

$$\Sigma_{ij}^{\text{in}}(t_1, t_2) = \left\langle S_k(t_1) S_l^\dagger(t_2) \right\rangle = \sum_{klpq} V_{ikl} V_{jpq}^* \left\langle x_k(t_1) x_l(t_1) x_p^\dagger(t_2) x_q^\dagger(t_2) \right\rangle$$

$$(11.46)$$

We now will factorize the product into binary averages $\sim \langle xx^\dagger \rangle$ (Born approximation), once pairing k with p and l with q, and the other time pairing k with q and l with p. This decomposition is allowed by Wick theorem as our unperturbed Hamiltonian is quadratic. The two pairings yield the same sign and give an overall factor of 2. Looking at Eq. (11.28), we see that the paired terms give $\langle x_k(t_1) x_p^\dagger(t_2) \rangle = \Pi_{kp}^n(t_1, t_2)$. At steady state the only quantity that matters is the time difference $\tau = t_1 - t_2$, whence we get the nonequilibrium equation

$$\Sigma_{ij}^{\text{in}}(\tau) = 2 \sum_{klpq} V_{ikl} V_{jpq}^* \Pi_{kp}^n(\tau) \Pi_{lq}^n(\tau) \qquad (11.47)$$

This equation could have been written down directly by inspection from the Langreth rule (Eq. (11.38)) for phonons, $\left[\Pi \otimes \Pi\right]^< = \Pi^< \Pi^<$. Since this gives us products in τ space, in frequency domain, we get a convolution

$$\Sigma_{ij}^{\text{in}}(\omega) = 2 \sum_{klpq} V_{ikl} V_{jpq}^* \int d\omega' \Pi_{kp}^n(\omega') \Pi_{lq}^n(\omega - \omega')$$

$$\Sigma_{ij}^{\text{out}}(\omega) = 2 \sum_{klpq} V_{ikl} V_{jpq}^* \int d\omega' \Pi_{kp}^p(\omega') \Pi_{lq}^p(\omega - \omega') \qquad (11.48)$$

We can describe the equations with Feynman diagrams (Fig. 11.6). We can then use $\Gamma = \Sigma^{\text{in}} + \Sigma^{\text{out}}$, $A = \Pi^n + \Pi^p$, and the Meir Wingreen equation (Eq. (11.17)) to calculate the effect of anharmonicity on the equilibrium phonon density of states as well as the nonequilibrium phonon current.

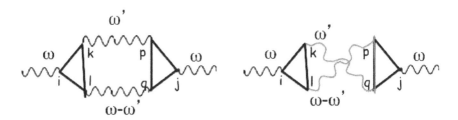

Fig. 11.6   *Three phonon Feynman diagrams.   The left corresponds to* $\Sigma^{in}_{ij}(\omega)$ *in Eq. (11.48), while the right amounts to swapping two indices.   The self energies have equal magnitude and give a factor of 2 in Eq. (11.48) once we use Langreth rules for the Green's functions.*

Chapter 12

Simple limits: the Landauer equation

In this and subsequent chapters, we simplify the NEGF equations to show their implications in various limits. In the coherent ballistic limit in absence of interactions, we get the Landauer theory that interprets conductance as the transmission at the Fermi energy. This is in contrast to the Kubo formula operative at low bias, where we interpret conductance as current noise. The Landauer viewpoint leads to quantization of electric and thermal conductance, and interference effects such as Fano oscillations. In dirty samples, the emergence of scattering within the channel leads to Ohm's law and the Drude theory for conduction. In presence of scattering, we also recover the Marcus theory for chemical reactions, Variable Range Hopping for amorphous semiconductors, and Frenkel Poole for molecular solids.

12.1 The Landauer viewpoint: conductance is transmission

The previous chapters laid out the mathematical foundations for quantum transport, and their relation to classical transport. Our aim in this chapter is to explore the physics of quantum transport at its various limits.

The most general equation we wrote down in the previous chapter was the NEGF Meir Wingreen equation for current flow (Eq. (11.17)). That equation works for an interacting system with scattering, and is limited by our ability to write down a suitable inscattering function $\Sigma^{\text{in}} = \left\langle SS^{\dagger} \right\rangle$ which may need to be non-perturbative in the interaction parameter (we discuss the challenges in the chapter on correlation). However for a simple noninteracting two terminal system, the maths simplifies considerably and we get the Landauer equation (Eq. (11.5)). The Landauer equation simply connects the low bias conductance with the quantum mechanical transmission function at the Fermi energy, $G \propto T(E_F)$. At higher bias, we integrate the transmission across an energy window over which the two contact Fermi functions differ (Fig. 8.2, between the shaded bars). The window arises from Pauli exclusion, which blocks current flow in and out of the contacts outside the energy window. This is easy to see. When an energy level lies within the window, one contact wants to fill it and the other wants to empty it and a current keeps flowing. For levels lying outside this window, the Fermi functions agree, the level stays filled or empty and there is no current. At zero temperature,

$$I = I_0 \int_{\mu_2}^{\mu_1} T(E)dE \qquad (12.1)$$

where $I_0 = q/h$ (Eq. (11.5)), and $\mu_{1,2}$ are the bias-separated quasi-Fermi levels inside the two contacts, $\mu_1 = \mu_2 + qV$.

The above equation is remarkable in that it connects a measured transport quantity, the conductance $G = \partial I/\partial V$, with a quantum mechanical quantity, its transmission probability T summed over a certain energy window. In other words, the equation describes the conductor much like a wave-guide with specific conducting modes. We will see shortly that this approach still manages to recover our familiar Ohmic behavior for large diffusive conductors. However it's main strength manifests for small systems where it generates interesting quantum results such as conductance quantization, resonances and antiresonances, tunneling and interference effects, all of which have been observed experimentally.

It is straightforward to generalize the Landauer equation to a multimoded conductor at finite temperature:

(a) Sum over the individual transmissions of each mode 'i' lying within the energy window;

(b) convolve with a thermal broadening function $F_T(E) = -\partial f/\partial E$

$$I = I_0 \int_{\mu_2}^{\mu_1} \sum_i T_i(E) \otimes F_T(E)dE \qquad (12.2)$$

where $f(E - E_F) = 1/\left(1 + e^{(E-E_F)/k_B T}\right)$ is the Fermi-Dirac distribution, which looks like a step function around the electrochemical potential E_F

at zero temperature, and gets smeared over an energy range $\sim k_B T \approx 25$ meV at room temperature. The convolution ends up creating a thermally broadened window set by a difference of Fermi functions, i.e.,

$$\int_{\mu_2}^{\mu_1} \ldots \otimes F_T(E) dE = \int_{-\infty}^{\infty} \ldots \left[f_1(E) - f_2(E) \right] dE \qquad (12.3)$$

where $f_{1,2}(E) = f(E - \mu_{1,2})$. We write the mode summed transmission as

$$\sum_i T_i(E) = \underbrace{\frac{\sum_i T_i(E)}{M(E)}}_{\bar{T}(E)} M(E) = \bar{T}(E) M(E) = \bar{T}_p M_p + \bar{T}_e M_e \qquad (12.4)$$

where $\bar{T}(E)$ is the mode-averaged transmission that lies between zero and unity. The subscripts 'p' and 'e' indicate that in general we need to sum over all propagating as well as evanescent (tunneling) modes in the system. So the final result can be cast in the form of **the Landauer equation** (Eq. (11.5))

Landauer Equation

$$I = I_0 \int_{-\infty}^{\infty} dE \bar{T}(E) M(E) \left[f_1(E) - f_2(E) \right] \qquad (12.5)$$

In evaluating the transmission through a clean ballistic channel, we will only look at the quantum evolution of the electron, and ignore any thermal effects, based on the reasoning we presented earlier, i.e., for small devices, there is not enough time to scatter inside. In contrast to our rather intricate mathematical juggleries for the classical Fokker-Planck and drift-diffusion equation, the formal derivations here are a lot simpler, with the thermodynamics existing merely in the contacts, and that too rather trivially. We simply assume that the contacts are unperturbed by the electron flow and stay pinned at a local Fermi-Dirac equilibrium distribution. In other words, we dictate the contacts act as 'reservoirs', which implicitly means they have a lot of modes, and a lot of inelastic scattering among their modes to thermalize and thus locally equilibrate the electrons.

The difference between a crystalline solid and an organic molecule is in the nature of the modes. For solids with minimal boundary effects, the modes are propagating Bloch states, perhaps zone-folded at their unstrained and passivated boundaries. For molecules, the suitable modes are their atomic or molecular orbitals. It is also worth mentioning that the modes that contribute to conduction are usually their propagating modes inside the bands. Evanescent modes inside the bandgap arise from tunneling and make negligible contributions to current, except in low bandgapped materials like graphene.

12.2 'Quantum' Ohm's law: conductance quantization

A significant achievement of the Landauer theory was in its explanation of the quantized two-terminal resistance (and its extensions such as the quantized thermal conductance). The quantized conductance arises trivially from the Landauer equation, considering a single mode with two spin channels ($M = 2$) and ballistic transport ($T = 1$),

$$I = \frac{q}{h} \int_{-\infty}^{\infty} dE \underbrace{T}_{=1} \underbrace{M}_{=2} \left[f_1 - f_2 \right] = \frac{2q}{h} \underbrace{(\mu_1 - \mu_2)}_{qV} = \frac{2q^2}{h} V \qquad (12.6)$$

which gives us the quantized conductance $G = \partial I / \partial V$

$$\boxed{\begin{array}{l} \text{Conductance Quantum} \\ G_0 = 2q^2/h \approx 77\mu A/V \end{array}} \qquad (12.7)$$

and a corresponding quantized resistance $R_0 = 1/G_0 \approx 12.9$ kΩ. The results undergo a minor change at finite temperature. As we discussed earlier, the Landauer equation (Eq. (12.5)) can be written in a slightly different way, as a convolution of the *zero-temperature conductance* with a thermal broadening function $F_T(E) = -\partial f(E)/\partial E$ evaluated around the equilibrium Fermi energy.

$$\boxed{I_1 = \frac{q}{h} \int_{\mu_2}^{\mu_1} dE \left[\bar{T}(E)M(E) \otimes F_T(E) \right]} \qquad \begin{array}{l} \text{(different form of Landauer} \\ \text{equation)} \end{array}$$

$$(12.8)$$

This equation can be deduced easily by writing the Fermi function as

$$f(E - \mu) = \int_{-\infty}^{\infty} dE' f(E' - \mu)\delta(E' - E)$$

$$= \underbrace{f(E' - \mu)\Theta(E' - E)|_{-\infty}^{\infty}}_{=0}$$

$$- \int_{-\infty}^{\infty} dE' \frac{\partial f(E' - \mu)}{\partial E'}\Theta(E' - E) \quad \text{(integrate by parts)}$$

$$= F_T(E) \otimes \Theta(E - \mu) \quad \text{(after variable change } E' \to E' + \mu)$$

$$\tag{12.9}$$

with $F_T = -\partial f/\partial E$. Plugging this into Landauer equation, we get

$$I_1 = \frac{q}{h} \int_{-\infty}^{\infty} dE \bar{T}(E)M(E)\Big[\Theta(E - \mu_1) - \Theta(E - \mu_2)\Big] \otimes F_T(E) \tag{12.10}$$

which leads to Eq. (12.8).

At zero temperature, the zero-bias resistance can be written as

$$R = \Big([\partial I/\partial V]_{V=0}\Big)^{-1} = \frac{h}{q^2 \bar{T} M_0} \tag{12.11}$$

where $\bar{T}$ is the average transmission per mode at the Fermi energy and M_0 is the number of modes. The equation can be rewritten in a more instructive way that separates the contact resistance from the resistance intrinsic to the channel (Fig. 18.8). We get the equation promised at the start of the book (Eq. (1.2))

$$R = \underbrace{\frac{h}{q^2 M_0}}_{\text{contact}} + \underbrace{\frac{h}{q^2 M_0}\Big(\frac{1 - \bar{T}}{\bar{T}}\Big)}_{\text{channel}} \tag{12.12}$$

The channel resistance arises from scattering processes inside the channel and contributes to the conventional Ohm's Law. The contact resistance, which is most prominent for near ballistic channels, arises from the mode narrowing as electrons from the numerous modes in the contact reservoirs need to squeeze through the fewer modes in the channel.

From the total resistance (Eq. (12.11)), we can then extract the two-terminal conductance $G_{2t} = G_0 M_0 \bar{T}$, while from the channel resistance in Eq. (12.12), we can get the four-terminal conductance $G_{4t} = G_0 M_0 \bar{T}/(1 - \bar{T})$, with $G_0 = q^2/h$.

How do we estimate the number of modes M_0? For a macroscopic object, they can be estimated in k-space, by chopping up the cross-sectional area πk_F^2, into little blocks of area $2\pi/L_x \times 2\pi/L_y = 4\pi^2/S$. Here $S = a^2$ is related to the size of the cross section, giving us $M_0 = Sk_F^2/4\pi$. This leads to the finite cross section **Sharvin conductance**

$$G_{\text{Sharvin}} = \frac{2q^2}{h}\left(\frac{k_F a}{2}\right)^2 \qquad (12.13)$$

This result assumes a circular cross section, and needs to be modified to account for other cross sectional geometries.

12.3   Quantum Fourier's law

Through an analogous process, we can calculate the thermal current in the Landauer formula, using the energy E as the equivalent thermal 'charge' carried primarily by phonons in insulators. Given a temperature differential across the insulator, each phonon mode carries an energy $E = \hbar\omega$ and has a nonequilibrium occupancy driven by the separate Bose-Einstein distributions $N_{1,2} = 1/\left[e^{E/k_B T_{1,2}} - 1\right]$ in the contacts. Through a process analogous to electronic current, we get the heat current

$$I_Q = \frac{1}{h}\int dE\, E\bar{T}(E)M(E)\left[N_1 - N_2\right] \qquad (12.14)$$

where the difference in N is because of the temperature difference ΔT across the contacts. Again for a single ballistic mode, $\bar{T} = M = 1$, we can simplify the equation significantly. $N_1 - N_2 = \left(\partial N/\partial T\right)\Delta T = \left(Ee^{E/k_B T}/\left[e^{E/k_B T} - 1\right]^2\right)\Delta T/k_B T^2$, which gives us the thermal conductance quantum $k_0 = I_Q/\Delta T$

$$\boxed{\begin{array}{c} \underline{\text{Thermal Conductance Quantum}} \\ k_0 = \pi^2 k_B^2 T/3h \approx 0.95\ pW/K^2 \end{array}} \qquad (12.15)$$

The extension of Landauer theory to heat flow, however, is a bit more delicate, because the natural separation between contacts and channels that we mandated for electrons, driven by a large difference in mode density, is harder to justify for thermal flow (I acknowledge Prof. Ali Shakouri for bringing this to my notice). Contrary to electrical conductivity that has a published range of 19 orders of magnitude, thermal conductivity spans only 4, meaning that the superficial separation between equilibrium reservoirs and nonequilibrium channel is harder to justify. The bosonic nature of phonons also means that there is no natural 'window' function of the quantity $N_1 - N_2$, making thermal transport inherently broad band and thus rather involved, barring a few special conditions. On top of that, thermal conductances depend on anharmonic phonon-phonon scattering (Section 11.7), impurity (Eq. (22.34)) and boundary scattering. This is typically captured by adding the scattering rates as parallel conducting channels to generate the popular Callaway model,

$$\tau^{-1} = \underbrace{vL^{-1}}_{\text{boundary scattering}} + \underbrace{BT^3\omega^2}_{\text{3 phonon}} + \underbrace{A\omega^4}_{\text{Klemmens impurity scattering}} \qquad (12.16)$$

12.4 Kubo formula: conductivity is noise

Physicists often focus on the low bias transport coefficients such as conductivity, optical response or susceptibility, and build a lot of interactions into the Hamiltonian to see their effect. From a device point of view, we primarily focus on the strongly non-equilibrium regime, but it is worth digressing for a second to derive the Kubo formula, which basically connects a transport coefficient with a correlation function. Given a Hamiltonian $\mathcal{H} = \mathcal{H}_O + \hat{V}(t)$ where the second part is the interacting Hamiltonian, we can use the interaction representation (Eq. (3.20)). Using the symbol $\langle \cdots \rangle_O$ to represent averaging over $\mathcal{H}_O$, defining $S = i \int dt' \hat{V}(t')/\hbar$, and employing the identity $e^{-B} A e^{B} = B + [A,B] + [A,[A,B]]/2! + \cdots$, we get

$$\left\langle O_I \right\rangle = \left\langle e^S O e^{-S} \right\rangle_O \approx \left\langle O \right\rangle_O + \left\langle [O,S] \right\rangle_O, \quad S = i \int dt' \hat{V}(t')/\hbar \qquad (12.17)$$

We now use the substitutions for current and electric field, $O \to j_\alpha(t)$, $H_O \to \hbar\omega$, $\hat{V}(t) \to \vec{j} \cdot \vec{\mathcal{E}}(t)$, to get the Kubo formula for conductivity

$$\underbrace{\sigma_{\alpha\beta}}_{\text{conductivity}} = \frac{\delta j_\alpha}{\delta \mathcal{E}_\beta} = \sigma_{\alpha\beta}^{(0)} + \frac{i}{\hbar} \int dt' \underbrace{\left\langle \left[j_\alpha(t), j_\beta(t') \right] \right\rangle_O}_{\text{correlation}} e^{i\omega(t-t')/\hbar} \qquad (12.18)$$

The double j arose because we are evaluating the current in presence of a small perturbation $\hat{V}$ that in turn is proportional to current. The term $\sigma_{\alpha\beta}^{(0)}$ comes from the magnetic field contribution in the current, i.e., the second term in $\vec{j} = q(\vec{p} - q\vec{A})/m$, and is the Drude term $inq^2/m(\omega + i/\tau)\delta_{\alpha\beta}$ (we encounter this shortly in Eq. (13.12)). Beyond that, Eq. (12.18) says that *the low-bias conductivity is the current-current correlator* $\sim \langle j_\alpha(t) j_\beta(t') \rangle$.

The equation, a manifestation of the so-called 'fluctuation-dissipation' theorem, is particularly useful for near equilibrium 'contactless' systems described by a Hamiltonian, e.g. when describing dipole radiation. It is however important to exercise due caution when dealing with quasi-ballistic transport. After all, what stops us from using a time-dependent Hamiltonian to calculate the AC conductivity $\sigma_{ac}(\omega)$ using Kubo formula, and then simply pull out DC transport by taking $\omega \to 0$? An AC current does not need to deal with contacts and can be established locally, as in an antenna. However in practice, AC results are extracted mainly for a bulk system with a simple local Hamiltonian where the operating wavelength is small enough to ignore the contacts and their self-energies, in other words, $\lambda < L$. But this inequality amounts to $\omega = 2\pi v/\lambda > 2\pi v/L$, above a low frequency cut off set by the transit time L/v, meaning that we cannot take its zero frequency limit and expect to recover the DC conductance for a *finite* and clean ballistic system, such as its conductance quantum.

The Kubo formula finds usage in infinitely large systems for linear response near zero bias, i.e., low voltages $\hat{V}$ (note all the higher order terms like $[O, [O, S]]$ that we just dropped !). In essence, we interpret conductivity as 'equilibrium current noise'. To go to strongly interacting currents in finite systems far from equilibrium, we will need to include higher order terms $\sim \langle j_\alpha(t) j_\beta(t') j_\gamma(t'') \rangle + \dots$, bring in boundaries with self energies $\Sigma(t, t')$ and extend our analysis to a Keldysh contour to account for an unknown final state by bringing in separate correlation Green's functions $G^{n,p}$. After all the substitutions, the Kubo formula $j \sim \langle [j, \hat{V}] \rangle$ gets replaced with $j \sim [G^n, H]$ and needs the full NEGF machinery.

12.5 Explicit quantum effects: Fano interference

For much of device physics, quantum mechanics enters implicitly through an effective potential that electrons swim in. The potential determines the effective mass that limits electron mobility, and semiconducting bandgaps that set the threshold voltage and optical absorption thresholds. In recent days quantum effects on devices have been growing, for instance through quantized energy levels at heterojunction interfaces. Many effects sit in the bandstructure and can be extracted separately at equilibrium. The real strength of the Green's function approach is its ability to include quantum effects in transport. The Fano interference is a notable example (Fig. 1.1(d)).

Consider a conducting channel with a molecular adsorbate or a quantum dot in a side coupled geometry. At low temperature and in the absence of decoherence, an electron wave can flow along two pathways, a direct path of continuum channel levels forming a band, and a secondary path through the discrete levels characterizing the dot (including phonon sidebands). What we will show is that the phase difference between the channel and the dot determines the resonant-antiresonant structure of the output conductance. Since the Fermi wavelengths of the electrons depends on the local doping, this conductance is gate tunable.

The Hamiltonian of the combined channel-dot system (d: dot, ch: channel)

$$H = \begin{bmatrix} \epsilon_d & \tau \\ \tau & \epsilon_{ch} \end{bmatrix} \tag{12.19}$$

assumed to be coupled to contacts at either end by a scalar self energy $-i\gamma_{d,ch}$ determining their respective level broadenings. The *isolated* channel and contact Green's functions (Eq. (11.11)) in absence of each other are

$$g_d(E) = \frac{1}{E - \epsilon_d + i\gamma_d}, \qquad g_{ch} = \frac{1}{E - \epsilon_{ch} + i\gamma_{ch}} \tag{12.20}$$

The channel Green's function G_{ch} in presence of the dot is given by inverting, $(EI - H)^{-1}$, and extracting the bottom right corner (i.e., channel) element. As expected, it satisfies the Dyson equation (7.31)

$$G_{ch}^{-1} = g_{ch}^{-1} - \tau^2 g_d \implies G_{ch} = \frac{g_{ch}}{1 - \tau^2 g_{ch} g_d} \tag{12.21}$$

The change in channel density of states $D_{ch} = -\text{Im}(G_{ch})/\pi$ (Eq. (11.16)) before and after adsorption of the dot, with and without τ

$$\delta D_{ch} = \frac{1}{\pi} Im\left[\frac{g_{ch}}{1 - \tau^2 g_{ch} g_d} - g_{ch}\right] = \frac{1}{\pi} Im\left[\frac{\tau^2 g_{ch}^2}{(g_d^{-1} - \tau^2 g_{ch})}\right] \tag{12.22}$$

We define the spectral functions, level broadenings and phase angles

$$a_{ch} = -\frac{1}{\pi}\text{Im}(g_{ch}), \qquad a_0 = a_{ch}\frac{\gamma_0}{\gamma}$$

$$\gamma_0 = \tau^2\pi a_{ch}, \qquad \gamma = \gamma_d + \gamma_0$$

$$q_0 = \frac{E - \epsilon_{ch}}{\gamma_{ch}} = -\frac{Re(g_{ch})}{Im(g_{ch})}, \qquad \xi = \frac{E - \epsilon_d - \gamma_0 q_0}{\gamma} \qquad (12.23)$$

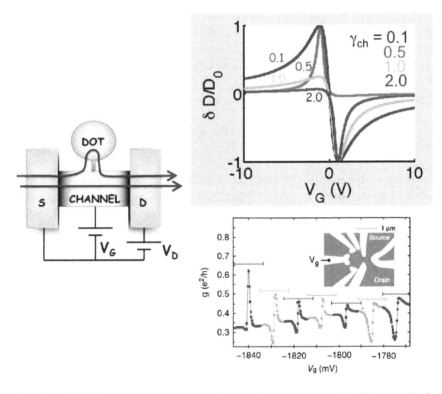

Fig. 12.1   *(Left) Fano interference occurs due to coherent superposition between a local-ized and a delocalized conducting state, such as at the interface between a quantum dot and a quantum wire in a side-coupled geometry. Depending on the ratio of the phase angles the conductance shows peaks, valleys or a resonant-antiresonant pair (top right), as seen in experiments (top right), reproduced from A. C. Johnson, C. M. Marcus, M. P. Hanson and A. C. Gossard, 'Coulomb Modified Fano-Resonance in a One-Lead Quantum Dot', Phys. Rev. Lett. 93, 106803 (2004).*

so that $g_{ch} = \pi a_{ch}(q_0 - i)$, Using $g_d^{-1} = (E - \epsilon_d) + i\gamma_d$, we then get

$$\delta D_{ch} = \tau^2 \pi Im \left\{ \frac{a_{ch}^2(q_0 - i)^2}{\underbrace{(E - \epsilon_d - \tau^2 \pi a_{ch} q_0)}_{\xi\gamma} + i\underbrace{(\gamma_d + \tau^2 \pi a_{ch})}_{\gamma}} \right\}$$

$$= \underbrace{\frac{\tau^2 \pi a_{ch}^2}{\gamma}}_{a_0} \left[1 - \frac{(q_0 + \xi)^2}{\xi^2 + 1} \right] \tag{12.24}$$

Equation (12.24) describes the resonant-nonresonant superposition of the two transport paths. Depending on the value of the renormalized phase angle ξ, the spectral function shows resonance, antiresonance or a combination. These features are gate tunable and observed experimentally (Fig. 12.1).

Chapter 13

Classical Ohm's law: the Drude formula

It is well recognized that the Landauer equation in its original form describes the coherent propagation of quantum waves. What is not as well appreciated is how it can also describe classical particulate transport. The relation between the two limits arises, rather elegantly, by the multiple scattering contributions to the mode averaged transmission $\bar{T}$. Indeed, we can play with the shape of the transmission function to capture all sorts of scattering processes and various ways of averaging over them. *We thus have a unified set of equations that work from quantal to classical limits.*

The fundamental equation for current flow that we invoked was $J = (\sigma/q)d\mu_n/dy$ (Eq. (10.3)). This equation is true regardless of whether the system is dominated by drift or diffusion. At equilibrium, the quasi-Fermi level or electrochemical potential must be spatially invariant. We then split this into a term that depends on the variation of $\mu_n - U$, the *chemical potential* (and thus variation of charge density), and a term proportional to the variation of $-U/q$, the *electrostatic potential*. Identifying the former as diffusion and latter as drift, we derived the generalized Einstein's relation for conductivity (Eq. (10.6)).

The equilibrium electron density in a material can be obtained by an equation that is almost as simple as the Landauer approach. At zero temperature, it is just the sum over the occupied states

$$n_0 = \int_{-\infty}^{E_F} D(E)dE \tag{13.1}$$

where $D(E)$ is the density of states in the channel, in units of $/eV/cm^3$. We can easily generalize to finite temperature with a Fermi-Dirac distribution

$$n_0 = \int_{-\infty}^{\infty} D(E)f(E - E_F)dE \qquad (13.2)$$

At non-equilibrium, we will have a position dependent quasi-Fermi energy $\mu_n(y)$, and a position dependent bandstructure varying with a potential $U(y)$ that locally shifts the densities of states, $D(E, y) = D(E - U(y))$.

$$n(y) = \int_{-\infty}^{\infty} D\Big(E - U(y)\Big)f\Big(E - \mu_n(y)\Big)dE$$

$$= \int_{-\infty}^{\infty} D\Big(E\Big)f\Big(E - \mu_n(y) + U(y)\Big)dE \quad \text{(change of variable)}$$

$$(13.3)$$

Comparing with Eq. (10.4) to identify F, i.e., $n = F(|\mu_n - U|)$, we can then rewrite the generalized Einstein's equation $\sigma = q^2 \mathcal{D} dF/d\mu_n$ to give us

$$\boxed{\sigma = q^2 \mathcal{D} \int_{-\infty}^{\infty} D(E)\Big(-\frac{\partial f(E)}{\partial E}\Big)dE} \qquad (13.4)$$

The significance of this equation is that it directly describes the transport properties, i.e., the conductivity of a sample, as a Fermi-surface property, since $-\partial f/\partial E$ is non zero only near the Fermi energy. It is not the total electron count proportional to f that matters, but rather the dynamics of the electrons near or above the Fermi energy that determines current. This would explain why the conductivity of glass and copper are vastly different. They have comparable electron densities, but for copper, many electrons lie near their Fermi energy and are free, while for glass, most of them lie far from the Fermi energy (which lies inside a large band-gap), so that all the electron occupied states see a constant Fermi energy $-\partial f/\partial E \approx 0$, and are not 'free' to conduct.

For a *single band effective mass model*, there is a one-to-one correspondence between what happens at the Fermi energy and what happens at the band-bottom, so we can recast the problem in terms of the total electron number rather than a Fermi surface property itself. For a non-degenerate semiconductor for example, $f(E) \approx \exp[-(E - E_F)/k_B T]$ so that $-\partial f/\partial E = f(E - E_F)/k_B T$. This means that we can rewrite the conductivity as

$$\sigma = q^2 \frac{\mathcal{D}}{k_B T} \underbrace{\int_{-\infty}^{\infty} D(E) f(E - E_F) dE}_{n} \quad \text{(non-degenerate)} \qquad (13.5)$$

which now makes the total n paradoxically relevant, but actually only so *because of the one-to-one relation between the total number of electrons, and the more relevant electron count at the Fermi energy.* Writing $\sigma = qn\mu$, where μ is the mobility, we get our text-book Einstein relation for non-degenerate semiconductors, $\mathcal{D}/\mu = k_B T/q$.

Note that classical drift-diffusion equation can be obtained directly from the Landauer equation, albeit with some care! Later in the chapter on scattering, we will prove the *additivity of the channel resistance* $\propto (1 - \bar{T})/\bar{T}$ in Eq. (22.5), writing it as $\Delta z/\lambda$, where λ is the mean free path of momentum scattering. If we now apply the Landauer equation between two adjacent segments of a channel separated by Δz, we must ignore the contact resistance and use the four-terminal channel resistance $(1 - \bar{T})/\bar{T} = \Delta z/\lambda$ instead (see discussions on G_{4t} after Eq. (12.12)). We also consider a variation in electrochemical potential from $\mu_n(z)$ to $\mu_n(z + \Delta z)$ imposed by scattering. We then get

$$I = I_0 \int_{-\infty}^{\infty} dE \; \underbrace{\frac{\lambda}{\Delta z}}_{\bar{T}/(1-\bar{T})} \; M(E) \left[f\left(E - \mu_n(z + \Delta z) \right) - f\left(E - \mu_n(z) \right) \right]$$

$$= I_0 \int_{-\infty}^{\infty} dE \lambda M(E) \frac{\partial f}{\partial \mu_n} \frac{\partial \mu_n}{\partial z}$$

$$= S \frac{\partial \mu_n}{\partial z} \underbrace{I_0 \int_{-\infty}^{\infty} dE \lambda \frac{M(E)}{S} \left(-\frac{\partial f}{\partial E} \right)}_{\sigma/q} \qquad (13.6)$$

where σ was defined from Eq. (10.3) using $J = I/S = (\sigma/q)\partial\mu_n/\partial z$. Using $I_0 = q/h$, and comparing with our definition of σ in Eq. (13.4), we get the relation $D(E)\mathcal{D} = (1/h)\lambda M(E)/S$. Both diffusion along the transport direction and mode count are related,

$$\mathcal{D} = \langle v_z^2 \rangle \tau$$

$$M = \underbrace{g}_{\text{degeneracy}} \times h \langle |v_z| \rangle D(E) S = g \frac{h \langle |v_z| \rangle}{L} D_T(E) \qquad (13.7)$$

where $\langle \cdots \rangle$ denotes an angular average over the relevant dimensions, and g is the degeneracy (e.g., 2 for spins, 2 more for valley, another 2 for bilayer graphene, assuming these degeneracies are not accounted for in the density of states $D(E)$ per unit volume, or the total volume integrated density of states $D_T(E)$).

It is worth emphasizing that the mode count, which can be shown to be an integer for a system with discrete sub-bands, depends not just on the total density of states D_T, but in fact on the states that have finite velocity v. This is critical, because *localized states, such as from traps, may contribute to the density of states, but will have no direct contribution to the transmitted current.*

For a 1-D ballistic channel, it is easy to show that the density of states per spin is $D_T(E) = L/2\pi\hbar v = L/hv$ and $\langle |v_z| \rangle = v$, so that $M = 1$. For a larger system that can be decomposed into 1-D subbands, we can then show readily that $M(E) = \sum_n \Theta(E - \epsilon_n)$.

We then have

$$\lambda = \frac{hDSD}{M} = \frac{D}{2\langle |v_z| \rangle} = \frac{\langle v_z^2 \rangle}{2\langle |v_z| \rangle}\tau \qquad (13.8)$$

For a large macroscopic object, the number of modes was estimated earlier (after Eq. (12.12)) as $M_0 \approx Sk_F^2/4\pi$. In addition, the transmission per mode was obtained as $\bar{T} = \lambda/(\lambda + L)$ as discussed before Eq. (13.6) (and also derived later in Eq. (22.5)). For a free electron, the Fermi wavevector can be estimated from the electron density that occupies the Fermi sphere,

$$N = \frac{\text{spin} \times \text{k sphere volume}}{\text{k space grid volume}} = \frac{2 \times 4\pi k_F^3/3}{(2\pi/L)^3} \qquad (13.9)$$

giving us $k_F = (3\pi^2 n)^{1/3}$, with n being the electron density $n = N/L^3$. From Eq. (13.8), we get

$$\lambda = (v_F\tau_{sc}/2) \int_{-1}^{1} \cos^2\theta d(\cos\theta) / \int_{-1}^{1} |\cos\theta| d(\cos\theta) = 4v_F\tau_{sc}/3 \quad (13.10)$$

in 3D. Finally, we can write $v_F = \hbar k_F/m$, so that channel resistance can be written in a form that maps onto classical Ohm's law

$$R_{\text{channel}} = \frac{L}{\sigma S} \qquad (13.11)$$

with the conductivity σ now a material-dependent parameter given by the Drude formula

$$\sigma = \frac{q^2}{h}\frac{\lambda}{\lambda_F^2} = \frac{nq^2\tau_{sc}}{m}, \qquad (13.12)$$

keeping in mind that this equation only applies to a nondegenerate material with n 'free' electrons (the more general equation being Eq. (13.4)).

13.1 Conductivity beyond diffusivity: Variable Range Hopping in amorphous semiconductors

It is tempting to say that Eq. (13.4) contains all we need to know about conductivity, but it is worth emphasizing that this still assumes through-band transport, where the transmission is given by $\lambda/(\lambda + L)$. This works for most diffusive systems, but there are notable examples where tunneling through band-gaps can still contribute to conductivity, where we will need to hark back to the original Landauer equation (Eq. (12.5)), extended to both propagating and decaying modes (Eq. (12.4)). Two examples worth mentioning are (a) variable range hopping, and (b) the minimum conductivity across graphene, both of which arise from the preponderance of tunneling over through-band transport. We will discuss graphene later in the chapter on scattering.

In amorphous semiconductors with a random onsite potential, we have in effect a random distribution of quantum wells with associated random energy levels that are usually not aligned (Fig. 13.1 left). This means that the electron will need to transport inelastically (i.e., allow a change in energy between initial and final states). Since the probability of such inelastic transport goes as $\exp(-\Delta E/k_B T)$, the electron would prefer to find levels as close to each other energetically as possible. For a random distribution, we can imagine the levels are uniformly distributed with density of states D over a band-width, $\Delta E \approx 1/(4\pi R^3/3)D$, so that over larger and larger volumes, it is likely to find levels closer to each other. This incentivizes the electron to tunnel over longer distances, but the probability of such tunneling in turn decreases according to the WKB formula (chapter on tunneling) as $\exp[-\alpha R]$, so we find a sweet spot where the product of the tunneling and inelastic probabilities is maximized. Maximizing the transmission

$$T_{\text{VRH}} = e^{-\alpha R - \Delta E/k_B T}, \qquad \Delta E = \frac{1}{(4\pi R^3/3)D} \qquad (13.13)$$

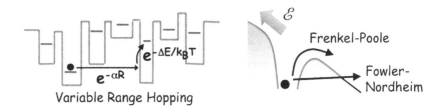

Variable Range Hopping

Fig. 13.1   *(Left) Tunneling assisted thermionic emission (Variable Range Hopping). (Right) Field assisted tunneling (Fowler-Nordheim) and Field-assisted thermionic emission (Frenkel-Poole).*

with respect to R, we get the optimized R^*, which when substituted back into the probability, gives a variable range hopping (VRH) conductivity

$$\sigma_{\mathrm{VRH}} = \sigma_0 \exp\left[-\left(T_0/T\right)^{1/4}\right] \tag{13.14}$$

The exponent is indicative of the nature of the hopping such as its dimensionality — for 2-D this becomes $1/3$, while for quasi-1D systems or systems with strong Coulomb interactions, we get $1/2$.

13.2   Frenkel Poole conduction in molecular solids

For organic molecules with a series of impurity traps, current flows by an intermittent process reminiscent of variable range hopping, except instead of a combination of tunneling and through band conduction, it arises by a combination of charge trapping and thermal emission over a barrier lowered by an electric field (Fig. 13.1). This is the thermionic analogue of Fowler Nordheim tunneling through oxides (Fig. 14.4) commonly used to write information onto the floating gates in a Flash memory or EEPROM. The average transmission over a distribution of impurities is given by

$$T = \left\langle e^{-U(r)/k_B T}\right\rangle, \quad U(r) = -\frac{q^2}{4\pi\epsilon_0 r} - q\mathcal{E}r \tag{13.15}$$

The minimum U is obtained for $r^* = \sqrt{q/4\pi\epsilon_0\mathcal{E}}$. Substituting for r^*, we get the Frenkel Poole conduction around the barrier $\phi_B = -q^2/4\pi\epsilon_0 r^*$

$$T_{\mathrm{Frenkel\ Poole}} = e^{-\left[\phi_B - \sqrt{q^3\mathcal{E}/4\pi\epsilon_0}\right]/k_B T} \tag{13.16}$$

The expression translates to a field dependent mobility of the form $\mu = \mu_0 e^{\sqrt{\mathcal{E}/\mathcal{E}_0}}$ commonly seen for organic conductors.

In both systems above (Fig. 13.1), we have a random distribution of scattering centers, and the transmission is calculated by exploiting the exponential sensitivity to evaluate the most likely dimensions for scattering, which is then substituted into the expression for conductivity. Let us now see how a specific scattering site manifests itself within the Landauer theory.

13.3 Current flow with inelastic scattering

Simple momentum scattering events such as from impurity centers can be included brute force into the Hamiltonian and evaluated with the Landauer equation. We can do a configurational averaging over various random impurity sites to get the role of disorder. To include more complicated scattering physics, such as phonons for instance, we will need to expand the scope of the Landauer equation. The full NEGF Meir Wingreen equation allows us to work out these processes. We will derive an expression for the scattering current in the chapter on scattering (Eq. (24.32)). That equation superficially looks like the Landauer equation, except it allows for virtual contacts separately vertically along the energy axis.

$$I_{sc} = I_0 \int dE M(E) \int dE' \left[T_{E \to E'} f_E \left(1 - f_{E'}\right) - T_{E' \to E} f_{E'} \left(1 - f_E\right) \right]$$
(13.17)

Note that if the transmissions were energy independent, we could just switch around the definitions of the dummy variables E and E' and the result would be zero! Similarly if the transmissions were energy symmetric, i.e., $T_{E \to E'} = T_{E' \to E}$, then we would be back to the coherent Landauer equation (Eq. (12.5)). For equilibrium phonons coupling with nonequilibrium electrons, the two processes of absorption and emission are related by a Boltzmann distribution set by the phonon temperature,

$$\frac{T_{E \to E'}}{T_{E' \to E}} = e^{-(E' - E)/k_B T_{ph}}.$$
(13.18)

Usually the energy difference is picked up by phonon absorption and emission at energy $\hbar\omega = E' - E$, so that

$$T_{E \to E'} \propto N_\omega = 1/\left[\hbar\omega/k_B T_{ph} - 1\right]$$
$$T_{E' \to E} \propto N_\omega + 1 = N_\omega e^{\hbar\omega/k_B T_{ph}}$$
(13.19)

The ratio of the backward and forward currents

$$\frac{f_{E'}\left(1 - f_E\right)T_{E' \to E}}{f_E\left(1 - f_{E'}\right)T_{E \to E'}} = e^{(E-E')\left(\frac{1}{k_B T} - \frac{1}{k_B T_{ph}}\right)} \tag{13.20}$$

which yields no scattering current where $T = T_{ph}$, i.e., when the electrons and phonons equilibrate with each other. We will use this equation to derive the inelastic phonon current later in the chapter on scattering. However, one result that is useful to derive is the Marcus transfer rate.

13.4   Marcus theory of electron transfer

When we look at the reaction rate for a chemical process, the rate constant plays the role of conductance in current flow. The Landauer theory modified for scattering gives us a natural way to estimate this process. What the equation above states is that the current flow with inelastic scattering, taking us across two different energy levels, is proportional to the probabilities of an initial filled state and a final empty state. We start with an initially filled state at high energy E and final empty state at low energy E', so that $f_E \approx 1$ and $f_{E'} \approx 0$. This occupancy differential alone should drive the transient current. Ignoring spontaneous emission, the current term in Eq. (13.17) then becomes proportional to

$$\underbrace{f_E\left(1 - f_{E'}\right)}_{\approx 1}\left[1 - \underbrace{\frac{f_{E'}\left(1 - f_E\right)}{f_E\left(1 - f_{E'}\right)}}_{e^{-(E-E')/k_B T}}\right] \tag{13.21}$$

In a chemical reaction, we consider the transient process driving us unidirectionally from reactants to products, which involves just one of the directed terms in the expression above. The scattering rate can be obtained from Fermi's Golden Rule (Eq. (11.14)) and is proportional to the scattering potential V^2 times the density of states renormalized by the polaron formation (the Franck Condon factor), giving us the electron transfer rate

$$k_{ET} \propto V^2 e^{-\Delta E/k_B T} \tag{13.22}$$

Marcus theory writes down this energy barrier in terms of the reorganization energy of a molecule, which we later identify as polaron energy. ΔE is the energy 'barrier' separating the original conformational potential of the molecule, and the reorganized structure corresponding to a displaced coordinate (we identify this as the polaronic energy later). This gives

$$k_{ET} = V^2 e^{-(\lambda - \Delta E)^2 / 4\lambda k_B T} D_{FC} \qquad (13.23)$$

where D_{FC} is the Frank-Condon density of states. A pictorial description of the entire inelastic process will be discussed later in the chapter on scattering (Fig. 24.7).

PART 4

Adventures and Applications:

Ballistic Quantum Flow

Chapter 14

The vanishing act: electron tunneling

In the eventful journey of the electron across molecules and solids, tunneling bears a distinct role as an oft perceived quantum effect. Most quantum phenomena, like interference for instance, require extreme delicacy, typically cryogenic setups so not to perturb the electron's phase. Tunneling however pervades nanoscale systems even under ambient conditions. In fact, standby power dissipation from room temperature oxide tunneling is a major limiter in transistor scaling. While many effects of quantum origin such as bandgaps, magnetism and effective mass can and often are included *ex post facto* within otherwise classical descriptions, tunneling warrants a thoroughly quantum treatment. The preponderance of tunneling in today's nanodevices makes our quantum kinetic description of current flow all the more relevant to present day industry. In this chapter, we will discuss quantum tunneling and the tunnel current across thin oxide films and molecular wires. In the next chapter, we will deal with the added complexity that arises when tunneling is further symmetry limited by a topological index. We discuss three examples where the molecular nature of the tunneling states matters. These include tunnel magnetoresistance in magnetic tunnel junctions, symmetry filtering across Fe/MgO interfaces, and finally the rich metamaterial behavior of electron waveguiding across graphene PN junctions.

14.1 Frustrated transmission and tunneling

If we plot the transmission probability of a classical object incident on a barrier, say a ball thrown at a wall, that probability plotted against energy looks like a step function — once the particle's total energy (kinetic plus gravitional potential) clears the barrier, it has a hundred percent probability of going over. If its energy lies below, it bounces back with absolute certainty. Waves, however, behave in a qualitatively different way. They tend to backscatter anytime they perceive an abrupt change in medium, whether a wall or a well. Water sloshes back when it moves from a deep sea to a shallow ocean shelf, doing so more aggressively for an abrupt rocky promontary than a graded sandy beach. In particular, if the energy of a wave falls below the wall, the wave can still penetrate into the barrier with finite probability. This is seen even with classical E-M waves, which reflect off of a metallic barrier, yet manage to propagate a small exponentially decaying tail into the metallic skin. For a thin enough metallic slab that is all skin, a thin sliver of light can make it to the other side and propagate, leading to the optical phenomenon known as frustrated transmission.

Tunneling is the quantum mechanical analog of frustrated transmission. Recall that the kinetic energy of the electron is connected to the curvature of the wavefunction, $\sim \nabla^2 \psi$. Since a finite potential barrier can only support a finite kinetic energy, the curvature must also stay finite on all space (in contrast to a box with infinitely high walls or a delta function scatterer). This means that the wave cannot vanish abruptly, but is constrained to decay gradually at a wall even if its energy turns out to be inadequate. If the barrier is thinner than this decay length, the wave emerges at the other side with finite though reduced amplitude, and continues propagating thereafter.

The seeping through of a wavefunction in itself is not surprising — as we saw above, frustrated transmission is well known in classical electromagnetism. What makes this process profound in quantum mechanics is that the probability of an electron is given by $|\psi|^2$, meaning that there is a finite non-zero probability of the entire indivisible, and to our knowledge, structureless electron, to emerge at the other side of the wall. And more intriguingly, this 'wall' could consist of atoms that are frozen at low temperature or even non-existent, like the vacuum gap in a scanning tunneling microscope, meaning that the dynamics of the barrier or its constituents is in no way complicit in the electron's act of tunneling.

It is hard to wrap our minds around the inherent randomness of each electron 'deciding' whether to tunnel or not, even while a collection of electrons conforms to an overall probability density. Not to mention the spooky act of tunneling itself, the entire undivided particle disappearing and reemerging at the other end, and the fact that the act of measurement itself influences the electron fundamentally. For instance, nailing down the location of an electron before or after tunneling, say with a flash of light, would require a high resolution and a corresponding wavelength shorter than the barrier thickness. But then the momentum $p = h/\lambda$ transferred from the photons to the electron is large enough to disrupt the measurement and promote it over the barrier even in the absence of tunneling.

And yet, technology is intimately connected to tunneling. Scanning tunneling microscopes rely on electron tunneling to image atomic surfaces. Tunneling through the gate oxide was the primary source of standby power dissipation in recent generation transistors. In fact it is possible that emerging devices like tunnel-field effect transistors may one day exploit quantum tunneling to our advantage.

The maths of transmission through a step and a barrier have been worked out in many textbooks. We will just outline the main steps here. We can write down the solutions to Schrödinger's equation in the various pieces of the potential (Fig. 14.1(a)), as superpositions of plane waves $\sim e^{\pm ikx}$ in the free regions where $U = 0$. Inside the barrier region, we use superpositions of $e^{\pm ik'x}$ when $E > U_0$ and exponentials $e^{\pm \kappa x}$ below, with $k = \sqrt{2mE/\hbar^2}$, $k' = \sqrt{2m(E-U)/\hbar^2}$ and $\kappa = \sqrt{2m(U-E)/\hbar^2} = ik'$. At the interfaces where the potential changes, we relate the superposition coefficients by matching the wavefunction and its derivative. Matching the wavefunction conserves charge density $\propto |\psi|^2$, while matching its derivative in addition conserves current density that goes as $\psi^* \nabla \psi - \psi \nabla \psi^*$. The resulting reflection coefficient, i.e., the ratio of reflected *current density* (Eq. (9.16)) to incident for a particle on a step of height U_0 is given by the mismatch between wavevectors at fixed energy E

$$R_{\text{step}} = 1 - T_{\text{step}} = \begin{cases} \left(\dfrac{k-k'}{k+k'} \right)^2 & \text{if } E > U_0 \\ \left| \dfrac{k-i\kappa}{k+i\kappa} \right|^2 = 1 & \text{if } E < U_0 \end{cases} \tag{14.1}$$

in fact, *the fractional change in velocity along the transport direction* (if effective masses were different across the step, we would be using ratios of k/m above). This is quite similar to what we get for geomet-

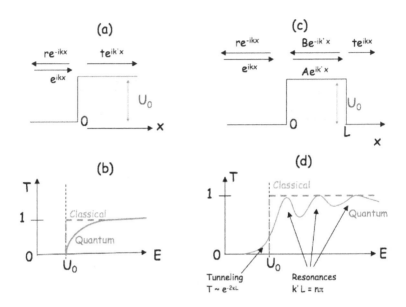

Fig. 14.1 **Left column**: *Schematic potential profile and wavefunction amplitudes (top), and corresponding transmission probability (bottom) for a particle incident on an abrupt step (T = 1 − R, with R from Eq. (14.1)). There is quantum backscattering (T < 1) even when the electron has enough energy to clear the step (i.e., E > U₀).* **Right column**: *Repeated for a particle incident on a barrier of finite width, for which the transmission probability (Eq. (14.3)) shows sweet spots ('resonances') above the barrier and a non zero tunneling probability below. The resonances arise when the width can accomodate exactly an integer number of half wavelengths (i.e., k' = nπ/L), which is the quantization condition for a particle in a box of the same width.*

rical optics, where the Fresnel equations obtained by matching the tangential component of magnetic field H and the normal component of displacement field D across a non magnetic interface give us a reflectivity $R = (Z_1 - Z_2)^2/(Z_1 + Z_2)^2$, with the electromagnetic impedance Z ultimately proportional to the velocity v (in 2 D, we get projections $Z \cos\theta$ which amounts to working just with the normal components of the velocity).

What is counter-intuitive about this equation, from a particulate or classical point of view, is that even when the energy of the electron is sufficient to clear the wall, it perceives a change in potential nonetheless and gives a finite probability of reflection. How would we make this exact quantum mechanical result consistent with our classical expectations, where $R = 0$ if $E > U_0$, and 1 if $E < U_0$? We may think that the classical result can be recovered by reducing the De Broglie wavelength, e.g. by increasing the mass m or somehow letting the Planck constant $\hbar \to 0$. However, one can easily verify that in the above equation, both m and $\hbar$ cancel out between numerator and denominator! Recovering the classical result is tricky because the lengthscale over which the potential varies here is zero. The potential is infinitely abrupt. To recover the classical result, we will need to make this potential vary no faster than the De Broglie wavelength. The concept is easily illustrated with a slightly different potential that looks like a step function smeared out over a distance x_0, similar to an inverse Fermi function $1 - f$, i.e., $U(x) = U_0/(1 + e^{-x/x_0})$, where (incredibly!) an exact solution exists, that can be simplified in the asymptotic limits to derive the following expression for reflection coefficient (the algebra is illustrative though tedious, but the result is what we focus on). See Landau-Lifschitz Vol. 3, problem 3, pages 93-94 (1981 edition).

$$R = \frac{\sinh^2 \left| \pi x_0 (k - k') \right|}{\sinh^2 \left| \pi x_0 (k + k') \right|} \tag{14.2}$$

It is easy to verify that if we now set $\hbar \to 0$ or $m \to \infty$ *before* making x_0 vanish, then the reflection coefficient R decays exponentially to zero as $\sim \exp\left[-2\pi x_0 k'\right] = \exp\left[-x_0/\lambda\right]$, thus correctly recovering our classical expectation above barrier (below barrier k' is imaginary and we get $R = 1$). Note also that this exponential has an isolated essential singularity at $\lambda = 0$. All its derivatives with λ vanish at $\lambda = 0$, meaning that we cannot do a simple Taylor expansion in λ/x_0 to extract the lowest order classical corrections.

We can extend this study to a barrier of finite width L, where the usual approach is to start with superpositions described earlier (plane waves and exponentials) and match wavefunctions and derivatives at the two interfaces. The resulting textbook expression for transmission (Fig. 14.2) is

$$T_{\text{barrier}} = \begin{cases} \dfrac{4k^2 (k')^2}{\left[k^2 + (k')^2\right]^2 \sin^2 k'L + 4k^2 (k')^2 \cos^2 k'L} & \text{if } E > U_0 \\[4mm] \dfrac{4k^2 \kappa^2}{\left[k^2 - \kappa^2\right]^2 \sinh^2 \kappa L + 4k^2 \kappa^2 \cosh^2 \kappa L} \approx \dfrac{16E(U_0 - E)}{U_0^2} e^{-2\kappa L} & \text{if } E < U_0 \end{cases}$$

$$\tag{14.3}$$

with k, k' and κ defined earlier. We get resonances for energies above the barrier height when the width can fit an integer number of half wavelengths, $k' = n\pi/L$, and tunneling (frustrated transmission) below. The

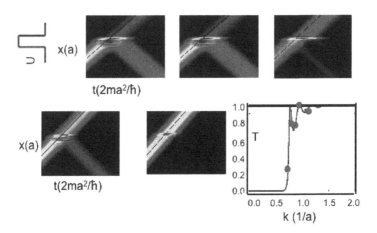

Fig. 14.2 *(Top left) One dimensional potential barrier. Colored panels show a colorplot of the electron probability density at various energies, obtained by solving the Schrödinger equation numerically. The chosen energies correspond to the k values marked in the last figure with red dots. The transmission, calculated by integrating over the transmitted current density and dividing by the incident current density, lies on the analytical expression derived in Eq. (14.3)*

exponential is easily generalized to a spatially varying potential by replacing $\kappa L \to \int_a^b \kappa(x)dx$ between the turning points of the potential, i.e., points a and b where $E = U(a) = U(b)$).

Once again, we can seek its classical limit by examining its smooth analogue, $U = U_0 / \cosh^2 (x/x_0)$. In this case, the expression for the transmission worked out (Landau-Lifschitz Vol. 3, problem 4, pages 94-95).

$T_{\text{smooth barrier}}$

$$
= \begin{cases}
\dfrac{\sinh^2\left[\pi k x_0\right]}{\sinh^2\left[\pi k x_0\right] + \cos^2\left[\dfrac{\pi}{2}\sqrt{(1 - 8mU_0 x_0^2/\hbar^2)}\right]} & \text{if } 8mU_0 x_0^2/\hbar^2 < 1 \\[4ex]
\dfrac{\sinh^2\left[\pi k x_0\right]}{\sinh^2\left[\pi k x_0\right] + \cosh^2\left[\dfrac{\pi}{2}\sqrt{(8mU_0 x_0^2/\hbar^2 - 1)}\right]} & \text{if } 8mU_0 x_0^2/\hbar^2 > 1
\end{cases}
$$

$$(14.4)$$

For $x_0 << \lambda = \sqrt{\hbar/8mU_0}$, we approach a delta function provided we scale the potential height to keep the area finite $U_0 = U_1/2x_0$ so that as $x \to 0$, $U(x) \to U_1\delta(x)$. In this case, we can Taylor expand in small x_0/λ and get the transmission

$$T_{\text{delta function}} = \underbrace{\lim_{x_0 \to 0}}\, T_{\text{smooth barrier}} = \frac{1}{1 + \dfrac{m^2 U_1^2}{\hbar^4 k^2}} \tag{14.5}$$

This equation is consistent with our expectation of scattering off of an abrupt defect. Transforming the delta function into a finite energy by using $\delta\epsilon = U_1 a$, we get the familiar Breit Wigner formula for scattering off of a point defect (see equation for total transmission of modes M from Eqs. (11.24) and (11.25))

$$\boxed{T_{\text{defect}} = T_{\text{delta function}} = \frac{1}{1 + \dfrac{\delta\epsilon^2}{\Gamma^2}}} \tag{14.6}$$

with $\Gamma = \hbar^2 k / ma = \hbar v / a$ (i.e., set by the transit time a/v), and $v = \hbar k / m$.

Unsurprisingly, we can get the same answer simply by our particle on a barrier prescription, but keeping track of a derivative discontinuity for the delta function potential.

In the opposite limit of a slowly varying potential, $x_0 \gg \lambda$, we recover the classical result

$$T_{\text{classical}} = \frac{1}{1 + \dfrac{\cosh^2 \pi k_U x_0}{\sinh^2 \pi k x_0}}$$

$$\approx \frac{1}{1 + \exp\left(2\pi x_0 k \left(\sqrt{U_0/E} - 1\right)\right)} = \begin{cases} 1 & \text{for } E > U_0 \\ 0 & \text{for } E < U_0 \end{cases} \tag{14.7}$$

where $k_U = \sqrt{2mU_0/\hbar^2}$, in the limit $x_0 \to \infty$. Note that the standard expression for a particle on a barrier (Eq. (14.3)) cannot be derived in a straightforward way out of this function, because we will need an analytically solvable smoothly varying potential with two independent length-scales, one defining the width of the barrier (which stays finite), and the other denoting the length over which the edges of the barrier gradually vary.

14.1.0.1 *Band-to-Band/Zener Tunneling*

Zener tunneling happens when an applied voltage aligns the conduction band in one region with the valence band in an adjacent region, separated by a narrow potential barrier that becomes more and more triangular and progressively transparent to tunneling. The tunneling probability through a triangular barrier can be exactly solved using Airy functions, but it is simpler to work directly with the WKB assumption. The WKB exponential term generalizes the under barrier result $T \sim e^{-2\kappa L}$ in Eq. (14.3) to a spatially varying potential $U(x)$

$$T(E) \sim e^{-2\int_a^b \sqrt{2m[U(x)-E]}dx/\hbar} \tag{14.8}$$

where a and b are the turning points between which $U(x) > E$. Linear expanding $U(x) - E \approx U_0(x-a)/(b-a)$, the integral becomes $2\sqrt{2mU_0}(b-a)/3\hbar$ and we then get

$$T_{\text{Zener}} \sim e^{-4\sqrt{2mU_0}(b-a)/3\hbar} = e^{-4\sqrt{2m}U_0^{3/2}/3\hbar\mathcal{E}_0} \tag{14.9}$$

where $\mathcal{E}_0$ is the electric field equal to $U_0/(b-a)$. The triangular barrier gives a tunneling contribution similar to a square barrier, albeit at two thirds of the maximum height, $2U_0/3$. For a silicon dioxide film of thickness 1 nm and an applied voltage 1 V across it, this tunneling probability equals 0.001.

14.2   Tunneling through a thin film

The transmission expressions for a barrier are mainstream in quantum mechanics textbooks, and the WKB exponential variation is used widely. However, what is usually not discussed is the actual current carried by these transmitting states. To make that connection, we will now invoke Landauer's theory, which essentially requires us to sum the transmissions over all modes within an energy window set by the quasi-Fermi levels in the contacts. The total current is highly relevant to gate controlled structures like MOSFETs, where the scaling of the oxide has historically limited the scaling of transistor technology.

14.2.1   *Current through a thin film: the Simmons model*

We mentioned earlier that a significant source of standby power dissipation in today's transistors is electron tunneling across thin oxides. To that end, we would like to calculate the current density across a thin oxide barrier of cross sectional area S. The WKB approach of the previous section gives us the transmission probability for a single electron incident from the source. The current can thence be extracted by summing this transmission over all the transverse current modes incident from the source.

Landauer equation decomposes nicely when we assume no potential variations across the cross section of the oxide perpendicular to its thickness. Under this assumption, relevant when the oxide film is much thinner than wide, it is convenient to separate the total electron energy E into a longitudinal component along the transport axis and a transverse component perpendicular to the axis, $E = E_L + E_T$, with $E_L = \hbar^2 k_\parallel^2/2m + U$ and $E_T = \hbar^2 k_\perp^2/2m$. *The simplification arises because only the longitudinal energy E_L is relevant for transmission across the barrier because of the uniform device cross section.* Summing over the independent transverse modes $\vec{k}_\perp$, we get

$$I = \frac{q}{h} \sum_{\vec{k}_\perp} \int dE_L \bar{T}(E_L)[f_1(E) - f_2(E)] \tag{14.10}$$

(the $\sum_k$ plays the role of M in the Landauer equation described earlier). The equation can be rewritten by separately summing over the transverse modes, to give us

$$I = \frac{q}{h} \int dE_L \bar{T}(E_L)[F_1(E_L) - F_2(E_L)] \tag{14.11}$$

where the two-dimensional Fermi functions are given for a free electron as

$$F_{1,2}(E_L) = \sum_{\vec{k}_\perp} f_{1,2}(E_L + E_T)$$

$$= \int dE_T \underbrace{D_T(E_T)}_{Sm/2\pi\hbar^2} f_{1,2}(E_L + E_T)$$

$$= \frac{Smk_BT}{2\pi\hbar^2} \ln\left[1 + e^{(E_L - \mu_{1,2})/k_BT}\right] \tag{14.12}$$

In deriving the above equation, we assumed closely spaced transverse modes that allow us to replace the sum $\sum_{\vec{k}_T}$ with an integral $\int dE_T D_T(E_T)$ over the transverse two-dimensional density of states $D_T(E_T) = mS/2\pi\hbar^2$ for a free electron of energy $E_T = \hbar^2 k_\perp^2/2m$. In effect, we have outsourced the transverse mode summation from the transmission function to the modified Fermi function in Eq. (14.11), *mapping this onto a one dimensional transmission problem.*

At this point, let us assume a WKB expression for the tunneling transmission, simplifying the sinh and cosh functions in Eq. (14.3) with exponentials. The exponential term for an arbitrary potential $\phi(x)$ can be simplified analytically by Taylor expanding the potential and ignoring integral contributions beyond quadratic (Gaussian) terms. In other words,

$$\mathcal{T}(E_L) \approx e^{-\mathcal{A}(E_F + \langle \phi \rangle - E_L)^{1/2}} \quad (14.13)$$

where the potential average $\langle \phi \rangle = \int_{s_1}^{s_2} \phi(x)dx$, and $\mathcal{A} = 2\Delta s\sqrt{2m}/\hbar$, $s_{1,2}$ being the turning points for the electron within the barrier. Substituting this transmission expression in Eq. (14.11) and after some algebra, we get the standard Simmons expression for the current density $J = I/S$,

$$
\boxed{
\begin{array}{c}
\text{Simmons Equation} \\
J = J_0 \left[\langle \phi \rangle e^{-\mathcal{A}\langle \phi \rangle^{1/2}} - \left(\langle \phi \rangle + qV \right) e^{-\mathcal{A}(\langle \phi \rangle + qV)^{1/2}} \right]
\end{array}
}
\quad (14.14)
$$

where $J_0 \approx q/2\pi h(\Delta s)^2$. The current depends on the effective tunneling mass sitting in $\mathcal{A}$, the barrier thickness Δs at the tunneling point, and the average barrier height $\langle \phi \rangle$.

The Simmons model deals with just the average barrier height $\langle \phi \rangle$ and mapping it onto a rectangular potential. We already presented the result of tunneling through a rectangular barrier of height V_0, by solving the 1-D Schrödinger equation for each segment individually, and then equating their wavefunctions and their current densities at the interfaces. The transmission (Eq. (14.3)) can be reexpressed as

$$\bar{T}(E) = \frac{1}{1 + \alpha_c^2 \sinh^2(\eta_c L)}, \quad (14.15)$$

where $\alpha_c = (1 + r^2)/2r$, with $r = (\eta_c/k)(m_c/m_d)$ representing the velocity ratio, $\eta_c = \sqrt{2m_c E/\hbar^2}$ and $k = \sqrt{2m_d(V_0 - E)/\hbar^2}$. m_c and m_d are the effective masses of the two identical contact regions and the device (barrier) region respectively. For above barrier, the sinh terms get replaced by sines.

A NEGF simulation of current flow through an insulator is relatively straightforward to set up. We start with a 1-D uniform grid spanning the transport direction, the other two directions subsumed into the F functions. The kinetic energy terms can be written as a tridiagonal matrix with diagonal entries $2t$ and principle off-diagonal entry $-t$, where $t = \hbar^2/2m^*a^2$

depends on the local effective mass m^* and grid spacing a. At the interfaces between different materials we impose continuity of current density. This actually amounts to adjusting the onsite term to make the sum of entries along the row vanish. To this tridiagonal kinetic energy matrix we add a diagonal voltage matrix that takes into account the band-offset at the material interfaces due to the barrier (the barrier is between the contact Fermi energy and the conduction band minimum of the insulator). We then add a linear voltage drop across the insulator under bias. The contact self-energy matrices are mostly zero, except at the entries corresponding to their interface with the channel. These terms are given by $-t \exp(ika)$, Eq. (11.22), where k is obtained by inverting the tight-binding contact E-k, taking into account the fact that under bias the drain E-k slips past the

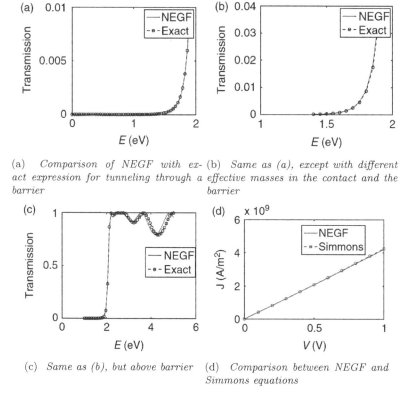

(a) *Comparison of NEGF with exact expression for tunneling through a barrier*

(b) *Same as (a), except with different effective masses in the contact and the barrier*

(c) *Same as (b), but above barrier*

(d) *Comparison between NEGF and Simmons equations*

Fig. 14.3 *Comparison between NEGF, analytical results (Eq. (14.15)) and Simmons' approximations for a tunnel barrier. The Simmons result needed to be scaled by a scalar quantity $\sim 3 - 4$ to account for a missing kinetic energy mismatch pre-factor.*

source E-k by the applied bias. The transmission is obtained by the NEGF prescription (Eq. (11.20)), while the current density is obtained by Eq. (14.11). Figure 14.3(a)-(c) compares the tunneling current computed using the analytical equation (14.15) and the numerical solution of NEGF. The barrier height is 2 eV and the width is 3.05 nm. The three subfigures correspond to tunneling with equal masses $m_c = m_d = m_0$ (m_0 being the free electron mass), tunneling with $m_c = m_0, m_d = 0.2m_0$, and above barrier transport, where mismatches can be directly traced back to finite grid issues.

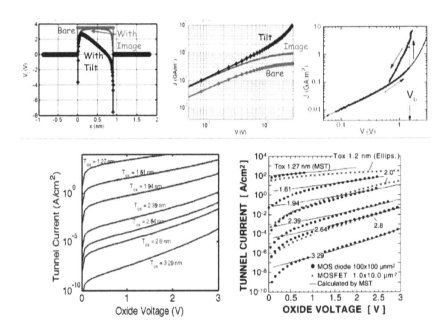

Fig. 14.4 *(Top Left) barrier geometry, (top middle) Theoretical and (top right) experimental tunneling currents through an Al_2O_3 barrier ($\Phi_B = 3.52$ eV, $\kappa = 8.4$, $m_{ox} = 0.4m_0$). The bare tunneling I-V as extracted from Simmons model (red) is enhanced by image-induced barrier lowering (blue) and shows a strong upturn when the field makes the barrier triangular, leading to Fowler-Nordheim tunneling. The experiment (reproduced from K. Gloos, P. J. Koppinen and J P Pekola, 'Properties of native ultrathin aluminium oxide tunnel barriers', J. Phys.: Condens. Matter 15, 1733 (2003), Fig. 9.) shows a hysteresis, possibly due to charge traps in the oxide. (Bottom left) Theory vs (Bottom right) experiment (reproduced from M Hirose, M Koh, W Mizubayashi, H Murakami, K Shibahara and S Miyazaki, 'Fundamental limit of gate oxide thickness scaling in advanced MOSFETs', Semicond. Sci. Technol. 15, 485 (2000)) with Simmons model fits for tunneling across a SiO2 barrier ($\Phi_B = 3.13$ eV, $\kappa = 3.9$, $m_{ox} = 0.5m_0$).*

Once we've benchmarked NEGF with analytical results at equilibrium, we then proceed to calculate an I-V and compare it with Simmons' expression. Figure 14.3(d) shows this comparison for a 3.52 eV, 0.912 nm barrier (the current looks linear since the voltage applied is much lower than the barrier height). Although the agreement looks exact, it is worth emphasizing that the Simmons current had to be scaled by a factor of $\sim$ 3.6, primarily to account for the WKB pre-factor $\sim 1/\sqrt{k}$ that the exponential term in Simmons drops. This factor captures the semiclassical scattering due to the difference in kinetic energy (effective mass) between the contact and the channel, a prefactor amounting to $16EV_b(m_c/m_d)/(E + V_b m_c/m_d)^2$ averaged over the energy bias window.

Figure 14.4 compares the Simmons J-V with experiments on a 0.912 nm Al_2O_3 barrier with barrier height 3.52 eV. Aside from the hysteresis in the experimental J-V (which is probably attributable to traps), the J-Vs look quite similar, especially when we complement the bare Simmons curve (red) with image corrections in the contacts (blue), and account for a field-driven Fowler Nordheim tunneling (black). The latter mechanism turns on because the source to drain bias makes the potential more triangular, making the tunneling exponentially more efficient through the narrowed part of the potential. The result is a sharp upturn in the $J - V$ curve, experimentally observed. Finally at the bottom, we show results for tunneling across SiO_2. *The point to emphasize is that the model captures not just the overall shape, but also the magnitude and quantitative details of the $J - V$ curve*

14.2.2   *From organic monolayers to single molecule tunneling*

The Simmons model was developed for an isotropic medium with a bandgap, and would apply to organic electronics with a random arrangement of molecules (top left). It is worth looking at a variant, where the transport is patently anisotropic. For an organized array (top right) of self assembled molecules (SAM) with stronger wavefunction overlap along the backbone than between molecules, we would expect this result to simplify to an expression proportional to the molecular packing density. In the former, the transport is three dimensional, but in the latter, we imagine several copies of a single isolated molecule and the current must be limited by the weak intermolecular hopping. The expression for current density must accordingly move from per unit area to per molecule.

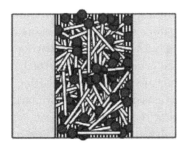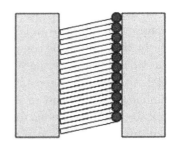

To capture the effect of ordering, and proceed accordingly from organic electronics to molecular electronics, we need to revisit the Landauer equation that generated Simmons' approximation. Specifically, we simplified the 2D Fermi functions as $F_{1,2}(E_L) = \sum_{\vec{k}_T} f_{1,2}(E_L + \hbar^2 k_T^2/2m_T)$ corresponding to the transverse effective mass m_T parametrizing hopping between two neighboring molecules. In the isotropic Simmons' formula, we set $m_T = m$ and replaced the sum by a integral over a continuum of closely packed, transverse modes. Once intermolecular tunneling is suppressed in an ordered SAM, $m_T \gg m$, while k_T is bounded by the transverse Brillouin zone, meaning that $E_L \gg E_T$, and E_T can in fact be dropped. The sum over k_T now has an argument that is independent of k_T, so it simply yields the number of k_T points in the transverse direction, equal to the number of molecules N_{mol}. We thus have a clear dichotomy

$$F_{1,2}(E_L) \approx \frac{Sm^* k_B T}{2\pi\hbar^2} \ln\left[1 + e^{(E_L - \mu_{1,2})/k_B T}\right]$$

(amorphous layer, isotropic conduction)

$$\approx N_{\text{mol}} f_{1,2}(E_L)$$

(SAM, low transverse current).   (14.16)

The Landauer equation for an insulating SAM thus differs from that of an oxide (Eq. (14.11)) and reads

$$I = \frac{2q}{h} N_{\text{mol}} \int dE_L \bar{T}(E_L) M(E_L)[f_1(E_L) - f_2(E_L)]$$ (14.17)

leading to the modified Simmons equation

Modified Simmons Equation

$$I = \frac{4q N_{\text{mol}}}{h \mathcal{A}}\left[\langle\phi\rangle^{1/2} e^{-\mathcal{A}\langle\phi\rangle^{1/2}} - (\langle\phi\rangle + qV)^{1/2} e^{-\mathcal{A}(\langle\phi\rangle + qV)^{1/2}}\right]$$

(14.18)

Notice that there is now an explicit dependence on the molecular packing density. In practice though, this packing dependence is observable only with pre-fabricated metal contacts. Depositing a metal top contact after the growth of a weakly packed monolayer would lead to pinholes and metal shorts that damage the device yield.

14.2.3   *Absolute values: resistance of an alkane molecule?*

Clearly the current is controlled largely by the contact injection rates, augmented by the intrinsic conductivity of the wires. Experiments reveal an encouraging pattern when it comes to these currents. Akkerman *et al.* pooled a wide range of alkane conduction data and showed that the current per molecule (extracted from an estimate of the molecular populations in various testbeds) roughly falls into three categories of resistances depending on the set up. Taking C12 as a prototypical molecule, the measured resistances give us (Fig. 14.5)

$$R_{2\text{chem}} \approx 10^4 M\Omega \quad \text{(Two chemisorbed contacts)}$$
$$R_{1\text{chem}} \approx 10^7 M\Omega \quad \text{(One chemisorbed contact)}$$
$$R_{\text{layer}} \approx 10^{10} M\Omega \quad \text{(Extra resistive layer)}$$

$$(14.19)$$

which amounts to a single-molecule current of 100 pA at 1-V for a well contacted molecule, 0.1 pA with a physisorbed contact, and 0.1 fA with an extra resistive layer. These numbers allow us to extract some intriguing features about the molecular contacts. The resistance per molecule can be calculated by taking I/N_{mol} in Eq. (14.17), differentiating with respect to voltage, and using a Lorentzian expression for the single-point transmission, augmented by the additional decay factor controlled by β. After some simple algebra we get

$$R = R_0 e^{\beta L}[(1+\alpha)^2/4 + (\phi_B/\gamma_1)^2]/\alpha \quad R_0 = h/2q^2 \approx 13K\Omega$$

$$\alpha = \frac{\gamma_2}{\gamma_1} \approx e^{-2\kappa_{\text{vac}}d}, \quad \kappa_{\text{vac}} = \sqrt{2m^*\phi_{\text{vac}}/\hbar^2} \qquad (14.20)$$

where in addition, we have already determined a microscopic expression for β. As we discuss later in the chapter on resonance, the typical level broadening with a S-Au coupling is $\gamma_1 \approx 0.15$ eV. Using a barrier height $\phi_B \sim 1.3$ eV, $\alpha = 1$ and a $\beta \approx 0.76/\text{C}$ extracted from the data for doubly chemisorbed contact, we get $R_{2\text{chem}} \approx 10^4 M\Omega$ for C_{12}, which is consistent with the data. For a singly chemisorbed contact, assuming a vacuum layer

$\sim 2 \overset{\circ}{A}$ on the physisorbed site which reduces its γ_2 by roughly two orders of magnitude through a WKB formula, i.e., $\alpha = \gamma_2/\gamma_1 = 0.01$, and using a $\beta \approx 1.05/C$ as from the data, we get $R_{1\text{chem}} \approx 3 \times 10^7 M\Omega$, which is again consistent with the data.

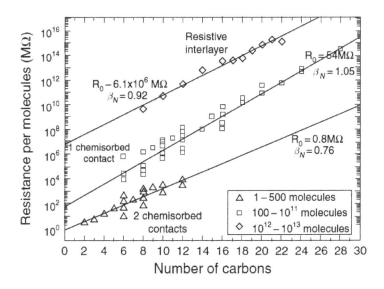

Fig. 14.5 *Compilation of experimental alkane resistances shows three classes of values per molecule, corresponding to doubly chemisorbed species, physisorption at one end and the presence of additional barriers at the weaker end (H. B. Akkerman and B. de Boer, 'Electrical conduction through single molecules and self-assembled monolayers', J. Phys.: Condens. Matter 20, 013001 (2008)). These values are consistent with a simple model for current flow through off-resonant tunneling (see text).*

Chapter 15

Tunneling with an index — molecular signatures of tunneling

Our treatment of tunneling in the previous chapter and in standard quantum mechanics textbooks requires just a continuum description of the barrier in terms of an overall effective mass. The underlying molecular nature and the 'bottom-up' viewpoint may seem like an overkill. However, there are notable examples where the orbital structure of the tunneling electrons imposes additional symmetry rules that constrain the transmission severely. We present three examples in this chapter. The tunnel magnetoresistance across a magnetic tunnel junction is governed by the alignment of majority spins across the oxide. This magnetoresistance can be further tweaked by aligning the orbital structures with the spin structures, so that certain orbital derived bands promote large spin-dependent tunneling and reduce the read current in magnetic memory devices. Finally, 2-D materials like graphene and topological insulators have a band index that we associate with pseudospin in the former and spin in the latter. The alignment of these band indices across a tunnel PN junction makes their transmission strongly angle-dependent and gate tunable, giving us an extra knob to control their device properties.

15.1 Tunnel magnetoresistance across a magnetic tunnel junction (MTJ) — a continuum description

If you follow memory technology, you may be aware of MTJs that consist of an oxide sandwiched between two ferromagnetic metals. Each ferromagnet has a lot of electrons with a majority spin (say, 'up') and a few minority spins that are opposite. In the absence of spin flip, the large number of up spins in one magnet would prefer to hop only to the majority up channel of a neighboring ferromagnet, and avoid the down channel. Thus when the two ferromagnets are parallel to each other the MTJ has a low resistance, while in the antiparallel state, the resistance is very high. The two resistance states can then be designated as bits '0' and '1', and form the basis of a non-volatile memory.

For tunneling across oxides and thin films, we usually start with the WKB formalism and expand that with a mode summation into the Simmons equation. We explicitly ignore the pre-factor $\sim 1/\sqrt{k}$ which gives us the kinetic energy mismatch across the junction. For an MTJ however, this prefactor is the critical quantity, as it is large when the magnets are parallel (large overlap between majority states) and small when they are antiparallel (small overlap between majority and minority states). Extracting this properly requires considerable algebra starting from a spinor representation of the electrons (analogous to the expression for pseudospins in graphene, Eq. (5.14)), and taking into account the slippage of the contact states under bias. Fortunately, we can use the results we already have. For a square barrier, we have already worked out the WKB expression including the pre-factor (Eq. (14.3)). The equation was

$$T \approx \frac{16E(U_0 - E)}{U_0^2} e^{-2\kappa L} = \frac{16k^2\kappa^2}{(k^2 + \kappa^2)^2} e^{-2\kappa L} \qquad (15.1)$$

where $k = \sqrt{2mE/\hbar^2}$, $\kappa = \sqrt{2m(U_0 - E)/\hbar^2}$.

The actual expression for the angle-dependent current is a mix of the transmission pre-factor from Eq. (15.1), the Simmons factor for a barrier Eq. (14.14), and the half-angle spinor components in Eq. (5.14). The equation

$$\boxed{\begin{array}{c} \text{Tunnel Magnetoresistance} \\[6pt] J_{\text{TMR}} = \underbrace{(J_\uparrow + J_\downarrow)}_{\text{spin term}} \times \underbrace{J_0 \left[\langle\phi\rangle e^{-\mathcal{A}\langle\phi\rangle^{1/2}} - \left(\langle\phi\rangle + qV\right) e^{-\mathcal{A}(\langle\phi\rangle + qV)^{1/2}} \right]}_{\text{simmons term (Eq. (14.14))}} \\[10pt] J_\sigma = \frac{16k_\sigma\kappa^2}{(k_\sigma^2 + \kappa^2)} \left[\frac{k_\sigma^+ \cos^2\theta/2}{k_\sigma^{+2} + \kappa^2} + \frac{k_{\bar\sigma}^+ \sin^2\theta/2}{k_{\bar\sigma}^{+2} + \kappa^2} \right], \quad (\sigma : \uparrow, \downarrow, \quad \bar\sigma : \text{opposite spin}) \end{array}}$$

$$(15.2)$$

where we now separate out the k-states between source and drain (local potentials 0 and $-qV$), between up and down (down state band-edge is lower than up state by ferromagnetic exchange energy Δ), and between metal contact and oxide electron effective masses (m_C and m_{ox}).

$$k_\uparrow = \sqrt{\frac{2m_C(E_F - qV)}{\hbar^2}}, \qquad k_\downarrow = \sqrt{\frac{2m_C(E_F - \Delta - qV)}{\hbar^2}}$$

$$k_\uparrow^+ = \sqrt{\frac{2m_C E_F}{\hbar^2}}, \qquad k_\downarrow^+ = \sqrt{\frac{2m_C(E_F - \Delta)}{\hbar^2}}$$

$$\kappa = \sqrt{\frac{2m_{ox}(U_0 - E_F - qV/2)}{\hbar^2}} \tag{15.3}$$

15.2 Orbital description of symmetry filtering — how MgO captured the oxide market in nonvolatile memory technology

The last section describes how the TMR across an MTJ depends on properties like effective mass in the metal contacts, tunneling effective mass, barrier thickness and height, and contact spin polarizations. The TMR is limited by a WKB tunneling expression. Here is where the molecular nature, specifically the orbital structure of the composite materials, can make the TMR much higher than the simple Bloch wave based analysis above, so that the resistance difference between 0 and 1 is very high. One way to do this is if we can play with the orbital chemistry of the oxide and ferromagnetic bands such that the up and down spins have highly different tunneling transmissions, even stronger than a conventional WKB would predict. This has now been accomplished with Fe/MgO/Fe MTJs. Inside the bandgap of MgO, there are decaying states designated as Δ_1 (Fig. 15.1), built out of s, p_z and $d_{3z^2 - r^2}$ orbitals, where z is the tunnel direction. These states share angular symmetry with the conduction and valence bands of MgO, and must turn around in the complex $E - \kappa$ plane to join the band-edges. Compared to a regular WKB tunnel state, such a turn-around reduces the decay constant κ, especially when the Fermi energy lies midgap. This is easy to see with a simple two band tunneling model, consisting of an alternating hopping parameter that oscillates between t_1 and t_2 and a common onsite energy ϵ_0. Implementing Bloch's theorem for the dimer array, we get the E-k which gives two branches (see Eq. (5.4))

$$E_\pm = \epsilon_0 \pm \sqrt{t_1^2 + t_2^2 - 2t_1 t_2 \cos kd} \tag{15.4}$$

that correspond to the valence and conduction bands of the insulator, d being the dimer distance. When the Fermi energy E_F lies in the bandgap,

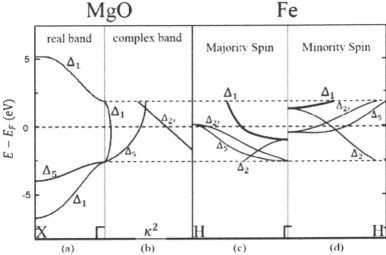

Fig. 15.1   *(Top) Fe/MgO/Fe MTJ, (bottom) bandstructure across junction explaining origin of high tunnel magnetoresistance (see text), because of preferential tunneling of Fe majority spin Δ_1 states that connect by orbital symmetry the MgO conduction and valence bands (Refs. 1–3 of Chapter 15).*

the k-values are complex (decaying instead of propagating waves). We can then write (for $t_1 > 0 > t_2$)

$$(E_F - \epsilon_0)^2 = t_1^2 + t_2^2 - 2t_1 t_2 \cosh \beta d/2 \tag{15.5}$$

with $\beta/2$ representing the imaginary part of k, so that the probabilities and electron transmissions go as $|e^{ikd}|^2 \sim e^{-\beta d}$. Noticing that the bandgap $E_g = 2|t_1 - t_2|$, the bandwidth $\Delta = 2|t_2|$ (assuming $|t_2| < |t_1|$), and the barrier height $\phi_B = E_F - (\epsilon_0 - |t_1 - t_2|)$, we get

$$\beta = \frac{2}{d} \cosh^{-1} \left[1 + \frac{2\phi_B(E_g - \phi_B)}{\Delta(E_g + \Delta)} \right] \tag{15.6}$$

For small barrier heights $\phi_B \ll E_g$, and in the limit $E_g \gg \Delta$, we can use a simple series expansion. Using $\cosh^{-1}(z) = \ln\left[z + \sqrt{z^2 - 1}\right] \approx \sqrt{2(z - 1)}$

for $z - 1 \gtrsim 0$, we then get $\beta \approx 2\sqrt{2m^*\phi_B/\hbar^2}$, where m^* signifies the mass of an electron conducting through the band ($\Delta = 2\hbar^2/m^*d^2$). The approximation leads to the familiar WKB tunneling in the small barrier limit, $T \sim \exp\left[-2\sqrt{2m^*\phi_B d^2/\hbar^2}\right]$. Symmetry filtering occurs for larger barriers, specifically when the Fermi energy is placed symmetrically near the middle of the gap, whereupon the two band effect kicks in. Setting $\phi_B = E_g/2$ for midgap in Eq. (15.6), the transmission simplifies to a weaker power law decay, $T \sim \Delta/2\phi_B = \hbar^2/m^*\phi_B d^2$, assuming the bandgap is significantly larger than the bandwidth. The anomalous long decay length significantly favors midgap states and filters out other decaying states.

In contrast to the Δ_1 states, the other evanescent states such as Δ_5 and Δ_2' have orbital components in the x-y plane, which do not need to join the bandedges because of their different symmetry. There is thus a pronounced difference in filtering properties of electrons tunneling through Δ_1 midgap vs other states, which is known as symmetry filtering.

But how does symmetry filtering translate to actual spin filtering? This is where the ferromagnetic contact comes into play. Fe grown epitaxially on MgO(100) also supports Δ_1 states, but only its majority spin sector has the Δ_1 states at the Fermi energy (i.e., this state acts as a 'half-metal'). Such a half metal then allows only up spins to tunnel through the Δ_1 states, while suppressing the down spins.

The combination of orbital and spin symmetry predicts a large ($\sim 1000\%$) tunnel magnetoresistance (TMR) for Fe/MgO/Fe, leading to its adoption as a key MTJ material stack. Experimental TMRs are still at a record high, though not as high as predicted. A possible origin is oxygen vacancies in MgO that depolarize the Fe spins. Expanding the search for other classes of half metals, such as in the half, full and inverse Heusler alloys, is an active and growing field of research in the field of high throughput computational materials science.

The decay constant β arising from the complex bands representing tunneling states has in fact been measured for various organic molecules. β has been established to lie between 0.76 to 1.07 per carbon (i.e., $0.84/\text{Å}$) in alkanethiols, about 1.7 per ring for aromatics and about $0.22/\text{Å}$ for carotenoid polyenes. The striking part is the relative constancy of these numbers across various material subclasses, for instance, within the alkanes or the aromatics. Our expression involves an inverse cosh term (i.e., a logarithm) and thus varies slowly with ϕ_B, essentially determined by the pre-factor, explaining why a variety of organic molecules have very similar values of β. For a range of selected barrier heights ϕ_B we get $\beta \approx 0.6 - 0.8 \text{Å}^{-1}$. Given a C-C distance of around 1.27 Å, this amounts to $\approx 0.76 - 1$ per C-C dimer. Note that the price we pay for the simplicity of a few parameters is the inability to capture realistic details of the band-structure, such as any non-parabolicity or multiplicity of the conduction and valence bands as well as any asymmetry between the two.

15.3   Graphene PN junctions — metamaterials, Klein tunneling, Veselago and all that jazz!

A second example where tunneling brings in 'molecular signatures' is electron tunneling across a graphene PN junction, created for instance by electrostatic 'doping' with two backgates to voltages $V_{G1,G2}$. As we saw with graphene, the electron wavefunctions (Eq. (5.14)) have a molecular (Bloch) part that is determined by the phase relation between the two dimer p_z basis sets involved. The states at zero and 180 degree are in fact orthogonal (Fig. 5.9), and can be designated as up and down pseudospins. The orthogonality prevents backscattering of low energy electrons at normal incidence, ensuring that across a PN junction, the transmission at low angle is pinned to unity. This is called Klein tunneling. The overall transmission is thus angle dependent (called chiral tunneling), and the transmission lobe narrows significantly with increase in voltage across the PN junction, while still maintaining perfect transmission at normal incidence.

Let us work out the tunnel transmission across the junction by redoing our particle on a step algebra (Eq. (14.1)), but including the 2D k-vector and the Bloch prefactor. The incident waves can be written as

$$\begin{pmatrix} 1 \\ e^{i\theta_1} \end{pmatrix} e^{i(k_x x + k_y y)} + r \begin{pmatrix} 1 \\ -e^{-i\theta_1'} \end{pmatrix} e^{i(-k_x' x + k_y' y)} \tag{15.7}$$

where the sign change in the second term happens when we flip the sign of k_x in the exponent, according to Eq. (5.14) for the upper branch, i.e.

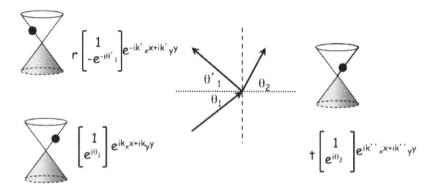

Fig. 15.2 *Component wavefunctions for an electron in graphene incident at a potential step from the left. The dot marks the electron wavevector. Matching the plane wave and matrix components at the step (dashed line) yields their trajectories and their angle-dependent transmission.*

$(-k'_x + ik'_y)/k'$. The transmitted wave is given by

$$t \begin{pmatrix} 1 \\ e^{i\theta_2} \end{pmatrix} e^{i(k''_x x + k''_y y)} \tag{15.8}$$

We need to match the wavefunctions at the interface, say at $x = 0$ for all y values, like we did for a 1D particle on a step earlier. Since we're matching both dimer atom contributions, in effect, we're matching the current density as well, decomposed onto the finite atomic grid.

Let us first match the phases from the plane waves at $x = 0$ for all ys. This gives us a k_y conservation, i.e., $k_y = k'_y = k''_y$. Assuming there is a gate voltage V_{G1} applied to the incident side relative to its Dirac-point (where the cones meet) and a voltage V_{G2} to the transmitted side, we get $k_{F1} = qV_{G1}/\hbar v_F$ and the same for the other side, which gives us Snell's law

$$\boxed{\begin{array}{c} \theta_1 = \theta'_1 \\ \dfrac{\sin\theta_1}{\sin\theta_2} = \dfrac{V_{G2}}{V_{G1}} \end{array}}$$ (Snell's law for electrons)

When we go across opposite dopings, e.g., n to p, then the ratio V_{G1}/V_{G2} is negative, meaning that *the transmission angle θ_2 is negative*. In other words, graphene acts like a **negative index metamaterial**. Under these conditions, the electrons will bend the 'wrong' way, and focus without a lens. This negative index lensing is sometimes called the 'Veselago' effect, after the scientist who first postulated negative index Snell's law (Fig. 15.3).

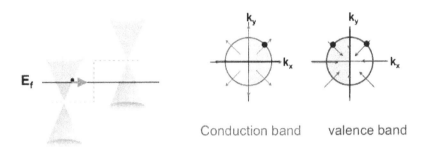

Fig. 15.3   *(Left) Dirac cones across an NP junction. (Right) Fermi circles with group velocities $\partial E/\hbar \partial \vec{k}$ pointing along $\vec{k}$ for electrons but opposite for holes. The $x-directed$ currents (shaded half circles) focus across the junction without a lens, acting as a negative index metamaterial (i.e., following Snell's law but bending the 'wrong' way!)*

For unequal doping, when we go from a denser to a rarer medium, we expect total internal reflection beyond a certain critical angle θ_C. This angle is determined by the conservation of k_y, i.e., quasimomentum component perpendicular to the interface. Since the maximum k is set by the Fermi surface on the lighter doped side, the electrons incident at wide angles from the heavier doped side must reflect, as they cannot conserve k_y. The result is two-fold (Fig. 15.4). Outside θ_C, they reflect, while inside a critical cone, they cross-over, and for PN junctions, they tend to focus (while for NN or PP, they defocus). If we now go back to Eqs. (15.7), (15.8) and match their (2×1) pseudospin pre-factors, we should get the actual transmission probabilities. This is analogous to matching suitable components of $\vec{E}$ and $\vec{H}$ to get Fresnel equations for EM waves at an interface. The pre-factor matching gives us the following equations

$$1 + r = t$$
$$e^{i\theta_1} - re^{-i\theta_1'} = te^{i\theta_2} \tag{15.9}$$

Using $\theta_1' = \theta_1$ from the first Snell's law (Eq. (15.3)), and eliminating t,

$$r = \frac{e^{i\theta_1} - e^{i\theta_2}}{e^{-i\theta_1} + e^{i\theta_2}} \tag{15.10}$$

where θ_1 and θ_2 are related by Snell's law. This gives us the transmission probability as

$$T = 1 - |r|^2 = |t|^2 \left(\frac{v_{x2}}{v_{x1}}\right) = |t|^2 \left(\frac{\cos \theta_2}{\cos \theta_1}\right) \tag{15.11}$$

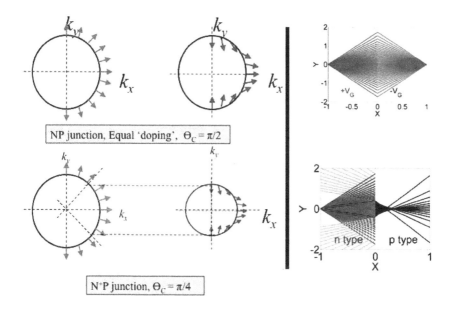

Fig. 15.4 *(Top) A symmetric pn junction focuses a point source of electrons. (Bottom) Unequal doping leads to total internal reflection as large angle electrons on the higher doped side cannot match k_y across the junction*

Keep in mind that for $\theta_1 > \theta_C$, θ_2 is in fact, complex! The result then is

$$T = \frac{2\cos\theta_1 \cos\theta_{2\mathcal{R}}}{\cosh\theta_{2\mathcal{I}} + \cos(\theta_1 + \theta_{2\mathcal{R}})} \tag{15.12}$$

where the subscripts $\mathcal{R}$ and $\mathcal{I}$ refer to Real and Imaginary parts. For all points above the critical angle, $\theta_1 > \theta_C$, θ_2 is complex, with $\theta_{2\mathcal{R}} = \pi/2$, so that $T = 0$ and we have total internal reflection. For all angles below, $\theta_{2\mathcal{I}} = 0$ and θ_2 is real. For homogeneous doping, $\theta_1 = \theta_2$, and $T = 1$, as expected. For symmetric PN junctions, $\theta_2 = -\theta_1$ and $T = \cos^2\theta_1$.

Most significantly, when $\theta_1 = 0$, θ_2 is also 0, whereupon Eq. (15.12) always gives $T = 1$. *We refer to this as Klein tunneling.* This is indeed a striking result! Note that EM waves also maximize their transmission at normal incidence, but that transmission is material-dependent, set by the ratio of refractive indices n_1/n_2 and is less than unity. For graphitic electrons however, the equivalent ratio V_{G1}/V_{G2} proves immaterial for the zero angle modes, which perfectly transmit. The difference arises because the transmission depends on the ratio of the x-directed velocity components (see statement immediately following Eq. (14.1)). Photons actually

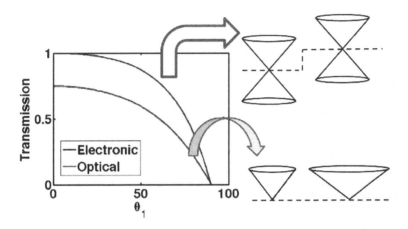

Fig. 15.5 *Optical transmission is governed by the reduction in speed and change in opening angle of the photonic band. Electron transmission across a graphene PN junction does not suffer a change in speed or opening angle, which is set by the $\pi - \pi$ bonding chemistry alone. The shift in Dirac point does not slow down the normally incident electrons, causing them to Klein tunnel perfectly.*

slow down or speed up between materials and their Dirac like cone changes opening angle and slope. The slope of the graphene Dirac cone is set by the $\pi - \pi$ bonding and does not change between regions with different gate potentials. Instead the cones shift vertically at an interface at constant opening angle. Since the projected velocity does not change across a junction for $\theta = 0$, the normally incident electrons transmit perfectly. More generally as we saw, the transmission is a sensitive function of angle θ_1, a behavior we refer to as *chiral tunneling*.

But is there any real tunneling here, in the absence of a band-gap? The answer is yes, in presence of a split between the two gates across the PN junction. The transverse modes see an angle-dependent tunnel barrier, while the normal mode is still perfectly transmitted following Klein tunneling. A WKB analysis for a slowly varying potential across the split gives us a slightly modified pseudospinor

$$\psi(x) \approx \frac{1}{\sqrt{k_x(x)}} \begin{pmatrix} 1 \\ \pm e^{i\theta(x)} \end{pmatrix} e^{i\left[\int_0^x k_x(x')dx' + k_y y\right]} \tag{15.13}$$

Going through the same exercise as before for a 2D particle at a step, and realizing that k_y gets quantized to $n\pi/W$ by the graphene width W, so that

the longitudinal propagation constant $k_x(x) = \sqrt{\left(\dfrac{E - U(x)}{\hbar v_F}\right)^2 - \left(\dfrac{n\pi}{W}\right)^2}$

becomes imaginary around the transition point, we get for a p-n junction

$$T(\theta_1) = \frac{2 \cos \theta_1 \cos \theta_{2\mathcal{R}}}{\cosh \theta_{2\mathcal{I}} + \cos (\theta_1 + \theta_{2\mathcal{R}})} e^{-2\pi d k_{F1} k_{F2} \sin \theta_1 \sin \theta_2 / (k_{F1} + k_{F2})}$$

(15.14)

There is a critical angle $\theta_C \approx 1/\sqrt{\pi k_{F\|} d}$ where $\|$ refers to the parallel combination of momenta in the exponent in Eq. (15.14). The current is obtained by summing over the incident modes, which amounts to an integration over θ_1 with a corresponding mode density $M(\theta_1)$. Assuming the dominant angle dependence comes from the exponent, the integrand can be replaced with a Gaussian, whereby the mode-averaged transmission works out to

$$\bar{T}(E) \approx \frac{1}{\pi |E|} \sqrt{\frac{U_0 \hbar v_F}{d}}$$

(15.15)

around the Dirac point (Fig. 15.6). The final transmission is explicitly gate voltage dependent, with the barrier $U_0 = q\Delta V_G = \hbar v_F(|k_{FI}| + |k_{FT}|)$. The absence of a band-gap makes the gate voltage dependence at this stage still modest. In the last chapter, we will see how we can make the dependence drastic by introducing an additional barrier.

Finally, it is worth pointing out that the equations here contain a fair amount of subtle physics. For instance, we argued that Klein tunneling is driven by the constancy of the velocity component v_x along $\theta = 0$. Indeed, the reflectivity of photons at an interface are given by

$$R_{EM} = |r|^2 = \left|\frac{v_{x1} - v_{x2}}{v_{x1} + v_{x2}}\right|^2 = \left|\frac{v_1 \cos \theta_1 - v_2 \cos \theta_2}{v_1 \cos \theta_1 + v_2 \cos \theta_2}\right|^2$$

(15.16)

But Eq. (15.10) gives us for the graphene PN junction

$$R_{GPNJ} = |r|^2 = \left|\frac{(\cos \theta_1 - \cos \theta_2) + i(\sin \theta_1 - \sin \theta_2)}{(\cos \theta_1 + \cos \theta_2) - i(\sin \theta_1 - \sin \theta_2)}\right|^2$$

(15.17)

While this supports our argument that for normal incidence where $\theta_1 = \theta_2 = 0$, $R = 0$ and $T = 1$ (Klein tunneling), the expression itself looks more complicated than Eq. (15.16). Where do the extra complex terms

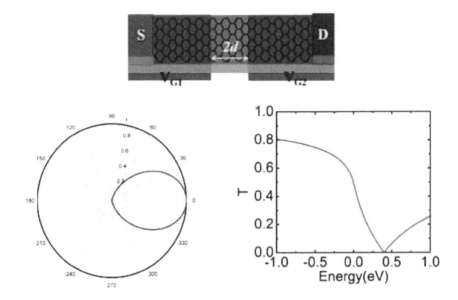

Fig. 15.6 *Strongly angle-dependent transmission T (bottom left) across a backgated graphene p-n junction (top), the angle dependence arising from the molecular pseudospin structure, i.e., the phase-coherent mixing of the two dimer p_z orbitals involved. The rotation of the pseudospins around the graphene Fermi circle (Fig. 5.9), makes the transmission (Eq. (15.14)) angle-dependent. Integrating over the transverse modes, the net transmission and ballistic current across the junction is energy and thus gate voltage dependent (bottom right)*

arise? It is noteworthy that EM-waves pick up a complex phase when we venture *out-of-plane* to perpendicular polarization, whereupon Eq. (15.16) swaps indices 1 and 2 for the angular terms, generating polarization dependent phenomena such as Brewster angles. Since our graphitic electrons are longitudinal rather than transverse waves, we expect to pick up a phase variation when we rotate in plane, as seen in Fig. 5.9. The added terms in Eq. (15.17) account for that in-plane phase variation.

Chapter 16

Through band conduction

The key element in designing the conductance of a material is the alignment of its energy bands relative to the contact Fermi energies. The Landauer theory gives us a convenient bottom up way to describe this current, and can be readily modified to include scattering effects (Chapter 10). The Landauer conductance is proportional to the number of modes lying within the conduction window opened up by the applied source to drain bias, times the average transmission probability of these modes. The physics is often dominated by the mode-averaged transmission function alone, such as a WKB function $T \sim e^{-\kappa L}$ for quantum mechanical tunneling through a bandgap (Eq. (14.3)), an Ohmic contribution for diffusive transport $T \sim \lambda/(\lambda + L)$ in a dirty material (Eq. (22.5)), or a sinusoidal term $T \sim 1/(A \sin^2 kL + B)$ for a coherent interferometer (see Fano interference earlier and Eq. (14.3)). We will see later how all these transmissions hail from a common origin (the $\sin^2 kl$ function analytically continues into a decaying exponential $\sinh^2 \kappa L$ outside a band, and phase averages to the Ohmic limit $\sim 1/L$ inside a band). However, a lot of physics, especially with high quality materials and interfaces, arises from bandstructure alone, and their biggest contribution is to the mode density $M(E)$. In what follows, we will focus on transport where the mode spectrum is the key player, in other words, ballistic ($T = 1$, no *velocity scattering*) resonant conduction through the middle of a band.

We will start with a simple system, namely graphene. Since its density of states and mode spectrum can be worked out easily, we will begin with the current voltage characteristic through bulk graphene. Our focus here will be the overall shape of the IV curve, which is ultimately what device engineers focus on. Physicists often worry about the details of the low bias Ohmic conductance, which often turn out to be more complicated and sensitive to morphology and interfacial chemistry. We will see this later for graphene when we discuss the role of charge puddles on low bias conductance. But it is easier instead to start with *just the overall shape of the I-V curve.*

16.1 Starting simple: current flow and band to band transfer in bulk graphene

The current for graphene shows a direct band to band transition at high bias. This is not conventional tunneling since there is no bandgap, but there is still an upturn in the current when the voltage is large enough to cross the Dirac point and excite charges in both bands. It is straightforward to calculate the ballistic current-voltage characteristic by integrating over the number of modes. Given the linear $E - k$ for graphene, the number of modes is given by

$$M(E) = \frac{\hbar|v_x|}{L_x} D(E) \qquad (16.1)$$

To be more precise, the prefactor of the D term equals the escape rate $\hbar/\tau$ satisfying Matthiessen's rule, effectively the sum of the rates of individual processes, i.e., the contact injection and removal rates $\gamma_{1,2}$ and the transit time $L_x/|v_x|$. We assume here the contacts are good enough that the transit time is the rate limiting step (Eq. (11.6)). The electron transit speed is obtained from the linear $E-k$ of graphene, $E = \pm\hbar v_F|k|$, leading to a group velocity $v = v_F$ and $|v_x| = 2v_F/\pi$. As explained earlier, the speed of the graphitic electrons is invariant and determined only by bonding chemistry and $\pi - \pi$ overlap. The effective mass of the electrons however change with applied bias.

The density of states $D(E) = \overbrace{2}^{\text{for spin}} \sum_{k} \delta(E - E_{\vec{k}})$ (Eq. (9.16)) evaluates to

$$D(E)dE = 2 \underbrace{\times \ 2}_{\text{for 2 valleys}} \times \underbrace{\left(\frac{dk_x dk_y}{\Delta k_x \Delta k_y}\right)}_{\text{evaluated over 1 valley}}$$

$$= 4 \times \frac{2\pi k dk}{(2\pi/L_x)(2\pi/L_y)}$$

$$= \frac{2L_x L_y}{\pi} \frac{|E|dE}{\hbar^2 v_F^2} \qquad \text{(using the graphene E-k)} \qquad (16.2)$$

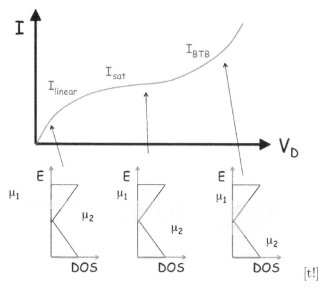

Fig. 16.1 *Schematic graphene ballistic I-V obtained by integrating the mode density (proportional to the linear density of states, DOS, shown below), over the energy window (gray slab) between contact electrochemical potentials $\mu_{1,2}$ separated by the applied drain bias. The integrand keeps increasing with drain bias (as does the current), but the rate of increase dI/dV_D slows down till μ_2 crosses the conduction band and reaches the Dirac point, then the rate increases again after we encounter the valence band.*

so that

$$
\begin{aligned}
D(E) &= \frac{2L_x L_y |E|}{\pi \hbar^2 v_F^2} && \text{density of States} \\
M(E) &= \frac{\hbar |v_x| D(E)}{L_x} = \frac{4 L_y |E|}{\pi^2 \hbar v_F} && \text{mode distribution}
\end{aligned}
\tag{16.3}
$$

The zero-temperature ballistic current is then given by Eq. (12.5)

$$
\begin{aligned}
I &= \frac{q}{h} \int_{\mu_2 = E_F - qV_D}^{\mu_1 = E_F} dE\, M(E) \\
&= \frac{2q}{h} \left(\frac{4 W E_F^2}{h v_F} \right) \left[1 \mp \left(1 - \frac{q V_D}{E_F} \right)^2 \right]
\end{aligned}
\tag{16.4}
$$

where $W = L_y$ is the device width, and the minus sign above is for $qV_D < E_F$, and the plus sign is for $qV_D > E_F$ (the sign flip occurs because of the absolute value of energy in the density of states). For a device of width $1\mu m$

and Fermi energy $E_F = 0.2$ eV, the pre-factor outside the square bracket is about 0.3 mA. The current can be separated into three clear regimes — a low bias linear regime of current (I_{linear}), a higher bias quasi-saturation current (I_{sat}) when μ_2 reaches the Dirac point, and a much higher bias quadratic rise due to band-to-band transport (I_{BTB}) once μ_2 crosses the Dirac point.

$$I = \begin{cases} I_{\text{linear}} = I_0 M E_F \left(\dfrac{qV_D}{E_F} \right) & \text{for } qV_D \ll E_F \\[2ex] I_{\text{sat}} = I_0 M E_F & \text{for } qV_D = E_F \\[2ex] I_{\text{BTB}} = I_0 M E_F \left(\dfrac{qV_D}{E_F} \right)^2 & \text{for } qV_D \gg E_F. \end{cases} \qquad (16.5)$$

where $I_0 = G_0/q = q/h$. The total number of modes including spin and charge is

$$M = 4 \times W/(\lambda_F/2) = 4W k_F/\pi = 4W E_F/\pi \hbar v_F \qquad (16.6)$$

where $E_F = qV_G \alpha_G$ with α_G the gate transfer factor. We can also write in terms of the electron density n_{2d} and the total electron number N_{2d}

$$N_{2d} = n_{2d} S = 4 \int \frac{d^2 \vec{k}}{\Delta k_x \Delta k_y} = 4 \frac{\pi k_F^2}{\left(\dfrac{2\pi}{L_x} \right) \left(\dfrac{2\pi}{L_y} \right)} \qquad (16.7)$$

so that

$$\boxed{k_F = \sqrt{\pi n_{2d}}} \qquad (16.8)$$

and $M = 4W \sqrt{n_{2d}/\pi}$

Several complications can arise at this point:

- the density of states can pick up a background due to charge puddles with a spatially varying onsite potential, averaging out and eliminating the strict Dirac point;
- various diffusive scattering processes can reduce the average transmission per mode;
- local doping of the contact regions can lead to further asymmetries in the current, due to the creation of a Schottky like potential profile across the channel;

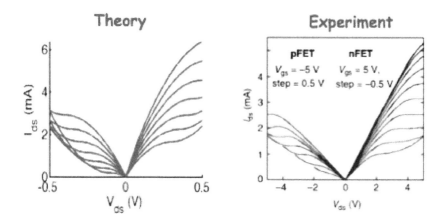

Fig. 16.2   *Theoretical and experimental I-V for a 1 μm wide graphene sample (data from P. N. First et al., MRS Bulletin 35, 296 (2010)). The calculations assume ballistic transport, Eq. (16.4). The gate voltages are ramped up from ±0.2 V upto ±0.5 V, assuming an initial doping in the sample that places the starting Fermi energy at $E_F = 0.1\,eV$. Note that the voltages plotted in the simulation are local potentials, which correspond to a much larger applied voltage.*

- the gate transfer factor α_G can become V_G dependent, so that applying one volt at the gate does not shift the Dirac point by the entire electron volt.

We will discuss the first two when we talk about scattering. The third and fourth are electrostatic issues related to the contact potential and the quantum capacitance of the graphene sheet. We will address them when we talk about molecular transistors.

16.2   From conductance to conductivity quantization in graphene

For ballistic graphene, the low bias conductance is quantized $G = I/V_D = q^2 M/h$ where M is the number of propagating modes, equal to 4 (for 2 valleys and 2 spins) times the number of Fermi half-wavelengths we can fit into the width (Eq. (16.6)). As we reach the Dirac point, however, the number of modes drops to zero (Fermi wavelength swells to infinity) and we expect to get zero conductivity. However, a really interesting twist appears that makes the conductivity finite, and in fact, quantized, a feature unique to the zero bandgapped feature in graphene. Specifically, for a sample with width much longer than length, we get a continuum of closely spaced subbands with near zero bandgaps, for which the tunneling contributions actually become non-negligible.

For a graphene segment of finite width, the quantized n^{th} subband has energy $E_n = \pm \hbar v_F \sqrt{k_x^2 + (n\pi/W)^2}$, with $k_y = n\pi/W$, at zero energy, leading to the decay constants $q_{x,n} = n\pi/W = k_y$. The contacts however are still zero bandgapped metallic graphene, so the wavefunctions for the three regions are (following the same approach as Eq. (15.7))

$$
\Psi_I = \begin{pmatrix} 1 \\ e^{i\theta} \end{pmatrix} e^{i(k_x x + k_y y)} + r \begin{pmatrix} 1 \\ -e^{-i\theta} \end{pmatrix} e^{i(-k_x x + k_y y)}, \quad x < 0, \text{ source}
$$

$$
= \begin{pmatrix} ae^{q_{x,n}x} \\ be^{-q_{x,n}x} \end{pmatrix} e^{ik_y y}, \quad 0 < x < L, \text{ nth subband}
$$

$$
= t \begin{pmatrix} 1 \\ e^{i\theta} \end{pmatrix} e^{i(k_x x + k_y y)}, \quad x > L, \text{ drain} \tag{16.9}
$$

So the matching equations are

$$
1 + r = a
$$
$$
e^{i\theta} - re^{-i\theta} = b
$$
$$
ae^{q_{x,n}L} = te^{ik_x L}
$$
$$
be^{-q_{x,n}L} = te^{ik_x L + \theta} \tag{16.10}
$$

Solving for r, we get the transmission for the nth subband, $T_n = 1 - |r|^2$. The final result is

$$
T_n(\theta) = \frac{\cos 2\theta + 1}{\cos 2\theta + \cosh(2q_{x,n}L)} \tag{16.11}
$$

Assuming normal incidence, $\theta = 0$, we get $T_n = 1/\cosh^2(q_{x,n}L)$, similar to WKB tunneling. Using Landauer theory, the conductance for a set of closely spaced modes (implying $W \gg L$) is

$$
G = \frac{2q^2}{h} \sum_{n=-\infty}^{n=\infty} T_n = \frac{2q^2}{h} \frac{1}{\Delta q_n L} \sum_n \frac{\Delta q_n L}{\cosh^2 q_N L}
$$

$$
= \frac{2q^2}{h} \frac{1}{(\pi/W)L} \int_{-\infty}^{\infty} \frac{dx}{\cosh^2 x} = \frac{4q^2 W}{hL\pi} \tag{16.12}
$$

Setting $G = \sigma W/L$, we get $\sigma = 4q^2/\pi h$.

For most materials, the evanescent tunnel modes are exponentially smaller than the propagating modes. For a wide sheet of graphene, a continuum of such exponentials with vanishingly small barriers effectively gives an Ohmic, $\sim 1/L$ dependence, while the number of modes scales with width, so that the end result is a *conductivity, rather than conductance quantization*. Experimentally, the surprise is that a similar universal conductivity is seen for dirty samples, except it is roughly a factor of 3 higher. One possible reason is charge puddles that can create added scattering centers and increase the number of modes near the Dirac points for a higher conductivity (Section 23.3).

16.3 From graphene to benzene: the nature of molecular IVs

While a bulk system such as graphene has a continuum band that leads to an easily integrable mode spectrum, the results get complicated near interfaces, such as metals. In general, the mode spectrum around interfaces is quite complicated (although we frequently treat it simply as a doped extension to regular graphene leading to a Schottky potential). The magnitude of the current in particular, depends on the metal carbon bonding chemistry. This complication becomes predominant for molecules attached to metal contacts, where one can argue that the entire heterojunction is 'all interface', compounding the morphology and interface specific problems for molecular conduction. Nonetheless, we can extract some physics out of the Lorentzian lineshapes for the molecular densities of states.

The current voltage characteristic of a molecular conductor has the generic shape shown in Figs. 1.2(a) and 16.3. Several features are noteworthy and a lot of effort has gone into understanding each. Each arises from a different aspect of the band alignment shown in Fig. 16.3 on the top, indicating the Fermi energy of the contacts, the barrier height separating the Fermi energy from an individual resonant level, and the broadening of that level by interactions with the contact. What we will argue is that the low bias conductance of the molecule depends on the barrier to broadening ratio ϕ_B/γ, while the saturation current depends on the broadening parameter γ alone. The intermediate open shell current, usually relevant for shorter molecules in the Coulomb Blockade limit, depends on the resistive coupling asymmetry across the contacts as well as the charging energy U_0. The

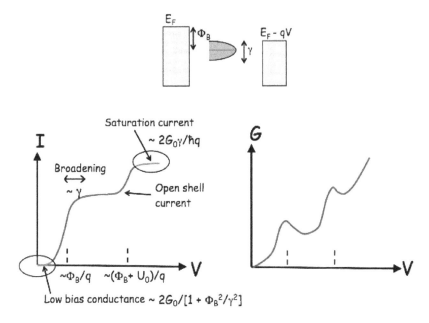

Fig. 16.3   *(Top) Molecular alignment with metal Fermi function, determined by the Fermi energy E_F, barrier Φ_B and broadening γ. (Bottom left) Current voltage characteristic show intermediate open shell plateaus and ultimately closed shell saturation currents. The low bias conductance depends on the normalized barrier ϕ_B/γ, while the saturation current depends primarily on the contact injection efficiency γ. The intermediate plateau looks Ohmic vs saturating, the slope depending on γ/U_0. (Bottom right) The inflexion points in the I-V correspond to peaks in the conductance, and in the absence of complex electrostatics, to peaks in the low bias transmission function.*

transitions at which we see the current plateaus (to be more accurate, the transitions where the conductance maximizes, in other words, the current shows a point of inflexion from concave upwards to convex) is given by the barrier height ϕ_B between the molecular level and the Fermi energy for the first plateau, and additional Coulomb costs $\sim U_0$ for the next (these voltages are the local bias values, and will in general be related to the actual applied drain bias values through a transfer factor determined by capacitance ratios). All of these conclusions can be obtained from a very simple implementation of the Landauer model.

The mode density is given by

$$M = \gamma D/2\pi \qquad (16.13)$$

where γ is the effective broadening of the levels inversely proportional to the total charge transport time (Eq. (11.4)). For bulk solids with good contacts, this time is dominated by the intrinsic transit time

$$t_{\text{transit}}^{\text{Bulk}} \approx L/v + L^2/2\mathcal{D} \tag{16.14}$$

(at least asymptotically), where v is the band velocity and $\mathcal{D}$ is the diffusion constant. However for a short molecule with typically poor contacts mismatched at the interfaces, the rate limiting steps are the injection and removal times, and γ is dominated by the individual contact induced broadenings given by Fermi's Golden Rule (Eq. (11.14)), i.e., determined by the orbital coupling across the contact interface and the contact density of states near the Fermi energy (we assume here the contact couplings are equal, $\gamma_1 = \gamma_2 = \gamma$. For unequal coupling, the γs add in parallel). The molecular level can be described as a Lorentzian $D(E) = 2\gamma/\left((E - \epsilon_0)^2 + \gamma^2\right)$ for two spins, and $M = \gamma D/2\pi$, so that the low bias conductance

$$G = \frac{q}{\hbar}M(E_F) = \frac{2G_0}{1 + (\phi_B/\gamma)^2} \tag{16.15}$$

where $G_0 = q^2/h$. In other words, the low bias conductance depends on the barrier height, which as we shall see shortly depends on a slew of factors from work function to interface structure and density of states. The quantity γ is also tricky to estimate, as it is also morphology dependent, sensitive to chemistry, dispersive (i.e., energy dependent) and orbital dependent (i.e., it needs a matrix representation to be accurate). The low bias conductance is thus not an intrinsic measure of the channel properties alone, but rely equally on the detailed microstructure of the contacts. In what follows, we will attempt to get a simple expression for ϕ_B and γ (Eqs. (17.15) and (17.17)).

The saturation current is obtained by integrating the Landauer equation (we assume zero temperature here, and can convolve this with the thermal broadening function if needed later on).

$$\begin{aligned} I &= \frac{q}{h} \int_{\mu_1}^{\mu_2} dE M(E) = \frac{q}{h} \int_{\mu_1}^{\mu_2} dE \frac{2\gamma^2}{(E - \epsilon_0)^2 + \gamma^2} \\ &= \frac{2q\gamma}{h} \left(\tan^{-1}(\mu_2/\gamma) - \tan^{-1}(\mu_1/\gamma) \right) \end{aligned} \tag{16.16}$$

If we assume the bias window completely straddles the level, we can then get the maximum saturation current by assuming $\mu_2/\gamma \to \infty$ and $\mu_1/\gamma \to -\infty$, which makes the term in the bracket equal to π, giving us the saturation current

$$I_{\text{sat}} = \frac{q\gamma}{\hbar} \qquad (16.17)$$

In other words, *the saturation current does not depend directly on the barrier height, and is primarily determined by the net broadening alone.*

As we will later see in the chapters on correlation, the intermediate plateaus are obtained by shell filling, and are scaled by degeneracy factors that arise from the combinatorics of many-body excitations driven by unscreened Coulomb interactions.

Chapter 17

Conduction in molecules

17.1 Where are the 'levels'?

The three ingredients we needed for the current calculation are the energy level spectrum of the active molecule closest to the contact Fermi energy (typically the Highest Occupied Molecular Orbital or 'HOMO' and Lowest Unoccupied Molecular Orbital or 'LUMO' levels), the Fermi energy placement in between those levels, and the broadening of the levels by the continuum of levels in the contacts. Let us start with the bare molecular levels. Bandgaps and energy levels in molecules need a lot of care in interpreting, as electron electron interactions get significantly amplified for localized orbitals. The precise locations of the measured levels depend on the nature of the experiments, whether photoemission and inverse photoemission or optical bandgaps or transport bandgaps. Analogously, the computed results depend on the details of the electronic structure method invoked, each enjoying its own domain of relevance. Figure 17.1(a) shows a range of calculated bandgaps as well as measured gaps between gas phase and multilayer samples. Figure 17.1(b) shows the computed frontier energy levels of benzene as extracted from GAUSSIAN98. Figure 17.1(c) shows the measured gas phase spectra compared with transport spectra for perylene-3,4,9,10-tetracarboxylic dianhydride (PTCDA). Figure 17.1(d) shows how the calculated bandgap for PTCDA systematically increases with self interaction correction from LDA DFT to Hartree Fock, where a theorem (Koopman's theorem) connects the calculated HOMO level with the gas phase ionization potential $-I = E_N - E_{N-1}$ (E_N: total energy of N electron system), and naturally leads to the identification of the LUMO as the electron affinity $-A = E_{N+1} - E_N$.

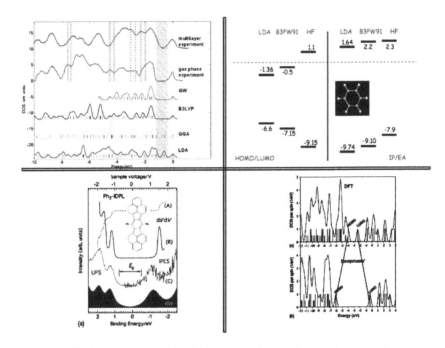

Fig. 17.1 *Clockwise from top left. (a) Variation in calculated and measured spectra of PTCDA (N. Dori et al., Phys. Rev. B 73, 195208, 2006). (b) Variation is larger for calculated HOMO LUMO levels of benzene in GAUSSIAN98, vs ionization potentials (IP) and electron affinities (EA). (c) Increase in bandgap with increasing interactions, from LDA/DFT to Hartree Fock ('Basic Theory of the Molecule-Metal Interface', F. Flores and J. Ortega, in The Metal Molecule Interface, N. Koch et al. (d) Experimental transport spectra and ionization potential compared with LDA/DFT results (K. Iketaki et al., J. Phys. Chem. C 113, 1515 (2009)).*

Let us focus on Fig. 17.1(b). We see that the bandgaps set by the difference between the Ionization Potential and Electron Affinity (IP/EA) are relatively invariant with bandstructure method between Hartree Fock (HF), Becke 3 parameter Perdew Wang hybrid functionals (B3PW91), and Density Functional Theory with the Local Density Approximation (LDA). In contrast, the HOMO-LUMO gaps obtained from the quasiparticle energies are sensitively method dependent. Within DFT itself, the gaps are 5.16 eV for the generalized gradient approximation (GGA) with the Perdew Burke Ernzerhof functional (PBE), 6.27 eV for the Heyd Scuseria Ernzerhof functional (HSE) and 10.51 eV for the GW approximation ('G': Green's function, 'W': Screened Coulomb potential).

The experimental optical bandgap of benzene in the gas phase is around 10.51 eV. This correlates well with the HF gap at 10.21 eV (and closer to a GW calculation), which can be related with the ionization potential (IP) and the electron affinity (EA), features that include single electron effects (Koopman's theorem guarantees that the Hartree Fock HOMO level matches the IP for a large enough basis set). In contrast, the bandgap of benzene in solution phase or on a metal substrate is significantly reduced as the Coulomb correlations are screened by counterions or image charges, and correlates closer with the LDA bandgap.

The inconsistency arises from the importance of electronic correlations, and from our inability to precisely handle them computationally. As we explain in the chapter on correlation, the issue lies in how adequately we can handle self interaction correction through a proper orbital dependent potential. Our ability to capture these corrections and incorporate them in transport will determine whether our calculated transport bandgap is set by the HOMO-LUMO gap or the IP-EA gap (Fig. 27.9). The exact total energy E_N for an N electron system, for instance the expressions worked out for H_2 using full CI in the chapter on correlation, show a derivative discontinuity at integer N, and linearly interpolate in between for fractional charges. In fact, the slopes on each side of the integer charge represent the IP and the EA for the N electron system,

$$\left.\frac{\delta E}{\delta n}\right|_{N+\delta} = -A = E_{N+1} - E_N$$

$$\left.\frac{\delta E}{\delta n}\right|_{N-\delta} = -I = E_N - E_{N-1} \tag{17.1}$$

where I is the ionization potential and A is the electron affinity. We can derive this discontinuity from basic statistical mechanics, tracking just the three states of occupancy N_0, $N_0 + 1$ and $N_0 - 1$ at energies E_0, $E_0 + I$ and $E_0 - A$. The equilibrium many body occupancy follows Boltzmann statistics

$$\left\langle N \right\rangle = \frac{\sum_N N e^{-\beta(E_N - \mu N)}}{\sum_N e^{-\beta(E_N - \mu N)}} = N_0 + \left(\frac{e^{\beta(A+\mu)} - e^{-\beta(I+\mu)}}{1 + e^{\beta(A+\mu)} + e^{-\beta(I+\mu)}}\right), \quad \beta = \frac{1}{k_B T}$$
$$\tag{17.2}$$

If we simply invert this equation to plot μ vs $\left\langle N \right\rangle$, the discontinuity jumps out at zero temperature (the chemical potential $\mu = \delta E/\delta N$ denoting the aforementioned derivative). We can see this analytically as well in our simple three state example. Clearly when $\langle N \rangle = N_0$, we get $\mu = -(I + A)/2$. However, in between for fractional occupancies, we can set $\langle N \rangle = N_0 + \Delta N$ and solve the quadratic equation for the corresponding $e^{\beta \Delta \mu}$ in terms of ΔN, giving us two complex conjugate roots. Assuming zero temperature, $\beta \to \infty$, and a convex function for $E(N)$, we then get

$$\mu = \begin{cases} -I & \text{if } N_0 - 1 < N < N_0 \\ -(I + A)/2 & \text{if } N = N_0 \\ -A & \text{if } N_0 + 1 > N > N_0 \end{cases} \qquad (17.3)$$

Clearly the slope $\mu = \delta E / \delta N$ is constant for noninteger occupancies and undergoes a discrete jump at the integer occupancies. This is the derivative discontinuity that a proper exchange correlation functional must then reproduce at zero temperature. For a noninteracting system, the kinetic

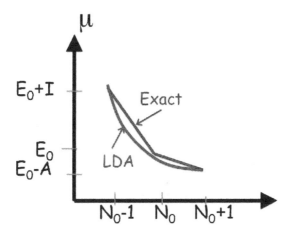

Fig. 17.2 *Derivative discontinuity in exact correlated energy, compared to model calculations. This discrepancy accounts for the underestimation of bandgap in LDA/DFT.*

energy alone contributes to the derivative discontinuity, which gives us the quasiparticle energies associated with the Highest Occupied Molecular Level (HOMO) and the Lowest Unoccupied Molecular Level (LUMO), ϵ_{HOMO} and ϵ_{LUMO}. For an interacting system however, an additional discontinuity should arise from the exchange correlation term, and account for the larger bandgaps (this added discontinuity amounts to an orbital dependent self interaction correction that we allude to later in the book). The Coulomb correlations alone create an additional difference between the IP and EA (one involves the attractive Coulomb effect of $N-1$ electrons upon removal, and the other the repulsion from N electrons upon addition). Unfortunately typical model DFT calculations such as the Kohn Sham approach to LDA give us exchange correlation functions that are smoothly varying with N, meaning that this extra bandgap due to self interaction correction is missing in these calculations. A slew of results have cropped up over the years to fix this omission. These range from the LDA + U approach that adds in a separate Hartree term for the inner correlated bands, to Quantum Monte Carlo (QMC) to hybrid DFT approaches such as B3PW91, B3LYP, HSE or exact exchange.

When computing the energy levels, care must also be taken to (a) map the bare gas phase levels with experimental spectra (e.g., Photoemission data), and (b) renormalize these levels by interaction with the contacts, which is known to significantly reduce the bandgap by screening out the Coulomb correlations responsible for the derivative discontinuity in the exchange correlation functional. A full out LDA calculation will underestimate the gap since it has no derivative discontinuity whatsoever (and thus applies to the fully screened limit). For a slowly varying charge density, we already extracted the Thomas Fermi exchange potential (Eq. (6.30)), $V_{TF} = -C_X \int n(R)^{4/3} d^3 R$, obtained by integrating the pair correlation function that is a modified Bessel function. For a homogeneous electron gas, we can then calculate the one electron exchange potential $E_X = \delta V_{TF}/\delta n = -C_X n^{1/3}$ solely in terms of the 'size' of the electron, defined as the dimensionless Wigner Seitz radius $(4\pi r_s^3/3)n = a_0^3$ where a_0 is the Bohr radius, giving us

$$E_X = -\frac{0.916}{r_s} \times \frac{q}{4\pi\epsilon_0 a_0} \qquad (17.4)$$

We can also work out the correlation term for the high density and the low density limits analytically. For low densities, the kinetic energy term $\propto \nabla^2 n \sim 1/r_s^2$ is unimportant compared to the Coulomb term $\propto 1/r_s$, as are the electron overlap and thus Pauli exclusion, so that the electrons tend to act classically. The classical Coulomb energy to assemble an isotropic sphere is a textbook problem in Electrostatics. We find the electric field using Gauss' Law and then the potential (which looks like a point charge $V(R) = Q(R)/4\pi\epsilon_0 R$), and then integrate over a sphere to get the total energy $E = 3/5r_s \times q/4\pi\epsilon_0 a_0$. The energy to bring an electron from infinity to the center of the sphere can also be obtained from the Gauss' law potential in presence of the sphere, so that we get the total energy for the low density classical limit. Subtracting the above exchange term, we then get the correlation term for a homogeneous electron gas

$$E_C = \left(-\frac{0.8836}{r_s} + \frac{3}{2r_s^{3/2}}\right) \times \frac{q}{4\pi\epsilon_0 a_0}, \quad \text{(low density)} \quad (17.5)$$

where in the second term, we corrected for the zero point energy in the harmonic potential acting inside the sphere. For the opposite limit of high electron density, the result can again be obtained analytically, evaluating the Feynman diagrams for the multiple electron–electron scattering processes and then selectively summing the most divergent terms to generate a logarithmic term arising from the long ranged, unscreened nature of the Coulomb interaction in a homogeneous electron gas.

$$E_C = \left(-0.0232r_s + 0.0622\ln r_s + 0.004r_s\ln r_s\right) \times \frac{q}{4\pi\epsilon_0 a_0}, \quad \text{(high density)} \quad (17.6)$$

The two limiting density expansions can be extended to intermediate electron densities using Quantum Monte Carlo calculations. We could accordingly start with the Hartree Fock calibrated with gas phase data and then add charge polarization and screening effects from the contacts. Alternately, we can start with the LDA limit and add correlations due to self interaction correction in the presence of contacts. The B3PW91 and B3LYP hybrid functionals do that by mixing exact exchange with LDA/GGA

$$E_{XC}^{\text{B3LYP}} = \underbrace{\left(E_X^{\text{LDA}} + E_C^{\text{LDA}}\right)}_{\text{LDA}} + \underbrace{\left[a_X\left(E_X^{\text{GGA}} - E_X^{\text{LDA}}\right) + a_C\left(E_C^{\text{GGA}} - E_C^{\text{LDA}}\right)\right]}_{\text{GGA corrections}}$$

$$+ \underbrace{a_0\left(E_X^{HF} - E_X^{LDA}\right)}_{\text{HF corrections}} \quad (17.7)$$

so that the net result happens to lie energetically closer to the partially screened limit relevant for transport spectra. The three Becke coefficients above are obtained from experimental data such as atomic energies and ionization potentials, while obeying a set of constraints such as the correct integrated sum of the exchange correlation holes.

17.2   Where is the Fermi level?

The Landauer equations earlier were written in their simplest, lumped parameter incarnations, but are nonetheless useful to get quick estimates (a more detailed model will need to calculate energy dependent matrices for the self energies). The broadening γ is given by Fermi's Golden Rule, $\gamma = 2\pi|\tau|^2 D$ (Eq. (11.14)), which depends on the local density of states D at the contact bonding site, and the components of the bond coupling matrix τ oriented perpendicular to the contact.

The barrier height, on the other hand, is given primarily by the difference in energy between the Fermi energy of the contact and the closest conducting level in the molecule, assuming their vacuum levels are aligned in the first place. This barrier height determines the onset of conduction. The physics, however, is a bit more involved because of charge transfer between the channel and the contact and the corresponding energy level adjustment in presence of a sizeable Coulomb cost.

For bulk semiconductors unencumbered by contacts, we can define an intrinsic energy level E_i where the mobile electron and hole densities are equal. For an undoped semiconductor, each electron thermally excited from valence to conduction band must leave behind a corresponding hole in the valence band, keeping the system charge neutral (for doped semiconductors, charge excitation leaves behind a corresponding charged dopant). The mobile electron density for a nondegenerate semiconductor at equilibrium is

$$n = \int_{E_C}^{\infty} D(E - E_C)f(E - E_F)dE \approx \int_{E_C}^{\infty} D(E - E_C)e^{-(E-E_F)/k_BT}dE$$

$$= e^{-(E_C-E_F)/k_BT} \underbrace{\int_0^{\infty} D(E)e^{-E/k_BT}dE}_{N_C} \qquad (17.8)$$

For an isotropic parabolic band, the lumped density of states can be written as

$$N_C \approx \frac{2}{\lambda^3} \qquad (17.9)$$

where λ is the thermal De Broglie wavelength, $\lambda = \sqrt{2\pi m^* k_B T/h^2}$. For materials with substantive bandgaps (several times the thermal energy k_BT), the charge neutrality at the intrinsic energy amounts to

$$n_i = \underbrace{N_C e^{-(E_C-E_i)/k_BT}}_{n} = \underbrace{N_V e^{-(E_i-E_V)/k_BT}}_{p} \qquad (17.10)$$

placing the intrinsic level *roughly midway between conduction and valence band, offset a bit towards the lower density of states* through a 'lever rule' akin to a center of mass.

$$E_i = \frac{E_C + E_V}{2} + \frac{k_B T}{2} \ln \left(\frac{N_V}{N_C} \right) \qquad (17.11)$$

The actual Fermi energy in the channel however can be displaced with doping. Assuming N_D fully ionized donors or N_A fully ionized acceptors for instance, we get

$$E_F - E_i \approx \pm k_B T \ln \left(N_{D,A}/n_i \right) \qquad (17.12)$$

where the intrinsic charge density $n_i = \sqrt{N_C N_V} e^{-E_G/2k_B T}$. The above equation functions just as expected. Donors shift the Fermi energy towards the conduction band and favor electrons, while acceptors shift it towards the valence band and favor holes.

How do these results change for an organic molecule like benzene? The equivalent of the intrinsic level E_i is called the *charge neutrality level* E_{CNL}. This level defines an *energy boundary such that filling until this level with electrons keeps the molecule neutral in situ, i.e., in presence of contact bonding.* For a short molecule, the broadening and density of states come from its interactions with the contacts, meaning that the placement of the neutrality level within its bandgap is a property not intrinsic to the molecule *per se*, but to its atomic environment including details of the contact surface chemistry at the bonding site. Analogous to above then, if the contact Fermi energy lies above the CNL, the channel is 'doped' by electrons flowing in from the contacts. The only catch now is that unlike bulk semiconductors with a fairly high charge density, the paucity of carriers in a molecule means that there is inadequate screening, so that a charging energy is associated with the act of doping in general.

Figure 17.3 shows the process of band alignment. Due to the disparity between the contact Fermi energy and the molecular CNL, a net charge δQ driven by the energy disparity flows between the two, in its quest to set up equilibrium and equate the Fermi energies. If the charging has no penalty, the molecular level stays put and the vacuum levels in the two materials stay aligned. In general however, there is a Coulomb cost for this charge transfer 'doping' that shifts the molecular level and misaligns the vacuum levels proportionally. Let us assume the levels move up by δU, driven by the net charge build up δQ, so that

$$\delta U = U_0 \delta Q / q, \qquad (17.13)$$

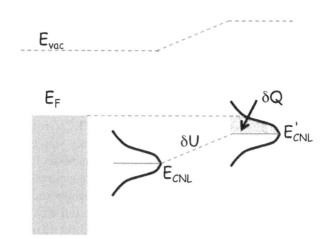

Fig. 17.3 *Alignment of molecular state with contact. Charge transfer is driven by the difference between the metal Fermi energy and the molecular charge neutrality level, balanced with the Coulomb cost for adding the charges and deviating from neutrality. The screening parameter S (Eq. (17.15)) controls the degree of 'state pinning'. Incidentally, S is the same ratio fronting the U_L term in Eq. (18.31), with $C_Q = q^2 D_0$ and $C_\Sigma = q^2/U_0$.*

where U_0 is the single electron charging energy (this self interaction energy is the difference between a Hartree Fock transport gap and a Kohn Sham LDA-DFT quasiparticle transport gap). The charge density build up is now given by the balance, proportional to the energy difference between the Fermi energy and the shifted charge neutrality level,

$$\delta Q = q \int_{E_{CNL}+\delta U}^{E_F} D(E)dE \approx qD_0 \times (E_F - E_{CNL} - \delta U), \qquad (17.14)$$

where $D_0 = D(E_F)$ is the density of states created by the metal at the molecular interface, referred to as Metal Induced Gap States (MIGS) or Induced Density of Interface States (IDIS). From the two coupled equations (Eqs. (17.13), (17.14)), we get the new shifted charge neutrality level and the barrier height in terms of the screening parameter S

$$S = 1/(1 + U_0 D_0)$$
$$E'_{CNL} = E_{CNL} + \delta U = (1 - S)E_F + SE_{CNL}$$
$$\phi_B = E_F - E'_{CNL} = S(E_F - E_{CNL})$$
$$\delta U = (1 - S)(E_F - E_{CNL})$$
$$\delta Q = qD_0 S(E_F - E_{CNL}) \qquad (17.15)$$

For a large trap density and corresponding charging energy $U_0 D_0 \gg 1$, the charge neutrality level stays rigidly pinned to the Fermi energy so that changing the metal work function has little effect on the molecule or the ability of the contact to dope the molecule. For the opposite limit, the level position is unpinned to the contact, so that changing the metal workfunction can change the doping of the molecule and allows contact engineering (relevant for organic LEDs, solar cells, FETs etc). These two limits are usually referred to as the 'Bardeen' limit and the 'Schottky Mott' limit. The single parameter S that determines the extent of pinning in our simple model is set by $U_0 D_0$, which more generally arises from $D_0 \rightarrow -\int dE D(E-U) \partial f / \partial E$ (even this is an approximation as it assumes the shape of the density of states is unaffected. The full NEGF treatment with self consistent Poisson would capture more subtle nuances as well). Note also that we are assuming the Fermi energy cuts right through the density of states, so we are using the degenerate expression D_0 instead of the Boltzmann averaged lumped density of states $N_C = \int D(E) e^{-E/k_B T}$ for nondegenerate semiconductors.

So far, we have only included the Hartree contribution from monopolar charges transferred between metal and molecule. Two higher order effects are also important at this stage:

- the Pauli repulsion between parallel spins in the metal and the molecule (equivalently, the exchange correlation hole) that pushes some of the escaping metal wavefunction tails back into the contact (this is called the '**pillow effect**'). The antisymmetrization of the molecular and contact orbitals $\phi_{a,b}(\vec{r})$ creates a molecular dipole just from the exchange correlation hole, given by $\vec{D} = -2q\vec{r}_{a,b}/(1-S_{ab}^2)$, where $\vec{r}_{a,b} = \int d^3\vec{r} \phi_a(\vec{r}) \vec{r} \phi_b(\vec{r})$ is the dipole, $\vec{r}$ is measured from the midpoint of the atoms, and $S_{ab} = \int d^3\vec{r}\phi_a(\vec{r})\phi_b(\vec{r})$ is the orbital overlap. This dipole moment creates a shift in the potential V^{pillow}.
- the role of any built in molecular dipoles, including bond dipoles at the interface as well as image charges set up by the molecule on the contacts. From the computed charge transfer δQ we can calculate this dipolar contribution as $V^{\text{dipole}} \approx n_{\text{mol}} \delta Q / \epsilon$ where n_{mol} is the areal dipole trap density, and ϵ is the permittivity of the interfacial layer. The image term can be approximated as $V^{\text{image}} \approx -qq'/\epsilon |z-z'|$ at distance z from the metal, where $q' = -q(\epsilon - 1)/(\epsilon + 1)$ for an image plane and z' is the location of the image charge obtained from the method of images.

The net effect of these added effects is a lowering of the Coulomb term,

$$\delta U' = (1 - S)(E_F - E_{CNL}) + Sq(V^{\text{pillow}} + V^{\text{dipole/image}}) \qquad (17.16)$$

17.2.1 A case study: Benzene Dithiol (BDT) on gold

For various reasons, primarily historical, benzene has ended up being the 'fruitfly' in molecular conduction studies. From a theoretical and computational point of view, this makes sense. The so-called 'aromaticity' of the π conjugated structure, a delocalized electron cloud spread across the alternating single and double bonds, argues for a non negligible conductivity, while the compactness of its structure suggests that it maybe computationally tractable. Experimentally however, the situation is a lot murkier, because of the extreme difficulty of making two covalent bonds to the ends of benzene with single crystal planar metal contacts.

The workfunction of gold at around 5 eV places its Fermi energy close to the HOMO level of BDT (the lowest energy molecular geometry is planar upon attachment to contacts, leaving a sulphur pz orbital to bond with the benzene HOMO level and pushing the BDT HOMO even closer to the gold Fermi energy). Attachment of gold leads to a small amount of charge transfer $\delta Q \approx 0.1q$. The final energy placement can be studied from an extended molecule involving BDT and a large cluster of gold, and identifying the molecular HOMO level by plotting its eigenmodes and looking for a sizeable overlap with the benzene HOMO wavefunction. The LDA barrier height is actually quite small (~ 0.5 eV), but the actual result should be much higher in the presence of correlation that is not fully screened out.

Thermoemf measurements indeed suggest that BDT is p type. The nature of the IV asymmetry in presence of known contact asymmetry also indicates p type, consistent with our initial energy level placement. The extracted barrier height however is much larger ($\phi_B \sim 1.3$ eV).

The contact broadening γ can be estimated from Fermi's golden rule. Near the Fermi energy, the density of states in gold is primarily s dominated but with a non negligible amount of p orbital contribution as well. The surface density of states D_S of Au(111) is estimated to be around 0.07/eV for s electrons (p orbitals give a similar factor, but few p orbitals of Au couple with the sulphur p orbitals). Next, we need the bond energy. The optimized structure for thiol places it above the centroid of an Au(111) triangle, with the Au-S bond energy ~ 2 eV. The HOMO wavefunction of BDT is fairly delocalized over the ring, so that only about 1/8 of the electron overlaps with the sulphur, making the coupling τ between contact and molecule that much smaller, and the broadening

$$\gamma = 2\pi|\tau|^2 D_S \approx 0.1 - 0.2 \text{ eV} \tag{17.17}$$

For benzenedithiol, this gives us a maximum current through a closed shell to be around 3-7 μA, consistent with experimental data on break junction measurements with well documented conductance histograms (note that the IV gets stretched out in the presence of a sizeable single electron charging energy U_0 that acts as an additional voltage dependent broadening, but charging doesn't increase the peak current). There are scattered alternate data on BDT that show smaller current magnitudes. Given the extreme size of BDT, this discrepancy could well be due to the inability of a

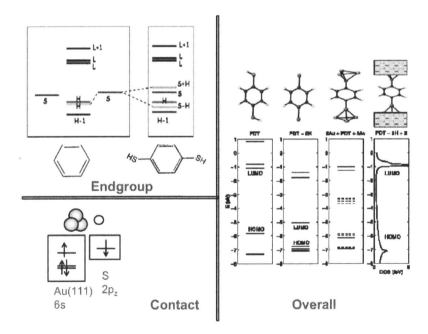

Fig. 17.4   *(Top left) Correlation diagram showing the frontier levels of sulphur and benzene, and their evolution to BDT. (Bottom left) The gold triad on Au(111) donates one unpaired 6s electron that pairs up with the pz orbital of the sulphur endgroup. (Right) B3PW91/LANL2DZ plots showing the evolution of BDT levels upon contacting with gold (Prashant Damle, PhD Thesis, Purdue University)*

single molecule to bridge the gap, instead having two separate molecules connected to each contact separately and interacting with each other through a herringbone configuration (this configuration would certainly explain, along with the low observed conductances, the symmetric nature of the observed IVs, in contrast to single bridging molecules that would tend to show polarization asymmetries). Furthermore, the Au-S bonding depends sensitively on attachment geometry. The radial wavefunction varies exponentially with perpendicular distance to the metal plane while the angular overlap of orbitals varies as a power law of a sine or cosine. This too can alter the current by multiple orders of magnitude. The bonding geometry itself is also unclear, because of step edges and polycrystalline phases on gold surfaces, and also because the energy difference between the various sites (on top of atom, in hollow vs on top of bond) lie within the thermal barrier.

The zero bias transmission was estimated earlier as $2G_0/(1+\phi_B^2/\gamma^2)$. Thermoemf measurements indicate a barrier height $\phi_B \sim 1.2 - 1.3$ eV, larger than typical $LDA-GGA$ estimates. This would lead to a predicted low bias conductance $G \sim 10^{-2}G_0$ (resistance of 1.3 MΩ), consistent with experiments. Experiments with xylyl dithiol show a conductance $G \sim 6 \times 10^{-4}G_0$ which argues for a much weaker Au-S coupling, as we expect in the presence of bulky methyl intervening groups between benzene and sulphur.

17.3 Molecular rectification

The historical interest in organic molecules started with the prediction of a diode like behavior in a D^+BA^- donor bridge acceptor structure by Aviram and Ratner. The IP of the donor must exceed the EA of the acceptor (loosely speaking, HOMO of D must be above the LUMO of A). The bridge prevents the immediate transfer of charges and reorganization of the levels, but maintains the energy disconnect. Applying a negative bias to the donor end lowers its levels, increases the energy disconnect and increases the donor to acceptor barrier, reducing the current. A negative bias on the other hand would bring the levels closer, lowering the barrier and thereby increasing the current. This is similar to a regular PN junction where forward bias lowers the built-in potential across it while reverse bias increases it. The only difference is in the electrostatics; since the potential varies perpendicular to the molecular axis, due caution must be exercised in drawing

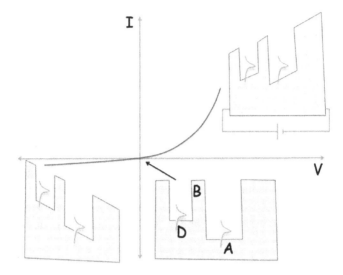

Fig. 17.5   *Current asymmetry due to in-built barrier asymmetry across a donor-bridge-acceptor (D-B-A) series within a molecule. For positive bias on the donor, the states are brought closer to resonance while for negative the states fly away from each other and the current plummets.*

1-D band diagrams in the absence of extended 2D sheet charges that could have acted as parallel plate capacitors with negligible fringing fields.

The Laplace potential suffices to bring out the role of an intrinsic structural asymmetry. One can think of this as a *barrier asymmetry*, for instance, if one has a pronounced vacuum barrier at one side or a Schottky barrier at an interface with a semiconducting contact. Under opposite bias voltages, the barrier potential tilts in opposite directions, enhancing the average barrier height in one direction and reducing it in the other (like the built-in potential of a p-n junction), giving rise to a corresponding asymmetry in the I-V. A more subtle *polarization asymmetry* arises when contacts come into play. The current through a resonant level is symmetric in the contact broadenings $I \propto \gamma_1 \gamma_2 / (\gamma_1 + \gamma_2)$. After all, it is determined by the sum of the two escape rates $\sim \hbar / \gamma_{1,2}$. Simply swapping the broadenings does not change the sum or the current, assuming the contacts are nondispersive and the molecule has no structural asymmetry. Regardless of voltage polarity, the weaker contact is always the rate limiting step. However, even in the absence of an inherent structural asymmetry, the system can still show an aforementioned polarization asymmetry arising simply from Poisson's equa-

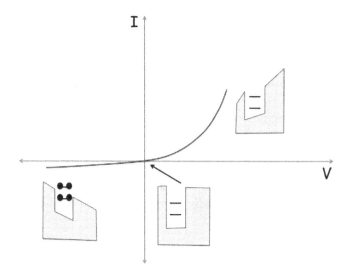

Fig. 17.6 *Polarization asymmetry arises when the molecule picks up a net charge for one polarity, as a consequence of barrier asymmetry. A positive bias on the thicker barrier makes the thinner one fill the levels faster than they are emptied, leading to a net addition of charge or shell filling for a LUMO level or n-type conduction. For negative bias the charges are depleted (shell tunneling for LUMO). The build up of charge in one direction drags out the I-V and makes the latter rectifying as well. For HOMO levels (p-type conduction) the asymmetry reverses.*

tion, i.e., the self consistent screening fields. While the current is symmetric in the contact broadenings, the nonequilibrium charge is not. The self consistent charge on a level (Eq. (11.2)) is given by $(\gamma_1 f_1 + \gamma_2 f_2)/(\gamma_1 + \gamma_2)$. Let us assume contact 1 is dominant $(\gamma_1 \gg \gamma_2)$, and also that the level is sharp and well contained inside the bias window. For positive bias on contact 2, $f_1 = 1$ and $f_2 = 0$, and the dominant contact 1 keeps it filled with the charge is given by $\sim q\gamma_1/(\gamma_1 + \gamma_2) \approx q$. For opposite bias though, $f_2 = 1$ and $f_1 = 0$ and the level is emptied by the dominant contact (charge $\sim q\gamma_2/(\gamma_1 + \gamma_2) \approx 0$. So one way, the level stays filled, the other way it stays empty.

The IV asymmetry depends on the original state of the level. If the level was a HOMO level that was filled to begin with (*p* type conduction), then in the first case contact 1 keeps it filled. In the second case however, contact 2 empties it and imposes an additional charging penalty $\propto U_0$, postponing the onset of conduction in and stretching out the current. We thus have an IV asymmetry just from the differential charging at opposite bias voltages. For a LUMO level instead, contact 1 imposes the charging cost and stretches out the IV in the opposite direction (for positive bias on contact 1). Thus, the IV picks up an asymmetry, and in fact the nature of the polarization asymmetry tells us whether conduction was p type or n type.

The polarization asymmetry can be understood with a many body picture as well. As before there is a charging cost associated with the action of the faster contact. In a many body picture, for one polarity the charge escapes immediately and there is no single electron charging, while for the opposite polarity the charge lingers and there is Coulomb Blockade. In the literature, the former is usually referred to as *shell tunneling* while the latter is known as *shell filling*.

17.4   Molecular dipoles on semiconductors

A sheet of molecular dipoles can influence the threshold voltage of a semiconducting channel via charge transfer 'gating'. Figure 17.7 shows experimental and theoretical I-Vs for alkanethiols on GaAs with a copper top contact. The workfunction difference between the metal top contact and the semiconducting substrate creates a Schottky barrier over which the charges in the GaAs valence band need to thermally jump and then tunnel through the molecule to reach the metal contact (Fig. 1.1(c)). This barrier clearly, is larger for n-type GaAs than p-type. An intervening molecular dipolar layer absorbs some of the voltage drop, reducing the Schottky barrier but increasing the tunnel barrier. The I-V (Fig. 17.7) shows a kink in the forward bias current when the charge is thermionically injected over the Schottky barrier into the molecular tunnel barrier. The I-V is strongly asymmetric; for forward bias, the Schottky barrier is reduced while for reverse bias it is increased.

The shape of the I-V is further controlled by voltage division – longer molecules offer larger voltage drops, reducing the GaAs Schottky barriers. While this reduces the current through the increasing tunnel resistance, the reduced Schottky barrier also reduces the onset voltage for tunneling seen from the kink, making the I-V less rectifying and more symmetric. The system acts as two nonlinear resistors in series. At low bias the voltage drops primarily across the Schottky barrier, while at high bias the molecu-

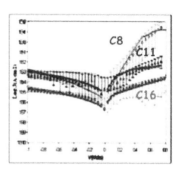

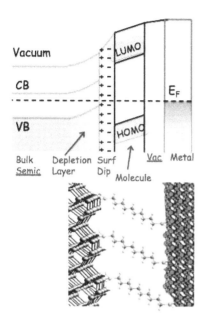

Fig. 17.7 *(Left) Cu-alkanedithiol-GaAs junction (bottom), and corresponding equilibrium band-diagram (top). The sign of the interfacial dipole determines the voltage division between semiconductor and molecule, and the corresponding I-V asymmetry. (Right) Experimental I-V (dots with error bars, courtesy Prof. Nathan Swami and Dr. Archana Bahuguna. Also see F. Camacho-Alanis, L. Wu, G. Zangari, N. Swami, 'Molecular junctions of 1 nm device length on self-assembled monolayer modified n- vs. p-GaAs', Journal of Materials Chemistry, 18, 5459 (2008).) while solid lines show theoretical simulations. The results show rectification, along with a kink signifying a crossover between low-bias thermionic emission over the semiconductor Schottky barrier, and high-bias tunneling through the molecular barrier. For longer chains, the tunnel current drops. Also longer chains support a larger fraction of the voltage drop, reducing the surface potential and Schottky barrier. As a result, the kink separating thermionic and tunnel emission shifts to lower voltages, making the I-V more symmetric*

lar tunnel barrier dominates. This means that one can increase the Schottky current by dropping more voltage at the molecule-GaAs interface, say using a sheet dipole created by the molecule-semiconductor workfunction difference, or engineered using molecular redox groups. Increasing the doping also reduces the Schottky barrier width, controlling part of the current.

PART 5

Controlling Electron Flow:

Switching and Gating

Chapter 18

Electrical gating

The multibillion dollar semiconductor industry relies on our ability to significantly alter the current flowing between two terminals, by the application of a voltage to a third, gate terminal. What has made this switching action evolve into a formidable industry is our ability to dope transistors with well controlled doses of trace impurities to make them electron or hole rich, and then to integrate multiple (currently billions) such complementary transistors onto a chip. In complementary metal oxide semiconductor (CMOS) technology, the central unit is an inverter consisting of an n doped and a p doped transistor placed in series between a voltage source and a ground. A common input signal is placed on their gates, so that when one (say the n FET) is turned on by pulling its Fermi energy closer to the majority band, its complement is turned off. This means that the output signal drawn from a point between the transistors shorts through the active FET to either the power supply or the ground, floating thereby to a value that is inverse to the input. What has made this inverter particularly attractive is that due to the serial placement of the complementary FETs, aside from the transient phase when the voltages are changing, the system stays perpetually in open circuit between supply and ground (as either the n or the p MOS is off), and the energy budget is thus quite manageable.

From an inverter, we can construct a NOR or NAND gate, then use those to construct higher order logic blocks such as an adder or a ring oscillator. At its heart is a transistor, which at the highest level of operational abstraction is simply a voltage controlled switch that toggles between ON and OFF. At a device level however, we need to make sure the switching occurs fast enough, reliably enough, and with as little static and dynamic power dissipation as possible. At a circuit level, we need to ensure there

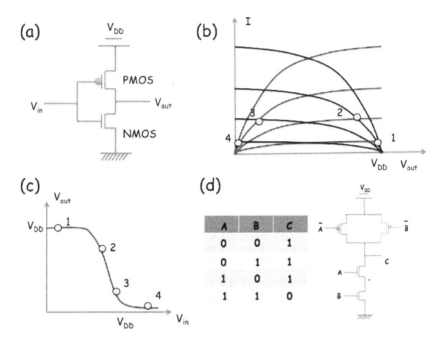

Fig. 18.1 *(a) An inverter is constructed by placing a PMOS and NMOS in series between the power supply voltage V_{DD} and the ground. A large input gate voltage V_{in} attracts electrons and turns the NMOS on while shutting off the PMOS, shorting out the lower branch and letting V_{out} equalize with ground. A small (or negative) input gate shorts out the PMOS branch and sets $V_{out} = V_{DD}$. The element thus acts as an inverter (high input gives low output and vice-versa). (b) To extract the inverter characteristic, we flip the PMOS I-V along the voltage axis to satisfy Kirchoff's voltage low. To obey the current law, we then match up the NMOS current (red curves) at $V_G = V_{in}$ with the PMOS current (black curves) at $V_G = V_{DD} - V_{in}$ and extract the output V_{out} from the intersection points. (c) The resulting $V_{in} - V_{out}$ curve shows an inverter characteristic. The steepness is the gain, and must exceed unity in order to combat circuit losses and restore a digital signal (high stays high, low stays low) that may tend to get corrupted and start drifting towards the middle. (d) A pair of inverters can be hooked up as shown to construct a not-and or 'NAND' gate (truth table shown), from which we can then construct any higher order Boolean function.*

is adequate 'gain' so that the device can reliably switch another one down-stream, while still retaining the amplification to restore signals that are otherwise corruptible to noise. We will discuss these aspects in greater detail down the road.

In this chapter, we will begin with ultimately scaled molecular CMOS and its gateability in a three terminal configuration. We will argue that such a transistor is performance compromised by a slew of problems ranging from parasitic slowness to poor gateability and rampant source to drain tunneling. Some of these features extend to other molecular systems, notably graphene. In the next chapter, we visit the relatively unexplored domain of conformational gating. We will argue that such a gate could have niche applications in designing sleep circuits. We will finally discuss two terminal switching, which underlies negative differential resistance (NDR), hysteresis and stochastic processes and telegraph noise. Such processes may have some applications (e.g., a latch in lieu of DRAM), but more importantly they carry useful spectroscopic information about *in situ* molecular species.

18.1 Current saturation in classical transistors

Textbook derivations of transistor current are based on classical transport, drift diffusion (Fig. 18.2). The channel starts off being doped opposite to the contacts (e.g., we have highly doped n^+ contacts around a p-type channel, creating a large barrier to electron flow). The transistor turns on by applying enough gate voltage to exceed the threshold to invert the p-channel and make it effectively n-type, thereby shorting the channel between the contacts. The inversion requires that the gate suck enough minority carriers to the surface (electrons for a p-type channel) to exceed the density of majority carriers at equilibrium. Under an applied drain bias, a current flows through the inversion layer, and is proportional to the inversion charge density (thus, to the overdrive of gate voltage over the threshold for inversion), times the average electron drift (roughly proportional to electric field and thus the voltage at the power source). The net result is the square law theory of MOSFETs. Initially current increases with drain bias as the backflow from the drain end over the source side barrier reduces (Fig. 18.2). This continues until backflow is completely suppressed and the current saturates (Fig. 1.1(b)). The saturation current varies quadratically with gate voltage above the inversion threshold, especially for transistors a few generations old. In these textbook devices, current saturates due to pinchoff, when the local channel potential at the drain end is large enough to pull its voltage difference with the gate to below the threshold for inversion. Such a saturating curve with a high output impedance is essential to ensure adequate gain (which is related to the product of the output impedance $Z_T = (\partial I/\partial V_D)^{-1}$ and the transconductance $g_m = \partial I/\partial V_G$), so that a small change in input gate voltage can cause an adequate change in output voltage across a target capacitor in order to circumvent noise and fully restore its signal.

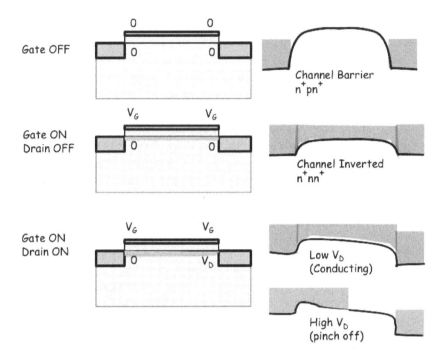

Fig. 18.2   *(Top) Transistor in OFF state, with electron distribution in pink showing n type source and drain and p type (electron deficient) channel. (Middle) A large enough positive gate bias above a threshold V_T sucks in enough electrons to invert the channel (dark green), shorting out the source to drain barrier and turning the FET ON. (Bottom) A drain bias gives directionality and creates an inversion current that rises linearly with V_D, until the latter pulls the channel at the drain end below inversion (i.e., the gate to channel voltage $V_G - V_D$ sinks below V_T). At this point, no more drain charges backflow over the peak of the barrier (which the gate capacitance must hold fixed), whereupon the current saturates.*

Critical to the saturation physics is the strong gate control that must hold the source side barrier fixed while the drain bias increases so that backflow is completely suppressed. Holding the barrier against slippage is getting harder to sustain with smaller channels where the drain fields tend to invade and cause barrier lowering. Various engineering designs are constantly being adopted to ensure that the gate capacitance stays dominant.

As transistors shrunk over the years, a new saturation mechanism came in. The electron speed ceased to grow with the increasing electric field as devices shrunk, saturating instead because its carrier mobility was compromised at high fields by scattering processes from phonons and electron

electron interactions. We moved accordingly from a square law variation to a linear variation of current with gate voltage, as the electron speed saturated under high bias through a lot of collision events.

The current density in any system is always decomposable into its charge carrier density times the electron velocity, whether drift or diffusion. To understand conventional transistor action, let us quickly go over the textbook equations before reverting to the Landauer formalism relevant to molecular processes. The areal charge density Q (in Coulombs/cm^2), drift velocity v and current I are given as a function of position z along the channel axis by

$$I(z) = \frac{Q(z)A}{t(z)} = \frac{Q(z)Wdz}{dz/v(z)}$$

$$Q(z) = -qC_{ox}\left(V_G - V_T - V(z)\right)$$

$$v(z) = \mu\mathcal{E}(z) = -\mu\frac{dV(z)}{dz} \qquad (18.1)$$

where μ is the mobility, V is the local channel potential. V_T is the threshold gate voltage required to invert the charge, set by the metal and semiconductor work function difference, the local band bending needed for strong inversion (a two step process — first bend enough to turn the channel intrinsic, and then dip further to compensate for the bulk majority carrier density), and finally the capacitive voltage drop across the oxide as well as impurity charges. To get the current, we integrate both sides of Eq. (18.1) from $z = 0$ to L with boundary conditions $V(0) = 0$, $V(L) = V_D$, keeping in mind that at steady state, $I(z) = I$ is independent of z.

Square Law Theory

$$I = \frac{1}{L}\int_0^L I(z)dz = \frac{\mu W C_{ox}}{L}\left[\left(V_G - V_T\right)V_D - \frac{V_D^2}{2}\right] \qquad (18.2)$$

Note that this equation predicts an inverted parabola for its I–V. Since it presupposes an inverted channel and ignores any depletion charges in Eq. (18.1), it can only be trusted upto the pinchoff point, i.e., the tip of the parabola reached at the saturation voltage $V_{sat} = V_G - V_T$. Beyond this saturation voltage where Q in Eq. (18.1) changes sign, we assume the current stays saturated to the maximum value $I_{sat} = I(V_{sat}) = \mu W C_{ox}\left(V_G - V_T\right)^2/2L$, giving us the square law theory. We can then add more complexity to this equation, such as the role of depletion charges, or the competing effect of the drain capacitor. The latter requires a body coefficient $m = 1 + C_D/C_{ox}$ inside the square bracket as a prefactor to the quadratic term in Eq. (18.2).

It is interesting to note that the only place where temperature explicitly enters our equations is in the threshold voltage V_T through a logarithmic expression for the surface potential (and to some extend, the number of fully ionized dopants). This will change as we go down this chapter, starting with temperature dependences entering the phenomenological mobility μ, to more explicit contact Fermi Dirac distributions for near ballistic FETs, and finally to thermal weighting factors when the electron has multiple configurational potentials to choose from.

Over the years, the square law theory reduced to a linear voltage variation with device shrinkage, as scattering preventing the velocity from increasing indefinitely with electric field $\mathcal{E} = -V/L$. Device engineers incorporate this effect with an empirical field dependent mobility

$$v = \mu \frac{\mathcal{E}}{\left[1 + \left(\mathcal{E}/\mathcal{E}_0\right)^\beta\right]^{1/\beta}} \longrightarrow \mu\mathcal{E}_0 \quad \text{when } \mathcal{E} \gg \mathcal{E}_0 \qquad (18.3)$$

so that the speed saturates at high electric fields due to increased scattering with impurities and phonons. Let us play with this expression for the special case of $\beta = 2$ where we can try something analytical. Plugging back this expression for v in the current equation as before and squaring both sides, and defining $x = V_G - V_T - V$, $x' = -V'$, we get

$$\frac{(xx')^2}{1 + (x')^2/\mathcal{E}_0^2} = \left(\frac{I}{\mu W C_{ox}}\right)^2 = i^2 \qquad (18.4)$$

where i is a position-independent constant. Collecting the x' terms together, we then get

$$i = x'\sqrt{x^2 - i^2/\mathcal{E}_0^2} \qquad (18.5)$$

The right hand side is now an exact differential and the left hand side is position independent by continuity. We can now integrate once again over channel coordinate z to get

$$iL = \int_0^L x'\sqrt{x^2 - i^2/\mathcal{E}_0^2}\,dy = \int_{V_G-V_T}^{V_G-V_T-V_D} \sqrt{x^2 - i^2/\mathcal{E}_0^2}\,dx \qquad (18.6)$$

Set $x = i/\mathcal{E}_0 \cosh\theta$, whereupon the right hand side becomes

$$iL = \frac{1}{2}\left(\frac{i}{\mathcal{E}_0}\right)^2 \left[\frac{\sinh 2\theta}{2} - \theta\right]_{\theta_2}^{\theta_1} \tag{18.7}$$

with $\cosh\theta_1 = (V_G - V_T)\mathcal{E}_0/i$ and $\cosh\theta_2 = (V_G - V_T - V_D)\mathcal{E}_0/i$.

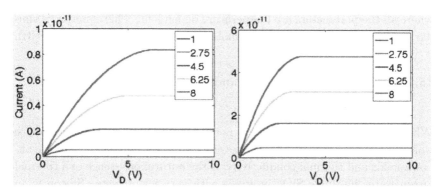

Fig. 18.3 *I-V characteristics for a device of length $L = 20\mu m$ (left) and $L = 2\mu m$ (right). The first shows conventional square law theory, but the second shows linear scaling because of field-dependent mobility that saturates the velocity prior to pinch-off (courtesy Yunkun Xie).*

Figure 18.3 shows how we progress from quadratic (left) to linear (right) scaling of the saturation current with gate voltage as we move from long to short channel. We discuss current saturation in the ballistic limit in Section 18.7.

A critical quantity in our analysis is the effective mobility of electrons and holes, which can depend on various parameters such as temperature, doping, electric field and substrate orientation. Takagi showed that the mobilities follow a 'universal' curve when plotted vs the effective transverse electric field which depends on the depletion and inversion charge densities, $\mathcal{E}_{\text{eff}} = q(N_{\text{dep}} + \eta N_{\text{inv}})/\epsilon_{Si}$, where η is an empirical parameter that is frequently taken to be $1/2$ for electrons and $1/3$ for holes. The equation can be fitted using Matthiessen's rule, i.e., the additivity of scattering rates, with the functional forms for each scattering mechanism (e.g., phonons or surface roughness) justifiable from Fermi's Golden Rule studies and eventually fitted to experimental data,

$$\mu_{\text{ph}} = A\mathcal{E}_{\text{eff}}^{-0.3}T^{-1.75}, \quad \mu_{sr} = B\mathcal{E}_{\text{eff}}^{-\gamma}, \quad \mu_{\text{Takagi}}^{-1} = \mu_{\text{ph}}^{-1} + \mu_{sr}^{-1} \tag{18.8}$$

with constants A_{el}, A_{hole}, B, γ fitted to data. Deviations from the universal curve, primarily at high doping densities N_I, are attributed to Coulomb scattering. An empirical way to capture these results for bulk materials is the Caughey-Thomas model,

$$\mu_{\text{Caughey-Thomas}} \approx \mu_{min} + \frac{\mu_0}{1 + (N_I/N_0)^\alpha} \qquad (18.9)$$

where all the parameters are temperature dependent. This equation is similar in form to the field dependent mobility we invoked earlier (Eq. (18.3)).

18.2 Ultra thin body FETs and FinFETs — interpreting effective mass

When it comes to speed, silicon is not the optimal material in the device world. The main strength of silicon, aside from its abundance, higher breakdown fields and thermal conductivity, is the natural occurence of a thermal grown water insoluble SiO_2 interface with very few defects. Silicon is an indirect bandgap material with a conduction band minimum along the Δ valley at $\vec{k}_0$ that is 0.85 times the Brillouin zone distance along the (100) or $\Gamma - X$ axis (Fig. 5.5). There are six equivalent constant energy ellipsoids along the $\pm x, \pm y, \pm z$ axes with longitudinal effective mass $m_l \approx 0.91\ m_0$ and transverse mass $m_t \approx 0.19\ m_0$,

$$E \approx E_C + \frac{\hbar^2(k_l \pm k_0)^2}{2m_l} + \frac{\hbar^2 k_T^2}{2m_t} \qquad (18.10)$$

with k_l and $\vec{k}_T$ cycling among $\pm k_x$, $\pm k_y$ and $\pm k_z$. If we now confine the silicon crystal along the z-axis, then the projected states onto the X-Y plane consist of four ellipsoids that retain their ellipticity with masses m_l and m_t, and two along the z-axis that project onto the 2-D Γ point with a circular constant energy contour and mass m_t (Fig. 18.4). The two Δ_2 ellipsoids pointing at the z direction quantize their subband bottoms to $\hbar^2(\pi/d)^2/2m_l$ by the larger effective mass m_l, creating the lower energy two-fold degenerate unprimed ladder. The four other ellipsoides quantize their subband bottoms to $\hbar^2(\pi/d)^2/2m_t$ by the smaller effective mass m_t, creating the four-fold degenerate higher energy primed subbands.

A convenient way to track the dynamics is to work out the rotation matrices that take us between the *device* coordinate system D (defined by the source-drain-gate-substrate), the *crystal* coordinate system C given by the

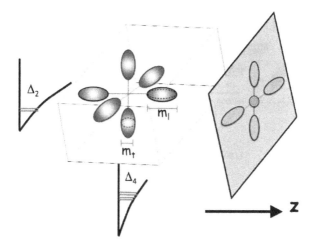

Fig. 18.4 *Constant energy ellipsoids around the bottom of the silicon conduction band (Δ valley in Fig. 5.5) project onto various directions upon confinement along the z axis, creating a sequence of primed and unprimed subbands when projected onto the x-y plane.*

unit cell axes (the $\langle 100 \rangle$ family assumed here) and the *ellipsoidal* coordinate system E which depends on the orientation of the Δ and Λ ellipsoidal valleys. In general, the kinetic energy for electron flow is quadratic in the quasimomenta (but not for holes, as we mention later)

$$H_K = \frac{\hbar^2}{2} \vec{k}_E^T M_E^{-1} \vec{k}_E \qquad (18.11)$$

We want to convert it to the device axes $\vec{k}_D$ where

$$\vec{k}_D = \begin{pmatrix} -i\partial/\partial x \\ k_y \\ -i\partial/\partial z \end{pmatrix} \quad \text{(along device axes)}$$

$$M_E = \begin{pmatrix} m_l & 0 & 0 \\ 0 & m_t & 0 \\ 0 & 0 & m_t \end{pmatrix} \quad \text{(along ellipsoidal axes)} \qquad (18.12)$$

We now make two transitions — one takes us from the device to the crystal (to $\langle 100 \rangle$), and the other takes us then from the crystal to the ellipsoid.

$$\vec{k}_E = R_{E \leftarrow C} \vec{k}_C, \quad \vec{k}_C = R_{C \leftarrow D} \vec{k}_D \qquad (18.13)$$

so that

$$M_D^{-1} = R_{E \leftarrow D}^T M_E^{-1} R_{E \leftarrow D}$$
$$R_{E \leftarrow D} = R_{E \leftarrow C} R_{C \leftarrow D} \qquad (18.14)$$

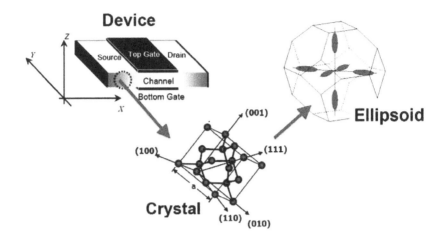

The rotation matrices can now be worked out for transitions from crystal axis to various wafer orientations D and various ellipsoidal valleys C, specifically the Δ valleys along the (100) axes (those of the cubic unit cell) and the Λ valleys along the (111) axes (along the cubic body diagonals).

$$R^{(001)}_{C\leftarrow D} = \begin{pmatrix} 1 & 0 & 0 \\ 0 & 1 & 0 \\ 0 & 0 & 1 \end{pmatrix} \quad R^{(111)}_{C\leftarrow D} = \begin{pmatrix} -\dfrac{2}{\sqrt{6}} & 0 & \dfrac{1}{\sqrt{3}} \\[2mm] \dfrac{1}{\sqrt{6}} & \dfrac{-1}{\sqrt{2}} & \dfrac{1}{\sqrt{3}} \\[2mm] \dfrac{1}{\sqrt{6}} & \dfrac{1}{\sqrt{2}} & \dfrac{1}{\sqrt{3}} \end{pmatrix} \quad R^{(110)}_{C\leftarrow D} = \begin{pmatrix} 0 & \dfrac{1}{\sqrt{2}} & \dfrac{1}{\sqrt{2}} \\[2mm] 0 & \dfrac{-1}{\sqrt{2}} & \dfrac{1}{\sqrt{2}} \\[2mm] 1 & 0 & 0 \end{pmatrix}$$

$$(18.15)$$

$$R^{\Delta_1}_{E\leftarrow C} = \begin{pmatrix} 1 & 0 & 0 \\ 0 & 1 & 0 \\ 0 & 0 & 1 \end{pmatrix} \quad R^{\Delta_2}_{E\leftarrow C} = \begin{pmatrix} 0 & 1 & 0 \\ 0 & 0 & 1 \\ 1 & 0 & 0 \end{pmatrix} \quad R^{\Delta_3}_{E\leftarrow C} = \begin{pmatrix} 0 & 0 & 1 \\ 1 & 0 & 0 \\ 0 & 1 & 0 \end{pmatrix} \quad (18.16)$$

$$R^{\Lambda_1}_{E\leftarrow C} = \begin{pmatrix} \dfrac{1}{\sqrt{3}} & \dfrac{1}{\sqrt{3}} & \dfrac{1}{\sqrt{3}} \\[2mm] -\dfrac{1}{\sqrt{2}} & \dfrac{1}{\sqrt{2}} & 0 \\[2mm] \dfrac{-1}{\sqrt{6}} & \dfrac{-1}{\sqrt{6}} & \dfrac{2}{\sqrt{6}} \end{pmatrix} \quad R^{\Lambda_2}_{E\leftarrow C} = \begin{pmatrix} \dfrac{1}{\sqrt{3}} & \dfrac{1}{\sqrt{3}} & \dfrac{-1}{\sqrt{3}} \\[2mm] \dfrac{-1}{\sqrt{2}} & \dfrac{1}{\sqrt{2}} & 0 \\[2mm] \dfrac{1}{\sqrt{6}} & \dfrac{1}{\sqrt{6}} & \dfrac{2}{\sqrt{6}} \end{pmatrix} \quad (18.17)$$

$$R^{\Lambda_3}_{E\leftarrow C} = \begin{pmatrix} \dfrac{-1}{\sqrt{3}} & \dfrac{1}{\sqrt{3}} & \dfrac{1}{\sqrt{3}} \\[2mm] \dfrac{1}{\sqrt{2}} & \dfrac{1}{\sqrt{2}} & 0 \\[2mm] \dfrac{-1}{\sqrt{6}} & \dfrac{1}{\sqrt{6}} & \dfrac{-2}{\sqrt{6}} \end{pmatrix} \qquad R^{\Lambda_4}_{E\leftarrow C} = \begin{pmatrix} \dfrac{-1}{\sqrt{3}} & \dfrac{1}{\sqrt{3}} & \dfrac{-1}{\sqrt{3}} \\[2mm] \dfrac{1}{\sqrt{2}} & \dfrac{1}{\sqrt{2}} & 0 \\[2mm] \dfrac{1}{\sqrt{6}} & \dfrac{-1}{\sqrt{6}} & \dfrac{-2}{\sqrt{6}} \end{pmatrix}$$

The final effective mass tensor M_D in the device coordinate system is a full matrix with components m_{ij}. It is now straightforward to work out the various mass components relevant to our transistor.

$$m_Z = m_{33} \quad \text{(confinement effective mass)}$$

$$m_X = \left(1/m_{11} - m_{33}/m_{31}^2 \right)^{-1} \quad \text{(transport effective mass)}$$

$$m_Y = \left(1/m_2' - m_X/(m_{12}')^2 \right)^{-1} \quad \text{(width effective mass)}$$

$$m_2' = \left(1/m_{22} - m_{33}/m_{23}^2 \right)^{-1}, \qquad m_{12}' = \left(1/m_{12} - m_{33}/m_{23}m_{31} \right)^{-1}$$

$$m_{DOS} = \sqrt{g m_X m_Y} \quad \text{(density of states effective mass)} \tag{18.18}$$

where g is the ellipsoidal degeneracy.

Each of the above masses has a clear role to play. The confinement effective mass m_Z determines the bottoms of the subbands, deciding in turn which ones are selectively populated and active at low bias and low temperature. Keeping the masses high and the subband bottoms low increases the degeneracy of the subbands and thus the carrier density and overall ON current. The carrier density is also determined by the density of states effective mass m_{DOS}, obtained by imagining a single ellipse with the same area as all the ellipses above put together. Finally, the transport effective mass m_X determines the carrier mobility. It also determines the ease of tunneling, which may affect the OFF current more substantially than the ON current for ultrascaled devices with source to drain tunneling.

We can now rewrite the simplified Schrödinger equation in the device coordinate system in terms of the new mass elements. Assuming a potential $V(x)$ along the transport direction varying slowly enough to eliminate Zener tunneling between subbands, we can treat the latter as decoupled modes and throw away the dipolar cross-coupling terms, giving us

$$\left[\frac{-\hbar^2}{2m_X} \frac{\partial^2}{\partial x^2} + V(x) \right] \psi_n(x) = \left[E - \epsilon_n - \frac{\hbar^2 k_y^2}{2m_Y} \right] \psi_n(x) \tag{18.19}$$

where the subband energies ϵ_n satisfy the confinement equation

$$\left[\frac{-\hbar^2}{2m_Z} \frac{\partial^2}{\partial z^2} + W\left(x + \frac{m_{33}}{m_{31}} z, z \right) \right] \psi_i(x, z) = \epsilon_i \left(x + \frac{m_{33}}{m_{31}} z, z \right) \psi_i(x, z)$$

(18.20)

where W is the confinement potential, sampled now along an angular axis in the x–z plane set by the ellipsoid orientations.

Using these equations, we can make estimates for how big the ballistic currents are for Si or Ge of various wafer orientations. The equations are more complicated for the holes. The latter are typically derived from the p-orbitals of the atoms and show significant non-parabolicity at low energies. The warped bands are described by eigenstates of parametrized multiband models, taking on empirical forms like that look like

$$E \approx E_V - Ak^2 \pm \sqrt{\left[B^2 k^4 + C^2(k_x^2 k_y^2 + k_y^2 k_z^2 + k_z^2 k_x^2) \right]}$$

(18.21)

For small k-values, it is often convenient to describe these as multiple parabolic bands with different masses, and designate them as heavy hole bands, light hole bands, and in the presence of spin-orbit coupling, additional split-off bands.

18.3   Molecular FETs: what are 'mobility' and 'inversion'?

The equations for current flow in organic FETs are similar to conventional FETs, i.e., classical drift diffusion. The complexity arises in the various hopping processes that influence the field dependent mobility, for instance the Frenkel Poole equation, Eq. (13.16). In contrast, molecular FETs (more precisely, *single molecule* FETs) consist of ordered monolayer arrays, at least in idealization. Hopping processes along their backbone, i.e., through-band transport channels, become more important than intermolecular hopping. The expectation is that such molecular FETs act as a ballistic channel. The mechanism for current saturation is qualitatively different in *ballistic* transistors, a concept that began as an idealization but is quite relevant today. 2014 transistors operate well above 50% of the ballistic limit (the observed current is above half the predicted ballistic ON current), while transistors based on nanowires and tubes have been clocked at 90 to 100% of their ballistic limit. Organic molecules are expected to operate in these quasiballistic regimes, although paradoxically, their bane has been their poor contacts with the interconnects. Nonetheless, it is worth studying a model molecular transistor, to identify its strengths and weaknesses.

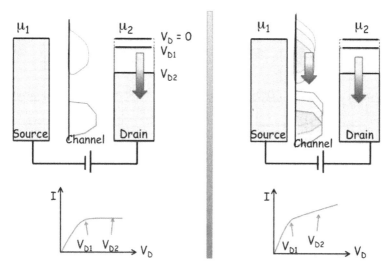

Fig. 18.5 *Current saturation is driven by the band alignment (i.e., the relative voltages) across the channel contact heterostructure, specifically, (left) the slippinge of the drain Fermi energy μ_2 past the channel density of states (DOS) into its bandgap. This requires that the channel states be held fixed relative to the slipping drain states by the gate, whose capacitance must therefore dominate the electrostatics. Aggressive scaling has caused the gate capacitance to lose its edge over the source and drain, causing (right) the channel states to slip as well and preventing the current from saturating. See Fig. 18.11.*

It is straightforward to understand how current would saturate in a ballistic 'molecular' transistor, and indeed, what the corresponding resistance and mobility mean in the absence of scattering. A source to drain bias (Fig. 18.5) pulls the drain's quasi Fermi energy towards the bandgap. Increasing the conduction window initially increases the current, but pretty soon the drain Fermi energy crosses the bandedge and into the gap, whereupon no further charge can be extracted from the band, and the current as a result saturates. The saturation current depends on gate bias which sets the initial Fermi energy location inside the band. In a ballistic transistor therefore, the main mechanism behind the flat current is *charge saturation* rather than velocity saturation. For a given gate bias, we reach the maximum charge we could draw from the band (or for extended transistors, the maximum we can pull out above the top of the barrier). We develop the mathematical relations shortly in Section 18.7.

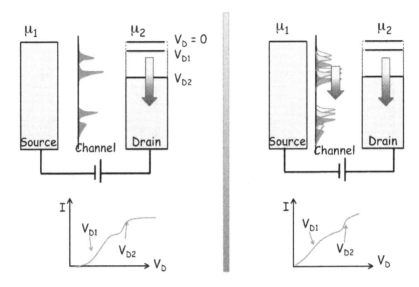

Fig. 18.6   *Analyses repeated for a molecular DOS with (left) good gate control, where the current shows saturating steps after each peak in the DOS is crossed by μ_2. (Right) With poorer gate control, common for short molecules, saturation is once again compromised.*

We can see the parallel between the saturation mechanism in a ballistic molecular FET vs that in conventional transistors if we just look at the band alignment in each case. For conventional doped FETs, we need to look at the surface potential in lieu of the lumped molecular density of states. For n type conduction (the case considered in Fig. 18.5), turn on occurs by pulling *up* the Fermi energy μ_1 from a 'barrier' state, from the HOMO LUMO gap in molecules, vs all the way from the valence band in a p channel FET (inverting the latter in the process). Current flows when we apply a drain bias to pull μ_2 *down* and open a window of conduction within the turned on states. Current saturates when the drain bias is large enough to neutralize the turn on, i.e., μ_2 slips back into the gap state for molecules, or back towards the valence band in a conventional FET, removing the inversion (pinch off). For a molecular FET, we simply look at the lumped density of states, while for a nanoFET, we look at the states at the top of the barrier where the gate control is maximum.

3D electrostatics plays a critical role in ensuring saturation in each case. The saturation works effectively if the channel states stay fixed while the drain Fermi energy slips past the bandedge. Allowing the channel states to slip in response to the drain bias compromises the saturation of the current, which in turn degrades the inverter gain (the non-ideal electrostatics is usually referred to as 'short channel effects'). To avoid channel slippage, the gate capacitance must exceed the source/drain capacitance significantly and keep the channel potential fixed. Historically, the capacitance ratio $C_{ox} = \epsilon_{ox}A/t_{ox}$ and $C \approx \epsilon_{Si}A'/L$ has been close to 13 (A, A' are the areas of the channel controlled by the gate and drain respectively). The capacitance ratio means that the silicon channel needed to be about $13(\epsilon_{Si}/\epsilon_{ox}) \sim 40$ times longer than the oxide thickness. On the other hand, the oxide thickness could not be pushed down below ~ 1 nm else the tunneling current across the oxide would dominate performance (the tunnel current can be understood in terms of the Simmons equation, Eq. (14.14)).

A compromise was reached with the advent of **high k** oxide technology, that allowed the careful insertion of a few monolayers of high dielectric constant materials such as HfOx inside the SiO2 oxide stack, increasing thus the gate capacitance without significant tunneling. Initial progress was hampered by the reduction in channel mobility from soft phonon modes, but with the advent of metal gates, these modes were dynamically screened out, leading to the successful onslaught of high k technology. Another progress happened with the scaling of the gate controlled channel area A, by placing multiple gates around a thinned down channel. The resulting structures ranging from dual gate to trigates, Ω-gates, FinFETs and finally wrap around gates (e.g., in a silicon nanowire) are all designed to enhance the role of the multigates relative to the source and drain capacitances.

The question we would like to address is how one interprets the phenomenological quantities such as μ or N for a small molecular conductor. Indeed, we need to invoke instead a different 'bottom up' approach and tie up with the top down way of thinking. To model current in a ballistic molecular conductor, we need to invoke Landauer equation (Eq. (12.5)). Treating the channel as a point, we can rewrite the equation as

$$I = \frac{q}{h} \int dE \underbrace{2\pi\gamma_{\text{eff}}D}_{M}[f_1 - f_2]$$

$$= q \int dE v(E) \frac{C_Q}{q^2 L}[f_1 - f_2] \tag{18.22}$$

where we replaced the mode count with $M = hvD/L$, and invoked the quantum capacitance $C_Q = q^2 D$ which we interpret in the next section. It is straightforward then to associate the transit time of electron flow with the RC time constant familiar to circuit theory, except we deal with quanta of resistance and capacitance in each case

$$M = h v D / L \implies \underbrace{\frac{L}{v}}_{t} = \underbrace{\frac{h}{q^2 M}}_{R_Q} \times \underbrace{q^2 D}_{C_Q} \implies \boxed{t = R_Q C_Q} \qquad (18.23)$$

18.4 Compact model for potential drop across a molecular conductor

To understand the relation between this equation developed for a ballistic ultimately scaled molecular FET vs a classical transistor, we need to add a diffusive resistance in series and an electrostatic capacitance also in series,

$$R \to R_Q + R, \quad C \to C_Q C_E / (C_Q + C_E)$$
$$t = RC \qquad (18.24)$$

so that for a small ballistic channel ($C_Q \propto D \ll C_E$, $R_Q \gg R$), we get the above result in Eq. (18.23), while for a large diffusive channel ($C_Q \gg C_E$, $R_Q \ll R$) we get the square law theory described earlier in Eq. (18.2). For a molecular FET, the relevant capacitance is set by the density of states, while the relevant velocity is set by the hopping parameters $\gamma = \hbar v / L$.

Why did we interpret the density of states as a capacitance, and that too a quantum one? To understand this, let us reexamine Poisson's equation:

$$\vec{\nabla} \cdot \left(\epsilon \vec{\nabla} \phi \right) = -q^2 \left[n(\vec{r}) - N_D \right] \qquad (18.25)$$

where N_D is the background dopant density, and $n(\vec{r}) = \int dE D(E, \vec{r}) f_0(E)$, depends on the local density of states $D(E, \vec{r})$ per unit volume. If we now resolve this equation onto a grid of points $\{i\}$ with uniform spacing a, and operate in the linear response regime where the above equation simplifies, $\delta n_i \approx D_i (\mu_i - \phi_i)$ evaluated at energy μ_i, then the left side of Eq. (18.25) can be written using finite difference as $[\epsilon_{i-1}(\phi_i - \phi_{i-1}) - \epsilon_i(\phi_{i+1} - \phi_i)]/a^2$. Thereupon the electrostatic equation (18.25), reexpressed in terms of the electrostatic capacitance per unit area $C_i = \epsilon_i / a$ reads

$$C_{i-1}(\phi_i - \phi_{i-1}) - C_i(\phi_{i+1} - \phi_i) = -C_i^Q(\mu_i - \phi_i) \qquad (18.26)$$

naturally identifying the quantum capacitance per unit area $C_i^Q = q^2 D_i a$. This circuit equation represents charge conservation at a node.

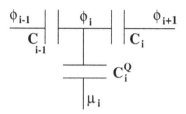

Fig. 18.7 *Capacitive voltage division in a small quantum object involves both the local electrostatic capacitances C_i and the semiconductor quantum capacitance C_i^Q set by the local density of states.*

A similar decomposition of the equation of continuity $\vec{\nabla} \cdot \left(\sigma \vec{\nabla} \mu \right) = 0$ amounts to

$$G_{i-1}(\mu_i - \mu_{i-1}) - G_i(\mu_{i+1} - \mu_i) = 0 \qquad (18.27)$$

describing Kirchhoff's law of current conservation at a junction of resistors, with conductance per unit area $G_i = \sigma_i/a$.

The above equations allow an elegant interpretation in terms of lumped circuit elements. The equation of continuity (18.27) can be visualized as a network of resistors characterized by individual conductivities σ_i, determining the local electrochemical potentials μ_i and current densities. The homogeneous (Laplace) part of the Poisson equation (18.26) can be expressed in terms of a network of capacitors characterized by individual dielectric constants ϵ_i, which determine the local electrostatic potentials ϕ_i. The two networks are connected at each point by a local quantum capacitance $C_i^Q = q^2 D_i a$ per unit area across which a voltage drop $(\mu_i - \phi_i)/q$ creates a local charge $qn_i = qD_i (\mu_i - \phi_i)$ that acts as the inhomogeneous driving term in Poisson's equation. It is straightforward to generalize this to a finite channel, as shown below.

We can solve the above circuit equation for the local electrostatic potential ϕ_m and the current I. The linear-response current is given by

$$I = \frac{V_D}{R_S + R_D} \qquad (18.28)$$

while the voltage division fraction, defined as the fraction of the applied potential that acts on the molecule, is given by

$$\frac{\phi_m}{V_D} = \frac{C_E}{C_Q + C_E} r_C + \frac{C_Q}{C_Q + C_E} r_R \qquad (18.29)$$

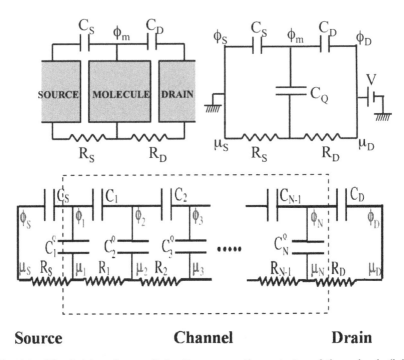

Source   **Channel**   **Drain**

Fig. 18.8   *The division of an applied voltage among the contacts and the molecule (left) is determined both by the resistance ratio and the capacitance ratio, weighted by the total electrostatic capacitance $C_E = C_S + C_D$ and the quantum capacitance C_Q. The average molecular potential ϕ_m can be obtained from the circuit diagram shown on the right, and amounts to Eq. (18.25). A distributed circuit model at the bottom generalizes to include the entire potential profile with variations in local μ and ϕ.*

where the capacitive division factor $r_C = C_D/C_E$ while the resistive division factor $r_R = R_S/(R_S + R_D)$. The electrostatic capacitance $C_E = C_S + C_D$ can be expanded to include a gate capacitance in parallel. *We see now that for a dot with a small density of states (insulating molecule, $C_Q \ll C_E$) the capacitive division simply described by Laplace's equation dominates, while for a dot with a large density of states (conducting molecule, $C_Q \gg C_E$) the resistive division dominates.*

18.5   Role of the potential profile

The electrostatic potential profile across a molecular conductor plays a critical role in determining its I-V. For a simple lumped circuit model of the molecule, the Poisson's equation can be recast as $U = U_L + U_0(N - N_0)$, whence for small bias voltages the NEGF equations for charge read as,

$$N_0 = \int dE D(E) f(E - E_F)$$

$$N = \int dE D(E - U) \frac{\gamma_1 f(E - \mu_1) + \gamma_2 f(E - \mu_2)}{\gamma_1 + \gamma_2}$$

$$\Delta N \approx \int dE D(E) \left[\frac{\gamma_1 (U - U_1)}{\gamma_1 + \gamma_2} + \frac{\gamma_2 (U - U_2)}{\gamma_1 + \gamma_2} \right] \left. \frac{\partial f}{\partial E} \right|_{E=E_F}$$

$$U = U_L + U_0 \Delta N \tag{18.30}$$

where $f(E) = 1/(1 + \exp(E/kT))$, while $U_{1,2}$ are the potentials at the source and drain contacts. The Fermi derivative $\sim \delta(E - E_F)$ pulls out just the Fermi energy from the integral expression for ΔN at low temperature, whereupon the equation set simplifies to

$$U = \frac{U_L C_E + U_N C_Q}{C_E + C_Q}. \tag{18.31}$$

Here the *neutrality potential* $U_N = (\gamma_1 U_1 + \gamma_2 U_2)/(\gamma_1 + \gamma_2)$ is obtained by setting $N = N_0$, while the quantum capacitance $C_Q = q^2 D_0$, where $D_0 = D(E = E_F)$ and the electrostatic capacitance $C_E = q^2/U_0$. The equation for U illustrates that the voltage division across a molecule interpolates between an *insulating limit* U_L where charges are freely added and removed from the system, and a *neutrality limit* U_N when the density of states is very high (e.g., in a metallic species or with abundant metal induced gap states) and the system is pinned near neutrality. The limits are isomorphic with the Bardeen limit and the Schottky limit for Fermi level pinning at metal-semiconducting interfaces. The actual potential lies in between at a value determined by the ratio $U_0 D_0 = C_Q/C_E$, which requires a careful evaluation of the local density of states and its charging cost.

18.6 Quantum capacitance: charge control vs energy level control

One of the less appreciated aspects in nano-electronics is the seminal role electrostatics plays in determining the band-alignment. While classical device engineering does pay attention to Poisson's equation and electron screening, albeit in a mean-field limit, what is not often recognized is that the electrostatics has radically different import for nanostructures with diminished densities of states. We can illustrate this by raising an apparent conundrum. Suppose we have a metallic top gate deposited on a suspended graphene sheet with an oxide insulator in between. If we apply 1 volt gate bias on this gate, which of the following equations is correct?

- $Q_{ind} = C_{ox}V_G$.
- $U = -qV_G$ where U defines the shift in energy level (e.g., Dirac point) of the gated region.

Prima facie, both look reasonable — after all, the Q equation is seminal in circuit theory and the starting point for analyzing MOS devices above threshold (we are assuming here that there are no competing electrodes). At the same time, if we apply a gate voltage V_G to a material and there is no other capacitive division, would not all energy levels feel the full force of this voltage and shift up by $-qV_G$? And yet, these two equations cannot both be correct — for one because the first depends on the geometry of the oxide while the second seems universal. Also, since graphene has a linear band-structure and a corresponding linear density of states, its charge density varies quadratically with separation between the Dirac point and the Fermi energy. If proposition 1 were correct, the gate voltage would also vary quadratically with shift in Dirac point, while proposition 2 argues for a linear variation. The standard graphene literature seems to gravitate towards the former. Which then is correct?

The answer, interestingly, is *both*, except in opposite limits! We can see this by looking at the self-consistent solution to Poisson-NEGF introduced in the previous subsection. From that analysis we derived two equations, for low drain bias (i.e., linear response)

$$U = \frac{U_L C_E + U_N C_Q}{C_E + C_Q}$$

$$Q = q\Delta N = \left(\frac{C_{ox}C_Q}{C_{ox} + C_Q}\right)\left[\frac{U_N - U_L}{q}\right] = \left(\frac{C_{ox}C_Q}{C_{ox} + C_Q}\right)\left[V_G - V_T\right]$$

$$(18.32)$$

where U_L is the Laplace solution (for a single gate, $U_L = -qV_G$, while for multiple gates it would be a capacitive average of the gate voltages), U_N is the neutrality potential, C_E is the total electrostatic capacitance (C_{ox} for a single gate electrode, a parallel sum of capacitances if there are many electrodes) and $C_Q = q^2 D$ is the quantum capacitance.

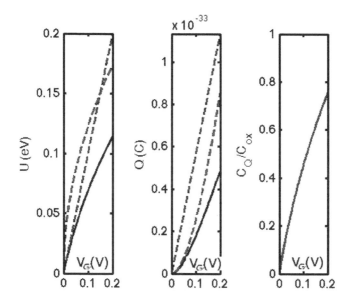

Fig. 18.9 *For a single gated graphene sheet with 1 nm oxide (left), the shift in Dirac point (black solid) starts linear (parallel to $-qV_G$, blue dashes), then bends around to give an additional $\sim \sqrt{V_G}$ dependence (parallel to red dashes). However, the magnitude of the shift is not the standard $\hbar v_F \sqrt{\pi C_{ox} V_G}$, but lower. (Middle) Similarly, the induced charge consistently stays less than $C_{ox}(V_G - V_T)$, blue dash, closer to $C_Q(V_G - V_T)$ (red). (Right) The reason is that for a 1nm oxide thickness, the quantum capacitance is distinctly smaller than the gate oxide capacitance and dominates the electrostatics.*

The equations clearly show the answer to the conundrum: when the Fermi energy is far enough from the Dirac point that the density of states is large and $C_Q \gg C_{ox}$ (or alternately we have very thick oxides), $U \approx U_N$ while $Q \approx C_{ox}[V_G - V_T]$ and proposition 1 is correct. If however we have very thin oxides or we are close enough to the Dirac point, $C_Q \ll C_{ox}$, $U \approx U_L = -qV_G$, while $Q \approx C_Q(V_G - V_T)$ and proposition 2 is correct.

In solid state devices in the form of films, the oxide capacitance is in series with a depletion capacitance along the thickness of the film, which reduces the overall gate capacitance. Truly 2-D material systems have an intrinsic advantage associated with the above equation, since they do not have enough material to support a depletion width, $U \approx U_L$, meaning that the density of states is inadequate to pin the levels and the 2-D channel then responds to the applied terminal voltages readily. This improved electrostatics is a key driver for 2-D material based devices.

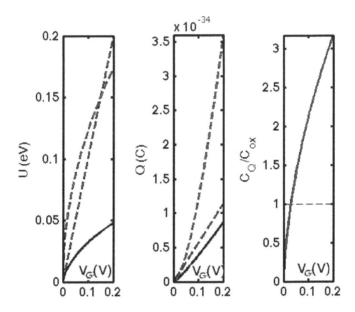

Fig. 18.10   *For thick oxides at 10 nm, the quantum capacitance is large, and the curves follow our expectations from classical electrostatics*

Let us work out some numbers for SiO_2, ignoring charge puddles

$$\frac{C_{ox}}{C_Q} \approx \frac{[\epsilon_{ox}S/t_{ox}]}{2q^2 S|E_F|/\pi\hbar^2 v_F^2} = \frac{0.15 \text{ eV-nm}}{|E_F|t_{ox}} \tag{18.33}$$

For graphitic samples with charge puddles, we replace E_F with $\sqrt{E_F^2 + 2\sigma_E^2/\pi}$ with σ_E typically around 50 meV.

18.7   I-V for a ballistic silicon MOSFET

For a 2-d electron gas described by single band with an isotropic effective mass m^*, $E = E_C + \hbar^2 k^2/2m^*$, area $S = WL$, we can use Eq. (9.16) for the 2-D density of states by state counting

$$D(E)dE = \frac{2 \text{ for spin}}{(2\pi/L)^2} \times \underbrace{2\pi k dk}_{\text{2D} - \text{k space volume}} \implies D(E) = \frac{m^* S}{\pi\hbar^2}\Theta(E - E_C)$$

$$\tag{18.34}$$

and the corresponding velocity $v(E) = \partial E/\hbar \partial k = \sqrt{2E/m^*}$. This gives us the ballistic current from Eq. (18.22), including a potential shift U in the velocity and density of states $D(E - U)$, which simplifies to

$$I = \frac{q\sqrt{2m^*}W}{\pi\hbar^2} \int_{E_C + U}^{\infty} dE(E - E_C - U)^{1/2}\Big[f(E - \mu_1) - f(E - \mu_2)\Big] \quad (18.35)$$

This equation can be re-expressed in terms of the Fermi integrals, $\mathcal{F}_{1/2}(\eta) = (2/\sqrt{\pi}) \int_0^{\infty} x^{1/2}dx/[e^{(x - \eta)} + 1]$ and solved numerically (with considerable in-built physics - see papers by Natori and Lundstrom referenced later). For a non-degenerate semiconductor (Fermi energy far from band-edges) these integrals reduce to e^{η}, in which case the current can be simplified as

$$I \approx I_0 e^{q(\alpha_G V_G + \alpha_D V_D)k_B T}\Big[1 - e^{-qV_D/k_B T}\Big], \quad \text{(Nondegenerate)}$$
$$(18.36)$$

using a capacitive voltage division $U = -q(\alpha_G V_G + \alpha_D V_D + \alpha_S V_S)$ with grounded source ($V_S = 0$) for the Laplace potential. We can immediately see that for superior gate control $\alpha_D \ll \alpha_G$ (strong input-output asymmetry), the current saturates at large drain bias - while for poor gate control ($\alpha_D \approx \alpha_G$), the saturation is compromised by the so-called drain-induced barrier lowering (DIBL) discussed earlier in Fig. 18.5.

Note that the saturation seems to trigger at an unphysically low voltage $V_D \sim k_B T/q$, which is an artifact because we can no longer assume non-degenerate semiconductors near saturation and replace the Fermi integrals by exponentials in Eq. (18.35). A way to resolve this without going into the Fermi integrals is to go to zero temperature, where the Fermi functions turn into step functions. Equation (18.35) then simplifies to the form

$$I = \frac{2q\sqrt{2m^*}W}{3\pi\hbar^2}\left[\Big(\mu_1 - U - E_C\Big)^{3/2} - [\max\Big(\mu_2 - U - E_C, 0\Big)]^{3/2}\right] \quad \text{(Zero T)}$$
$$(18.37)$$

where we assume $\mu_1 = E_F$ is inside the conduction band (NMOS is ON), so that when the drain electrochemical potential $\mu_2 = E_F - qV_D$ slips past the sliding bandedge $E_C + U$, with $U = -q(\alpha_G V_G + \alpha_D V_D)$, the current starts to saturate, with the degree of saturation set by the drain control parameter α_D (Fig. 18.11). Note also that the equation predicts an abrupt transition at threshold that will be smoothened out at finite temperature.

Classical transistors follow a similar trend, but the details differ - the electron velocity is set by the electric field and mobility, which in turn has

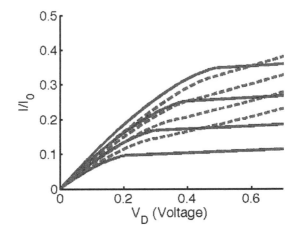

Fig. 18.11 *I-V for a ballistic MOSFET obtained by solving Eq. (18.37) for a material with Fermi energy 0.2 eV above conduction band-edge, plotted for a gate voltage varying from 0 to 0.3 V in steps of 0.1 V. For strong gate control ($\alpha_G = 0.9$, $\alpha_D = \alpha_S = 0.05$, blue solid lines) the currents saturate while for weak gate control ($\alpha_G = 0.5$, $\alpha_D = \alpha_G = 0.25$, red dashed lines) the current is quasi-linear (low output impedance) and shows a weaker gate-dependance (low transconductance).*

additional voltage-dependences (e.g. Frenkel-Poole) that deviate from the above three-halves power law. We can capture these deviations by the modifications in Eq. (18.24). Equation (18.36) for a non-degenerate ballistic MOSFET is nonetheless quite useful in its compactness and simplicity. We can use it to estimate the gain by cascading a PMOS and an NMOS transistor in series between power supply voltage V_{DD} and ground (Fig. 18.1). For one of the two transistors, we use the above equation with the input and output voltages given by $V_{in} = V_G$, $V_{out} = V_D$, while for the other we use $V_{in} = V_{DD} - V_G$, $V_{out} = V_{DD} - V_D$, and equate the currents since the transistors are in series. Dropping the second term inside the square brackets in Eq. (18.36) for voltages fairly large compared to $k_B T/q$, the inverter equation then amounts to a straight line in the middle of the inverter curve

$$\alpha_D V_{out} + \alpha_G V_{in} = 0.5(\alpha_G + \alpha_D)V_{DD} \qquad (18.38)$$

with the inverter gain given by the slope $|dV_{out}/dV_{in}| = \alpha_G/\alpha_D$. Clearly we get a better gain for $\alpha_G \gg \alpha_D$ when the gate controls are superior and the currents saturate. We also need this gain and the current saturation to persist over a fairly large voltage range, meaning we need a sizeable bandgap with negligible density of states within it. These two criteria - good electrostatics and low midgap density of states, are precisely where small organic molecules tend to suffer, purely for geometrical reasons.

18.8   Current non-saturation in aromatic molecules

Experiments on short organic molecules show very weak gating action. The current changes by a modest ratio (compare a CMOS ON-OFF of 10^3-10^4), and the gate voltage required to bring about that change is substantive. The origin of this poor gateability is not hard to fathom in the light of the previous discussions. First is the loss of capacitive control that inevitably sets in when we shrink the channel size down to a short molecule. For an organic molecule a few nanometers long, the effective oxide thickness must be a few Angstroms or less (recall the historical capacitance ratio of 13 in Section 18.3), making it impossible ultimately to sustain its job as an insulator - an electron would instantly tunnel out. The invariable trade off between maintaining high capacitance and high resistance extends to all insulators, whether SiOx, a high k dielectric or even vacuum. This means realistically much of the channel charge will also be controlled by the drain, causing the channel states to slip with drain bias. The slippage prevents the drain Fermi energy from crossing its bandedge abruptly, thereby preventing saturation (we mentioned in Fig. 18.2, saturation requires that the gate capacitance holds the top of the barrier fixed). Indeed, most molecular IVs show very modest gate control (compared to the orders of magnitude desired for logic applications) and very poor saturation.

More importantly, most molecular measurements involve metal contacts to take advantage of self assembly. For short molecules electron wavefunctions near the metal Fermi energy tunnel through and short out the source to drain path even when we expect the molecule to turn off. This causes the molecule to effectively act metallic. We are familiar with weak gateability for semimetals like graphene, but the surprise (in retrospect probably not all that surprising) is that even traditional molecules like hydrogen become metallic on absorption onto contacts like platinum, as their bandgaps get inundated with electronic states from metals. The metal induced gap states (MIGS) decay exponentially with distance; but the molecular lengths are typically smaller than the decay lengths and become pervious to tunneling, so that depending on the molecular barrier, their IVs either look Ohmic or activated, but rarely saturate and are only weakly gate dependent. In other

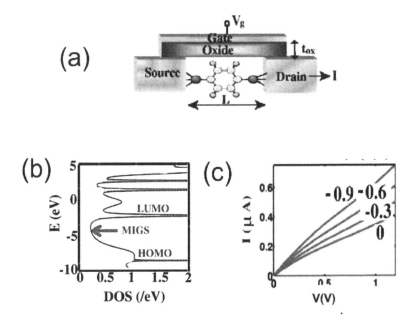

Fig. 18.12   *(a) Schematic molecular FET. (b) Calculated density of states for benzene dithiol (BDT) adsorbed on Au(100) contact surfaces. The presence of metal induced gap states (MIGS) makes the molecule act somewhat metallic, erasing its bandgap. (c) The resulting current changes very little with gate voltage and does not saturate. The poor OFF current, subthreshold swing and output impedance imply large voltages for switching, high leakage and dissipation, and poor gain if fashioned into an inverter.*

words, both their output impedance $Z_T = (\partial I / \partial V_D)^{-1}$ and transconductance $g_m = \partial I / \partial V_G$ are low, their product $\partial V_D / \partial V_G = \partial V_{\text{out}} / \partial V_{\text{in}}$ thus indicating a fundamentally weak inverter gain.

The dual attributes of direct tunnel bands from the metal contacts and poor electrostatic integrity make most organic molecular wires unsuitable for conventional high performance switching applications. It is worth emphasizing that this is somewhat independent of the molecular chemistry. The problem lies here in the geometry, specifically the size and the relative proximity of the competing electrodes. It may be possible to ameliorate the situation with very high dielectric constant insulators (e.g., ionic gating), but the tradeoffs nonetheless remain. It behooves us therefore to look for other innovative ways to gate a molecular channel.

Chapter 19

Configurational gating

Much of the gating we encounter in device physics is electrostatic, associated with depleting or accumulating charges near the gate terminal at the surface of the channel. Such a process is fast, because of the low effective mass of the electrons. However, it makes it hard to fully turn off the channel, especially when the oxide is very thin, leading to considerable leakage and stand-by power dissipation. Cantilevers and relays allow us to physically turn off the channel by moving pieces out of the conduction path. These processes are relatively slower, but clearly consume considerably less power. Such a gated geometry could vary from an entire cantilever to a fluctuating bond, a rotating molecular ring or a relay protein in a gated sodium channel.

The key mechanism behind a configurational gate is the electrostatic manipulation of its potential landscape in configuration space in order to eliminate the barrier between two wells, (for instance, using a cantilever in Fig. 19.1). The barrier removal happens by moving the local maximum continuously into a minimum, together with a bias forward to drive a sharp transition. At the point of inflexion, we expect an abrupt change in configuration, such as the conformation of a relay, magnetization in a multiferroic assembly, polarization in a negative capacitance FET or conductivity in a VO_2 switch. In cases where one of the wells is conductive and the other insulating, such a transition creates a sharp sub-thermal switch, as we shall see towards the end of the book. However, the recovery of the original state requires us to bias backward into the original state. In a later chapter, we will argue why hysteresis arises naturally in these switches.

Our purpose in this chapter is to lay out the mathematical machinery to understand how to gate a configurational degree of freedom and the point of destabilization for the associated transition. We will use these equations later to quantify the influence of this abrupt transition on power dissipation in a configurationally gated transistor.

19.1   Case study: a molecular relay

An attribute that clearly distinguishes organic molecules from conventional solid state switches is mechanical flexibility. This opens up the option for reconfigurable logic, and more generally, conformational gating. Such gating processes are likely to be slower than purely electronic switching, but they can reduce the energy cost substantially by physically moving the channel away from the current path when turned off. Besides the possibility of a reduced overall energy-delay product, a molecular relay provides a compelling case study for understanding how mechanical gating can couple with electronic transitions. It is also believed that ion channels in axonal systems operate subthermally under the action of a gate voltage. While the equations underlying such a system, such as the Hodgkin-Huxley equation or the Fitzhugh-Nagumo model have been purely empirical (discussed next chapter), exploring molecular relays provide a microscopic view of how subthermal switching and hysteresis can arise.

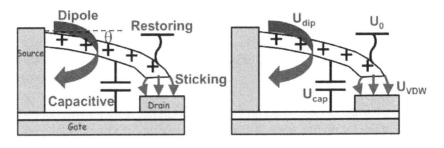

Fig. 19.1   *(Left) Schematic picture of cantilever and the various forces operating on it. (Right) Various terms contributing to the electronic potential (Eq. (19.4))*

A conventional switch operates around an energy barrier separating two distinct states cohabiting the same conformational potential landscape (Fig. 19.2(a)). The role of the gate voltage is to lower the barrier at an opportune moment and drive a current in presence of a drain bias designed to introduce directionality, resulting in an abrupt (and thus dissipative) transition downhill. At the end of an entire charging-discharging cycle, the energy dissipated is CV_G^2, as we discussed towards the beginning of the book.

A molecular relay enjoys an additional gating mechanism besides electronic gating, namely its conformational orientation. Device scaling has made it incredibly hard to reduce the OFF current in the presence of tunneling processes. Conformational switches solve the problem by brute force, moving the channel physically away from the current conduction pathway, as seen in the continuous shift in the stable conformation (Fig. 19.2(b)). This switching is sped up by two somewhat independent attributes — the fact that several charges sitting on the cantilever rotate together as a single giant charge, and the fact that the cantilever configuration undergoes an abrupt phase transition when the destabilizing electrostatic forces overcome the mechanical restoring forces. The hard turn off enables a steeper switching profile than the Boltzmann thermal limit (see last few chapters), while the forming and breaking of adhesive bonds with a substrate sharpens the pull-in. The reduced OFF current and the low subthreshold swing are potentially useful for applications during their sleep cycles where a slow but sharp turn off can save significantly on power dissipation. Let us discuss how we can make some small changes to our Landauer approach to capture these conformational effects.

We rewrite the Landauer equation as a convolution between the *zero-temperature* current and a thermal broadening function $F_T(E) = -\partial f(E)/\partial E$, derived in Eq. (12.8)

$$\boxed{\begin{array}{c} \underline{\text{Conformationally Averaged Landauer Equation}} \\[1em] I = \dfrac{q}{h} \int_{\mu_1}^{\mu_2} dE M(E) \left\langle \bar{T}(E, V_G) \right\rangle \otimes F_T(E, V_G) \\[1.5em] \left\langle \bar{T}(E, V_G) \right\rangle = \dfrac{\int d\theta \bar{T}(E, \theta) e^{-U(\theta, V_G)/k_B T}}{\int d\theta e^{-U(\theta, V_G)/k_B T}} \end{array}}$$

(19.1)

where $M(E)$ is the mode count in the cantilever. The V_G dependence in F_T arises from the shift of the Fermi energy E_F with V_G with a capacitive gate transfer factor α_G. $\bar{T}(E, V_G)$ is the mode and conformation averaged

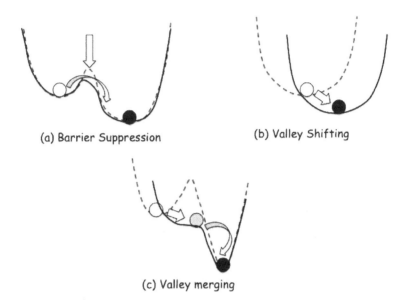

(a) Barrier Suppression   (b) Valley Shifting

(c) Valley merging

Fig. 19.2   *The movement of the most likely conformational angle, θ^*, depends on the gate modification of the potential profile. In conventional digital transistors (a), the gate suppresses a central barrier allowing a thermal jump between two fixed minima. In a conformational transistor (b), the gate shifts a single valley continuously in an analog fashion. The absence of a barrier makes the process low energy but also makes it corruptible with noise. A compromise is accomplished in a relay (c) with the presence of a global minimum into which a local minimum shifts and eventually merges. This dynamics allows a low power analog process to pick up an eventual pseudodigital transition. The second minimum can arise from capacitive destabilization or from short-ranged Van der Waals attraction (second and fourth terms in Eq. (19.4)).*

transmission function controlled by the gated molecular orientation. $\langle ... \rangle$ represents the angular average of the cantilever angle with respect to the horizontal θ, weighted by the Boltzmann probability distribution, assuming the mechanical modes are at equilibrium.

Let us work out the physics behind the abrupt transition. We will work out the associated reduction in subthreshold swing later in the book. For a long stiff cantilever, the fluctuations about the mean can be ignored, so that $P(\theta) \propto e^{-\beta U(\theta)} \sim e^{-\beta[U(\theta^*)+(\theta-\theta^*)^2 U''(\theta^*)/2]}$ becomes a delta function around the potential minimum at θ^*, whereupon the angle averaged current becomes

$$I = \frac{q}{h} \int_{\mu_1}^{\mu_2} dE M(E) \bar{T}\Big(E, \theta^*(V_G)\Big) \otimes F_T(E, V_G) \qquad (19.2)$$

evaluated at a single gate-dependent, potential minimizing angle $\theta^*(V_G)$. The transmission is limited by electron tunneling into the drain from the edge of the cantilever through the vacuum barrier of width $t_{\text{air}}(\theta) = h - L\sin\theta$, h being the bare height above the drain before bending and L being the length of the cantilever. The transmission, using a WKB approximation, is $\bar{T}(E,\theta) \propto \exp[-2\kappa t_{\text{air}}(\theta)]$.

Aside from the tunnel transmission, the physics of the relay is governed by its conformational potential $U(\theta, V_G)$, which can be separated into a flexural potential U_0 representing the tendency of the cantilever to snap back to its horizontal orientation, a gate driven field $\vec{\mathcal{E}}$ which couples with the dipole moment $\vec{\mu}$ to destabilize the cantilever, a capacitive coupling U_{cap} with the back gate also to bend the cantilever, and finally a Van Der Waals 'pull-in' U_{VDW} that makes the cantilever 'stick' to the drain when close enough instead of a snap back

$$U[\theta, V_G] = U_0(\theta) + U_{VDW}(\theta) + U_{\text{dip}}[\theta, V_G] + U_{\text{cap}}[\theta, V_G] \qquad (19.3)$$

where

$$U_0(\theta) \approx \frac{1}{2} U_0^{\text{Bend}} \theta^2 \quad \text{(Bending energy)}$$

$$U_{VDW}(\theta) = 4\epsilon \left[\left(\frac{\sigma}{t_{\text{air}}(\theta)} \right)^{12} - \left(\frac{\sigma}{t_{\text{air}}(\theta)} \right)^6 \right] \quad \text{(Adhesion to contact)}$$

$$U_{\text{dip}}[\theta, V_G] = -\vec{\mu} \cdot \vec{\mathcal{E}} = -\mu \left(\frac{\epsilon_{\text{air}} V_G \sin\theta}{t_{\text{air}}(\theta)} \right) \quad \text{(Field dipole coupling)}$$

$$U_{\text{cap}}[\theta, V_G] = -\frac{1}{2} \left(\frac{\epsilon_{\text{air}} A}{t_{\text{air}}(\theta)} \right) V_G^2 \quad \text{(Capacitive coupling)} \qquad (19.4)$$

The physics of the evolution of θ^* is easily seen from the potential profile in Fig. 19.2. In contrast to the usual two well one barrier form characterizing purely electrostatic gating in conventional FETs (Fig. 19.2(b)), gating the conformational degree of freedom amounts to shifting the minimum of the flexural potential (i.e., the equilibrium orientation of the cantilever) continuously in an analog format (Fig. 19.2(a)). This proceeds until there is an added destabilization that happens due to a phase transition from the capacitive term and a similar one from the Van Der Waals sticking term, whereby the most likely cantilever configuration transitions to the deeper well. In other words, we start with the two well one barrier problem, but instead of barrier lowering the two wells merge laterally into each other.

The transition from multivalued to single valued minima can be seen easily by solving the transcendental equation corresponding to the wells, $dU(\theta^*)/d\theta^* = 0$, visualized in Fig. 19.3 as the intersection of two sets of curves, one of which shifts vertically with increasing V_G. For low values of V_G, we have three intersection points corresponding to three extrema — the two well bottoms and the bump in between. At high enough V_G however, the wells merge and there is only one point of intersection to the right, the global minimum at the bottom of the Van der Waals well, and

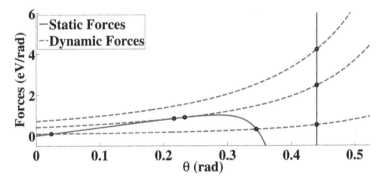

Fig. 19.3   *The transcendental equation obtained using $dU(\theta^*)/d\theta^* = 0$ generates the intersection of two sets of curves, one of which (dashed red) shifts up vertically as gate voltage V_G increases. For small values of V_G, we see three intersection points corresponding to the three extrema: the local minimum from the cantilever, the global minimum from Van der Waals, and the local maximum between the two wells. At high V_G, the two minima merge into a point of inflexion $d^2U(\theta^*)/d(\theta^*)^2$, followed thereafter by a single intersection point and thus a single, global minimum. The merging of minima (Fig. 19.2(c)) drives the sharp phase transition in relays.*

this is what drives the phase transition. The physics of the evolution of θ^* is easily seen from the potential profile earlier (Fig. 19.2 and Eq. (19.4)).

To recap, the phase transition occurs from the electrostatic (capacitive + Van Der Waals) potential that makes the two wells, the elastic and the electrostatic, merge laterally into each other. The transition occurs when the potential valley corresponding to the first metastable state reaches a point of inflexion, corresponding to the dual condition

$$\frac{dU(\theta^*)}{d\theta^*} = 0$$

$$\frac{d^2U(\theta^*)}{d(\theta^*)^2} = 0 \qquad (19.5)$$

We can simplify the algebra considerably by replacing the Van der Waals potential in Eq. (19.4) with a stiff parabola around its minimum, and ignoring the oxide thickness for the moment. We can then write

$$U(\theta) \approx \frac{1}{2}U_0^{\text{Bend}}\theta^2 + \frac{1}{2}k_{VDW}\left(\theta - \theta_0\right)^2 - \frac{\mu V_G \theta}{L(\theta_0 - \theta)} - \frac{\epsilon_{\text{air}}AV_G^2}{2L(\theta_0 - \theta)} \qquad (19.6)$$

where $k_{VDW} = 57.14\epsilon L^2/\sigma^2$ obtained by fitting a parabola around the Van Der Waals minimum. Setting its derivative equal to zero to extract the first metastable minimum, we get

$$(U_0^{\text{Bend}} + k_{VDW})\theta^* - k_{VDW}\theta_0 = \frac{2\mu V_G\theta_0 + \epsilon_{\text{air}}AV_G^2}{2L(\theta_0 - \theta^*)^2} \qquad (19.7)$$

which gives us the gate voltage dependence of the cantilever coordinate $\theta^*(V_G)$. Setting the second derivative to zero as well to get the inflexion condition gives as an additional equation, and from the pair of equations we can then extract the destabilization point

$$(\theta^*)_{\text{destab}} = \left[\frac{U_0^{\text{Bend}} + 3k_{VDW}}{3(U_0^{\text{Bend}} + k_{VDW})}\right]\theta_0 \qquad (19.8)$$

We get a destabilization near $\theta^* \approx \theta_0/3$ when the bending forces dominate (far from the Van der Waals potential). However, near $\theta \approx \theta_0$ to within a correction term $\Delta\theta \approx 2/\beta U''$, we will get an additional pull in as k grows in strength.

The aim of this chapter was to show how one can include an added configurational averaging on the Landauer approach to capture conformational gating. Such a conformational gating can describe the flipping, breaking, bridging or rotating of bonds, rings and sidegroups in molecular assemblies, nanowires and tubes - invoked at times to describe mechanical sensors, switches and resistive random access memory elements. The potential landscape differs from one kind of motion to another, as do the current equations that depend on the detailed interface geometry. For tunneling at the ends of cantilevers for instance, the current follows an exponential dependence dominated by the radial part of the wavefunctions, while for

rotational displacements along the contact surface we expect more complicated trigonometric dependencies arising from the varying angular overlap
between molecular and contact metal orbitals.

For a given molecular displacement, we can estimate the ON and OFF
currents at these end points with frozen configurations. However, what is
critical to these analyses is the role of thermal fluctuations that tend to dis-
place these assemblies from those on and off states, and the voltage required
to connect these end points while over-riding thermal fluctuations. A full
thermodynamic analysis involving the conformationally averaged Landauer
equation described in this chapter, is needed to evaluate the switchability
and gateability of these conformational transistors.

Towards the end of the book, we will use the equations developed in this
chapter to show how the sudden conformational phase transition can on occasion lead to subthermal switching at a voltage much less than the Boltzmann limit of $k_B T \ln 10/q$, consistent with experimental data on cantilevers
as well as ion channels. The analysis also shows the origin of hysteresis in
the output characteristics, an effect that tends to compromise the overall
performance.

Chapter 20

Two terminal switching

Let us move now from gated structures to purely two-terminal switching phenomena in molecular devices. The terminology 'switching' is used loosely, in that we will cover (a) *reproducible switching without memory* (where the curve is perfectly retraced upon a reverse swing), (b) *reproducible switching with memory* (i.e., a hysteretic curve), and (c) *stochastic switching* such as due to occupation/deoccupation of trap states. The detailed underlying mechanisms behind specific switching systems are still under debate, and the absence of systematic experimental data exploring various facets of the same molecule in the same device set-up makes it hard to validate such a model. However, our intent is to discuss the general physics behind such phenomena, and leave it to active researchers to figure out what specific sources of switching are operative in a specific example.

Where can we use such two-terminal switches? Conventional switching devices require a third terminal, a 'gate', which modulates the conductivity of the switch under the application of an external control signal (e.g., a gate voltage). Two-terminal switches are operative for instance in memory devices such as DRAM or FLASH, where we use a certain voltage to 'write' information onto the device. Read is usually a separate operation, often involving a much lower bias measurement of the device resistance. In particular, if the switching shows NDR (i.e., a current that has a peak when plotted against voltage), then we can use two back-to-back devices to create a GOTO pair, which acts like a latch and stores the memory bit in a non-volatile manner with very little leakage, alleviating the need for refresh.

We will discuss in sequence, how molecular states can give NDR, hysteresis and random telegraph noise.

20.1   Origin of Negative Differential Resistance (NDR)

The Landauer approach connects the measured current with the total transmission summed over all propagating modes lying within a bias window.
Since the bias window keeps increasing monotonically with drain bias, it
may not be clear how the integral could go down with increasing bias; but
that is exactly what happens for an NDR, namely, a *negative slope* in a
certain domain of the I-V characteristic. And the reason that happens
is because the transmission function itself can also alter shape under the
electrostatic impact of the changing drain bias. Under certain cases, the
transmission reduces significantly to create such an NDR. In many cases,
this involves expelling a level from the conducting 'window' set by $f_1 - f_2$,
the expulsion driven by electrostatics, specifically the Laplace potential of
the drain or the Coulomb repulsion from another nonconducting level.

Our aim here is to see how such an NDR, i.e., a negative slope, arises in the
I-V_D curve based on our formalism. From the Landauer current expression
(Eq. (12.5)), we can extract the slope and write down the conductance

$$G = \frac{\partial I}{\partial V_D} = \frac{q^2}{h} M(E_F) \bar{T}(E_F) + \frac{q}{h} \int_{E_F}^{E_F + qV_D} \frac{\partial \bar{T}(E, V_D)}{\partial V_D} M(E) dE \quad (20.1)$$

While the first term on the right is always positive, the second can sometimes be overwhelmingly negative if *the transmission function changes
shape by shrinking in area* upon application of a drain voltage.

The actual mechanism of the shrinkage of average transmission can be quite
varied. A common origin is the drain voltage driven sweep of multiple sharp
levels past each other (Fig. 20.1), so that the transmission increases upon
their alignment and decreases upon misalignment. Another mechanism is
the act of 'falling out of the band' (Fig. 20.2), for instance in a resonant tunneling diode, whereby a quantized level, capacitively shifted by the drain
voltage, sweeps past a slower shifting band-edge and plummets into the
band-gap (not to be confused with the mechanism of current saturation
in FETs where a *Fermi energy* slips past a channel bandedge). A similar
expulsion can be Coulomb driven (Fig. 20.3), repelled by a nonconducting
(trap) state. A fourth mechanism is the coupling between electronic and
mechanical variables so that the applied drain bias could lead to a conformational change in a molecule such as the breaking of bonds, or perhaps

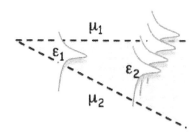

Fig. 20.1 *NDR arises when the Laplace potential under a drain bias sweeps a level sitting closer to it relative to another, making the average transmission go up near alignment and reduce away from alignment. The figure is plotted along the V_D axis and shows the relative slippage of ϵ_2 in and out of alignment with ϵ_1*

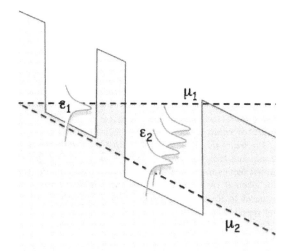

Fig. 20.2 *A sharp cutoff and NDR with an abrupt drop happens in a resonant tunneling diode (RTD) when the drain induced sweep pushes a level not just past another, but actually past a bandedge in a solid, for instance here past the bottom of the left quantum well*

the rotation of an intermediate ring that disrupts conjugation across the chain — mechanisms that drastically alter the overall transmission *function* and not just its value at a given energy, to its possible detriment.

In generic language, we can describe the NDR as driven by a 'dark' (i.e., non-conducting) state that shifts spectral weight away from the original conducting pathway and thus diminishes the overall transmission function.

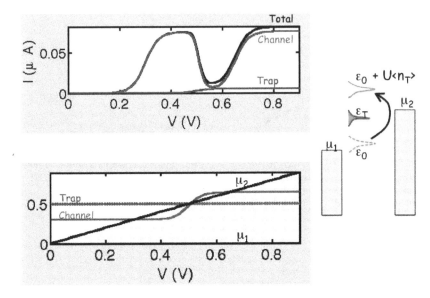

Fig. 20.3 *(Top) Current voltage for a channel and a trap shows NDR. (Right) Once the trap is accessed near 0.4 V, the channel is expelled from the conduction window by Coulomb repulsion, reducing the total current to the background trap contribution (red curve on top). Eventually at higher bias, the channel will be accessed again and the current will show a rise to the saturated value $\propto G_0\gamma_1\gamma_2/(\gamma_1 + \gamma_2)$. Comparing the IV with (Bottom) the energy levels of the trap and of the channel with Coulomb corrections $(\epsilon_0 + U \langle n_T \rangle)$ shows that the channel conducts when it enters the bias window (yellow wedge), then stops conducting when it leaves the bias window simultaneously as the trap is filled by μ_2, and conducts again when it reenters the window at a higher drain bias. Self interaction corrections are critical to this picture, because the trap must repel the channel upon filling, without being repelled by itself.*

A sharp NDR for instance, can be precipitated by expelling a level from the conducting window through Coulomb correlations. Figure 20.3 shows the calculated $I-V_D$ for a two-state system, one of which (a conducting level ϵ_0) has strong couplings $\gamma_{1,2}$ with the contacts, while the other (a non-conducting trap or 'dark' state at ϵ_T) has a weak coupling $\gamma'_{1,2}$ but a strong Coulomb cost of filling, say a single electron charging energy U. The current through the conducting level gets initiated when the voltage bias window straddles its peak by a few $\sim k_BT$. When the second non-conducting level enters the conduction window at higher bias voltage, even as it contributes mildly to the current (through its weak couplings $\gamma'_{1,2}$), it simultaneously repels the conducting state by imposing an energy shift U, evicting it in the process from the conduction window.

**The NDR occurs when the current loss from evicting the conduct-
ing state from the bias window exceeds the current gain through
the activation of the non-conducting level.** This amounts to the following two conditions

- the gain in non-conducting current, $\sim \gamma_1' \gamma_2'/(\gamma_1' + \gamma_2')$ must be lower than
  the loss in conducting current, $\sim \gamma_1 \gamma_2/(\gamma_1 + \gamma_2)$,
- provided the charging U is sufficient to produce a decent enough shift
  so that the conducting level actually slides past the now resonant non-
  conducting level, thereby escaping the bias window, and ceasing to
  conduct.

Thus, the two criteria we need, approximately stated (the exact maths
involves some inelegant inverse tangent functions).

$$\frac{\gamma_1 \gamma_2}{\gamma_1 + \gamma_2} < \frac{\gamma_1' \gamma_2'}{\gamma_1' + \gamma_2'}$$
$$\epsilon_0 + U > \epsilon_t \tag{20.2}$$

in other words,

$$\boxed{\begin{aligned} \frac{1}{\gamma_1'} + \frac{1}{\gamma_2'} &> \frac{1}{\gamma_1} + \frac{1}{\gamma_2} \\ \epsilon_0 + U &> \epsilon_t \end{aligned}} \tag{20.3}$$

Figure 20.3 arises out of the Landauer equation for two Lorentzian densities of states (a broadened channel and an unbroadened trap, $\gamma' = 0.1\gamma$),
Coulomb correlations included in the channel level as $\epsilon_0 + U \langle n_T \rangle$, with
trap occupancy $\langle n_T \rangle$ also obtained from NEGF (weighted integral of trap
density of states upto μ_1 and μ_2 at zero temperature).

The 'usual' version of NDR in a resonant tunneling diode also arises
from an act of *defenestration*, a conducting level falling out of a window
(Fig. 1.2(b)), except the level is driven by the Laplace potential (drain capacitance) rather than the Poisson potential (Coulomb repulsion), and the
window happens to be a neighboring band rather than that set by the voltage bias window.

There is a more general quantification we can make, if we describe the
entire scenario in terms of many-body states. We will see in the chapter

on correlation that all the blocking processes can be attributed to the activation of a generic 'dark' state that shuts off conduction (Fig. 27.2 and Eqs. (27.12)). The details behind the dark state could be involved; the state could arise from phonon bottlenecks or charge trapping or other sundry processes. The main upshot of the many body description is that we will find a similar inequality as above, except spelt out in many body language. We distinguish between *shell tunneling*, where the charges zip through the channel without lingering (when the barriers are asymmetrically disposed so that the charges escape faster than they enter), and *shell filling*, where the charges build-up progressively because their escape is slower than their entry. NDR happens during shell filling when the built-up charges block any further flow of charge and cause the current to collapse. What we will see in the many-body dark state description is that the shell filling happens when the lifetime of the charging states, typically a HOMO-1 or LUMO+1 state, is longer than the conducting states, such as a HOMO or a LUMO respectively.

20.2 Switching with memory requires intermediate scan rates

Hysteresis arises if the voltage is scanned on a time scale faster than the lifetime of the dark state, but slower than the response time of the active conducting state. In other words, we need a double inequality

$$\left(\frac{1}{\gamma_1'}, \frac{1}{\gamma_2'} \right) > \frac{1}{\hbar\omega} > \frac{1}{\gamma_1} + \frac{1}{\gamma_2} \tag{20.4}$$

where ω is the characteristic frequency denoting the scan rate.

Let us rationalize this inequality. As long as the system and its surrounding environment are stuck at equilibrium, we expect a unique solution to the I-V curve. The solution however splits into multiple branches if we depart from equilibrium and make forays into excited states, which would require a fast scan rate. The hysteresis is thus scan rate dependent, as we expect for an intrinsically non-equilibrium process. If however the slowest time constant is significantly larger than the time duration for an experiment or an observation, the scan rate dependence becomes weak and the hysteresis resembles a true steady-state memory.

We will now present a simple physical model for a hysteretic loop. The main requirement as we shall see is a multibranched N-shaped curve intersecting with a load line, which yields multiple solutions. Here, we will present a specific model that requires three time scales: (i) a fast scale which will represent in this case the electronic variable, (ii) a slow time scale that represents the environmental 'recovery' variable setting the surrounding potential which the electron resides in, and (iii) a third intermediate time scale that lies in between and represents the scan rate. The analysis appears in the literature in the context of 'memristors', spawning speculation as potential neuromorphic circuit elements, and as the purported fourth fundamental circuit element. We will skirt these issues and debates surrounding their true memristive properties (some of which are perhaps pedantic) and treat it instead as a simple model for hysteresis.

In an Ohmic regime, we will use

$$V = IR(I) \tag{20.5}$$

where the resistance is current-dependent. A simple model for the current dependence is a metallic filament that forms between source and drain, with the interface between the high resistance insulating part (length $L-w$) and low resistance conductive filament (length w) driven by the current flow,

$$R = R'_H[L - w] + R'_L w$$
$$\dot{w} = v = I/Anq = \beta_0 I \tag{20.6}$$

where R' (Ohms/meter) is the specific resistance. We can see readily that the resistance depends on w, which involves the time integral of the current (i.e., total charge), and *thus its history*. The history dependence makes the resistance a function of its memory, and thus leading to the name memristor. Upon integrating w, the equation can be rewritten as

$$V = I\left[R'_H L - \Delta R' \beta_0 \int_0^t I(\tau)d\tau\right] \tag{20.7}$$

where $\Delta R' = R'_H - R'_L$. The nonlinear term on the right leads to a frequency summation, so that the voltage loops much faster over time than the current and thus the interface mobility. In fact, as we loop cyclically through the voltage, its conductance will reach multiple values due to the tardiness of the recovery variable w, leading to the hysteretic I-V loop. More specifically, over time the charge $Q = \int_0^t I(\tau)d\tau$ builds up to cancel the term in the box, whereupon the resistance vanishes and the current shows an upswing to a higher conducting state. On the reverse swing, however, the voltage reduces with the high conductance state, and follows a different path downwards, all the way to negative voltage where the charge gets purged out of the system and the process resumes.

The final equation for the charge becomes

$$V = \dot{Q}\left[R'_H L - \Delta R' \beta_0 Q\right] = \frac{d}{dt}\left[Q R'_H L - \Delta R' \beta_0 \frac{Q^2}{2}\right] \tag{20.8}$$

Assuming a cyclical voltage variation $V = V_0 \sin \omega t$ with $\omega = 2\pi/T$ and integrating both sides over time,

$$\frac{Q^2}{2}\beta_0 \Delta R' - L R'_H Q + \frac{V_0}{\omega}\left[1 - \cos \omega t\right] = 0 \tag{20.9}$$

The quadratic equation generates two time-dependent roots and a current

$$Q(t) = \frac{L R'_H \pm \sqrt{\left| L^2 (R'_H)^2 - 2\beta_0 \Delta R' V_0 [1 - \cos \omega t]/\omega \right|}}{\beta \Delta R'}$$

$$I(t) = dQ/dt = \mp \frac{V_0 \sin \omega t}{\sqrt{\left| L^2 (R'_H)^2 - 2\beta_0 \Delta R' V_0 [1 - \cos \omega t]/\omega \right|}} \tag{20.10}$$

Figure 20.4 shows a set of hysteretic IVs for various parameters. If the critical charge that cancels the resistance, $Q_C = R'_H L/\Delta R' \beta_0$ happens within the voltage bound $|V_0|$, i.e., the process is slow enough that $\omega < 4\beta_0 \Delta R' V_0/L^2 (R'_H)^2$ (so that as V varies from 0 to V_0 the term under the square root switches sign), but is fast enough to complete a cycle during the measurement time, we see a kink in the IV leading to a pinched hysteresis loop. Of course, the model we used is simplified, and can be tweaked to include realistic effects, such as a nonlinear ionic conduction $G(V)$ (e.g., Eq. (13.16) or $\sim \sinh bV$ from tunneling), coupled with a 'window function' $F(w)$ to capture the ionic slow down at the ends, $\dot{w} = \beta_0 I F(w) G(V)$.

We can readily see that the hysteresis arose from the different initial conditions and parameters for forward vs reverse voltage sweeps, e.g., when the slow ionic coordinate w doesn't instantly recover on the reverse sweep

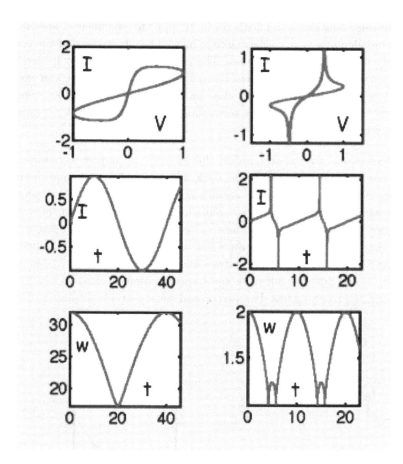

Fig. 20.4 *(Left column, top to bottom): Hysteresis curve, time trace of current I and a slow recovery variable w, the position of a conducting interface that determines the series resistance and thus the current, while in turn depending on the current that sets its speed. Parameters $R'_H = 1.6$, $R'_L = 1.5$, $T = 40$, $\beta_0 = 1$. (Right column, top to bottom): Analogous plots for $R'_H = 6$, $R'_L = 0$, $T = 10$, $\beta_0 = 1$. If the net charge accrued over time is enough to cancel the resistance within the voltage bounds, an NDR is seen in the $V - I$ curve leading to a pinched hysteresis curve (top right), else (top left) we get a regular featureless cyclical curve. It is easy to see that on the reverse swing, I and w are not in sync, and the different environment values w create different resistances for forward and reverse sweeps and generate the above hysteresis.*

(Fig. 20.4). In a molecule on its forward sweep for instance, the charge may inject into the trap at a voltage and block the channel (Fig. 1.3(c)) through a change of configuration (a change $M \to M^-$ where the charge is not easily extracted on the reverse sweep — similar to a flash memory). Let us for instance consider a dot with contacts that are asymmetric, specifically their broadenings $\gamma_2 \gg \gamma_1$ for negative bias on the drain, so that the trap is quickly filled and slowly emptied. The resulting shell filling will block the channel and create an NDR. On the reverse scan, however, we can assume $\gamma_2 \ll \gamma_1$ so that the trap stays filled on the time scale of the experiment, and the channel stays blocked throughout the voltage window, creating a hysteresis.

Here is one way to accomplish this reversal of contact asymmetry, in fact, one where the lifetime of the dark state is much longer than the measurement time so that the hysteresis actually seems scan rate-independent and steady state. Consider a resonant tunneling diode consisting of a quantum dot with asymmetric contacts $\gamma_1 \gg \gamma_2$, for instance a doped n^+ silicon channel on the left and an STM on the right. Under positive bias on the STM, a LUMO level ϵ_0 stays filled (fast electron entry, slow exit) but it is also dragged down by the Laplace potential drop across it. As it starts approaching the band-edge and gets partially depleted of elec-

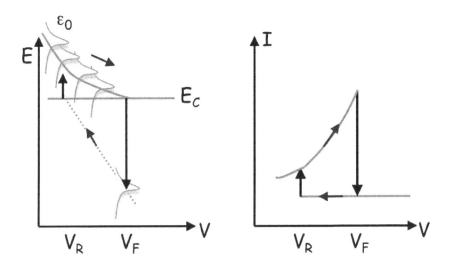

Fig. 20.5 *Hysteresis during trap induced NDR. (Left) Blue dotted line shows the Laplace solution that the molecular level would follow under drain bias in the absence of charging. The Coulomb cost of charge depletion slows down the level to the blue curved line, until it finally reaches the bandedge at V_F, and drops down to the Laplace solution upon complete charge removal inside the gap. On the reverse scan, the level goes up the Laplace solution till it hits the bandedge, and reaches the original position, albeit at a different reverse voltage V_R. (Right) The I-V picks up the hysteresis signature as well.*

trons, the Coulomb cost tries to push the level up and the LUMO follows a curved path in the energy space (Fig. 20.5). This Poisson driven path deviates from the linear Laplace path (dotted line) due to the charging costs. Once the level enters the silicon bandgap at voltage V_F, the injection from silicon γ_1 drops abruptly to zero and the LUMO gets abruptly emptied by the positive biased STM. Along with current blockade and an NDR, the LUMO drops abruptly in energy onto the Laplace driven trajectory. On the reverse bias scan however, the LUMO stays empty and follows the Laplace path and encounters the band-edge at a lower voltage V_R and the current recovers with a prominent hysteresis loop. Notably, this hysteresis is more or less scan-rate independent, because the presence of the band-edge acts as a charge 'trap' with a near infinite lifetime. In practice, it is compromised by higher order multiphonon processes to purge the excess hole charges dumped onto the level in the bandgap.

20.3   Ionic conduction and neuronal gating

We saw how nonlinearity and the finite scan rate create pinched hysteresis loops in memristor $I - Vs$, generating a sequence of asymmetric spikes in time. Biological neurons often encode information between such spikes. Each axon contains a series of transmembrane proteins or ion channels between extra and intercellular media (Fig. 2.3b). The ion channels are normally closed and are activated by ligand binding, mechanical gating (e.g., touch sensitive receptors) or voltage gating. In mechanical gated relays for instance, the channel opens abruptly (in fact, subthermally) when it picks up several ionic charges. Above a certain threshold (-55 mV), the sodium channels open and allow extracellular Na^+ ions to flow down gradient. During this polarization phase the membrane potential shoots up to ~ 30 mV. The increasing positive intercellular potential triggers an inhibitory mechanism that slows down the Na flow, and also opens a counterbalancing K-channel that moves K^+ ions in the opposite direction, depolarizing the channel. The result is an asymmetric spike that moves at a speed of $\sim 0.1 - 1m/s$. The positive charge flow tends to open neighboring channels as well, but since the K ions move slower than Na, there is an overshoot past threshold and the channel 'hyperpolarizes', leading to a refractory period of no spiking at the trailing edge of the spike. The result is the required input-output isolation and unidirectionality. The battery (Na^+-K^+ ATPase) eventually pumps the Na out of the channel and the

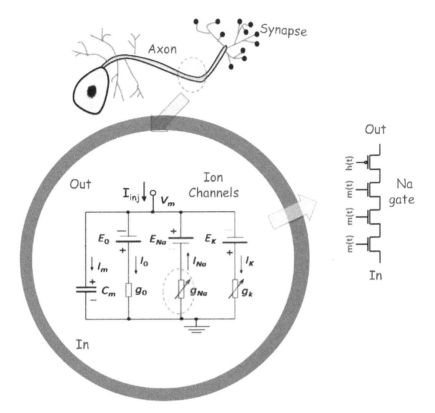

Fig. 20.6   *Top: Neural axon consists of ion channels and pumps separating intra and extracellular media (detail in Fig. 2.3b). Each channel can be written as a sequence of voltage-controlled resistors described by the Hodgkin-Huxley equation (bottom left) with considerable internal dynamics - e.g., the Na channel (right) performs an AND operation with three open and one inhibitory gates triggered by a mechanical relay.*

K back inside, resetting the ion concentrations to their equilibrium values. The pump operates against gradient, energized by the hydrolysis of adenosine triphosphate (ATP) molecules, by some accounts driven by a Brownian ratchet.

The dynamics across the ion channels is described by the Hodgkin-Huxley (HH) equation, $i = Na, K, 0$ that consists of a set of parallel voltage-controlled nonlinear resistors (Fig. 20.6). Based on the dynamics we will shortly develop for Coulomb Blockade (Eqs. (26.16), (26.17)), I find it tempting to interpret each gate as a blockaded dot that turns on abruptly at a critical charge threshold.

Hodgkin-Huxley

$$C_m \frac{dV}{dt} = I_{inj} - \sum_i g_i(V - E_i)$$

$$g_{Na} = (m^3 h)g_1, \quad g_K = n^4 g_2$$

$$m(V,t) \longrightarrow \frac{dm}{dt} = -\left[\frac{m(V,t) - m_\infty(V)}{\tau_m(V)}\right] = \alpha_m(1 - m) + \beta_m m$$

$$\alpha_m(V) = \underbrace{\int dE D(E) f_E \left[1 - f_{E-qV}\right]}_{\text{Blockade Rate}} = \frac{b(V - V_0)}{\left[1 - e^{-a(V-V_0)}\right]}$$

$$\alpha_m(V) \longrightarrow \beta_m(V), \quad f \longrightarrow 1 - f$$

Similar equations for n, h

$$(20.11)$$

where the number of open channels m_i, n_i for Na and K and inhibitory channels h_i represent the recovery variable 'w'. The parameters a, b, V_0 for each channel are fitted to experimental data. The capacitance of the lipid bilayer with relative dielectric constant $\epsilon \sim 5.6$ and thickness ~ 5nm is $C_m \approx 10^{-2} F/m^2$. The axoplasm conductivity is $\sim 2\Omega$-m. Assuming an axon 1 mm long and 5μm radius, the resistance is $\approx 2.5 \times 10^7 \Omega$.

Figure 20.7 shows the action potential spikes obtained by solving the HH equations with an injected current I_{inj}. In presence of that current, the spikes move unidirectionally along parts of the axon (the ones not covered by a myelin sheath, such as the nodes of Ranvier). Along any myelinated regions (e.g., in vertebrates requiring long distance communication), the larger thickness $\sim 1\mu$m and resulting lower capacitance reduces the $R - C$ time constant and makes the action potential signal jump quickly at ~ 100 m/s (also known as 'saltatory conduction'). During the entire transmission process, critical information is stored in the analog time intervals between spikes. When they arrive at the synapse, a weighted sum of pulses is compared against the threshold depolarization voltage of -55 mV to generate subsequent all-or-none spikes (often described by an 'integrate and fire' model). The synaptic weights adjust according to the interval between pre

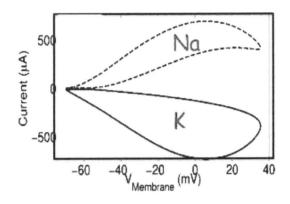

Time vs. Membrane Potential

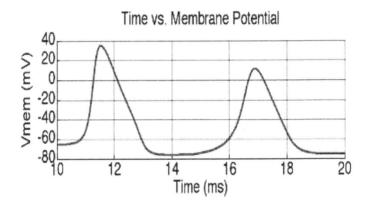

Fig. 20.7 *(Top) A "NEURON" code simulation of the Hodgkin-Huxley circuit equation (courtesy Dincer Unluer) shows pinched hysteretic I-Vs that translate into voltage spikes (bottom) that travel unidirectionally along each axon to the synapse. The synapse is an integrator and comparator — it performs a weighted sum over all incoming spikes, triggering another spike in a post synaptic neuron once the summed voltage exceeds the −55mV depolarization threshold. In neural models the synaptic weights can be set dynamically by the latency between pre and post neural spikes — causal spikes (pre before post) are potentiated while acausal spike weights are depressed.*

and post-synaptic spikes which sets their signal overlap. When pre-neuron fires causally before post-neuron (e.g., hearing a bell ring immediately before smelling the food served), the weight is exponentially augmented, connecting the two neurons. Conversely if pre comes after post, its weight is decreased and the acausal connection is suppressed. Such a spike-timing dependent plasticity enforces an associative Hebbian learning process where neurons that 'fire together wire together'.

20.4 How is a sequence of spikes generated?

A sequence of spikes needs a nonlinear resistor to generate a separation of time scales. Let us first establish a way to get a periodic array of spikes, which correspond to a limit cycle in phase space (an oscillator for instance). A regular oscillator is obtained by solving the phase space equations

$$\dot{x} = -\mu y, \quad \dot{y} = x/\mu \tag{20.12}$$

The solutions to this equation, obtained by taking the ratio of the two equations and interpreting $\dot{y}/\dot{x}$ as dy/dx, is a family of ellipses satisfying $x^2/\mu^2 + y^2 = C^2$.

If we now add a rotation to this equation, so that the rate of change of x depends linearly on x itself, we superpose an exponential increase or decrease.

$$\dot{x} = \mu(\alpha x - y), \quad \dot{y} = x/\mu \tag{20.13}$$

The solution is obtained by setting $(x, y) = (A, B)e^{\lambda t}$, and then extracting the two Lyapunov exponents $\lambda_{1,2} = (\mu\alpha \pm \sqrt{\mu^2\alpha^2 - 4})/2$. For positive α, this gives us a growing solution while for negative α it is damped. To make the solution self-sustaining, we will thus need a system with a positive α for small x and negative α for large x — in other words, a *nonlinear damping*.

Let us now include such a nonlinear damping. Consider a tunnel diode to represent the nonlinear damping, in parallel with an LC oscillator plus a current source I_0 to control the amplitude (Fig. 20.8).

$$L dI/dt = V, \quad C dV/dt = I_0 - I_{\text{diode}}(V) - I, \quad I_{\text{diode}}(V) \approx \gamma V^3 - \alpha V \tag{20.14}$$

We can make the substitution

$$V\sqrt{\frac{3\gamma}{\alpha}} \to x, \quad \alpha\sqrt{\frac{L}{C}} \to \mu, \quad \frac{t}{\sqrt{LC}} \to t, \quad \frac{L\sqrt{3\alpha\gamma}(I_0 - I)}{C} \to y \tag{20.15}$$

to give us the Van der Pol equations

$$\dot{x} = \mu\left(x - x^3/3 - y\right), \quad \dot{y} = x/\mu \tag{20.16}$$

which as desired has negative damping for small x and positive damping for large x. The fixed points of the curve where the derivatives vanish are between $y = x - x^3/3$ and $x = 0$, in other words, between an N-shaped curve that shifts vertically with I_0 and a vertical line $y = 0$. If we now make

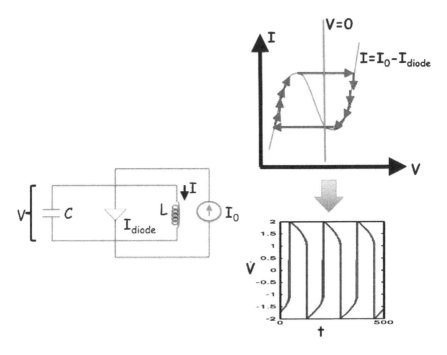

Fig. 20.8　*(Left) LC circuit plus a diode (nonlinear resistor) gives us phase space plots that are given by (top right) the intersection between a vertical straight line and an N-shaped I-V that shifts vertically over time with injection from a current source I_0. The result is a slow evolution along the stable branches of the N-shaped curve, and fast horizontal skips between them, which translates (bottom right) to a sequence of spikes in $V - t$ with sharp rise and fall times and slower evolution in between.*

μ very large, then it is easy to see that along the trajectory $y = x - x^3/3$, the deviations from the fixed point have very small $\dot{x}$ and evolve rather slowly. However, at the extrema of the nonlinear curve when we are forced to depart from the curve, the deviations $\dot{x}$ evolve rapidly while $\dot{y}$ changes little, and we get a fast horizontal jump to another sector of the curve. The phase space evolution thus toggles between a fast horizontal jump and a slow evolution back towards the extrema along the curve. Translated to a V–t plot, this gives us a slow inhibited evolution in x followed by a sharp rise, and a similar slow plateau followed by a fast drop. In other words, a series of spikes reminiscent of the action potential. More importantly, we reach a stable limit cycle, with the trajectory making quick jumps between different legs in order to stay entrained on the cycle.

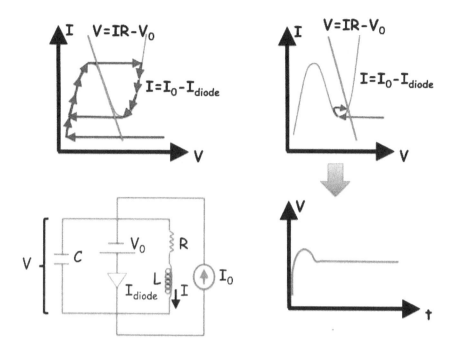

Fig. 20.9 *(Bottom Left) The Fitzhugh-Nagumo model amounts to an LCR circuit plus a diode (nonlinear resistor) and a battery that give us phase space plots set by (top left) the intersection between an inclined straight line and an N-shaped I-V that shifts vertically over time with injection from a current source I_0. When the line intersects the central negative sloped branch of the N-curve, the result is a slow evolution along the stable positive sloped branches, and fast horizontal skips between them, translating as before to a sequence of spikes in $V - t$. When the lines intersect on the positive sloped ends of the N-curve, the phase space evolution and the corresponding $V - t$ shows inhibited behavior with no spikes. In other words, we are getting an 'all-or-none' behavior.*

But how do we terminate the sequence so it has a clear beginning (in fact, one with a threshold) and an end? With a little bit of tweaking, we can add the physics of excitation and inhibition to this equation, generating the so-called Fitzhugh-Nagumo model. All we need is to add a sloped load line by including a battery and a linear resistor (Fig. 20.9).

$$LdI/dt = V - IR + V_0, \quad CdV/dt = I_0 - I_{\text{diode}}(V) - I \qquad (20.17)$$

The fixed point is now between the N-shaped curve and a sloped line with slope $1/R$ and intercept V_0/R. This means as the N-shaped curve shifts vertically with I_0, the load line changes intercept position over time. Eventually, that intercept will slip from the central branch with positive slope generating the oscillations, to either of the side branches with negative slopes. Once it does, the sequence will terminate. We can see it by introducing small variations around the fixed point

$$\delta \dot{I} = \delta V - R\delta I, \quad \delta \dot{V} = (1 - 3V_0^2)\delta V - \delta I \tag{20.18}$$

As long as we intersect the stable central branch, we get limit cycles that toggle between fast horizontal switches and slow transitions along the negative sloped branches, creating a series of spikes along the time axis. Once the load line reaches either of the negative sloped regions, for $|V_0| > 1/\sqrt{3}$, we get an inhibitory response and the spikes terminate at the beginning and the end of the train.

The Hodgkin-Huxley equation is structurally very similar to the equations above, except it extends the variables into multidimensional space to describe the sodium, potassium and leakage channels. It does not capture the 'all-or-none' excitation physics that Fitzhugh-Nagumo model captures, but provides a much better fit to the quantitative details of the spike. At the opposite end of the spectrum, we can simulate an artificial neuron network (ANN) with a slight change to the Fitzhugh-Nagumo equations (Eq. (20.17)) to generate the so-called Izhikevich model

$$
\begin{array}{c}
\text{Izhikevich-Model}\\[2mm]
\underbrace{v'}_{CdV/dt} = \underbrace{0.04v^2 + 5v + 140}_{-I_{diode}(V)} + \underbrace{I}_{I_0} - \underbrace{u}_{I}\\[4mm]
\underbrace{u'}_{LdI/dt} = a(bv - u) = \underbrace{abv}_{V} - \underbrace{au}_{IR}
\end{array} \tag{20.19}
$$

where v is the membrane potential, u is the recovery variable representing the K^+ channel. In addition, we impose an upper threshold at $+30$ mV

$$v \geq 30 \ mV, v \to c, u \to u + d \tag{20.20}$$

$a = 0.02$ is the scale of the recovery variable u, $b \approx 0.2$ is the sensitivity of u to subthreshold fluctuations, c sets the after-spike reset value -65 mV and $d = 8$ sets the limit on u.

Why do we need spiking neurons in the first place? It has been argued that the optimal shape of the spikes is given by maximizing the number of information bits per joule. Neurons encode information in the analog time intervals between spikes, i.e., pulse position (rather than pulse amplitude) modulation - capitalizing on the energy efficiency of packing information into the dead times between pulses. We can pack in more information by spacing the spikes closer together but they get slowed down by the refractory tails and also start to get corrupted by overlap. We can also try to squeeze each pulse, but that increases the energy cost.

20.5 Stochastic switching and Random Telegraph Signals (RTS)

While memristive spikes are generated at an optimal frequency, many processes in nature involve a thermodynamic sum over frequencies. Such a sum imparts a stochastic element to the switching, turning the spikes into random telegraph signals (RTS) that can be monitored with a fine time resolution (Fig. 1.3(d)). For instance, the steady-state occupancy of a trap state at equilibrium with the surroundings is given by the distribution function (g_s denoting the spin-degeneracy)

$$F(E) = \left[1 + g_s \exp\left[\frac{E - E_F}{k_B T}\right]\right]^{-1}, \qquad (20.21)$$

At high bias, we will generalize using the Beenakker analogue (Eq. (26.12)). In a time-dependent formalism we can then write down the electron capture and emission times by the trap assuming detailed balance, i.e., the ratio of capture and emission times follows a Boltzmann distribution

$$\begin{aligned}
\hbar/\tau_c &= \gamma_t F(\epsilon_t - \alpha_G V_G - \alpha_D V_D), \\
\hbar/\tau_e &= \gamma_t [1 - F(\epsilon_t - \alpha_G V_G - \alpha_D V_D)].
\end{aligned} \qquad (20.22)$$

where γ_t is the trap coupling parameter, and α_G and α_D are the capacitive gate and drain transfer factors. We can then go back to our model for trap induced blockade (Fig. 20.3) but solve away from steady state using Monte Carlo to capture the capture/emission times. Over a time-step dt, the probability for high current state is defined through a capture rate dt/τ_c, while that for low current state as dt/τ_e. The expression for γ_t may include inelastic processes that allow the trap to be filled and emptied within the band-gap, for instance, the Huang Rhys expressions for multi-phonon processes that we encounter later (Eq. (24.59)). The resulting simulations are shown in Fig. 20.11, and resemble action potential spikes generated along a neural axon (Fig. 20.6).

This is one of the domains where organic molecules could play a defining role. In conventional applications, they do not bring any obvious performance advantages compared to their inorganic counterparts — as interconnects, insulators or as transistor channels. However, short molecules do function better as quantum dots rather than as quantum wires, in other words, as charge storage centers and memory elements instead of charge carrying conduits. Organic molecules have large single electron charging energies due to their small sizes and low dielectric constants, as well as large electron-phonon coupling due to their confined sizes, conformational flexibility and poor coupling with the macroenvironment. In other words, they naturally support strong electronic and vibronic correlations, rendering them particularly useful as engineered *scattering centers*. Molecules may indeed function better *ON* the channel rather than *IN* it.

In a scattering configuration, strongly correlated dots can impose their characteristic fingerprints onto the current of an underlying gated channel. Figure 20.11 shows a possible characterization scheme whereby the attachment of a molecule on a backgated channel (step 1) would introduce characteristic trap levels inside its bandgap (step 2), identified computationally by the eigenfunctions sitting primarily on the molecular moiety. Sweeping a gate voltage past those traps would activate them at resonance and lead to a flicker in the output current through stochastic trapping/detrapping coupled with Coulomb Blockade of the channel. From the extracted frequency spectrum of the current (step 3), one can multiply by the frequency (step 4) to get rid of the overall $1/f$ noise that arises from a superpo-

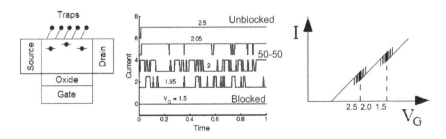

Fig. 20.10 *(Left) Gating across a trap level leads to RTS noise near each resonance where the occupation probability deviates from 0 or 1. Simulated time traces and schematic transfer characteristic of scattering current show inactivity below resonance, $50-50$ trapping and detrapping (noisiest) at resonance, and complete blockade above resonance. The ratio of trapping-detrapping times is set by the occupation probability.*

sition of Lorenzian densities of states, and then replot vs gate voltage (step 5) to get a snapshot of the Fermi surface dynamics in the channel. Each trap shows up as a peak in the frequency spectrum, and the peak 'corner' frequencies themselves maximize on resonance (Fig. 20.10) along the gate voltage axis, allowing us to read off the spectral information directly.

The normalized power spectrum Sf plotted against frequency and gate voltage could *serve as a unique fingerprint of the molecular states* in an electronic current noise measurement. In fact, the shape factors for different molecules (such as aniline and nitrobenzene) are calculated to be sig-

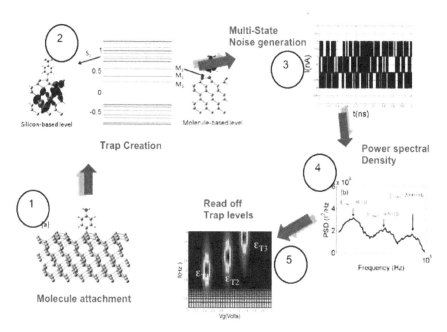

Fig. 20.11 *Schematic of a room temperature molecular 'bar-code' detector that reads signals from its gateable current noise. (1) Attachment of a molecular adsorbate on a clean silicon substrate leads to (2) traps in the silicon band-gap created by the molecular states (red lines). The states are characterized by their wavefunctions that cluster around the molecular orbitals away from silicon. (3) When the gated Fermi energy resonates with these states we get multilevel noise that translate into characteristic Lorentzian peaks (4) in the power spectral density. (5) Multiplying by frequency and replotting vs gate bias, we can read off the spatial and energetic locations of the individual traps as they each cross the Fermi energy under bias and generate an RTS sequence. The peaks in the corner frequency vs gate voltage give us the trap energies, while the ratio of capture to emission times gives us their gate and drain voltage dependencies and thus their spatial locations.*

nificantly different. Furthermore, they can provide insights into the Fermi surface crossings by various levels, notably, interactions among them that tend to smear out the peaks by the inter-trap charging energies. Experiments on carbon nanotubes show a suppressed conductance window in the transfer characteristic, with separate RTS signatures leading in and out of it. One way to explain the window is the presence of multiple interacting traps that activate and de-activate each other, leading to the creation and subsequent suppression of a blockade region with RTS signatures fringing it. A corner frequency scan shows how the trap signatures can split up due to interactions and can serve as a way to parametrize the correlation dynamics at the Fermi energy (Fig. 20.12).

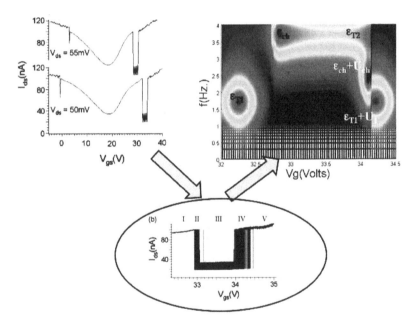

Fig. 20.12   *(Top left) Measured transfer characteristics in a nanotube FET (data from last two references of this chapter) shows a drain bias dependent window of suppressed current, with RTS signatures fringing it (zoom at the bottom). One way to explain these windows is the activation of a first trap that blocks the channel and the subsequent activation of a second trap that deactivates the first trap and unblocks the channel. Analyzing the RTS noise in Fourier space shows corner frequencies (top right) that give us a snapshot of the trap dynamics at the Fermi energy. The corner frequencies show a smear and a split (unlike Fig. 20.11), suggesting the activation and abrupt repulsion of traps from the Fermi window.*

Chapter 21

Magnetotransport

Magnetic fields inject additional dynamics onto electrons on their journey, tightening their rotational orbits and inducing them to cancel each other except at boundaries, where they skip merrily along the perimeters. The channel density of states collapses onto a ladder of equally spaced Landau levels representing bound states in the magnetic oscillator potentials. The levels are broadened by interactions with impurities and contacts (i.e., a self-energy), and occupied by electrons upto a Fermi energy. The placement of the Fermi energy on or off resonance with the Landau levels leads to dips and plateaus in the Hall voltage and show quantization effects. On the other hand, the interaction between the Landau levels and a commensurate background lattice potential yields a quasi-periodic phase plot known as the Hofstadter butterfly. Finally, collections of spins, exchange coupled to form a magnet, can impose their spin polarization onto electron currents passing through. When these polarized spins impinge on a soft magnet misaligned with them, angular momentum is exchanged between the two groups. The torque flips the magnet and allows us to write information onto it as a scalable, non-volatile memory. Once again, our dynamical equations allow us to explore these effects readily.

21.1 Semiclassical magnetotransport

Let us review the motion of a charge in a magnetic field. In presence of a magnetic field $\vec{B}$, a charge feels a Lorenz force

$$\vec{F} = q\vec{v} \times \vec{B} \qquad (21.1)$$

In presence of such a force, the electrons will form closed circular orbits around the lines of magnetic flux. Classically, the radius of the orbit can take any value, the corresponding speed set by force equality

$$\underbrace{\frac{mv^2}{r}}_{\text{centrifugal}} = \underbrace{qvB}_{\text{Lorenz}} \qquad (21.2)$$

with cyclotron frequency $\omega_c = v/r = qB/m$. Semiclassically, similar to the Bohr model of the atom, we need to postulate that the angular momentum is quantized to units of $\hbar = h/2\pi$, so that we have an additional equation

$$mvr = \hbar \qquad (21.3)$$

for the ground state (lowest energy) solution (we can interpret the above equation also as saying that the circumference of the orbit $2\pi r$ must be equal to integer multiples of the De Broglie wavelength h/mv). From the two equations, we get the radius of the orbit, called the cyclotron radius r_c

$$r_c = \sqrt{\frac{\hbar}{qB}} \qquad (21.4)$$

The flux contained in such an orbit is given by the flux quantum

$$\Phi_0 = B\pi r_c^2 = \frac{h}{2q} \approx 2.067 \times 10^{-15} Wb \qquad (21.5)$$

Finally, the coordinates of the oscillator centers are obtained by solving Newton's law with Eq. (21.1)

$$m\ddot{x} = F_x = q\dot{y}B$$

$$m\ddot{y} = F_y = -q\dot{x}B \qquad (21.6)$$

which upon integrating over time gives us

$$y - y_0 = \frac{m\dot{x}}{qB} = \frac{mv_x}{qB}$$

$$x - x_0 = -\frac{m\dot{y}}{qB} = -\frac{mv_y}{qB} \qquad (21.7)$$

21.2　Landau levels

To solve the problem quantum mechanically, we will need to make the Peierls substitution

$$\vec{p} \rightarrow \vec{p} - q\vec{A} = \vec{\Pi} \qquad (21.8)$$

with $\vec{p} = -i\hbar\vec{\nabla}$ being the canonical momentum and $\vec{\Pi} = m\vec{v}$ being the kinematic momentum. This substitution is necessary to account for the additional term in the potential, $q\vec{A} \cdot \vec{p}/\hbar$ that generates the Lorenz force upon spatial differentiation.

We will now use the *Landau gauge*, $\vec{A} = By\hat{x}$, one of several choices that gives the correct field $\vec{B} = \vec{\nabla} \times \vec{A}$. The Hamiltonian is now given by

$$H = \left[\frac{\Pi^2}{2m} + U(y)\right] = \left[\frac{p_y^2}{2m} + U(y) + \frac{(p_x - qBy)^2}{2m}\right]$$

$$= \left[\frac{p_y^2}{2m} + U(y) + \frac{1}{2}m\omega_C^2\left(y - y_k\right)^2\right] \qquad (21.9)$$

where $y_k = \hbar k_x/qB$ is the center of the cyclotron orbit. Note that $p_x = \hbar k_x$ is a constant of motion since there is no explicit x-dependence of the Hamiltonian (see Eq. (10.15)). The expression for y_k can be simplified by using the relation between canonical and kinematic momentum from Eq. (21.8), $k_x = (mv_x + qA_x)/\hbar$, and the Landau gauge $A_x = By$, giving us $y_k = (mv_x/qB + y)$, in agreement with the classical analysis (Eq. (21.7)). Also, $\omega_C = qB/m$ is the cyclotron frequency from the previous section. The Hamiltonian describes a series of shifted 1-D harmonic oscillators centered around $\hbar k_x/qB$.

The eigensolutions to the harmonic oscillator in the absence of a confining potential have the form (see caption in Fig. 3.1)

$$E_n = (n + 1/2)\hbar\omega_C$$
$$\Psi_n(x, y) = \chi_n(y - y_k)e^{ik_x x}$$
$$\chi_n(y) \propto H_n[y/r_C]e^{-y^2/2r_C^2} \qquad (21.10)$$

since the extent of the harmonic oscillator wavefunction $a_0 = \sqrt{\hbar/m\omega_C} = r_C$. H_n is the nth Hermite polynomial. These states do not carry any current, because the group velocity along the x-axis, $v_x = \partial E_n/\hbar\partial k_x = 0$.

In presence of some confining potential $U(y)$ which is zero in the middle but non-zero at the ends, we will get some 'skipping orbits' at the edges that do not cancel out, but create oppositely directed currents that ultimately carry the conductance (Fig. 21.1). We can use perturbation theory to estimate the energies in presence of the confining potential,

$$E_n \approx (n + 1/2)\hbar\omega_C + \langle n|U(y)|n\rangle$$

$$\approx (n + 1/2)\hbar\omega_C + U(y + y_k) \qquad (21.11)$$

since $|n\rangle$ is a shifted oscillator state (Eq. (21.10)). Thereafter

$$v_x = \frac{\partial U(y + y_k)}{\partial(\hbar k_x)}$$

$$= \frac{\partial U(y)}{\partial y}\frac{\partial(y_k)}{\partial(\hbar k_x)}$$

$$= \frac{1}{qB}\frac{\partial U(y)}{\partial y} \qquad (21.12)$$

In the middle of the sample, U varies weakly, but the confinement U becomes a large positive number outside the sample, so that at the upper edge we get $\partial U/\partial y$ positive and negative on the lower edge, and the states have positive and negative (opposite) group velocities, as well as separate orbital centers, y_k. Thus each Landau level now has localized states in the bulk, but oppositely directed extended states at the sample edges. The density of states is given by

$$D(E) = \sum_n A_n \delta(E - E_n) \qquad (21.13)$$

Since the states were merely redistributed by the presence of the magnetic field, we expect each delta function to be multiply degenerate. In fact, all the levels of the original 2DEG of DOS D_0 lying between two LLs (separated by $\hbar\omega_c$ in Fig. 21.1) coalesce into a delta function, so that the degeneracy

$$A_n = \hbar\omega_C D_0 = \hbar\omega_C \frac{mL_xL_y}{\pi\hbar^2}$$

$$= \frac{2qBL_xL_y}{h} = \frac{\Phi_B}{\Phi_0} = \frac{L_xL_y}{\pi r_C^2} \qquad (21.14)$$

where the flux is $\Phi_B = BL_xL_y$, while $\Phi_0 = h/2q$ is the flux quantum. The degeneracy is thus the number of flux quanta we can fit into the total flux, or equivalently the number of cyclotron orbits we can fit into the sample area. We can rigorously derive this degeneracy by noting that the

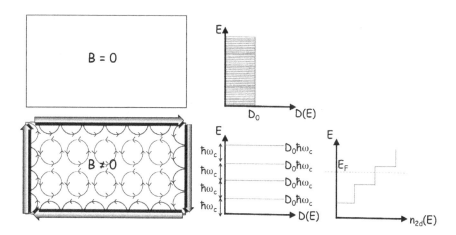

Fig. 21.1 *Left: Top view of 2D device without and with magnetic field, showing semi-classical electron trajectories. In the middle of the channel the orbital states cancel each other while those at the edges skip along, carrying a net edge current. Middle: The 2D density of states breaks into equally spaced Landau levels in a B-field, and (right) the electron density per unit area is quantized in multiples of $D_0 \hbar \omega_c / A = 2qB/h$ so that varying the magnetic field B changes the number of occupied Landau levels and the corresponding 2D electron density.*

k states are quantized in units of $\Delta k = 2\pi/L_x$, which makes the centers of the cyclotron orbits discrete by $\Delta y_k = \hbar \Delta k / qB = 2\pi r_C^2 / L_x$, so that the number of states $= 2L_y/\Delta y_k = L_x L_y / \pi r_C^2$

21.3 Hofstadter butterfly

The energy separation between the levels varies with applied magnetic field. If we superpose on top of this a periodic lattice potential, then the interaction between Bloch states characterized by lattice constant a and Landau levels characterized by cyclotron radius r_c generates a complex set of eigenvalues that form a self-similar, fractal pattern known as the Hofstadter Butterfly (Fig. 21.2). The key notion arises from the realization that the hopping terms are given by $-t = \int \Psi_i^* \hat{H} \Psi_j dx$ (Eq. (7.1)). Each wavefunction Ψ has a wavelike term along with the atomistic Bloch pre-factor. In presence of a magnetic field, we can continue as before with the Peierls substitution, $\vec{p} \to \vec{p} - q\vec{A}$, which gives us an extra factor $\exp(-i\frac{q}{\hbar} \int \vec{A} \cdot d\vec{l})$. We use the gauge $\vec{A} = (By, 0, 0)$, which modifies the x-components of the

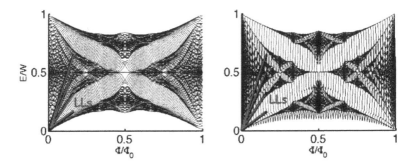

Fig. 21.2   *Hofstadter butterfly (Left: 11 × 11 square lattice in real space with hard wall bcs, right: 45 real space points with pbcs, and 145 k-space points), listing the energy eigenvalues normalized by bandwidth $8t_0$, vs the applied magnetic flux normalized with the flux constant, $\Phi/\Phi_0 = a^2/\pi r_c^2$. At low fields, the cyclotron radius is much larger than the lattice constant, and the electrons do not see the periodic lattice and act as free particles that quantize into Landau levels (LLs) in a B-field, shown superposed. At larger fields, the lattice makes an appearance and the levels split into the fractal structure called the Hofstadter butterfly, described by the Harper equation (21.17).*

coupling terms. The new Hamiltonian for an $N \times N$ square lattice then looks like

$$
H =
\begin{pmatrix}
4t_0 & -t_0 & 0 & 0 & \ldots & \ldots & -t_1 & \ldots & \ldots & 0 \\
-t_0 & 4t_0 & -t_0 & 0 & \ldots & \ldots & 0 & -t_2 & \ldots \\
0 & -t_0 & 4t_0 & -t_0 & & \ldots & \ldots & \ldots & -t_3 & \ldots \\
0 & 0 & -t_0 & 4t_0 & -t_0 & \ldots \\
\ldots & \ldots & \ldots \\
-t_1 & 0 & \ldots \\
0 & -t_2 & 0 & \ldots \\
0 & 0 & -t_3 & 0 & \ldots \\
\ldots \\
\ldots
\end{pmatrix}
=
\begin{pmatrix}
A & B & 0 & 0 & \ldots & 0 \\
B^\dagger & A & B & 0 & \ldots & 0 \\
0 & B^\dagger & A & B & \ldots & 0 \\
\ldots \\
\ldots \\
\ldots \\
\ldots \\
\ldots \\
\ldots \\
\ldots & \ldots & B^\dagger & A & B \\
\ldots & \ldots & B^\dagger & A
\end{pmatrix}
$$

$$(21.15)$$

where $t_m = t_0 \exp\left[-2\pi i m\phi/\phi_0\right]$, and A and B are $N \times N$ square matrices

$$
A =
\begin{pmatrix}
4t_0 & -t_0 & 0 & \ldots & \ldots & 0 \\
-t_0 & 4t_0 & -t_0 & 0 & \ldots & \ldots \\
0 & -t_0 & 4t_0 & -t_0 & \ldots & \ldots \\
\ldots & \ldots & \ldots & \ldots & \ldots & \ldots \\
\ldots & \ldots & \ldots & -t_0 & 4t_0 & -t_0 \\
\ldots & \ldots & \ldots & \ldots & -t_0 & 4t_0
\end{pmatrix},
\quad
B =
\begin{pmatrix}
-t_1 & 0 & 0 & \ldots & \ldots & 0 \\
0 & -t_2 & 0 & 0 & \ldots & \ldots \\
0 & 0 & -t_3 & 0 & \ldots & \ldots \\
\ldots & \ldots & \ldots & \ldots & \ldots & \ldots \\
\ldots & \ldots & \ldots & 0 & -t_{N-1} & 0 \\
\ldots & \ldots & \ldots & \ldots & 0 & -t_N
\end{pmatrix}
$$

$$(21.16)$$

We can implement periodic boundary conditions by filling the top right and bottom left corner elements of A with $-t_0$, and also the top right and bottom left $N \times N$ blocks of H with $B^\dagger$ and B respectively. Looking at the nth row, we get the Harper equation

$$4t_0\phi_n - t_0[\phi_{n-1} + \phi_{n+1}] - 2t_0[\phi_{n+N_1} + \phi_{n-N_1}]\cos(2\pi n\Phi/\Phi_0) = E\phi_n \tag{21.17}$$

Using the periodic Bloch solution, $\phi_n = \phi_0 e^{ikna}$, we get

$$E_{m,n} = 4t_0 - 2t_0\cos ka - 4t_0\cos(N_1 ka)\cos(2\pi n\Phi/\Phi_0) \tag{21.18}$$

with $k = 2\pi m/N_1 a$. Anytime the ratio Φ/Φ_0 reaches a rational number p/q, then when n reaches a multiple of q, the last cosine drops out and we recover a regular 2-D band.

21.4   Integer quantum Hall effect

The classical Hall effect arises by driving a current through a sample in presence of an applied magnetic field $\vec{B}$. The Lorentz force swivels the electrons to one side, creating an excess of electrons at one lateral end and a net positive charge on the other, in other words, a Hall voltage V_H. Quantum mechanics says that this voltage is quantized.

Consider a Hall bar with a current driven in the $x - y$ plane and B-field along $\hat{z}$. The voltage is generated along the y-direction and creates an electric field E_y. We expect current to flow until there is a voltage established along the y-direction, whereupon the Lorenz forces $q\vec{E}$ and $q\vec{v} \times \vec{B}$ cancel at steady-state. This means we get

$$qV_H/W = qvB \tag{21.19}$$

where V_H is the Hall voltage. The current density per width is $J = qn_{2d}v$, so that the current $I = JW = qn_{2d}Wv$. Thus, we get the Hall resistance

$$R_H = \frac{V_H}{I} = \frac{B}{qn_{2d}} \tag{21.20}$$

As we vary the Fermi energy, the longitudinal resistance drops to zero anytime we are in between two Landau levels, $\rho_{xx} = \sigma_{xx} = 0$, as all states in the bulk are localized, while those at the edges stay orthogonal and proceed ballistically. If at this time there are M Landau Levels filled at the Fermi energy (equal to the number of edge states that provide the conducting modes), then the 2-D electron density $n_{2d} = MA_n/(L_x L_y) = M/\pi r_C^2 =$

$2MqB/h$ (Fig. 21.1). Substituting, we then get a non-zero transverse (Hall) resistance

$$R_H = \frac{h}{2q^2 M}, \quad M = 1, 2, \dots \qquad (21.21)$$

which implies that the Hall resistance is quantized with a high degree of precision accorded by the lack of scattering between the spatially separated edge states.

21.5   Case study: anomalous quantum Hall for graphene

As an exercise, let us work out the quantum hall spectra for graphene. We start by writing down the 2x2 Hamiltonian as before (Eq. (5.11))

$$H = \begin{bmatrix} 0 & \hbar v_F \Pi_- \\ \hbar v_F \Pi_+ & 0 \end{bmatrix} \qquad (21.22)$$

where $\Pi_\pm = k_\pm - qA_\pm/\hbar$, with $k_\pm = k_x \pm ik_y = k_x \pm \partial/\partial y$ and $A_\pm = A_x \pm iA_y$. In the Landau gauge, $\vec{A} = By\hat{x}$, so that $A_\pm = By$. Defining the spinor components (Eq. (5.14)) as

$$\Psi = \begin{pmatrix} u_1(y) \\ u_2(y) \end{pmatrix} e^{ik_x x} \qquad (21.23)$$

we get from $H\Psi = E\Psi$, the equations

$$Eu_1 = \hbar v_F \left(k_x - \frac{\partial}{\partial y} - \frac{qBy}{\hbar} \right) u_2$$

$$Eu_2 = \hbar v_F \left(k_x + \frac{\partial}{\partial y} - \frac{qBy}{\hbar} \right) u_1 \qquad (21.24)$$

Left multiplying both sides of the first equation by $\hbar v_F \left(k_x + \partial/\partial y - qBy/\hbar \right)$, and using the second equation, we get

$$E^2 u_2 = \hbar^2 v_F^2 \left(k_x + \frac{\partial}{\partial y} - \frac{qBy}{\hbar} \right) \left(k_x - \frac{\partial}{\partial y} - \frac{qBy}{\hbar} \right) u_2$$

$$= \hbar^2 v_F^2 \left[-\frac{\partial^2}{\partial y^2} + \left(k_x - \frac{qBy}{\hbar} \right)^2 - \frac{qB}{\hbar} \right] u_2 \qquad (21.25)$$

where the last term came from the action of $\partial/\partial y$ from the first bracket on the $-qBy/\hbar$ term in the second bracket. This extra term arises because the canonical momentum Π_x does not commute with Π_y, but gives instead a curl of A which is the B term here.

The equation can be rewritten in reduced form as

$$\left[-\frac{\partial^2}{\partial y^2} + \left(\frac{y - y_k}{r_C^2} \right)^2 \right] u_2 = \epsilon u_2 \qquad (21.26)$$

where

$$\epsilon = \left(\frac{E}{\hbar v_F} \right)^2 + \frac{qB}{\hbar} = \left(\frac{E}{\hbar v_F} \right)^2 + \frac{1}{r_C^2} \qquad (21.27)$$

Note the similarity with a 1-D harmonic oscillator equation

$$\left[-\frac{\hbar^2}{2m} \frac{\partial^2}{\partial y^2} + \frac{1}{2} m\omega^2 (y - y_k)^2 \right] u_2 = \epsilon u_2 \qquad (21.28)$$

The energy eigenvalues of the 1D harmonic oscillator are $\epsilon = (n + 1/2)\hbar\omega$, with the factor $\hbar\omega$ equal to twice the square root of the product of the prefactors of the grad squared term and the parabolic term. The oscillator size is given by the fourth root of the ratio of those two pre-factors, in this case, $a_0 = \sqrt{\hbar/m\omega}$. By analogy then, for graphene, the solutions are

$$\epsilon_n = (n + 1/2)\frac{2}{r_C^2} = \frac{(2n+1)}{r_C^2} \qquad (21.29)$$

which substituted into the definition of ϵ gives us the energy eigenvalues which are the Landau levels

$$\left(\frac{E_n}{\hbar v_F} \right)^2 + \frac{1}{r_C^2} = \frac{(2n+1)}{r_C^2} \qquad (21.30)$$

so that

$$E_n = \left[\frac{\hbar v_F}{r_C} \right] \text{sign}(n) \sqrt{2|n|} \qquad (21.31)$$

with $n = 0, 1, 2, \ldots$. We can interpret v_F/r_C as a 'cyclotron frequency' ω_C. *Note that there is an energy at zero!*

The size of the oscillating wavefunction is set by the fourth root of the pre-factor ratio, which in this case becomes r_C. Thus

$$u_2 = H_n[(y - y_k)/r_c]e^{-(y-y_k)^2/2r_C^2} \qquad (21.32)$$

Plugging into the earlier equation for u_1, we get after some algebra

$$u_1 = u_2 \times \left[-\sqrt{2n} \left(\frac{H_{n-1}(z)}{H_n(z)} \right) \right] \qquad (21.33)$$

where $z = (y - y_k)/r_C$, with $y_k = \hbar k_x/qB$. Thus the nice pseudospin phase term $e^{i\theta}$ (Eq. (5.14)) gets messed up in presence of a B-field.

As before, adding in a confining potential $U(y)$ will endow these states with some curvature and a group velocity at the edges. Also, the k-states are once again quantized in units of $\Delta k = 2\pi/L_x$, giving us $\Delta y = \hbar \Delta k/qB = 2\pi r_C^2/L_x$, so that the degeneracy once again is $2L_y/\Delta y = L_x L_y/\pi r_C^2 = \Phi_B/\Phi_0$. We can still interpret this as the original graphene 2D densities of states integrated between two Landau levels, but since both the spacing and the graphene density of states are energy-dependent, the result is not that easy to see.

Since the weights of the Landau-level delta functions are unchanged, crossing M filled levels will once again give a Hall resistance $R_H = h/4q^2M$ with the added two coming from the two valleys. But remember that we also have a Landau level even at zero energy, so we occupy half-of it and get an additional Hall resistance $R_0 = h/2q^2$. The resistances work in parallel, so

$$R_H = \frac{h}{4q^2(M + 1/2)} \tag{21.34}$$

21.6 Case study: anomalous quantum Hall for bilayer graphene

Let us continue our exercise to look at bilayer graphene to see its Landau-level structure. We use the Hamiltonian from Eq. (5.21)

$$H = \frac{\hbar^2}{2m^*} \begin{bmatrix} 0 & \Pi_-^2 \\ \Pi_+^2 & 0 \end{bmatrix} \tag{21.35}$$

whose solution will be once again

$$E^2 u_2 = \frac{\hbar^4}{4m_*^2} \Pi_+^2 \Pi_-^2 u_2 \tag{21.36}$$

The commutator $[\Pi_+, \Pi_-] = -2qB/\hbar = -2/r_C^2$, so that

$$\Pi_+^2 \Pi_-^2 = \Pi_+(\Pi_+\Pi_-)\Pi_- = \Pi_+\left(\Pi_-\Pi_+ + [\Pi_+, \Pi_-]\right)\Pi_-$$

$$= \Pi_+\left(\Pi_-\Pi_+ - \frac{2}{r_C^2}\right)\Pi_-$$

$$= (\Pi_+\Pi_-)^2 - \frac{2}{r_C^2}\Pi_+\Pi_-$$

$$= \left(\Pi_+\Pi_- - \frac{1}{r_C^2}\right)^2 - \frac{1}{r_C^4} \tag{21.37}$$

We thus get

$$\sqrt{\frac{4m_*^2 E^2}{\hbar^4} + \frac{1}{r_C^4}} = \left(\Pi_+ \Pi_- - \frac{1}{r_C^2}\right) \tag{21.38}$$

which gives us the differential equation, following the steps earlier,

$$\left[-\frac{\partial^2}{\partial y^2} + \left(\frac{y - y_k}{r_C^2}\right)^2\right] u_2 = \epsilon u_2 \tag{21.39}$$

where $\epsilon = \dfrac{2}{r_C^2} + \sqrt{\dfrac{4m_*^2 E^2}{\hbar^4} + \dfrac{1}{r_C^4}}$. The solutions were as before, $\epsilon = (2n + 1)/r_C^2$, which gives us

$$E_n = \hbar \omega_C \sqrt{n(n-1)} \tag{21.40}$$

This too vanishes at $n = 0$. The extra degeneracy from the two layers compared to monolayer graphene gives us $R_H = h/2q^2(M + 1/2)$.

21.7 From spintronics to magnetoelectronics

The idea of encoding information onto the spin of an electron instead of charge has been around for a while. Instead of using the absence and presence of a charge to encode the 1 and 0 bits in conventional digital logic, we can use the up and down orientations of the spins. On the face of it, replacing charge with spin makes a lot of sense. Spins tend to couple weakly with their environment compared to charges, making them relatively robust with respect to noise. Furthermore, the energy required to flip a spin is miniscule (the Lorenz force on a spin acts perpendicular to its motion, so that the kinematic work is zero). It is natural then to expect that semiconductor spintronics is a way to low power computation.

Metal based spintronics, such as giant magnetoresistance (GMR) read heads, is now mainstream technology. A little thinking however illustrates where the challenges of semiconductor spintronics lie. Despite their isolation, single spins are fairly fragile entities that can still be easily corrupted with thermal noise. Proposals for solid state quantum computation typically rely on qubits such as precisely placed nuclear spins that need exceptional degrees of protection. Such isolation in turn prevents the spins from seeing their control fields, which now need to be large enough to circumvent thermal fluctuations. Large fields usually arise from large currents, which implies sizeable energy dissipated in the overhead circuitry. To put another way, *spins are cheap, but fields are expensive.*

Another problem is that most sources of spins, such as polarizing magnets, tend to generate mixtures of up and down electrons. The injection of majority spins is further diluted by the highly resistive semiconducting channels which themselves do not favor any one spin over the other. Accordingly, many physics experiments reliant on manipulating individual spin states employ circularly polarized light from tunable lasers. Finally, many propositions for spin based devices, such as the Datta-Das 'transistor' (Section 5.7), latch the spins onto the backs of mobile electrons, generating a highly dissipative charge current as well.

Significant strides have been made in mitigating at least some of these effects. The spin injection efficiency has been ramped up by putting a tunnel barrier that exponentially amplifies the polarization (recall the spin enhancing properties of symmetry filtering MgO, Section 15.2). Instead of exchange coupled spins, device concepts are migrating towards dipole coupled magnets (Fig. 2.5). The strong exchange coupling among spins within a magnet makes them rotate as one, reducing the energy dissipation and increasing robustness. A combination of crystal structure and shape engineering is then used to fabricate single domain magnets with a sizeable anisotropy, separating the up and down spins with an energy barrier substantially larger than $k_B T$.

Instead of charge currents, magnetic logic proposals now rely on spin currents that do not dissipate energy. The conversion efficiency from charge to spin currents has been amplified by using materials with large spin-orbit couplings, Section 5.7 (the so-called Giant Spin Hall Effect materials such as beta-tantalum), and also by geometrical scaling techniques that give us a spintronic 'gain'. In parallel, the polarization of the magnets needed for low write and read currents (Section 15.1) is now being pushed upwards through materials design. There is a worldwide search for half-metallic magnets that use clever chemical thumb rules (e.g., the Slater-Pauling rule) to open a bandgap selectively in one spin sector. This has its own challenges, as the half-metallicity tends to suffer when we move from ideal bulk systems to realistic heterostructures with strained surfaces and interfaces, finite bias voltages, and thermally driven ferromagnetic excitations.

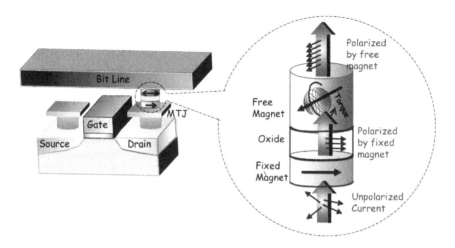

Fig. 21.3 *Schematic of a spin torque based random access memory (STTRAM) cell. Current is spin-polarized by transmission through a hard (fixed) magnet whose magnetization is pinned. The current is then repolarized when transmitting through a non-collinear soft (free) magnet. During repolarization the incident spin components perpendicular to the free magnet get absorbed. The addition of angular momenta generates a torque that flips the free magnet from a configuration antiparallel to parallel to the fixed magnet (i.e., high to low resistance). In effect, the free magnet must swivel the incident electrons to align with itself. In the process it exerts a torque on the electrons, and feels a counter-torque that in turn rotates it towards alignment with the incident electrons and the hard magnet. To flip back to the high resistance antiparallel state, current must be driven along the opposite direction, so that the removal of angular momenta from the free magnet and the associated swiveling flips it back to the antiparallel state.*

Finally, we have learned multiple ways to torque a magnet without actually using external magnetic fields with large current overheads. Examples include torques exerted by spin polarized currents, thermal torques across magnetic interfaces set at different temperatures, strain induced torques generated across a multiferroic material and torques created by spin-orbit coupling. Of these approaches, spin torque based devices are at this time the most mature (Fig. 21.3). More relevant to this book is the fact that *spin torque is inherently a non-equilibrium phenomenon*, so we will spend the next section discussing how to calculate such STTs.

21.8 Torquing magnets

Since this book is primarily about quantum transport, the inherently nonequilibrium concept of spin torque is a good candidate to employ our theoretical tools. Specifically, we will see how we can use NEGF to calculate the torque felt by a magnet. Based on our discussion above, we can see that the amount of torque is proportional to the component of the injected spin current density perpendicular to the magnetization $\vec{M}$ of the soft magnet. Accordingly, it depends on the dilution of current across the tunnel barrier separating the two magnets (the barrier height, width and effective mass that go into a Simmons-WKB expression). It also depends on the contact polarization that converts charge current into spin current. We thus expect the torque, i.e., the rate of change of soft magnetization, to have the form

$$\vec{\tau} = \partial \vec{M}/\partial t = a(V)\vec{M} \times \left(\vec{M} \times \vec{M}_F\right) + b(V)\vec{M} \times \vec{M}_F \qquad (21.41)$$

where $\vec{M}_F$ is the magnetization of the fixed magnet, and a, b are voltage-dependent coefficients. The above equation is the most general equation for a torque perpendicular to $\vec{M}$. The first component lies in the plane defined by $\vec{M}$ and $\vec{M}_F$ and is the most relevant current-driven torque that tends to rotate $\vec{M}$ into alignment or mis-alignment with $\vec{M}_F$ (Fig. 21.4). The second term rotates the soft magnet perpendicular to the plane of $\vec{M}$ and $\vec{M}_F$, and is often called the 'field-like' term (Figs. 21.4, 21.5).

Let us use our NEGF equations to calculate the torque on the soft magnet. We need the rate of change of angular momentum, i.e., magnetization $\vec{M}$,

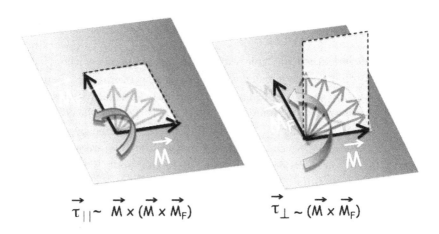

Fig. 21.4 *(Left) Non-equilibrium spin torque lying in the plane of the fixed and free magnetizations, driven by current polarized along the free magnet $\vec{M}_F$. (Right) Torque lying perpendicular to the plane, driven by a field-like term coming from direct ferromagnetic exchange between the two magnetizations. This term exists even at equilibrium.*

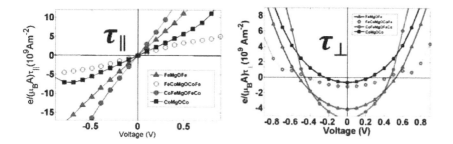

Fig. 21.5 *Calculated DFT-NEGF torques showing (left) the current driven component parallel to the $\vec{M} - \vec{M}_F$ plane, as well as (right) exchange-driven component
perpendicular to the plane, for Fe/MgO/Fe, FeCo/MgO/CoFe, CoFe/MgO/FeCo and
Co/MgO/Co magnetic multilayer junctions. Incorporating these torques into stochastic LLG allows us to predict circuit level write-error rates. Results are adapted from
'Spin Transfer Torque: A Multiscale Picture', Y. Xie, K. Munira, I. Rungger, M. T.
Stemanova, S. Sanvito and A.W. Ghosh, invited book chapter in Nanomagnetic and
Spintronic Devices for Energy Efficient Memory and Computing, ed. S. Bandyopad-
hyay and J. Atulasimha, Wiley 2016.*

which requires averaging the spin operator $\vec{S} = \hbar\vec{\sigma}/2$. $\vec{\sigma}$ is the Pauli spin
matrix in the local spin z basis set (Eq. (5.12) for $\sigma_{x,y}$, Eq. (21.42) for σ_z)

$$\vec{M} = \text{Tr}\left\langle \psi^\dagger \vec{S}\psi \right\rangle = \text{Tr}\left\langle \vec{S}\psi\psi^\dagger \right\rangle = \text{Tr}\left(\vec{\sigma}G^n \right)\hbar/2, \quad \sigma_z = \begin{pmatrix} 1 & 0 \\ 0 & -1 \end{pmatrix} \quad (21.42)$$

We can then use the Schrodinger equation $\partial\psi/\partial t = \mathcal{H}\psi/i\hbar$, $\partial\psi^\dagger/\partial t = -\psi^\dagger\mathcal{H}/i\hbar$, to get the torque

$$\vec{\tau} = \frac{\partial\vec{M}}{\partial t} = \frac{i}{2}\text{Tr}\left[\vec{\sigma}\left(G^n H - H G^n\right)\right] = \frac{-i}{2}\text{Tr}\left(G^n\left[\vec{\sigma}, \mathcal{H}\right]\right) \quad (21.43)$$

We can extract a spin current operator from this equation, describing the
spin current between sites i and j,

$$\vec{I}_i^S = \sum_j \frac{i}{2}\vec{\sigma}\left[G_{ij}^n H_{ji} - H_{ij}G_{ji}^n\right] \quad (21.44)$$

The net spin torque in Eq. (21.43) is then simply the divergence of spin
current at a point. For metallic layers, we often consider the incoming spin
current at the oxide-magnet interface and drop the outgoing spin current at
the other end of the magnetic layer, implying that all spins perpendicular
to the free magnet are completely absorbed and turned into torque.

To apply a torque, the Hamiltonian must be spin-dependent, as in a ferromagnet. For a spin Hamiltonian $\mathcal{H} = -\xi \vec{\sigma} \cdot \vec{B}$, the commutator can be evaluated from the easily verifiable commutative property of Pauli matrices

$$\left(\vec{\sigma} \cdot \vec{a}\right)\left(\vec{\sigma} \cdot \vec{b}\right) = \vec{a} \cdot \vec{b} + i\vec{\sigma} \cdot \left(\vec{a} \times \vec{b}\right) \qquad (21.45)$$

applied to one component of a Pauli matrix, $\sigma^\alpha = \vec{\sigma} \cdot \hat{x}^\alpha$, where $\hat{x}^\alpha = \hat{x}, \hat{y}, \hat{z}$

$$
\begin{aligned}
\left[\sigma^\alpha, \mathcal{H}\right] &= -\xi\left[\left(\vec{\sigma} \cdot \hat{x}^\alpha\right), \left(\vec{\sigma} \cdot \vec{B}\right)\right] \quad \text{(using } \mathcal{H} = -\xi\vec{\sigma} \cdot \vec{B}) \\
&= -2i\xi\vec{\sigma} \cdot \left(\hat{x}^\alpha \times \vec{B}\right) \quad \text{(using Eq. (21.45))} \\
&= -2i\xi\hat{x}^\alpha \cdot \left(\vec{B} \times \vec{\sigma}\right) \quad \text{(cyclical symmetry of scalar triple product)}
\end{aligned}
$$

$$(21.46)$$

so that $\left[\vec{\sigma}, \mathcal{H}\right] = 2i\xi\vec{\sigma} \times \vec{B}$. Eq. (21.43) then simplifies to

$$\frac{\partial \vec{M}}{\partial t} = \xi(\vec{\sigma} \times \vec{B})G^n = \frac{2\xi}{\hbar}\vec{M} \times \vec{B}, \quad \text{(perpendicular to } \tilde{M}) \qquad (21.47)$$

It is straightforward to see that $d|\vec{M}|^2/dt = 0$ by dot producting with $\vec{M}$, meaning that the magnetization stays constant in value and can only change orientation. In other words, the magnetization tends to precess around the magnetic field $\vec{B}$. Such a field can be generated externally as in a traditional magnetic random access memory (MRAM). It could come from the application of a strain across a magnetostrictive material that alters its potential landscape. For an STTRAM, the spin arises from the polarizing action of a fixed magnet $\tilde{M}_F$ on an injected current (Fig. 21.3) and will in general be of the form Eq. (21.41). We can work this out explicitly using plane wave eigenstates, the magnets imposing different $\vec{k}$ values for up and down spins at the same Fermi energy. It is straightforward to show (Ref. 1 for Ch. 15) that the current driven term simplifies at low bias

$$\tau_\parallel = \frac{4q^2\hbar^2}{m^2}\frac{\kappa_L\kappa_R k_{\uparrow L}k_{\uparrow R}}{[\kappa_L^2 + k_{\uparrow L}^2][\kappa_R^2 + k_{\uparrow R}^2]}\sin\theta e^{-2E_b(d)} \sim J_0 P_F \sin\theta V \qquad (21.48)$$

P_F the polarization of the fixed layer, θ the angle between magnets, and the subscripts L, R corresponding to left and right contact Fermi wavevectors k and decay constants κ through the barrier. We can readily generalize to an arbitrary spin-dependent Hamiltonian

$$\mathcal{H} = \frac{p^2}{2m} + V + g\mu_B\vec{\sigma} \cdot \vec{B} + \frac{\alpha}{\hbar}\hat{z} \cdot \left(\vec{\sigma} \times \vec{p}\right) \qquad (21.49)$$

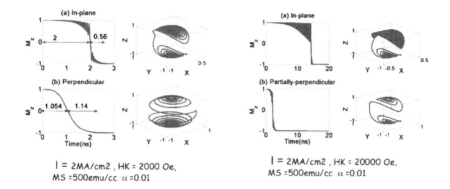

Fig. 21.6 *Magnetization switching (in units of saturation magnetization M_s) over time calculated for in-plane and perpendicular magnets, with (left) small vs (right) large anisotropy fields H_K and Gilbert damping α. See paper by Xie et al., referenced in Fig. 21.5.*

Following the steps before (an useful exercise!), we find that

$$\boxed{\frac{\partial \vec{M}}{\partial t} = \underbrace{-\vec{\nabla} \cdot \overleftrightarrow{\mathcal{J}}_S}_{\text{spin translation}} + \underbrace{\vec{J}_\omega}_{\text{spin rotation}}} \tag{21.50}$$

where the symmetrized torque operators

$$\overleftrightarrow{\mathcal{J}}_S = \left(\frac{\vec{v}\vec{M} + \vec{M}\vec{v}}{2}\right)G^n, \quad \vec{J}_\omega = \left(\frac{\vec{\omega} \times \vec{M} - \vec{M} \times \vec{\omega}}{2}\right)G^n$$

The currents will end up relating to velocities through Hamilton's equations

$$\vec{v} = \frac{\vec{p}}{m} + \frac{\alpha}{\hbar}\left(\hat{z} \times \vec{\sigma}\right) = \partial\mathcal{H}/\partial\vec{p}$$

$$\vec{\omega} = \frac{2}{\hbar}\left[g\mu_B\vec{B} + \frac{\alpha}{\hbar}\left(\vec{p} \times \hat{z}\right)\right] = \partial\mathcal{H}/\partial\vec{S}, \quad \vec{S} = \hbar\vec{\sigma}/2 \tag{21.51}$$

To the above spin torques, we add contributions from external magnetic fields, thermal fluctuations, Gilbert damping (that tends to bring spins back to equilibrium), as well as internal fields — determined by the material's potential landscape $\vec{H} = -\vec{\nabla}_{\vec{M}}U(\vec{M})$. U typically includes contributions from magnetocrystalline anisotropy (crystal-field) and shape anisotropy. The final $\partial\vec{M}/\partial t$ including all these equilibrium and non-equilibrium contributions gives us the macrospin Landau-Lifschitz-Gilbert (LLG) equation that we solve numerically to extract the switching dynamics (Fig. 21.6), energy dissipation, write time and bit flip errors — quantities critical to nonvolatile memory technology. We can also transfer it into an equivalent magnetic Fokker-Planck equation like we did in Chapter 9. With the advent of reliable DFT functionals as well as Materials 'Genome' approaches, we can do this 'first principles' i.e., predictably, all the way from atoms to circuits (Fig. 21.7).

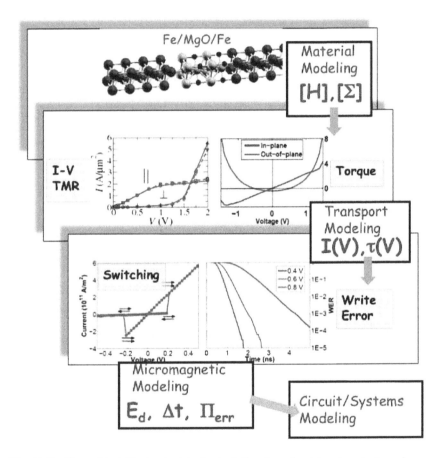

Fig. 21.7   *Hierarchical 'first principles' calculation from bandstructure to tunnel currents I (Eq. (15.2)) and torques $\vec{\tau}$ (Eqs. (21.41), (21.43)), thence to switching characteristics and bit flip errors using the stochastic LLG equation, leading ultimately to energy dissipation E_d, switching time Δt and write error rates Π_{err}. Results are adapted from 'Spin Transfer Torque: A Multiscale Picture', Y. Xie, K. Munira, I. Rungger, M. T. Stemanova, S. Sanvito and A.W. Ghosh, invited book chapter in Nanomagnetic and Spintronic Devices for Energy Efficient Memory and Computing, ed. S. Bandyopadhyay and J. Atulasimha, Wiley 2016.*

PART 6

A Smattering of Scattering: Resistance, Decoherence and Dissipation

Chapter 22

Simple models for scattering — combining transmissions

Along the electron's journey, it is likely to encounter multiple scattering events, potholes and road blocks, bridges and bus stops, that hinder, scramble and sometimes even accelerate its flow. The most familiar are momentum scattering events that create an electron backflow at an impurity site and generate a net resistance. As we saw earlier, a contact resistance appears even for a ballistic channel, as electrons from numerous contact modes M struggle to coalesce into a few channel modes m. The contact resistance, curiously, is a universal constant $R = h/2q^2$ for a single moded double spin channel, because the current is set by $\bar{T}_{M \to m} M$ (Eq. (12.5)), but the transmission probability itself goes as $\bar{T}_{M \to m} = m/(m+M) \sim 1/M$ for $m = 1 \ll M$, as we expect near a good contact. The resistance is akin to the buildup of a large incoming traffic merging from a six lane highway to a single lane. The opposite process creates no bottleneck and thus no resistance for electron escape into the contacts.

As far as scattering processes go, more ubiquitous are incoherent processes that destroy the electron's phase 'memory' while preserving momentum, washing out interference patterns and making the electron look classical. Finally, there are inelastic scattering processes that alter the electron's energy, heating it up or cooling it down by shifting energies between the electron and other scattering centers (for instance, setting an atom vibrating) and ultimately leading to energy dissipation. In some cases as we shall see, such processes can also help with transport. In our traffic flow analogy when going from a single lane to six, allowing the cars to change lanes (shift energies) would help them fan out and speed up as a result.

The actual process by which electrons equilibrate in a reservoir is very complicated, and involves a mixture of these processes that systematically rob the electron of its past history and in effect 'reset' the electron configuration. A proper theory of transport should be able to move effortlessly among these alternatives, depending on the structure and internal dynamics of the scattering center.

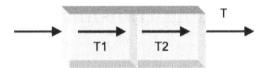

(a) *Direct transmission through two cascaded scatterers gives $T = T_1 T_2$. The electrons that do not make it to the end are redirected to the source.*

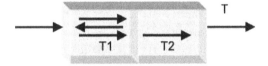

(b) *Electrons that are reflected at the source and enter the drain after a second try. $T = T_1 R_2 R_1 T_2$*

Fig. 22.1

We will try to see how scattering events can be included into the Landauer equation, specifically, in the mode-averaged transmission $\bar{T}$. Let us consider two segments with individual transmissions $\bar{T}_1$ and $\bar{T}_2$. What is the net transmission through the two segments cascaded in series?

A simple-minded theory would consider the two segments as independent, and wager that the net transmission is given by

$$\bar{T}_{\text{eff}} \approx \bar{T}_1 \bar{T}_2 \tag{22.1}$$

following the common statistical concept that for two independent uncorrelated events, their joint probability is simply the product of their individual probabilities. However, this would only be partially correct because the electron can reflect at the second interface with reflection probability $\bar{R}_2 = 1 - \bar{T}_2$, then turn back again at the first interface with probability $\bar{R}_1 = 1 - \bar{T}_1$, and then upon a second attempt go through the second interface. The probability of this second order process will then be

$\bar{T}_1 \bar{R}_2 \bar{R}_1 \bar{T}_2$. By the same logic, the electron could try a third attempt, fourth attempt and so on, the probability for emerging at the Nth attempt being $\bar{T}_1 (\bar{R}_2 \bar{R}_1)^{N-1} \bar{T}_2$. Summing over the entire spectrum of possibilities available to the electron, the final probability ends up as a geometric series

$$\bar{T}_{\text{eff}} = \bar{T}_1 \left(1 + \bar{R}_2 \bar{R}_1 + \bar{R}_2^2 \bar{R}_1^2 + \ldots \right) \bar{T}_2 = \frac{\bar{T}_1 \bar{T}_2}{1 - \bar{R}_1 \bar{R}_2} \qquad (22.2)$$

Since $\bar{R} = 1 - \bar{T}$, this can be written as

$$\bar{T}_{\text{eff}} = \frac{\bar{T}_1 \bar{T}_2}{1 - (1 - \bar{T}_1)(1 - \bar{T}_2)} = \frac{\bar{T}_1 \bar{T}_2}{\bar{T}_1 + \bar{T}_2 - \bar{T}_1 \bar{T}_2} \qquad (22.3)$$

or equivalently,

$$\frac{1}{\bar{T}_{\text{eff}}} - 1 = \left(\frac{1}{\bar{T}_1} - 1 \right) + \left(\frac{1}{\bar{T}_2} - 1 \right) \qquad (22.4)$$

meaning that $1/\bar{T} - 1 = \bar{R}/\bar{T}$ is additive. $\bar{R}/\bar{T}$ is thus the intrinsic, 4-terminal resistance of the channel, sans contacts (e.g., the resistance imposed by an impurity), and was the original focus of the Landauer equation when it was first formulated. If we have several segments cascaded together to create a length L, then according to this principle, $1/\bar{T} - 1$ is proportional to L and can be written as L/λ. This gives us the mode-averaged transmission

$$\boxed{\bar{T}_{\text{eff}} = \frac{\lambda}{\lambda + L}} \qquad (22.5)$$

$L \ll \lambda$ gives us $\bar{T}_{\text{eff}} = 1$, the ballistic limit. It is therefore straightforward to interpret λ as a mean-free path for backscattering. For a given microscopic scattering mechanism, we can calculate it using Fermi's Golden Rule for the electron scattering time. We will encounter a concrete example when we look at charge puddle scattering in graphene later in the next chapter.

22.0.0.1 *Including quantum coherence*

The transmissions $\bar{T}_{\text{eff}}$ describe the probabilities $|\psi|^2$ of the transmitted electrons, compiled for each possible history of the emergent electron. This kind of simple addition of probabilities is valid as long as there are no interferences between the waves. However, one lesson we learn from quantum mechanics is that the wavefunctions ψ themselves have independent physical meaning and the probability amplitudes for individual parallel processes need to be summed first before extracting the overall probability by mod-squaring. Instead of doing $\sum_i |\psi_i|^2$, the proper course of action should be to evaluate $|\sum_i \psi_i|^2$. This means we should be dealing with *reflection and transmission amplitudes*, described by r and t, which are in general complex, and whose mod-squared gives us $\bar{R}$ and $\bar{T}$. The sum over history should then read

$$t_{\text{eff}} = t_1 t_2 + t_1 r_1 r_2 t_2 e^{2ik\Delta x} + t_1 r_1^2 r_2^2 t_2 e^{4ik\Delta x} + \ldots$$

$$= \frac{t_1 t_2}{1 - r_1 r_2 e^{2ik\Delta x}} \tag{22.6}$$

and then doing $T_{\text{eff}} = |t_{\text{eff}}|^2$. In the above, we have also retained the phases picked up by the wave during each set of two passes through the segment of length Δx, not to mention any phases hidden inside r and t. Writing $r_i = \sqrt{\bar{R}_i} e^{i\phi_i}$, and a similar expression for t, we get (contrast with Eq. (22.2))

$$\bar{T}_{\text{eff}}(\theta) = \frac{\bar{T}_1 \bar{T}_2}{1 - 2\sqrt{\bar{R}_1 \bar{R}_2} \cos\theta + \bar{R}_1 \bar{R}_2} \tag{22.7}$$

where the phase $\theta = \phi_1 + \phi_2 + 2k\Delta x$.

It is straightforward to verify this relation with a concrete example, say, a particle incident on a barrier. This is a problem we encountered earlier. To recap, we matched the wavefunction Ψ and its derivative $d\Psi/dx$ (and thus the charge density $n \sim |\Psi|^2$ and current density $J \sim \Psi^* d\Psi/dx - \Psi d\Psi^*/dx$) across each interface. It is easy to check that the resulting barrier transmission (Eq. (14.3)) can be written analogously in terms of the transmissions and reflections across each step interface (Eq. (14.1)) as

$$T_{\text{barrier}} = \frac{T_{\text{step},1} T_{\text{step},2}}{1 - 2\sqrt{R_{\text{step},1} R_{\text{step},2}} \cos 2k'L + R_{\text{step},1} R_{\text{step},1}} \tag{22.8}$$

consistent with Eq. (22.7)

The equation is highly instructive, because in devices which maintain their phase coherence (e.g., samples at low temperature) the output transmission and thus the conductance shows oscillations with θ. As long as the phase θ can be tuned externally (e.g., varying the electron density with a gate and thus varying k_F which sits in θ), and as long as we are at low temperature where there are no thermally activated stochastic processes to randomize this phase (static impurities are fine since they impart shifts that are still time independent), we then reach a strange interference driven result, familiar to optics but certainly not to electronic circuit theory. This result states that the overall resistance of two resistors in series is tunable, no longer equal simply to their sum, and can in fact be smaller than an individual resistor! Such a phase coherent reduction in transmission is certainly familiar to microwave theory and transmission lines, but is not mainstream in large scale integrated circuit theory, which largely relies on the classical Ohm's law and its additivity.

If the electron is incident on the barrier at an energy below the barrier height, k' becomes imaginary, and we set $k' = i\kappa$, which then gives us the tunneling expression we have seen in the past (Eq. (14.3))

$$T_{\text{barrier}} = \frac{4k^2\kappa^2}{4k^2\kappa^2 \cosh^2 \kappa L + (k^2 + \kappa^2)^2 \sinh^2 \kappa L} \qquad (22.9)$$

which reduces to the WKB limit $\sim e^{-2\kappa L}$ for thick barriers.

To recover the classical Ohmic result (Eq. (22.2)), we need to average over the oscillations in Eq. (22.7) by integrating over their phase distribution $P(\theta)$. Assuming a uniform distribution over the phase angle, $P(\theta) = 1/2\pi$, we get

$$\bar{T}_{\text{eff}} = \frac{1}{2\pi} \int_0^{2\pi} \bar{T}_{\text{eff}}(\theta) d\theta \qquad (22.10)$$

Using the integral identity, $\int_0^{2\pi} dx/(A - B\cos x) = 2\pi/\sqrt{A^2 - B^2}$, we recover Eq. (22.2), valid for incoherent samples with a lot of phase averaging, where the phase relaxation length L_ϕ lies above the period L (justifying the use of Eq. (22.9)) but below the overall device length L_{dev} (i.e., $L < L_\phi < L_{\text{dev}}$), justifying the averaging in Eq. (22.10).

The standard way of including incoherence is to introduce phase-randomizing processes such as thermalized phonons. The phase information imposed on the phonon gets randomized in the process of thermalizing the phonons coupled with a heat bath. However, the above equation shows that we could also stick with coherent pieces and simply do an incoherent sum at the end. For instance, we chose to integrate the transmissivities (rather than the transmission amplitudes) over the phase angle, and in doing so, recovered an incoherent result rather trivially from a coherent one.

Note that the actual way of doing the ensemble averaging is important! What we just did is to extract the configurationally averaged transmission. The intrinsic resistance of the various segments are given by

$$R_{1,2} = \frac{h}{2q^2M}\left(\frac{1 - T_{1,2}}{T_{1,2}}\right) \tag{22.11}$$

and the resistance of the composite is

$$R = \frac{h}{2q^2M}\left(\frac{1 - \bar{T}_{\text{eff}}}{\bar{T}_{\text{eff}}}\right) \tag{22.12}$$

from Eq. (22.7). We could do the average by removing the phase information upfront as we did in Eq. (22.10), adding transmissions leading to a simple addition of resistances

$$\langle R \rangle_{\text{inc}} = \frac{h}{2q^2M}\left(\frac{1 - \langle \bar{T}_{\text{eff}} \rangle}{\langle \bar{T}_{\text{eff}} \rangle}\right) = R_1 + R_2 \tag{22.13}$$

This addition law ultimately gives us a linear dependence on length (i.e., Ohm's law). Indeed, we can apply the above equation to a segment of length $x + dx$ composed of a segment of length x in series with a differential segment of length dx and specific resistance α, we get

$$\underbrace{R(x + dx)}_{\langle R \rangle_{\text{inc}}} = \underbrace{R(x)}_{R_1} + \underbrace{\alpha dx}_{R_2} \tag{22.14}$$

which gives us $dR/dx = \alpha$ and thus $R = R_0 + \alpha x$.

If however we did this a slightly different way, assuming $L_{\text{dev}} < L_\phi$,

$$\langle R \rangle_{\text{coh}} = \frac{h}{2q^2M}\left\langle\frac{1 - \bar{T}_{\text{eff}}}{\bar{T}_{\text{eff}}}\right\rangle = R_1 + R_2 + 2R_1R_2 \quad (\text{verify}) \tag{22.15}$$

keeping all the phase information till the last step, we then get

$$\underbrace{R(x + dx)}_{\langle R \rangle_{\text{coh}}} = \underbrace{R(x)}_{R_1} + \underbrace{\alpha dx}_{R_2} + \underbrace{2R(x)\alpha dx}_{2R_1R_2} \tag{22.16}$$

in other words, $dR/dx = \alpha(1 + 2R)$, meaning $R = R_0 + \dfrac{e^{2\alpha x} - 1}{2}$. The main difference between the earlier classical diffusive (disorder and dephasing) and the current quantum diffusive limits (disorder without dephasing) is that we've now averaged over resistances $\sim 1/T_{\text{eff}}$ instead of conductances $\sim T_{\text{eff}}$, leading to an exponentially aggressive Anderson localization.

22.1 Simple models for interfacial scattering — Diffuse Mismatch Model (DMM)

In many systems we are looking at transport across heterostructures. Our intuition tells us that the amount of transmission must depend on how 'similar' the two look, and how well their modes match over the relevant energy window. This intuition can be translated into a deceptively simple expression for transmission using the diffusive mismatch model (DMM) originally derived for heat flow. The assumption behind this derivation is that we have two separately ballistic segments with mode spectra $M_1(E)$ and $M_2(E)$, and a diffusive process localized at the interface that randomizes the phase as a *one-shot* process.

We begin by writing the Landauer current expression as

$$I = I_0 \int \bar{T}_{12} M_1 \left[f_1 - f_2 \right] dE = I_0 \int \bar{T}_{21} M_2 \left[f_1 - f_2 \right] \qquad (22.17)$$

where we invoked a reciprocity (total conductance is the same regardless of which way the steady-state current flows). This gives us an equation

$$\bar{T}_{12} M_1 = \bar{T}_{21} M_2 \qquad (22.18)$$

We now invoke the loss of phase coherence at the interface. If an electron is moving to the right, it has either emerged from the left contact and crossed the interface with probability $\bar{T}_{12}$, or emerged from the right contact and reflected back at the interface with probability $1 - \bar{T}_{21}$. The lack of phase coherence means the electron has no 'memory' of its origin once it encounters the interface, and we can thus dictate that

$$\bar{T}_{12} = 1 - \bar{T}_{21} \qquad (22.19)$$

The two equations then give us

$$\bar{T}_{12} M_1 = \bar{T}_{21} M_2 = \frac{M_1 M_2}{M_1 + M_2} \qquad (22.20)$$

so that the total mode-averaged transmission, obtained by dividing the above by the total number of modes $(M_1 + M_2)/2$, normalized to the same area

$$\bar{T}_{DMM} = \frac{2M_1 M_2}{(M_1 + M_2)^2} \qquad (22.21)$$

The result differs by a factor of two from a purely ballistic transport model (also called the Acoustic Mismatch Model or AMM) with interference effects included (since the maximum transmission for two modes is $(1 + 1)^2 = 4$ rather than $1 + 1 = 2$). The AMM result gives us a transmission

$$\bar{T}_{AMM} = \frac{4M_1 M_2}{(M_1 + M_2)^2} \qquad (22.22)$$

which then gives us a reflection coefficient $1 - \bar{T}_{\text{eff}} = 1 - 4M_1 M_2/(M_1 + M_2)^2 = [(M_1 - M_2)/(M_1 + M_2)]^2$, i.e., the fractional difference in mode count at that energy (to be precise, the AMM only couches the result in terms of acoustic impedance Z, and $T_{AMM} = 4Z_1 Z_2/(Z_1 + Z_2)^2$, and what we are writing about is simply a conjectured generalization of that low-frequency impedance mismatch).

22.2 Simple model for incoherence — an isolated Büttiker probe

Earlier we discussed how to include simple momentum scattering events in the Landauer theory using a transmission probability $\lambda/(\lambda + L)$, allowing us to recover classical drift diffusion and eventually Ohm's law. We can further include phase-breaking scattering events by modeling each scattering center as a virtual contact that acts as a noninvasive voltage probe and draws no net current. Defining a local electron distribution function $f_s(E)$, in general not Fermi-Dirac, and a coupling γ_s for the scatterer, we can write down a current equation that generalizes the Landauer approach to multiple contacts. The Landauer Büttiker equation at any terminal is

$$I_i = \sum_j \frac{q}{h} \int dE T_{ij} [f_i - f_j] \qquad (22.23)$$

We assume that the transmission functions between any two contacts, real or virtual are known (in an independent level model, we can approximate $T_{ij}(E) = (2\pi\gamma_i\gamma_j/\gamma_T)D(E)$ (Eq. (11.5)) where $\gamma_T = \sum_i \gamma_i$ is the total broadening and D is the density of states). In addition f_1 and f_2, the contact Fermi Dirac functions are set at the two ends, while we wait for the scattering distributions f_s to float to their respective voltage values under nonequilibrium conditions. Our goal then is to determine these distributions and thence the net current between contacts 1 and 2 in presence of all these scatterers.

We start by setting each scattering current I_s at each probe to zero. This gives us a series of integral equations connecting the local distribution functions to the source/drain Fermi-Dirac functions. Since these are energy integral equations, they do not lead to a unique specification of the $f_s(E)$ functions, but rather, a constraint on their net energy sum. To get the distributions themselves out of this underspecified problem, we will need to make some additional model assumptions.

If we focus only on dephasing and momentum scattering but ignore inelastic scattering, so that there is no 'vertical flow' of electrons between different energy channels, then each current integrand individually vanishes at the virtual contacts. Solving these linear equations yields the local distribution functions readily. For a two-terminal channel with a single scattering center, setting $I_s = 0$ gives a non-thermalized distribution function

$$f_s(E) = \frac{T_{1s}(E)f_1(E) + T_{s2}(E)f_2(E)}{T_{1s}(E) + T_{s2}(E)}. \tag{22.24}$$

Substituting this back into Eq. (22.23) for the current, we can recast it in a Landauer form with an equivalent transmission

$$T_{\text{elastic}} = T_{12} + \frac{T_{1s}T_{s2}}{T_{1s} + T_{s2}}. \tag{22.25}$$

One can interpret this as a parallel combination of a scattering free channel (resistance proportional to $1/T_{12}$), and a scattering dominated channel given by two series resistances proportional to $1/T_{1s}$ and $1/T_{2s}$ (Fig. 22.2). The resulting distribution function f_s is patently non-thermalized. It can, of course, be forced into a Fermi-Dirac form, albeit with an energy dependent chemical potential $\mu_s(E)$.

We can now explore the opposite limit, where very strong inelastic scattering tends to scramble the various energy channels and bring the electrons to local thermal equilibrium. In other words, we posit a precise Fermi-Dirac form for each $f_s(E) = 1/[1 + \exp{(E - \mu_s)}/k_B T)]$ with *energy-independent parameters such as μ_s and temperature T*. The energy scattering means that we can no longer look at just the integrand. However, the prespecified form of $f_s(E)$ helps us solve the problem, with the only unknown now being μ_s. That electrochemical potential μ_s is obtained by adjusting its value till the numerically computed current I_s at the scattering point 's' vanishes.

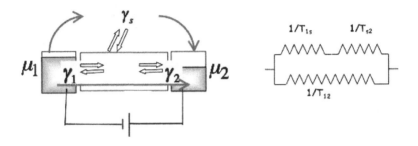

Fig. 22.2 *(Left) Schematic picture of scattering sources from contacts and a localized Büttiker probe. (Right) For elastic scattering, the scattering acts as a voltage divider and can be interpreted in terms of a simple circuit model.*

The Büttiker probe approach outlined above gives a simple description of scattering. By limiting ourselves to a phenomenological and predominantly energy-independent scalar form for γ_s, it ignores much of the rich microscopic dynamics of the scatterers. A realistic scattering process would involve a complicated $\gamma_s(E)$, with a complex energy dependence, in general elevated to become a matrix. We will deal with a more general matrix scattering later, but here we assume γ_s is a constant scalar.

22.3   A series of probes — D'Amato Pastawski model

We can extend the scattering model to the entire channel by considering a probe attached to each site and requiring that each of them individually draw a net zero current. We know the voltages at the two terminal sites and the currents (equal to zero each) at all intermediate sites. The equation can be written in matrix form as

$$
\begin{pmatrix} I_1 \\ I_2 \\ I_3 \\ \cdots \\ \cdots \\ I_{N-1} \\ I_N \end{pmatrix} = \int dE
$$

$$
\times \begin{pmatrix} \sum_i T_{1i}(E) & \cdots & -T_{12}(E) & \cdots & \cdots & -T_{1N}(E) \\ -T_{21}(E) & \sum_i T_{2i}(E) & \cdots & \cdots & \cdots & -T_{2N}(E) \\ \cdots & \cdots & \sum_i T_{3i}(E) & \cdots & \cdots & -T_{3N}(E) \\ & \cdots & -T_{43}(E) & & & \\ & \cdots & \cdots & & & \\ & \cdots & \cdots & & & \\ & \cdots & \cdots & & & \\ -T_{N1}(E) & \cdots & -T_{N3}(E) & \cdots & \cdots & \sum_i T_{Ni}(E) \end{pmatrix} \begin{pmatrix} f_1(E) \\ f_2(E) \\ \cdots \\ f_3(E) \\ \cdots \\ \cdots \\ f_{N-1}(E) \\ f_N(E) \end{pmatrix}
$$

$$(22.26)$$

Assuming a small drain bias, we can use $f_i(E) = F_T(E)\delta\mu_i$, where $F_T = \partial f_i/\partial \mu_i = -\partial f_i/\partial E \approx \delta(E - \mu_i)$, whereupon the equation simplifies to

$$
\begin{pmatrix} I_1 \\ I_2 \\ \boxed{I_3 = 0} \\ \cdots \\ \cdots \\ \cdots \\ I_{N-1} = 0 \\ \boxed{I_N = 0} \end{pmatrix} = \begin{pmatrix} \sum_i T_{1i} & \cdots & -T_{12} & \cdots & \cdots & -T_{1N} \\ -T_{21} & \sum_i T_{2i} & \cdots & \cdots & \cdots & -T_{2N} \\ \cdots & \cdots & \boxed{\sum_i T_{3i} \cdots \cdots -T_{3N}} \\ & \cdots & -T_{43} & & \\ & \cdots & \cdots & \mathbf{W} & \\ & \cdots & \cdots & & \\ -T_{N1}(E) & \cdots & -T_{N3} & \cdots & \cdots & \sum_i T_{Ni} \end{pmatrix}
$$

$$
\times \begin{pmatrix} \delta\mu_1 \\ \delta\mu_2 = 0 \\ \delta\mu_3 \\ \cdots \\ \cdots \\ \cdots \\ \delta\mu_{N-1} \\ \delta\mu_N \end{pmatrix}
$$

$$(22.27)$$

where we referenced all voltages relative to the right contact with $\delta\mu_2 = 0$ and evaluate all transmissions at $E = \mu_2$. We can then invert

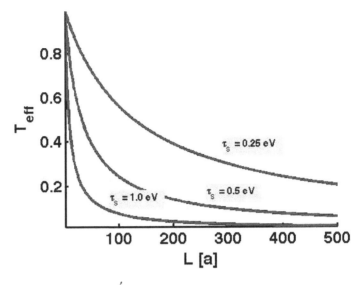

Fig. 22.3   *Transmission vs length for a series of Büttiker probes along a 1D channel for varying probe couplings τ_s. The coupling strengths can be connected to an overall mobility μ to represent the efficiency of scattering (courtesy Carlos Polanco).*

these equations to get

$$\delta\mu_i = \sum_{j=1}^{N} \left(W^{-1}\right)_{i,j} T_{j,0}\delta\mu_1 \qquad (22.28)$$

Substituting this back in the transport equation, we get

$$T_{\text{elastic}} = \underbrace{T_{1,2}}_{\text{coherent}} + \underbrace{\sum_{ij} T_{1,i}\left(W^{-1}\right)_{i,j} T_{j,2}}_{\text{incoherent}} \qquad (22.29)$$

as the generalization of the elastic single Büttiker probe result, Eq. (22.25).

22.4   Case study: scattering in a 1d wire

The ballistic transmission across a one dimensional wire is given by the Landauer transmission within the Fisher Lee formula,

$$T = \text{Tr}\left(\Gamma_1 G \Gamma_2 G^\dagger\right) = \gamma_1\gamma_2|G_{1N}|^2 \qquad (22.30)$$

assuming only nearest neighbor couplings, i.e., $\Gamma_1(i,j) = \gamma_1\delta_{i1}\delta_{j1}$ and $\Gamma_2(i,j) = \gamma_2\delta_{iN}\delta_{jN}$. For a homogeneous N-atom 1d wire with onsite energy

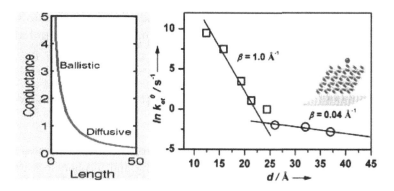

Fig. 22.4 *(Left) Transition from ballistic to diffusive based on Eq. (22.32). (Right) Electron transfer rates in helical peptides showing a similar behavior (H. S. Mandal and H B. Kratz, J. Phys. Chem. Lett. 3, 709 (2012).*

ϵ_0 and hopping $-t_0$, the Green's function can actually be calculated analytically (see paper by Mujica and book by Zimbolskaya). Using $\lambda = \epsilon_0 - E$ and $\zeta = \sqrt{\lambda^2 - 4t_0^2}$, this amounts to

$$G_{1N}(E) = \frac{2^{N+1}t_0^{N-1}\zeta}{(\lambda + \zeta)^{N-1}(\lambda + \zeta + i\Gamma)^2 - (\lambda - \zeta)^{N-1}(\lambda - \zeta + i\Gamma)^2}$$

$$\approx \frac{\sin ka}{t_0 \sin k(N+1)a} \quad \text{(if } \Gamma = 0, \quad E = \epsilon_0 - 2t_0 \cos ka) \qquad (22.31)$$

This equation takes into account hard-wall boundary conditions. We can simplify it if we ignore the edges and use plane wave solutions, in which case $G_{1N} = \psi_N/\psi_1 \propto e^{ikNa}$ from Bloch's theorem. k is complex outside the band, with $k = k_R + i\beta/2$, so that $T = \gamma_1\gamma_2 e^{-\beta Na}$. The equation tells us that the ballistic conductance reduces exponentially with channel length (as we argue after Eq. (22.16)). This is sensible, because the coherent 'one shot' transport of an electron must be driven ultimately by orbital overlap between the contacts, which decays exponentially as we separate the two.

If we now insert a series of Büttiker probes however, their resistances sit in series and transmissions add in parallel (generalizing Eq. (22.25)), so that the net incoherent transmission is $T_{\text{incoh}} = \left(\sum_i 1/T_i\right)^{-1} = T_i/N$ with T_i the transmission between any two neighboring probes. When we add this term to the coherent term, the resulting conductance then has the form (Fig. 22.4)

$$G = G_{\text{coh}}e^{-\beta Na} + G_{\text{incoh}}/N \qquad (22.32)$$

In other words, creating scattering 'islands' allows the electron to take a breather at each step and traverse a wider channel instead of negotiating a one-shot coherent jump.

22.5   Simple model for impurity scattering: Klemmens' equation

In the Landauer NEGF treatment of transport, the easiest process to include is elastic impurity scattering. Dephasing and incoherent scattering require constructing an added self energy, usually within the Born approximation, and going well beyond Landauer theory. But impurity scattering, due to a difference in material property at the defect site, is pretty straightforward to include directly in the device Hamiltonian. Let us try to get some simple estimates as well for the impact of such a scattering event.

The reduction in transmission at a scattering site due to an energy difference δE has already been shown (Eq. (14.6)) to take the form of a Breit Wigner distribution (we will see it again shortly). We can postulate a similar behavior for a mass difference as well, leading to the Klemmens' formula (originally derived for impurity scattering on phonon transmission)

$$T = \frac{1}{1 + \dfrac{\Delta m^2}{\Gamma_m^2}} \qquad (22.33)$$

Equating this with the standard expression, $T = 1/(1 + L/\lambda)$, we get the scattering time

$$\frac{1}{\tau} = \frac{v}{\lambda} = \frac{v \Delta m^2}{L \Gamma_m^2} \qquad (22.34)$$

which is the Klemmens formula for scattering. This equation can be derived independently from Fermi's Golden Rule, using $\hbar/\tau = |M|^2 D$, with the 1D density of states equal to $\hbar v/L$, and the matrix element given by $\Delta m / \Gamma_m$.

Typical impurity scattering processes involve a random distribution of such impurity masses. In many cases, such a distribution is well known, for instance (what else!) — a Gaussian. We can then average over such a Gaussian distribution to get a sense for the overall impact of impurity scattering on conductance. We will discuss a particular case study in the next chapter when we calculate the impact of charge puddles on the minimum conductivity in graphene.

22.6 Going smoothly from coherent to fully incoherent

Let us go back to the example at the beginning and see how we could switch smoothly between coherent and incoherent transmission. We will accomplish this by putting *two* Büttiker probes between two elastic scatterers. We need two probes to separately relax electrons coming in from either side (Fig. 22.5), so we have the flexibility of destroying time reversal symmetry completely. Note also that a single probe will simultaneously relax momentum and phase interdependently (we will see in a couple of chapters that momentum scattering requires us to relax components of the Green's function $G(x, x')$ along the $x + x'$ axis while spatial dephasing proceeds along the $x - x'$ axis, and these two are distinct only if the Green's function matrix and corresponding scattering matrix are at least 2×2 in size). For the D'Amato-Pastawski model with the 2-site Büttiker probe, it is straightforward to simplify Eq. (22.27) and show that

$$T = T_{LR} + \frac{\left(T_{L, \phi_1} + T_{L, \phi_2}\right)\left(T_{\phi_1, R} + T_{\phi_2, R}\right)}{2\left(T_{\phi_1, \phi_1} + T_{\phi_2, \phi_2} + T_{\phi_1, \phi_2} + T_{\phi_2, \phi_1}\right)} \quad (22.35)$$

analogous to Eq. (22.25). The transmission matrix components are obtained from the S-matrix connecting the outgoing and incoming electrons

$$\begin{pmatrix} a_L \\ a_R \\ a'_{\phi1} \\ a'_{\phi2} \end{pmatrix} = \begin{bmatrix} S \end{bmatrix} \begin{pmatrix} a'_L \\ a'_R \\ a_{\phi1} \\ a_{\phi2} \end{pmatrix}, \quad T_{ij} = |S_{ij}|^2 \quad (22.36)$$

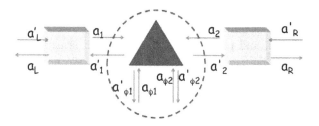

Fig. 22.5 *Incoherent Büttiker probes placed between two elastic scatterers. By controlling the probability of incoherent scattering P_ϕ, we can go smoothly between coherent and incoherent scattering.*

We can write down the scattering ('S') matrix components by connecting the input-output currents at each scatterer by inspection:

$$\begin{pmatrix} a_L \\ a_1 \end{pmatrix} = \begin{pmatrix} r_L & t_L \\ t_L & r_L \end{pmatrix} \begin{pmatrix} a'_L \\ a'_1 \end{pmatrix} \qquad \begin{pmatrix} a_R \\ a_2 \end{pmatrix} = \begin{pmatrix} r_R & t_R \\ t_R & r_R \end{pmatrix} \begin{pmatrix} a'_R \\ a'_2 \end{pmatrix}$$

$$\begin{pmatrix} a'_1 \\ a'_2 \\ a'_{\phi 1} \\ a'_{\phi 2} \end{pmatrix} = \begin{pmatrix} 0 & \sqrt{P_B} & \sqrt{1-P_B} & 0 \\ \sqrt{P_B} & 0 & 0 & \sqrt{1-P_B} \\ \sqrt{1-P_B} & 0 & 0 & -\sqrt{P_B} \\ 0 & \sqrt{1-P_B} & -\sqrt{P_B} & 0 \end{pmatrix} \begin{pmatrix} a_1 \\ a_2 \\ a_{\phi 1} \\ a_{\phi 2} \end{pmatrix} \qquad (22.37)$$

where P_B is the probability of the electron moving ballistically without scattering into the probe. The transmission/reflection coefficients satisfy (r is usually imaginary)

$$t_L^2 = 1 + r_L^2, \quad t_R^2 = 1 + r_R^2, \quad t_L t_R = 1 + r_L r_R \qquad (22.38)$$

from time-reversal symmetry and current conservation. The structure of the 4×4 matrix above is dictated by overall current conservation at the Büttiker probes. From Eq. (22.37), we write all a_Ls, a_Rs in terms of a_1s, a_2s from first line, re-express a_1s, a_2s in terms of a_ϕs from line 2, and reorganize in the form of Eq. (22.36) to get the transmission from Eq. (22.35)

$$\boxed{T = \frac{t_L t_R \left(1 + P_B\right)\left[1 + P_B r_L r_R\right]}{2\left[1 - P_B r_L r_R\right]^2}}$$

$$T_{\text{coherent}} = T(P_B = 1) = \frac{t_L^2 t_R^2}{(1 - r_L r_R)^2} \qquad \text{(see Eqs. (22.6), (22.7))}$$

$$T_{\text{fully incoherent}} = T(P_B = 0) = \frac{t_L t_R}{2} \qquad (22.39)$$

In the absence of elastic scattering where $t_{L,R} = 1$, we get $T = 1$ for completely coherent and $T = 0.5$ for fully incoherent, analogous to the AMM and DMM models. Complete incoherent scattering removes all memory of the electron origin and provides equal chance of it scattering forward and backward, much like the DMM model predicts. We now have a mechanism for smoothly going between the two limits by simply varying P_B. By extending to a series of such probes, we can now convert a resonant I-V (Figs. 1.2(a), 16.3) into a washed out, quasi-Ohmic one (Fig. 1.1(a)).

Chapter 23

Is scattering necessary?

23.1 Boltzmann and irreversibility

The Boltzmann distribution lies at the heart of the physics of relaxation and irreversibility. All of us are familiar with the effects of friction. If I drop a ball on the floor, it stops bouncing after a while. Why does it stop bouncing? We all know that the ball's energy gets dissipated into the floor and as a result the ball comes to rest while the floor heats up a bit (perhaps imperceptibly so). But what prevents the floor from cooling down, focusing all its thermal energy into a single consolidated upward motion and propel the ball towards the ceiling? After all, energy conservation allows that!

This brings us to the second law of thermodynamics and irreversibility and its counter-intuitive nature (unlike the first law, in fact unlike most 'laws', the second law actually involves an inequality!). The counterintuitive part is that viewed microscopically, i.e., two atoms colliding and scattering off, the reverse process looks equally likely (if one runs a film backward of two billard balls colliding, one would be hard pressed to know the difference). But viewed as an ensemble, we immediately see the problem created by running a scattering process backward — smoke would billow into a chimney or a cigarette, or the aforementioned ball would soar magically from the ground like flubber. The only difference seems to be that there are numerous smoke particles compared to our two billiard balls, and the associated 'tyranny of large numbers'.

One way to think of this is to compare the phase space for various modes. If both the ball and the floor consisted of just one atom coupled together with a spring, then exciting the ball would transport its energy to the floor atom, but the energy of the ball would recover shortly thereafter (there is a popular toy based on coupled pendula, known as 'Newton's Cradle'). If we now increase the number of atoms in the floor (i.e., the reservoir), and allow multiple modes in the reservoir, then the time taken for the entire system to migrate back to its exact initial phase space point (the so-called *Poincare recurrence time*) grows quickly with system size. For a true reservoir, this Poincare time must exceed any reasonable observation time, so that the above counter-intuitive processes, while not outright impossible, are rendered highly unlikely, with the recurrence time possibly greater than the age of the universe.

The macroscopic view is that the entropy of a closed system must not diminish. The microscopic view says that this entropy is given by the log of the number of distinct microstructures, $S = k_B \ln \Omega$, so that the entropy of two uncorrelated systems, $S = k_B \ln \Omega_1 \Omega_2 = k_B (\ln \Omega_1 + \ln \Omega_2)$, is additive. Most well-behaved reservoirs have Ω increasing with energy (higher energy typically generates higher momentum and thus a bigger phase-space), making S *an increasing function of energy*. In fact in classical thermodynamics, we interpret dS/dE as $1/T$, where T is the temperature — a positive quantity (reversing this trend momentarily away from equilibrium, for instance in the metastable states of a laser under population inversion, is often described as 'negative temperature'). Because of this increase in Ω, the corresponding probability of occupying any one microstructure goes down, $P \sim 1/\Omega$. In the reservoir, if we compare the ratio of probabilities at higher energy E_2 vs a slightly lower energy $E_1 = E_2 - \Delta E$, we get

$$\frac{P_2}{P_1} = \frac{\Omega(E_1)}{\Omega(E_2)}$$

$$= e^{[S(E_1)-S(E_2)]/k_B} \quad \text{(follows from S} = k_B \ln \Omega)$$

$$= e^{-\Delta E(dS/dE)/k_B} \quad \text{(Taylor expansion)}$$

$$= e^{-\Delta E/k_B T} \quad \text{(thermodynamic definition, T} = 1/(dS/dE) \quad (23.1)$$

which we recognize instantly as Boltzmann. Left to itself, a reservoir would therefore prefer to keep the energy of its microstates as low as possible, and deviate only to the extent that Boltzmann allows.

When a device is connected to a reservoir, such as a ball to an extended floor, its coupling with the reservoir excites its higher energy unoccupied modes and thereby propels the device preferentially towards its ground state with the same Boltzmann probability. The reservoir however reaches its ground state quickly because of the Boltzmann statistics (i.e., it 'thermalizes'), and in doing so, erases any information imparted onto it by the device. This information erasure (strictly speaking, only transient, but potentially spanning a long time due to the sheer size of the mode space in the reservoir) is at the heart of irreversibility. The timescale involved for this dissipative process is often called the Q-factor, and is ultimately a measure of the broadening $\gamma = \hbar/\tau$ that we invoke in quantum transport theory.

In classical devices, this thermalization shows up in the main evolution equations themselves, for instance, through the white noise added to Lángevin or the diffusion term entering Fokker-Planck. Recombination-generation are afterthoughts needed to take the electron away or reinject it into the system, and are not essential to the thermalization story. For ballistic quantum devices however, the thermodynamics is clearly separated from the dynamics. The latter is implicit in the Fermi-Dirac distributions maintained in the reservoirs, and the detailed microscopic processes that are invoked to enforce such a thermalization do not need to be dwelt upon (except the assumption that they are 'instantaneous' on the time scales of the electron transport). This is why temperature only enters the source term (the equivalent of generation) for quantum transport, but not the evolution itself. Indeed, that is what makes the Landauer equation such a nice convenient starting point — it neatly avoids diffusive processes that while seemingly omnipresent for non-ballistic devices, frequently obscure the essential physics of band-alignment and transport.

There are, however, notable cases where inelastic scattering *inside the device* is essential to consider, not merely as a realistic albeit inconvenient 'detail', but as the main kernel of the transport physics. This happens whenever we need to complete a circuit that is not otherwise completed. Thermalization in the contacts in itself may not be essential to transport *per se* — we might just need to deal with a non-thermalized, non-Fermi distribution. However, what would we expect to happen if the channel states are aligned, say, with a band-gap in the contact, or if the channel states themselves do not line up into a contiguous distribution at a fixed energy? (Think of a pn junction). The only way current can flow under such circumstances, indeed the only way we can complete the circuit, is to allow 'vertical transitions' along the energy level diagram, whereby an electron actually rises and falls in energy to bridge the gaps, emitting or absorbing photons or phonons in the process. Even the most elementary toy model must then include an inelastic scattering term to explain how current flows in such devices.

23.2 Does scattering help or hurt?

The simplest Landauer theory for current gives the zero-temperature, low-bias conductance as

$$G = G_0 M \bar{T} \tag{23.2}$$

where M is the number of modes at the Fermi energy and $\bar{T}$ is their mode-averaged transmission probability. Scattering is normally instrumental in reducing conductance. The presence of defects can cut down the transmission probability. It also typically reduces the effective number of conducting modes. If two segments of a device have modes M_1 and M_2, we saw that the effective mode density is given by $M_1 M_2/(M_1 + M_2)$ in the diffuse mismatch model, and this quantity, besides being smaller than both M_1 and M_2, clearly thrives only in the overlap band region where *both* modes exist (e.g., if we're outside the bandwidth of either segment, one of M_1 or M_2 goes to zero and so does the overall mode count).

The usual reduction in transmission with scattering can be described with the expression $\bar{T} = \lambda/(\lambda + L)$, alluded to earlier by combining multiple scattering events over their history and ignoring phase coherences. For instance, a site energy disorder $\delta\epsilon$ in a perfect 1-D chain will give a Lorentzian transmission of the Breit-Wigner form (Eq. (14.6)),

$$\bar{T} \approx \frac{1}{1 + \dfrac{\delta\epsilon^2}{\Gamma^2}} \tag{23.3}$$

i.e., the transmission reduces regardless of whether the defect is higher or lower in energy than the perfect channel. We can derive this equation from the expression for transmission over a barrier (Eq. (14.3)), by Taylor expanding around the resonances assuming $T_{\text{step}} \approx 1$, $\cos\theta \approx 1$ and replacing the deviation $\delta\theta \approx (d\theta/dE)\delta E$,- giving us an energy-dependent broadening

$$\Gamma \approx T_{\text{step}} \left(\frac{d\theta}{dE} \right)^{-1} \tag{23.4}$$

This expression can be readily rationalized when the phase angle arises from transit, so that $\theta = kL$ and $\Gamma = T_{\text{step}}(\hbar v/L)$ where v is the band velocity $v = \partial E/\hbar \partial k$.

If we now assume there is a random distribution of scattering disorder over a segment of length L, then the values of $\delta\epsilon$ take a random walk with average zero and variance proportional to the length, $\delta\epsilon^2 = CL$, where $C = W^2 N'$ with W being the bandwidth of the disorder and N' is the number of impurities per unit length. Equation (23.3) then gives $\bar{T} = \lambda/(\lambda + L)$, with $\lambda = \Gamma^2/C = \Gamma^2/W^2 N'$, related to Fermi's Golden Rule (Eq. (11.14)). Clearly if the disorder is narrow band (i.e., the energies fluctuate less), or if the impurity concentration is low, the scattering length is high.

Scatterers, however, can also open new conducting channels. We will shortly encounter inelastic tunneling spectroscopy as a concrete example, where electrons can enjoy an additional indirect scattering current by emitting phonons and witness an increase in conductance that shows up as characteristic peaks in $d^2 I/dV^2$. Another example we discuss in the next section is graphene, which has zero density of states at its Dirac point ideally; but charge puddles from impurities contribute to a spectral averaging of this Dirac point and escalate the minimum density of states, giving thereby not just a minimum conductivity, but one that can in fact increase with impurity concentration, at least initially. In such cases, we have an actual increase in modes, $M \to M_0 + \delta M$. The new conductance (Eq. (23.2)) in presence of scattering can then be written as

$$G = G_0(M_0 + \delta M)\frac{\lambda}{\lambda + L} \qquad (23.5)$$

for weak scattering. We readily see that the presence of these additional centers will escalate the conductance from its ballistic value $G_0 M_0$ if

$$\delta M/M_0 > L/\lambda \qquad (23.6)$$

in other words, if we have very few modes to begin with, $M_0 < \lambda \delta M/L$. This is intuitively sensible. Scattering centers hurt electron flow in highly conductive systems, but help in systems where there were very few modes for conduction to begin with. The Kondo effect is a perfect example, where the polarization of contact electrons by a spin impurity leads to a resonant state at the Fermi energy. In metals, such a state reduces conduction while in semiconducting quantum dots, it increases conduction (Section 27.6).

We can get a sense of this tradeoff is by invoking an additional energy-independent broadening Γ for the scattering mechanism. In a Lorentzian distribution, the broadening preserves the overall area under the distribution curve, and it does so by reducing the peak conduction while spreading some spectral weight to the wings. The conduction as a result reduces near resonance, and increases far away from it.

23.3 A case study: minimum conductivity in graphene in presence of charge puddles

Let us see how scattering can enhance the minimum conductivity in graphene worked out in Section 16.2. At low bias, charge puddles created by adsorbates and substrate 'doping' tend to misalign and average the Dirac points over a distribution of background charges, creating a non zero density of states. These puddles provide a viable route to conduction by injecting charge into the system. At higher bias or impurity concentrations, the reduction in mobility μ overrides the increase in charge density n.

Assuming a Gaussian distribution of dopants that in turn create a Gaussian distribution of shifts E_c around the Dirac point with average shift E_0,

$$D_{\text{puddle}}(E) = \sum_{E_c} D(E - E_c)P(E_c) \tag{23.7}$$

where

$$P(E_c) = \frac{1}{\sigma_E\sqrt{2\pi}}e^{-(E_c-E_0)^2/2\sigma_E^2} \tag{23.8}$$

with two defining parameters E_0 and σ_E (the standard deviation is typically ~ 50 meV).

Integrating the linear density of states (Eq. (16.3)) over the Gaussian distribution of shifted Dirac points, we get the modified density of states as

$$D_{\text{puddle}}(E) = \frac{2L_xL_y}{\pi\hbar^2v_F^2}\left[\sqrt{\frac{2}{\pi}}\sigma_E e^{-(E+E_0)^2/2\sigma_E^2} + erf\left(\frac{|E+E_0|}{\sigma_E\sqrt{2}}\right)|E+E_0|\right]$$

$$\approx \frac{2L_xL_y}{\pi\hbar^2v_F^2}\sqrt{(E+E_0)^2 + \frac{2\sigma_E^2}{\pi}} \quad \text{(fits above equation well)} \tag{23.9}$$

controlled by a doping shift and a smearing parameter. Near the Dirac point $E = -E_0$, the puddles now give us a non-zero density of states per unit area $\left(\sqrt{8/\pi^3}\right)\sigma_E/\hbar^2v_F^2$. At low voltages therefore, charge puddles increase the minimum conductivity by increasing the mode count, but at higher bias values, they degrade conduction by reducing scattering length.

Let us estimate the strength of the Coulomb scattering in 2-D using Fermi's Golden Rule (Eq. (11.14)), and extract a scattering length that we will include in the transmission (Eq. (22.5)). The Thomas Fermi equation (Eq. (4.16)) gives us the screened Coulomb potential

$$V_C(\vec{r}) = \frac{q^2}{4\pi\epsilon_0 r}e^{-\kappa r} \tag{23.10}$$

Using the eigenstates $\Psi_{i,f}(\vec{r}) = \frac{1}{\sqrt{2S}}\begin{pmatrix} 1 \\ e^{i\theta_{i,f}} \end{pmatrix}e^{i\vec{k}_{i,f}\cdot\vec{r}}$ from Eq. (5.14), normalized over area, we get the scattering matrix

$$V_{if} = \int d^2\vec{r}\,\Psi_f^*(\vec{r})V_C(\vec{r})\Psi_i(\vec{r})$$

$$= \frac{1}{2S}\int d^2\vec{r}\,e^{i\Delta\vec{k}\cdot\vec{r}}\left[1 + e^{i\Delta\theta}\right]\frac{q^2}{4\pi\epsilon_0 r}e^{-\kappa r}, \quad \Delta\vec{k} = \vec{k}_f - \vec{k}_i, \quad \Delta\theta = \theta_f - \theta_i$$

$$= \frac{1}{2S}\left[1 + e^{i\Delta\theta}\right]\int_0^\infty rdr\frac{q^2 e^{-\kappa r}}{4\pi\epsilon_0 r}$$

$$\times \underbrace{\int_0^{2\pi} d\theta e^{i\Delta kr\cos\theta}}_{2\pi J_0(\Delta kr)} \quad \theta : \text{angle between } \Delta\vec{k} \text{ and } \vec{r}$$

$$= \frac{q^2}{4\epsilon_0 S}\left[1 + e^{i\Delta\theta}\right]\underbrace{\int_0^\infty dr J_0(\Delta kr)e^{-\kappa r}}_{1/\sqrt{\Delta k^2 + \kappa^2}} = \frac{q^2}{4\epsilon_0 S\sqrt{\Delta k^2 + \kappa^2}}\left[1 + e^{i\Delta\theta}\right]$$

$$\tag{23.11}$$

Converted to energy for elastic scattering, where $|\vec{k}_f| = |\vec{k}_i| = E/\hbar v_F$,

$$(\Delta k)^2 = |\vec{k}_f - \vec{k}_i|^2 = k_f^2 + k_i^2 - 2k_f k_i\cos\Delta\theta$$

$$= 2E^2(1 - \cos\Delta\theta)/\hbar^2 v_F^2 \tag{23.12}$$

we get

$$|V|^2 = \frac{q^4\hbar^2 v_F^2(1 + \cos\Delta\theta)}{8\epsilon_0^2 S^2\left[2E^2(1 - \cos\Delta\theta) + \hbar^2 v_F^2\kappa^2\right]} \tag{23.13}$$

Assuming impurity density n_{imp} and cross sectional area S (impurity number $n_{\text{imp}}S$), we use the above matrix element in Fermi's Golden rule to give the momentum scattering time, related to the overall scattering time

(Eq. (11.14)) with an extra angular factor $1 - \hat{k} \cdot \hat{k}' = 1 - \cos \theta_k$ since forward scattering events do not degrade current (Section 23.4).

$$\frac{\hbar}{\tau_{sc}} = \sum_{\vec{k_f}} |V|^2 \delta(E - E_k)(1 - \cos \theta_k) n_{\text{imp}} S$$

$$= \frac{q^4 \hbar^2 v_F^2 n_{\text{imp}}}{16 \epsilon_0^2 \pi} \int D_{\text{puddle}}(E_k) dE_k \delta(E - E_k)$$

$$\times \int d\Delta\theta \frac{1 - \cos^2 \Delta\theta}{2E_k^2(1 - \cos \Delta\theta) + \hbar^2 v_F^2 \kappa^2}$$

$$= \frac{q^4 \hbar^2 v_F^2 n_{\text{imp}}}{16 \epsilon_0^2 S \pi} D_{\text{puddle}}(E) \frac{\pi}{2E^4} \left[2E^2 + \hbar^2 v_F^2 \kappa^2 - \hbar v_F \kappa \sqrt{4E^2 + \hbar^2 v_F^2 \kappa^2} \right]$$

$$(23.14)$$

with D_{puddle} defined earlier. For $E \ll \hbar v_F \kappa$, the term in square brackets expands to $2E^4/\hbar^2 v_F^2 \kappa^2 + O(E^6/\hbar^4 v_F^4 \kappa^4)$. The 2-D screening length is obtained using $\kappa = q^2 D_{\text{puddle}}/\epsilon_0$ (Eq. (4.16)), giving us $\hbar/\tau_{sc} = n_{\text{imp}}/16 D_{\text{puddle}}$. Using the Einstein relation from Eq. (10.6) (diffusion coefficient $\mathcal{D} = v_F^2 \tau_{sc}/2$), we get

$$\sigma = q^2 D_{\text{puddle}} \mathcal{D} = \frac{8q^2 v_F^2 \hbar}{n_i} D_{\text{puddle}}^2 \qquad (23.15)$$

where at high impurity density, $D_{\text{puddle}}^2 \approx 8\sigma_E^2/\pi^3 \hbar^4 v_F^4$.

There is a fair bit of physics involved in the variances of the charge distribution, typically connected with their influence on screening the potential. This can be worked out with the 2-D Poisson's equation to get us a self-consistent expression for σ_E which depends on the impurity concentration n_{imp}. We use an approximate formula for variance, $\sigma_E^2 \approx 2\hbar^2 v_F^2 n_{\text{imp}} + C$ with $C \sim 0.027$ eV^2, which matches with the self-consistent calculation fairly well in the dirty limit, giving us

$$\lim_{n_{\text{imp}} \to \infty} \sigma \approx \frac{128q^2}{\pi^3 h} = 4.12 \frac{q^2}{h} \qquad (23.16)$$

The equation predicts that even for dirty samples there is an asymptotic saturation of conductivity towards a value that is almost aspect ratio independent and weakly dependent on impurity density. More importantly, there is a reversal in trend at a critical aspect ratio and mode count whereby the conductivity increases with impurity density from the ballistic limit of $4q^2/\pi h$ by a factor $\sim \pi$, once their positive contribution to enhancing the overall mode count outdoes their deleterious contribution to mobility.

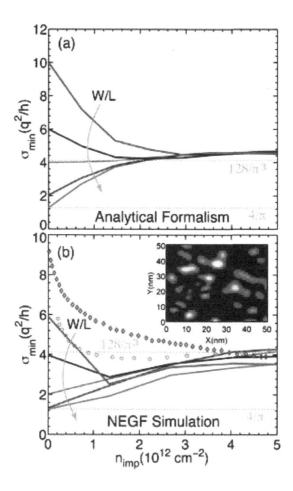

Fig. 23.1 *Minimum conductivity 'fan diagram' for graphene. (a) Quasianalytical treatment of NEGF with scattering included using Fermi's Golden Rule (FGR). (b) Numerical NEGF simulation with a Monte Carlo treatment of scattering by a random distribution of charge puddles (inset). The experimental data are (purple diamonds: J. H. Chen et al., Nat. Phys. 4, 377 (2008); orange circle, S Adam et al., PNAS 104, 18329 (2007)). For narrow samples with few modes, each mode needs to conduct well to get appreciable conductivity. Charge puddles hurt conductivity by reducing their transmission through scattering, causing σ_{min} to reduce with n_{imp} in the upper half plane of the fan diagram, as reflected by the experimental data as well. For wide samples with ample modes, scattering does not make a big difference, but the change in the mode count with the addition of impurity states at the Dirac point increases σ_{min} with n_{imp}, connecting with the quantized conductivity 4q²/πh in the lower half plane of the fan diagram. The theoretical model is published in 'Quantum transport at the Dirac point: Mapping out the minimum conductivity from pristine to disordered graphene', R. N. Sajjad, F. Tseng, K. M. Masum Habib and A. W. Ghosh, Phys Rev B 92, 205408 (2015).*

23.4   A variety of scattering mechanisms

Throughout the book, we employ a variety of scattering times — a conventional one given by Fermi's Golden Rule (Eq. (11.14)), a *momentum relaxation time* with an additional forward scattering term $f_k - f_{k'} \propto [1 - \hat{k} \cdot \hat{k}']$ (Eq. (23.14)), and an *energy relaxation time* with an exclusion principle factor $\propto [f_k(1 - f_{k'})N_q - f_{k'}(1 - f_k)(N_q + 1)]$ described earlier in Eq. (13.17). Aside from the occupancy terms, there is physics in the matrix elements $M_{kk'}$, which depend both on the perturbing potential as well as the 'size' and confinement of the wavefunctions. In device models, we typically work out the scattering times from different processes and add their contributions in parallel by summing τ_i^{-1} (Matthiessen's Rule), sometimes replacing the sum with an empirical expression, like the Takagi and Caughey-Thomas models discussed earlier (Section 18.1).

It is important to understand the scattering processes and identify the underlying potentials. Take decoherence in a spintronic device for example. Here we must evaluate matrix elements between different spin states with a potential $V = -J\vec{S} \cdot \vec{B}$ where $\vec{B}$ is a fluctuating magnetic field. Such a fluctating field can arise from exchange coupling of conduction electrons with rapidly thermalized holes and magnetic impurities (processes known as Bir-Aronov-Pikus and Overhauser effects). A fluctuating k-dependent internal field arises in materials with spin-orbit coupling, where $\vec{k}$ and thus any spin coupled with it is randomized by momentum scattering (this is called the Elliott-Yafet mechanism). Finally, we have the Dyakonov-Perel mechanism. The lack of inversion symmetry in materials like III-Vs shows up as a k-dependent magnetic field. An ensemble of electrons with a distribution of momenta see a corresponding distribution of magnetic fields, creating a wildly varying set of precessing electrons whose spins rapidly go out of alignment until they are reset by momentum scattering.

In the next chapter, we see how incoherent scattering processes enter NEGF. The equivalent of Fermi's Golden Rule is the self-consistent Born approximation (SCBA). $|M|^2$ is replaced by the deformation potential $\mathcal{D}_{ijkl} = \langle \lambda_{ij} \lambda_{kl} \rangle$ calculated from the incoherent, bilinear average of the scattering potentials λ (Eq. (24.21)). While we focus on vibrations next, it is relatively simple to generalize these principles. For spin dephasing for instance, the deformation potential will be given by the correlation $\langle B_i B_j \rangle$ of the fluctuating internal field from the thermalized magnetic scatterers.

Chapter 24

Vibrational scattering

24.1 Scattering in NEGF: the self-consistent Born approximation (SCBA)

The Büttiker probe approach described earlier gives a simple way to treat scattering. By limiting ourselves typically to an energy-independent, phenomenological scalar form for γ_s, we ignore much of the rich microscopic dynamics of the scatterers that would otherwise manifest themselves through the entire complex energy dependent matrix Σ_s. At the heart of electronic dephasing is an interaction between electronic degrees of freedom, and an external object (e.g., a photon, phonon or spin) whose dynamics is independently controlled by a thermal environment. The phase information transferred from the electronic subsystem to the scattering object is eliminated rapidly by coupling to the many modes of this large thermal bath, which imposes instead a thermalized distribution on the object (much like real metallurgical contacts were so far assumed to impose Fermi-Dirac distributions on local electrons). This information erasure is responsible for the loss of phase-coherence. Since we are dealing with an irreversible loss of information, at least on the time scales of the measurements in question, *dephasing cannot be captured trivially using a Hamiltonian. We need to invoke a non-Hermitian self-energy matrix to describe the non-unitary (irreversible) information flow.* (This comment should actually be qualified.. in principle, we can live with a Hamiltonian and do an incoherent sum over individually coherent calculations, but that sum over probabilities needs to be done by hand by throwing out interfering cross terms).

Let us look at scattering of an electron by a vibrational wave in a lattice, quantization of which yields 'phonons'. Upon electron phonon

scattering, the combined electron-phonon system evolves coherently according to a many-body Hamiltonian (Eq. (6.17)), describing in sequence an
electronic system with eigen-energies ϵ_i, a vibrating system decoupled into
normal modes (phonons) of frequencies ω_α, and an interaction term given
by $\sim \psi^*(r)U(r-R)\psi(r)$, which Taylor expanded gives a linear electron-
phonon coupling proportional to the electron number and the vibrational
amplitude (Eq. (7.16)). In other words,

$$\hat{\mathcal{H}} = \sum_i \epsilon_i n_i + \sum_\alpha \frac{1}{2}\left(\frac{p_\alpha^2}{m} + m\omega_\alpha^2 x_\alpha^2\right) + \sum_{i\alpha} \lambda_{i\alpha} n_i x_\alpha \tag{24.1}$$

Expanding the many-body wavefunctions in a basis of orthogonal single
particle electron states, i.e., $\Psi(x,t) = \sum_i c_i(t)\phi_i(x)$, the first electronic
term in the Hamiltonian with n becomes quadratic in the operators c,
$c^\dagger$. Similarly, expanding the coordinate and momentum terms using the
ladder operator representation diagonalizes the oscillator sub-element of
the Hamiltonian (Eqs. (7.11))

$$x_\alpha = \sqrt{\frac{\hbar}{2m\omega_\alpha}}\left(a_\alpha + a_\alpha^\dagger\right), \quad p_\alpha = \sqrt{\frac{\hbar m\omega_\alpha}{2}}i\left(a_\alpha - a_\alpha^\dagger\right) \tag{24.2}$$

The coefficients of the Hamiltonian operator in this $\{\phi\}$ basis set are

$$\hat{\mathcal{H}} = \sum_i \epsilon_i c_i^\dagger c_i + \sum_\alpha \hbar\omega_\alpha a_\alpha^\dagger a_\alpha + \sum_{i\alpha} \lambda_{i\alpha} c_i^\dagger c_i (a_\alpha^\dagger + a_\alpha). \tag{24.3}$$

with the first two sums acting like noninteracting particles (quadratic in
the operators and thus diagonalized in matrix notation), while the third
term introduces non quadratic elements. We introduced the electron and
phonon creation and annihilation operators $c, a, c^\dagger, a^\dagger$ to allow us to keep
track of the symmetry properties of the electron and phonon many body
wavefunctions. Indeed, the fermionic/bosonic nature of electrons/phonons
reflect themselves in the antisymmetric/symmetric nature of their many-
body states, so that the operators satisfy the conditions

$$[a, a^\dagger] = \{c, c^\dagger\} = 1 \tag{24.4}$$

between one creation and one destruction operator, and

$$[a, a] = [a^\dagger, a^\dagger] = \{c, c\} = \{c^\dagger, c^\dagger\} = 0 \tag{24.5}$$

between similar operators, with $[P, Q]$ denoting the commutator $PQ - QP$
and $\{P, Q\}$ denoting the anti-commutator $PQ + QP$. The above rules
enforce the symmetry properties of the wavefunction. For instance, the
fourth equality in Eq. (24.5) represents Pauli exclusion by implying that

$(c^\dagger)^2 = 0$, i.e., we cannot create two electrons at the same place (more generally, $\{c_\alpha, c_\beta\} = 0$, meaning that $\Psi_\alpha \Psi_\beta = -\Psi_\beta \Psi_\alpha$.)

Let us simplify our approach to a single phonon mode.

$$\hat{H} = \sum_i \epsilon_i c_i^\dagger c_i + \hbar\omega_0 a^\dagger a + \sum_i M \underbrace{(a + a^\dagger)}_{A_1} c_i^\dagger c_i \qquad (24.6)$$

and drop the 'i' index. Our goal will be to solve for the phonon degrees of freedom and use them as inputs to the electronic equation to extract the electron current in presence of phonon scattering.

Solving the Heisenberg equation $i\hbar d\hat{O}/dt = [\hat{O}, \hat{H}]$ first, where $\hat{O}$ stands for the phonon operators a and $a^\dagger$, we get

$$i\hbar \frac{da}{dt} = \hbar\omega_0 a + Mn$$

$$-i\hbar \frac{da^\dagger}{dt} = \hbar\omega_0 a^\dagger + Mn \qquad (24.7)$$

so that the equation for the displacement $A_1(t) = a + a^\dagger$ becomes

$$\frac{d^2 A_1}{dt^2} + \omega_0^2 A_1 = -2 \frac{M\omega_0 n}{\hbar} \qquad (24.8)$$

with solution

$$A_1(t) = A_0(t) - 2\frac{M\omega_0}{\hbar} \int dt' \Pi_\omega(t, t') n(t') \qquad (24.9)$$

where A_0 is *the solution in the absence of electron-phonon coupling*, i.e., the currents we have been exploring in the previous chapters. The phonon Green's function Π_ω (first introduced in Eq. (11.31)) satisfies the solution with just an impulse (delta function) perturbation, which forms the kernel of the particular integral because any function can be constructed out of such delta function terms.

$$\left(\frac{d^2}{dt^2} + \omega_0^2\right)\Pi_\omega(t, t') = \delta(t - t') \implies \Pi_\omega(\omega) = \frac{1}{(\omega_0^2 - \omega^2)} \qquad (24.10)$$

Substituting Eq. (24.9) in the Heisenberg equation for the electron operators with $\hat{H}$ (Eq. (24.6)), we get

$$\left[i\hbar\frac{d}{dt} - \epsilon_i\right]c_i(t) = \underbrace{MA_0(t)c_i(t)}_{\text{phonon scattering}} \underbrace{-2\frac{M^2\omega_0}{\hbar}\int dt' \Pi_\omega(t, t') n(t')}_{\text{polaronic shift}} \qquad (24.11)$$

If the number of electrons n were a constant, then the last term would be quite easy to evaluate in the frequency domain, and would amount to $-2M^2 n\omega_0/\hbar(\omega_0^2 - \omega^2)$, which has a DC value equal to $-2M^2 n/\hbar\omega_0$. This last term is a polaronic shift that can be interpreted as the change in zero-point energy of a stretched harmonic oscillator (the linear electron-phonon coupling acts effectively as an electric field that stretches the mean position of an oscillator). In other words, it is the shifted oscillator we will get from Eq. (24.1) by completing squares for an oscillator with a linear coupling, $m\omega^2 x^2/2 + Mxn/2 \to m\omega^2(x + 2Mn/m\omega^2)^2/2 - \underbrace{2M^2 n^2/m\omega^2}_{\text{polaronic shift}}$. (When comparing with the result from Eq. (24.11), keep in mind that the electron number n is zero or one, i.e., $n^2 = n$, when the electrons have no added dynamics).

Let us ignore this polaronic shift for now. We will discuss it in terms of the reorganization energy of an oscillator later (Eq. (24.53)). It is quite easy then to identify the rest of the right hand side of Eq. (24.11), the phonon scattering term, as a source, and construct the inscattering functions, just like we outlined in our introduction to NEGF equations. Indeed, defining $G = [i\hbar d/dt - H]^{-1}$, we get from Eq. (24.11)

$$c_i(t) = \sum_k M \int dt' G_{ik}(t, t') A_0(t') c_k(t') \qquad (24.12)$$

One can now write down the expression for the electronic correlation function using Eq. (11.17)

$$\begin{aligned} G_{ij}^n(t, t') &= \langle c_j^\dagger(t') c_i(t) \rangle \\ &= \sum_{kl} \int dt_1 dt_2 G_{ik}(t, t_1) \Sigma_{kl}^{\text{in}}(t_1, t_2) G_{lj}^\dagger(t_2, t') \qquad (24.13) \end{aligned}$$

where

$$\begin{aligned} \Sigma_{kl}^{\text{in}}(t_1, t_2) &= M^2 G_{kl}^n(t_1, t_2) \Pi^n(t_1, t_2) \\ \Pi^n(t_1, t_2) &= \langle A_0(t_2) A_0(t_1) \rangle = \langle [a + a^\dagger]_{t_2} [a + a^\dagger]_{t_1} \rangle \\ &= \underbrace{N_\omega e^{-i\omega(t_2 - t_1)}}_{\text{phonon absorption}} + \underbrace{(N_\omega + 1) e^{i\omega(t_2 - t_1)}}_{\text{phonon emission}} \qquad (24.14) \end{aligned}$$

where $N_\omega = \langle a^\dagger a \rangle = 1/[\exp(\hbar\omega/k_B T) - 1]$ is the equilibrium Bose-Einstein distribution of the phonons, obtained by averaging the phonon number operator $n = a^\dagger a$ over a Boltzmann distribution of energies, $\sum_{n=0}^\infty n e^{-n\hbar\omega/k_B T} / \sum_{n=0}^\infty e^{-n\hbar\omega/k_B T}$. This is also where the fundamental asymmetry between phonon emission and absorption processes manifests itself. The off diagonal terms connecting the N and $N + 1$ phonon sectors involve creation and annhiliation operators that do not commute, representing the symmetry of the phonon wavefunctions and their pair correlation, so that absorption involves the average number operator $\langle a^\dagger a \rangle = N_\omega$, while emission involves the reverse process $\langle a a^\dagger \rangle = \langle a^\dagger a + \underbrace{[a, a^\dagger]}_{=1} \rangle = N_\omega + 1$. The '1' term coming from the commutator here, independent of N_ω, gives rise to spontaneous phonon emission, while the other two terms in Eq. (24.14) represent stimulated emission and absorption.

Note that the electron phonon system evolves coherently according to the many body Hamiltonian (24.3), so we need to postulate a separate coupling, typically between the phonons and an external thermal reservoir, i.e., added Hamiltonian terms like $\sum_{k,\alpha=L,R} \epsilon_{k\alpha} a^\dagger_{k\alpha} a_{k\alpha} + \sum_{k\alpha} \tau_{k\alpha}(a^\dagger_{k\alpha} a + a^\dagger a_{k\alpha})$, involving contact variables $a_{k\alpha}$, to equilibrate the phonons and replace the dynamically evolving $\langle a^\dagger a \rangle$ phonon number operator with an equilibrium Bose Einstein distribution. Imagine we had a way to keep the phonons coherent and prevent thermalization, for instance, by interrogating the electrons and phonons using ultrafast pump and probe techniques in weakly coupled environments with large equilibration times. We would then need to actually calculate averages such as $\langle a^\dagger a \rangle$ properly using the evolution with the rest of the Hamiltonian, indeed even extract averages like $\langle a \rangle$, $\langle a^\dagger \rangle$, $\langle a + a^\dagger \rangle$ (mean displacement of the oscillators) for possibly coherent phonons, instead of simply setting them to zero. We discuss phonon coherence at the end of this chapter.

The Fourier transform (consistent with Eq. (11.34))

$$\Pi^n(E) = 2\pi[N_\omega \delta(E - \hbar\omega) + (N_\omega + 1)\delta(E + \hbar\omega)] \tag{24.15}$$

Since the product in time-space for Σ^{in} above becomes a convolution in Fourier space, it is easy to see from Eq. (24.14) (and independently from Langreth rule $\Sigma^< \sim G^< \Pi^<$) that

$$\Sigma^{\text{in}}_{kl}(E) = M^2 \left[G^n_{kl}(E - \hbar\omega)N_\omega + G^n_{kl}(E + \hbar\omega)(N_\omega + 1) \right] \tag{24.16}$$

Implicit here is an integration over ω with a phonon density of states. The first term shows phonon absorption while the second indicates emission, the term independent of N_ω being spontaneous emission.

The equations connecting the phonon contributions form two groups —
a set of dynamical equations, and a set of static equations. The *dynamical
equations* describe the non-equilibrium filling and emptying of states

---

Dynamical equations

$$G^{n,p}(E) = G(E)\left(\Sigma_1^{\text{in,out}}(E) + \Sigma_2^{\text{in,out}}(E) + \Sigma_{ph}^{\text{in,out}}(E)\right)G^\dagger(E)$$

$$\Sigma_{ph}^{\text{in,out}}(E) = \mathcal{D}_0(\omega) \otimes \left[N_\alpha(\omega)G^{n,p}(E \mp \hbar\omega) + \left(N_\alpha(\omega)+1\right)G^{n,p}(E \pm \hbar\omega)\right]$$

$$(24.17)$$

There is an alternate, elegant way to derive the equations above. Going
back to the original Hamiltonian Eq. (24.3), and the properties of the creation and annihilation operators on the N phonon subsectors derivable from
the phonon commutation relations in Eqs. (24.4), (24.5)

$$a|N\rangle = \sqrt{N}|N-1\rangle$$
$$a^\dagger|N\rangle = \sqrt{N+1}|N+1\rangle \qquad (24.18)$$

We can then write down the many body Hamiltonian in the many particle
basis set as a matrix. Using a basis set of direct products between the ith
N electron and the αth N_{ph} phonon state (the combined state represented
by the bra $|i,\alpha\rangle$), the Hamiltonian matrix Eq. (24.3) for the ith electronic
energy sub-block can be written as

$$\begin{pmatrix} \cdots\cdots & \cdots & & \cdots & \cdots\cdots \\ \cdots\cdots & E^i_{N-1} & \lambda_{i\alpha}\sqrt{N_\alpha} & \cdots & \cdots\cdots \\ \cdots\cdots & \lambda_{i\alpha}\sqrt{N_\alpha} & E^i_N & \lambda_{i\alpha}\sqrt{N_\alpha+1} & \cdots\cdots \\ \cdots\cdots & \cdots & \lambda_{i\alpha}\sqrt{N_\alpha+1} & E^i_{N+1} & \cdots\cdots \\ \cdots\cdots & \cdots & & \cdots & \cdots\cdots \end{pmatrix} \qquad (24.19)$$

where $E^i_N = \epsilon_i + N_\alpha \hbar\omega_\alpha$. One way to think of this is to imagine several
near-identical copies of the electronic subsystem varying only in the number
of phonons interacting with it (Fig. 24.1).

*In other words, the electrons act as their own contact, albeit with a different
energy argument displaced by the emitted or absorbed phonon energy. The*

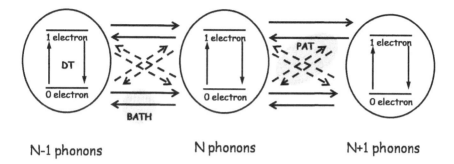

N-1 phonons　　　　N phonons　　　　N+1 phonons

Fig. 24.1　*Diagrams showing transport processes in electron-phonon Fock space, which can be thought of as copies of the electronic Fock space with transitions driven by Direct tunneling (DT), phonon-assisted tunneling (PAT) and thermalization processes driven by an external bath (BATH).*

many body Hamiltonian in Eq. (24.19) has the same form as a one-electron Hamiltonian that looks like

$$\begin{pmatrix} H_L & \tau_L^\dagger & 0 \\ \tau_L & H & \tau_R \\ 0 & \tau_R^\dagger & H_R \end{pmatrix} \tag{24.20}$$

The contact self energies after thermal averaging can be written down readily by analogy with Eq. (11.10)

$$\Sigma_L^{\text{in}} = \langle \tau_L g_L \tau_L^\dagger \rangle = \underbrace{\langle \lambda_{i\alpha}^2 \rangle}_{\mathcal{D}_0(\omega)} N_\alpha(\omega) G(\underbrace{E_{N-1}^i}_{E-\hbar\omega})$$

$$\Sigma_R^{\text{in}} = \langle \tau_R g_R \tau_R^\dagger \rangle = \underbrace{\langle \lambda_{i\alpha}^2 \rangle}_{\mathcal{D}_0(\omega)} \Big(N_\alpha(\omega) + 1 \Big) G(\underbrace{E_{N+1}^i}_{E+\hbar\omega}) \tag{24.21}$$

which leads to the second line of Eq. (24.17). The λ terms in the expression for τ combine together to generate the deformation potential $\mathcal{D}_0$, in other words, $(\mathcal{D}_0)_{ij} = \langle \lambda_i \lambda_j \rangle$ in a real space basis. In general, λ is a non-diagonal matrix in an arbitrary basis set, so that $\mathcal{D}_0$ is a fourth-rank tensor, $\mathcal{D}_{ijkl}^\alpha = \langle \lambda_{ik}^\alpha \lambda_{lj}^{\alpha\dagger} \rangle$ for the αth phonon mode. The electron-phonon scattering processes are often spatially localized, so that in a real space basis the λs become diagonal, $\lambda_{pq} = \lambda_p \delta_{pq}$. In that case, we can define $\mathcal{D}_{pq} = \langle \lambda_p^\dagger \lambda_q \rangle$ with only two indices, and the self-energy becomes $\Sigma_{kl}^{\text{in}}(E) = \mathcal{D}_{kl} G_{kl}^n(E \pm \hbar\omega)(N_\omega + 1/2 \pm 1/2)$. In matrix notation, this becomes

$\Sigma^{\text{in}}(E) = \mathcal{D} \otimes G^n(E \pm \hbar\omega)(N_\omega + 1/2 \pm 1/2)$, where $\otimes$ denotes an element-by-element as opposed to matrix multiplication, i.e., $(A \otimes B)_{ij} = A_{ij}B_{ij}$.

Equations (24.17) are referred to as the self-consistent Born approximation (SCBA), the self-consistency arising because $\Sigma^{\text{in,out}}_{\text{ph}}$ and $G^{n,p}$ depend on each other in the above equations (the Feynman diagrammatic version is in Fig. 24.2). Moreover, the equations conspire together to enforce current conservation by ensuring that no current leaks out at the scattering center. This is evidenced by writing out the Keldysh equation for the scattering current (Eq. (11.17)), re-expressed conveniently using the relations $A = G^n + G^p$ and $\Gamma = \Sigma^{\text{in}} + \Sigma^{\text{out}}$,

$$I \sim \int dE Tr(\Sigma^{\text{in}}_1 A - \Gamma_1 G^n) = \int dE Tr(\Sigma^{\text{in}}_1 G^p - \Sigma^{\text{out}}_1 G^n) \qquad (24.22)$$

and extended to the phonon scattering current

$$I_{\text{ph}} \sim \int dE Tr(\Sigma^{\text{in}}_{\text{ph}} G^p - \Sigma^{\text{out}}_{\text{ph}} G^n) \qquad (24.23)$$

The substitution from Eq. (24.17) with $\Sigma^{\text{in,out}}_{\text{ph}} \propto G^{n,p}$ makes the current drawn by the scattering center vanish (a condition explicitly used in the Büttiker probe approach). But replacing G with the non self-consistent Green's function, $\Sigma^{\text{in,out}}_{\text{ph}} \propto G^{n,p}_0$ maintains a scattering current leakage $\sim \int dE Tr(G^n_0 G^p - G^p_0 G^n)$ and violates current conservation.

Note that Eq. (24.17) above are for a single phonon mode at frequency ω, and will need to be integrated over a phonon density of states for a continuous distribution of phonons.

The *static equations* describing the states themselves are given by

$$\boxed{\begin{array}{c} \underline{\text{Static equations}} \\[1em] \Gamma_{\text{ph}}(E) = \Sigma^{\text{in}}_{\text{ph}}(E) + \Sigma^{\text{out}}_{\text{ph}}(E) \\[1em] \Sigma_{\text{ph}}(E) = \mathcal{H}\left(\Gamma_{\text{ph}}(E)\right) - i\frac{\Gamma_{\text{ph}}(E)}{2} \\[1em] G(E) = \left[EI - H - \Sigma_L - \Sigma_R - \Sigma_{\text{ph}}\right]^{-1} \end{array}}$$

$$(24.24)$$

where $\mathcal{H}$ denotes a Hilbert transformation

$$\mathcal{H}\big(\chi(\omega)\big) = \frac{1}{\pi} \int_{-\infty}^{\infty} d\omega' \frac{\chi(\omega')}{\omega - \omega'} \qquad (24.25)$$

We need to go this roundabout way because of the non-locality of the self energy matrices; they depend on Green's functions calculated at a different energy separated by $\hbar\omega$. Alternatively, we could directly use the last Langreth equation (Eq. (11.38)), $\Sigma \sim G^n\Pi + G\Pi^n + G\Pi$. Finally, the current is calculated from Eq. (11.17).

Note that at equilbrium, the *static equations* can be extracted directly from the self-consistent Born approximation (Eq. (24.14)) and written in terms of Feynman diagrams. At nonequilibrium, however, the relation is

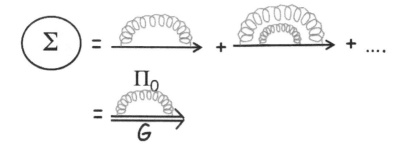

Fig. 24.2 *The self-consistent Born approximation, described as a sum of Feynman diagrams showing the underlying electron-phonon scattering processes. Self-consistency is needed since the diagrammatic equation needs to be evaluated along with Fig. 7.2.*

more complicated because of the different energy arguments of G^n and G^p that do not simply add up (Eq. (24.17)) and can only be captured from the SCBA equations indirectly upon Hilbert transforming. It is however instructive to notice that for *incoherent but elastic* scattering, where we simply randomize phase (so Σ stays complex) but don't relax the energy ($\omega = 0$ in the equations for $\Sigma^{\text{in,out}}$), we can then simplify

$$\Gamma_{\text{elastic}}(E) = \Sigma^{\text{in}}(E) + \Sigma^{\text{out}}(E) \propto \langle\lambda\rangle^2 \big(G(E) - G^\dagger(E)\big) \sim \langle\lambda\rangle^2 A(E) \qquad (24.26)$$

which is in effect, a NEGF based derivation of Fermi's Golden Rule (Eq. (11.14)).

24.2　Beyond SCBA: higher order terms

Most treatments of phonons in solid-state device end with the self-consistent Born approximation, which is indeed the dominant term for low electron-phonon coupling. As the electron number or the coupling constant increases, higher order processes start getting signficant. We can include higher order Feynman diagrams to capture second or third order electron-phonon coupling effects perturbatively. Two of the higher order processes are shown below. Following Feynman rules described earlier, we can write

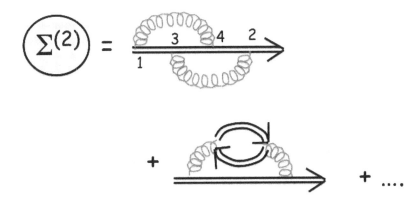

Fig. 24.3

down the self-energies by inspection. The first diagram gives us

$$\Sigma_{DX}(\tau_1,\tau_2) = \frac{\left\langle \lambda^4 \right\rangle}{3} \int d\tau_3 d\tau_4 G(\tau_1,\tau_3)\Pi_0(\tau_1,\tau_4)G(\tau_3,\tau_4)\Pi_0(\tau_3,\tau_2)G(\tau_4,\tau_2)$$
(24.27)

which we can extend to the Keldysh contour using Langreth rules. This term can also be cast as a conventional SCBA term, but including all the interesting dynamics into a 'dynamical vertex correction', $\lambda \to \tilde{\lambda}$, with the new coupling now frequency dependent from the time indices it carries.

$$\Sigma_{DX}(\tau_1,\tau_2) = \frac{\left\langle \lambda^4 \right\rangle}{3} \int d\tau_3 d\tau_4 \underbrace{G(\tau_3,\tau_4)\Pi_0(\tau_3,\tau_2)G(\tau_4,\tau_2)}_{\Gamma(\tau_3,\tau_4;\tau_2)} \Pi_0(\tau_1,\tau_4)G(\tau_1,\tau_3)$$
(24.28)

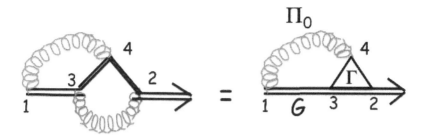

Fig. 24.4

The big question probably is whether these higher order terms are relevant. They could well be, as long as the couplings are large enough to warrant inclusion of such terms, yet small enough to allow perturbative treatments. One can use standard tricks like selectively summing the most divergent terms in the perturbation expansion that may still be summable, say as a geometric series. However, it is important to remember that what made Feynman diagrams so useful in QED is the presence of a small fine structure constant $\alpha \sim 1/137$. For electrons and phonons, this number may or may not be small depending on the regime of operation, and one may well need clever non-perturbative techniques to make things work, especially out of equilibrium, not to mention a compelling reason to do so.

24.3 The information content in the scattering matrix

The self-consistent equations introduced earlier allow us to explore various kinds of scattering — resistance, dissipation and decoherence, by simply varying the information content in the deformational potential D. The most general equation above captures momentum, energy and phase scattering. To get rid of energy scattering, we set $\omega = 0$, which makes each $\Sigma^{\text{in}}(E)$ relate directly to the energy-dependence of $G^n(E)$ as independent energy channels. The matrix $\mathcal{D}_0$ is diagonal in real space, but its diagonal entries are, in general, different. If we now go one step further and make those quantities equal so that $\mathcal{D}_0 = D_0 I$ where I is the identity matrix, then the *spatial relations* between the components of G^n are also preserved when we go to Σ^{in} and thence back to G^n. These processes do not mix spatial components, and thus eliminate momentum scattering (the latter comes from broken translational symmetry in real space), and capture

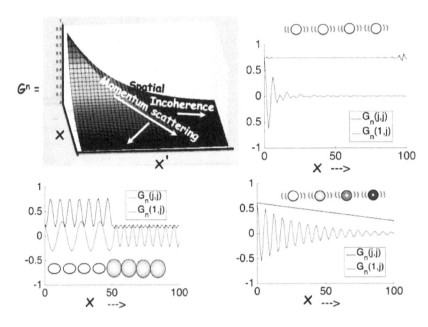

Fig. 24.5   *(Top left) Diagonal elements of $G^n(x, x')$ capture momentum scattering while off-diagonal elements capture spatial decoherence (temporal analogues are inelastic and time incoherent scattering). Impact of scattering terms on the diagonal and off-diagonal G^n elements for a 1-D chain: dephasing only (top right), momentum scattering only (bottom left) and both (bottom right). Simulations by Jingjie Zhang.*

dephasing alone. Figure 24.5 shows the impact of scattering on the correlation Green's function $G^n(x, x', t, t')$ for a 1-D chain with dephasing $(\mathcal{D}_0 = D_0 I)$, momentum scattering (change in mass) and both scattering mechanisms. Writing the Green's function explicitly in terms of the various coordinates, we can see that momentum scattering events disrupt spatial periodicity and cause decays along the $x + x'$ axis while spatially incoherent scattering influences matrix components along the $x - x'$ axis. Analogously, inelastic (energy) scattering disrupts temporal periodicity along $t + t'$, while temporal incoherence flows along $t - t'$.

24.4   When scattering helps: non-resonant Inelastic Electron Tunneling Spectroscopy (IETS)

The expected role of phonons is to create a relaxation channel to impede electron flow. However, phonons (or even localized vibrons) can also open channels where there previously was none and thereby assist flow. Inelastic Electron Tunneling Spectroscopy (IETS) is a prime example. When conduction happens through the tail of an electronic state, i.e., in the tunneling regime, electron-phonon coupling can open an additional channel for conduction when the applied drain bias exceeds the phonon energy $\hbar\omega$. The electron can then relax its energy through phonon emission before escaping into the lead (Fig. 24.6), the inequality $qV > \hbar\omega$ instrumental in removing any Pauli blocking from the leads. The added modes lead to a jump in conductance at $qV = \hbar\omega$, which becomes a prominent peak when one measures the second derivative d^2I/dV^2 with a lock-in amplifier tuned to twice the resonant frequency (Fig. 1.2(c)). Such peaks are catalogues for various phonon modes and can serve as a spectroscopic signature of a specific molecule.

IETS arises in the non-resonant limit of scattering. In the next section, we will show that a resonant electronic level produces phonon sidebands that can be thought of as a set of delta functions in conductance. The IETS jump in conductance can be viewed as a *continuous sequence of phonon sideband peaks (Fig. 24.6) that merge into a step*, the integral of delta functions resulting in a step function. Let us see how such an increase in current can come out of our full NEGF model within the SCBA approximation.

Invoking the full NEGF equations we presented, including scattering, the terminal current for an interacting electron system is given by Eq. (11.17)

$$I_1 = \frac{2q}{\hbar} \int dE \left(\Sigma_1^{\text{in}} D - \gamma_1 n \right), \tag{24.29}$$

simplified for a one level system where all matrices have become scalars (the inverse mapping process shown in Eq. (11.26)). The spectral function A in Eq. (11.17) is replaced by the density of states D, and the electron (hole) correlation function G^n (G^p) is replaced by the electron density n (hole density p), in addition to an overall factor of 2π in each case that has been absorbed in $\hbar = h/2\pi$. For a single phonon mode at zero temperature with coupling $\mathcal{D}_0$, we can use Eq. (24.17) with $N_\alpha = 0$ and Eq. (24.24) to simplify and get the in and outscattering functions as well as the net broadening:

$$\Sigma^{\text{in}} = \Sigma_1^{\text{in}} + \Sigma_2^{\text{in}} + \Sigma_s^{\text{in}} = \gamma_1 f_1 + \gamma_2 f_2 + \mathcal{D}_0 n(E + \hbar\omega)$$

$$\Sigma^{\text{out}} = \Sigma_1^{\text{out}} + \Sigma_2^{\text{out}} + \Sigma_s^{\text{out}} = \gamma_1(1 - f_1) + \gamma_2(1 - f_2) + \mathcal{D}_0 p(E - \hbar\omega)$$

$$\gamma = \Sigma^{\text{in}} + \Sigma^{\text{out}} = \gamma_1 + \gamma_2 + \underbrace{\mathcal{D}_0 \tilde{A}}_{\gamma_s} \tag{24.30}$$

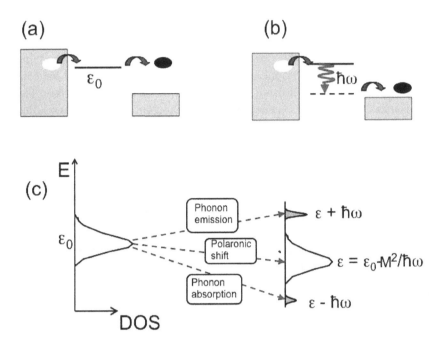

Fig. 24.6 *(a) Elastic scattering through a level leaves a hot electron in the drain and a cold hole in the source, and these need to rapidly thermalize in order to restore the Fermi Dirac distribution in the reservoirs (the excess electron in the drain thermalizes, sends out an electron to the battery, which in turn replenishes the one lost from the source and thermalizes it as well. (b) If the drain bias exceeds a phonon frequency, there is an extra channel for transport that opens up through phonon assisted tunneling, which shows up as IETS signatures off resonance. (c) Near an electronic resonance, the emission or absorption of phonons creates distinct sidebands with heights set by the Boltzmann ratio, and an overall shift (reorganization) of the oscillator coordinate that reduces the energy of the main peak by setting up a polaron.*

where we define $\tilde{A}(E) = n(E + \hbar\omega) + p(E - \hbar\omega)$, equal to the spectral function if $\hbar\omega = 0$. The zero-temperature Fermi functions are defined as $f_{1,2}(E) = \Theta(E_F - E + qV_D/2 \pm qV_D/2)$.

The Green's function technique allows us to calculate the scattering processes immediately. For a single electronic level ϵ_0, we can write down the Green's function as $G = 1/(E - \epsilon_0 + i\gamma/2)$. The density of states D and electron density n are obtained from Eqs. (11.24), (11.16), (11.17). Let us try to extract these results for weak coupling, i.e., to leading order in $\mathcal{D}_0$. This means to leading order in $\mathcal{D}_0$ we can ignore the γ_s term in γ from

Eq. (24.30) when substituting in the denominator of the $|G|^2$ term, and set $|G|^2 \approx |G_0|^2 = 1/[(E - \epsilon_0)^2 + (\gamma_1 + \gamma_2)^2/4]$. Following these substitutions and simplifications, the current (Eq. (24.29)) partitions conveniently into an elastic term and an inelastic term:

$$I_1 = I_1^{\text{el}} + I_1^{\text{inel}}$$

$$I_1^{\text{el}} = \frac{2q\gamma_1\gamma_2}{h} \int dE |G_0|^2 \left(f_1^E - f_2^E \right)$$

$$I_1^{\text{inel}} = \frac{2q\gamma_1\mathcal{D}_0}{h} \int dE |G_0|^2 \left[f_1^E p^{E-\hbar\omega} - \bar{f}_1^E n^{E+\hbar\omega} \right] \qquad (24.31)$$

where $\bar{f} = 1 - f$, and the energy arguments are written as superscripts for notational simplicity.

Since the inelastic current I_1^{inel} is already proportional to the small deformational potential $\mathcal{D}_0$, we can ignore the small, current conserving contribution of phonons in n and replace it with Eq. (11.24). A similar simplification can be made for p to leading order in $\mathcal{D}_0$. Since we are looking in the non-resonant tunneling regime, the level ϵ_0 can be assumed to be sitting far from the Fermi energy, making $|G_0|^2$ almost dispersionless. We can then take the terms $|G_0(E \pm \hbar\omega)|^2$ sitting in the n and p terms above, and yank them out of the integral assuming each is a slowly varying function of energy. After all these simplifications, we get

$$I_1^{\text{inel}} \approx \frac{q\gamma_1\gamma_2\mathcal{D}_0}{\pi h} |G_0|^4 \int_{-\infty}^{\infty} dE \left[f_1^E \bar{f}_2^{E-\hbar\omega} - \bar{f}_1^E f_2^{E+\hbar\omega} \right]. \qquad (24.32)$$

We can now see how scattering introduced Pauli exclusion terms $\sim f^E(1 - f^{E\pm\hbar\omega})$ *that were redundant in the coherent NEGF/Landauer approach.* If we look at elastic incoherent scattering by setting $\hbar\omega \rightarrow 0$, then the contribution of this 'inelastic' term has an integrand proportional to $f_1\bar{f}_2 - \bar{f}_1 f_2 = f_1 - f_2$, so that this term added back to the elastic current in Eq. (24.31) again looks like a Büttiker probe with a modified transmission function (Eq. (22.25)). Furthermore, if we went to finite temperature so that $N_\alpha(\omega) > 0$, we would get four exclusion terms corresponding to phonon emission and absorption going from left to right and vice-versa, $f_1^E \bar{f}_2^{E\pm\hbar\omega}(N_\omega + \frac{1}{2} \mp \frac{1}{2})$, $-\bar{f}_1^E f_2^{E\pm\hbar\omega}(N_\omega + \frac{1}{2} \pm \frac{1}{2})$, consistent with Eq. (13.17).

We now differentiate the inelastic current expression to obtain the conductance $\partial I_1^{\text{inel}}/\partial V_D$, in the process picking up multiple delta functions arising from differentiating the theta step functions sitting in $f_1(E)$ at zero temperature. We get

$$\frac{\partial I_1^{\text{inel}}}{\partial V_D} = \frac{q^2 \gamma_1 \gamma_2 \mathcal{D}_0}{\pi h} |G_0|^4 \left[1 - \Theta(\hbar\omega - qV_D) \right] \qquad (24.33)$$

assuming a positive qV_D. This expression therefore predicts a step function **increase** in conductance at $qV_D = \hbar\omega$ and a corresponding **positive peak** in the second-derivative

$$\boxed{\frac{\partial^2 I_1^{\text{inel}}}{\partial V_D^2} = \frac{q^3}{\pi h} \gamma_1 \gamma_2 \mathcal{D}_0 |G_0|^4 \delta(qV_D - \hbar\omega)} \qquad (24.34)$$

The equation predicts that the relative peak heights are directly proportional to the deformation potentials $\mathcal{D}_0$.

Note that in order to be consistent, we need also to evaluate the contribution due to vibrations from the elastic term. So far, we just replaced G with G_0, but we should ideally expand this term also to first order in $\mathcal{D}_0$ to be consistent with the inelastic contribution. If we go through this exercise, we find that IETS near resonance can actually produce a valley instead of a peak. Such an anomaly has indeed been observed experimentally.

24.5 Near resonant vibronic signatures — polarons and sidebands

Near a sharply defined electronic energy level, the emission or absorption of phonons creates additional discrete channels for conduction. These channels show up as sidebands in the measured density of states with suitable sum rules, and describe phonon assisted tunneling (Fig. 1.2(d)). The presence of phonons redistributes the spectral weights, so that the integral of the density of states with and without phonons stays the same (the main peak decreases in size to allow the sidebands to grow). In addition, we expect multiphonon peaks at separations that are multiples of the phonon frequency. The overall envelope of the phonon sidebands should show a decay away from the electronic resonance as multiphonon processes are increasingly less likely. At the same time, phonon emission must stay consistently stronger than absorption, since the former includes spontaneous emission as well.

Our focus in this section is to put all this on a quantitative footing using the NEGF formalism. We will accomplish this by computing the retarded electronic Green's function $G^R(t) = -i\Theta(t)\langle\{\bar{c}(t), \bar{c}^\dagger(0)\}\rangle$ to extract the density of states. We will use the Hamiltonian in Eq. (24.6) and assume a thermalized Bose Einstein distribution of phonons. We can pull off this calculation within the SCBA, but in this case (a single dispersionless phonon and a single electron mode with linear coupling), this can actually be done exactly. The trick is to do a polaronic transformation, a basis change (the equivalent of 'completing squares') that removes the linear electron phonon coupling and partitions the Hamiltonian into two decoupled groups by redefining the c and a variables.

24.5.0.1 *SCBA treatment of phonon sidebands*

Let us study the emergence of sidebands with a single isolated level and a single phonon mode coupled to it. The isolated Green's function is

$$G_0(E) \approx 1/(E - \epsilon_0 + i\gamma/2) \qquad (24.35)$$

which using Eqs. (24.17), (24.24) and making γ very small gives us

$$\Gamma_{\text{ph}}(E) = 2\pi g^2 \Big[A_{ab}\delta(E - \epsilon_0 - \hbar\omega_0) + A_{em}\delta(E - \epsilon_0 + \hbar\omega_0) \Big]$$

$$A_{ab} = N_{\omega_0} f(\epsilon_0) + (N_{\omega_0} + 1)\Big[1 - f(\epsilon_0)\Big] = N_{\omega_0} + 1 - f(\epsilon_0)$$

$$A_{em} = N_{\omega_0}\Big[1 - f(\epsilon_0)\Big] + (N_{\omega_0} + 1)f(\epsilon_0) = N_{\omega_0} + f(\epsilon_0) \qquad (24.36)$$

where $g = (M/\hbar\omega_0)^2$. At high bias values where all the fs reach unity and $1 - f$s reach zero, we get an emission peak proportional to $N_{\omega_0} + 1$ and an absorption peak proportional to N_{ω_0}. The phonon scattering between two states can then be estimated using Fermi's Golden Rule with the above exclusion terms, the matrix element g given by the spatial derivative of the electronic potential (Eq. (6.19)), the latter typically including screened Coulomb and spin-orbit coupling effects (Eq. (5.30)).

24.5.0.2 *Exact treatment with polaronic transformation*

The essence of the linear coupling in Eq. (24.6) is a shift in the mean position of the oscillator by an amount $x_0 = Mn/\hbar\omega_0$. The shifted wavefunction can be written as a Taylor expansion

$$\Psi(x - x_0) = \Big(1 - x_0\partial/\partial x + \dots\Big)\Psi(x)$$

$$= e^{-x_0\partial/\partial x}\Psi(x) = e^{-ix_0p/\hbar}\Psi(x) \qquad (24.37)$$

so that the translation operator can be written as an unitary transformation using Eq. (24.2)

$$X = e^{-ix_0 p/\hbar} = e^{-M(a-a^\dagger)c^\dagger c/\hbar\omega_0} \tag{24.38}$$

which satisfies $X^\dagger X = XX^\dagger = I$, the requirement for a reversible (unitary) evolution operation. Under this operation, the operators evolve just like a basis transformation

$$\bar{O} = XOX^\dagger \tag{24.39}$$

as would their products, seen easily by inserting the $X^\dagger X = I$ condition in between each pair,

$$X(ABCD)X^\dagger = (XAX^\dagger)(XBX^\dagger)(XCX^\dagger)(XDX^\dagger) = \bar{A}\bar{B}\bar{C}\bar{D} \tag{24.40}$$

so that the modified Hamiltonian becomes

$$XHX^\dagger = \bar{H} = \epsilon_0 \bar{c}^\dagger \bar{c} + \hbar\omega_0 \bar{a}^\dagger \bar{a} + M(\bar{a} + \bar{a}^\dagger)\bar{c}^\dagger \bar{c} \tag{24.41}$$

Using the relation (proved by expansion)

$$e^S O e^{-S} = O + [S, O] + \frac{1}{2!}[S, [S, O]] + \ldots \tag{24.42}$$

we then get

$$\bar{c} = X^\dagger c X = cX \quad (\text{since } [n, c] = -cn)$$
$$\bar{c}^\dagger = X^\dagger c^\dagger$$
$$\bar{n} = \bar{c}^\dagger \bar{c} = c^\dagger c = n$$
$$\bar{a} = a - Mc^\dagger c/\hbar\omega_0$$
$$\bar{a}^\dagger = a^\dagger - Mc^\dagger c/\hbar\omega_0 \tag{24.43}$$

This canonical transformation creates a shift in the oscillator coordinate $\sim a + a^\dagger$ but keeps the electron number unchanged. Substituting these transformations, we find that the Hamiltonian neatly decouples as

$$\bar{H} = \underbrace{(\epsilon_0 - \Delta_0)c^\dagger c}_{\bar{H}_{\text{el}}} + \underbrace{\hbar\omega_0 a^\dagger a}_{\bar{H}_{\text{ph}}} \tag{24.44}$$

where $\Delta_0 = M^2/\hbar\omega_0$ is the polaronic shift described earlier. In effect, we have removed the linear electron phonon coupling, and 'renormalized' the electronic energy levels with the polaronic shift. In other words, we accomplished the completing of squares at an operator level.

The decoupling of the redefined electronic and phononic sectors helps us extract the Green's function, and this is relatively straightforward since we have here an isolated system without added terms in the Hamiltonian (for instance, coupling of electrons with leads, or of phonons with a thermal bath). The new retarded Green's function, whose imaginary part gives us the desired density of states upon tracing, is given by

$$G^R(t) = -i\Theta(t)\left\langle \{\bar{c}(t), \bar{c}^\dagger(0)\}\right\rangle$$

$$= -i\Theta(t)\left\langle \{c(t), c^\dagger(0)\}\right\rangle_{\text{el}} \left\langle X(t)X^\dagger(0)\right\rangle_{\text{ph}} \qquad (24.45)$$

The first bracket is a trivial average over the electronic part of the Hamiltonian, and gives us the bare electron Green's function unrenormalized by phonons. The second term, however, is the critical bit of physics we are after, and involves the separated phonon part of the Hamiltonian averaged over the phonon density matrix. The term inside involves $X(t) = e^{-M[a(t) - a^\dagger(t)]/\hbar\omega_0}$. Our aim will now be to calculate these terms and simplify the angular average, and see how that leads to the sidebands as described pictorially in Fig. 24.6.

Since the Hamiltonian in the new variables is now separate, the evolution of the phonon annihilation operator and the electron operators are actually quite simple,

$$i\hbar\dot{\bar{a}} = [\bar{H}_{\text{ph}}, \bar{a}] = \hbar\omega_0\bar{a} \qquad (24.46)$$

so that

$$\bar{a}(t) = \bar{a}e^{-i\omega_0 t}$$

$$\text{Similarly } \bar{c}(t) = c(0)e^{-i(\epsilon_0 - \Delta_0)t/\hbar}, \qquad (24.47)$$

and by the polaronic transformation from Eq. (24.43), $a(t) = [\bar{a} + M/\hbar\omega_0]e^{-i\omega_0 t}$, i.e., a simple shift (readers can verify this rigorously by going through the maths step by step). Plugging this evolution of $\bar{a}$ back into the second angular bracket for G^R in Eq. (24.45), and taking care that the product of two exponentiated quantities $e^A e^B$ is not just an exponential involving their sum, but also involves their commutator (the Suzuki/Trotter formula or the Campbell Baker Hausdorff expansion), i.e.,

$$e^{tA}e^{tB} = e^{t(A+B)}e^{t^2[A,B]/2} + O(t^3) \qquad (24.48)$$

we get, with a slight redefinition $\tilde{M} = M/\hbar\omega_0$,

$$X(t) = e^{\tilde{M}[a(t) - a^\dagger(t)]}$$

$$= e^{-\tilde{M}^2/2}e^{-\tilde{M}a^\dagger(t)}e^{\tilde{M}a(t)} \qquad (24.49)$$

so that

$$X(t)X^\dagger(0) = e^{-\tilde{M}u} e^{a^\dagger u^*} e^{au}$$

$$\text{where } u = \tilde{M}\left(1 - e^{-i\omega_0 t}\right) \tag{24.50}$$

At zero temperature, only the ground state (zero phonon) states $|0\rangle$ are populated, so that $a|0\rangle = 0$ and $\langle X(t)X^\dagger(0)\rangle = e^{-\tilde{M}u} = e^{-\tilde{M}^2\left(1 - e^{-i\omega_0 t}\right)}$, so that

$$G^R(t) = -i\Theta(t)e^{-i\epsilon_0 t/\hbar - \tilde{M}^2\left(1 - i\omega_0 t - e^{-i\omega_0 t}\right)} \tag{24.51}$$

where we replaced the electronic part of Eq. (24.45) using Eq. (24.47) with $\langle\{\bar{c}(t), \bar{c}^\dagger(0)\}\rangle = e^{-i(\epsilon_0 - \Delta_0)t/\hbar}$ with $\Delta_0 = M^2/\hbar\omega = \tilde{M}^2\hbar\omega_0$.

In terms of the renormalized coupling strength $g = \tilde{M}^2 = (M/\hbar\omega_0)^2$, we can Fourier transform the Green's function and simplify using the expansion $\exp(ge^{-i\omega_0 t}) = \sum_{l=0}^\infty g^l e^{-il\omega_0 t}/l!$, leading us to the zero-T spectral function $A(\omega) = i\left(G(\omega) - G^\dagger(\omega)\right)$

$$A_{T=0}(\omega) = 2\pi \sum_{l=0}^\infty \underbrace{\frac{e^{-g}g^l}{l!}}_{A_l} \delta(\omega - \epsilon_0/\hbar + \Delta_0/\hbar - l\omega_0) \tag{24.52}$$

This equation suggests a Poisson distribution for the heights A_l of the sidebands (note that we only have a sum over positive values of l, i.e., sidebands for phonon emission, since we are at zero temperature). Such a Poisson process is known to be a characteristic of *coherent phonons* (Eq. (24.70)).

The Poisson term can also be understood in terms of the molecular reorganization energy. The polaronic shift $\Delta_0 = g\hbar\omega_0$ amounts to a shift in the molecular conformational potential by x_0 and the energy by $m\omega_0^2 x_0^2/2 = \Delta_0$. We can once again invoke the fast-zipping-electron-slow-moving-atom approximation (we called it Born Oppenheimer when evaluating the electronic states in a static conformational field; we will now call it **Franck Condon approximation** assuming the electrons are fully relaxed to ground state, when evaluating the reorganization). As seen from Fig. 24.7, we get a vertical transition that connects the ground state ($n = 0$) oscillator wavefunction of the undeformed molecule with the lth excited state of the deformed molecule (solutions outlined in the caption of Fig. 3.1,

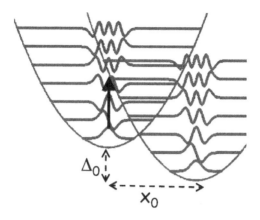

Fig. 24.7 *Polaron formation creates a shift in the underlying oscillator potential by Δ_0. Since electrons move much faster than the phonon frequencies, an electronic transition through a polaron formation involves a near vertical transition on the conformational potential, coupling in effect two states separated by a discrete phonon number (in this case, 1st to 6th phonon mode).*

$\sim H_l(x/a_0)e^{-x^2/2a_0^2}$), with a small polaron formed. The delta function in Eq. (24.52) connects the two energies, while the Poisson prefactor arises from the matrix element connecting these two initial and final states, and Eq. (24.52) then simply represents Fermi's Golden Rule

$$
\begin{aligned}
|M_{if}|^2 &= \left| \int \Psi_0(x)\Psi_l(x-x_0)dx \right|^2 \\
&= \left| \int_{-\infty}^{\infty} \frac{e^{-x^2/2a_0^2}}{\sqrt{a_0\sqrt{\pi}}} \frac{e^{-(x-x_0)^2/2a_0^2}}{\sqrt{2^l l! a_0 \sqrt{\pi}}} H_l((x-x_0)/a_0) dx \right|^2 \\
&= \frac{e^{-g}g^l}{l!}
\end{aligned}
$$ (24.53)

The integral above can be done by observing that the product of two Gaussians is a Gaussian

$$
\frac{e^{-(x-X_A)^2/2\sigma_A^2}}{\sigma_A\sqrt{2\pi}} \times \frac{e^{-(x-X_B)^2/2\sigma_B^2}}{\sigma_B\sqrt{2\pi}} = \frac{Ce^{-(x-X_C)^2/2\sigma_C^2}}{\sigma_C\sqrt{2\pi}}
$$

where

$$
X_C = \frac{X_A\sigma_B^2 + X_B\sigma_A^2}{\sigma_A^2 + \sigma_B^2}, \qquad \sigma_C^2 = \frac{\sigma_A^2\sigma_B^2}{\sigma_A^2 + \sigma_B^2},
$$

$$
C = \frac{e^{-(X_A-X_B)^2/2(\sigma_A^2+\sigma_B^2)}}{\sqrt{2\pi(\sigma_A^2 + \sigma_B^2)}}
$$ (24.54)

and then using the identity (Ref. Gradshteyn and Ryzhik)

$$\int_{-\infty}^{\infty} dx e^{-(x-y)^2} H_n(x) = (2y)^n \sqrt{\pi} \qquad (24.55)$$

If we kept the temperature dependences during the thermal averaging, we would get a slightly different result describing incoherence that is brought in during equilibration. We will now need to postulate a Bose Einstein distribution of the phonon population,

$$\langle X(t)X^\dagger(0)\rangle = \frac{\sum_n \langle n|e^{-n\hbar\omega_0/k_B T} X(t)X^\dagger(0)|n\rangle}{\sum_n e^{-n\hbar\omega_0/k_B T}} \qquad (24.56)$$

with $X(t)X^\dagger(0)$ written down in Eq. (24.50). Using the Taylor expansion of exponentials, and the action of annihilation operators (Eq. (24.18)), we get

$$e^{-au}|n\rangle = \sum_{l=0}^{\infty} \frac{(-u)^l}{l!} a^l |n\rangle = \sum_{l=0}^{n} \frac{(-u)^l}{l!} \sqrt{\frac{n!}{(n-l)!}} |n-l\rangle$$

$$\langle n|e^{a^\dagger u^*} e^{-au}|n\rangle = \sum_{l=0}^{n} \frac{(-|u|^2)^l}{(l!)^2} \frac{n!}{(n-l)!} = L_n(|u|^2) \qquad (24.57)$$

where L_n is a Laguerre Polynomial. Using the generating function of a Laguerre polynomial, $\sum_{n=0}^{\infty} L_n(|u|^2)x^n = \exp\left[|u|^2 x/(x-1)\right]/(1-x)$ to simplify the Boltzmann averaging in Eq. (24.56), we finally get

$$\langle X(t)X^\dagger(0)\rangle_{\text{ph}} = e^{-\Phi(t)}$$

$$\Phi(t) = |g|^2 \left[(N_{\omega_0} + 1)(1 - e^{-i\omega_0 t}) + N_{\omega_0}(1 - e^{i\omega_0 t}) \right] \qquad (24.58)$$

Substituting this into Eq. (24.45) and taking the imaginary part, we get the *Huang Rhys* formula for the sideband weights

$$A_T(\omega) = 2\pi \sum_{l=-\infty}^{\infty} \underbrace{e^{\hbar\omega_0 l/2k_B T - g(2N_{\omega_0}+1)} I_l\left(2g\sqrt{N_{\omega_0}(N_{\omega_0}+1)}\right)}_{A_l} \delta(\omega - \epsilon_0/\hbar + \Delta_0/\hbar - l\omega_0)$$

$$(24.59)$$

where N_{ω_0} is the equilibrium Bose Einstein population of phonons. The result is a set of phonon side-bands separated from the main electronic peaks by $l\hbar\omega_0$, in addition to the overall polaronic shift Δ_0 (Figs. 24.6, 24.8). The I_ls represent modified Bessel functions which skew the distribution in favor of emission processes over absorption, as expected.

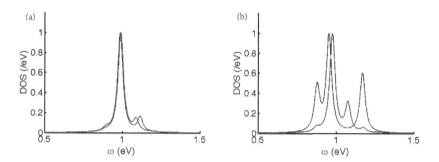

Fig. 24.8 *Comparison of non self-consistent Born approximation within NEGF (blue) for g = 0.2 eV, vs exact polaronic treatment (red). (b) Repeated for g = 0.5 eV.*

24.6 Current saturation from phonon emission

The above treatment ignores cross-counting between different phonon numbers and assumes that the phonons are at equilibrium. The sideband heights can be significantly altered by driving the system far from both electronic and phononic equilibrium: the first by applying a large bias with weak coupling, the second by preventing the phonons from equilibrating with the surroundings. Effects such as phonon bottleneck and negative temperature (emission weaker than absorption) can be seen in suspended carbon nanotubes. Under normal circumstances, phonon scattering tends to reduce device current and induces it to saturate. In regular MOSFETs, electrons at the drain end can relax their energy through phonon emission and avoid backscattering to the source, improving current saturation. In a carbon nanotube, optical phonons excited at ~ 0.19 eV can relax the electron energy and momentum and thereby saturate current for channels exceeding the mean-free path. We can capture this effect (Fig. 24.9) with the modified Landauer equation (Eq. (22.5)) for a single nanotube mode,

$$I = \frac{4q}{h} \int_0^{qV} dE \left(1 + \frac{L}{\lambda}\right)^{-1} \tag{24.60}$$

where we assume $\lambda = 10$ nm above the phonon energy (0.19 eV) and infinite below within a 15 meV bandwidth (i.e., using a Fermi rather than a step function). When the tube is suspended over a trench however, the low group velocities of the optical phonons keep them away from the contacts and create a phonon bottleneck (self-heating) that leads to a negative differential resistance in experiments, i.e., increasing voltage creates a build-up of hot

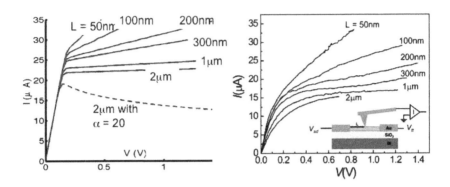

Fig. 24.9   *(Left) Phenomenological treatment of optical phonon scattering in a metallic carbon nanotube (Eq. (24.60)) and (Right) Experiments (Ji-Yong Park et al. Nano Lett. 4, 517 (2004)) show current saturation when the voltage exceeds the breathing mode energy ≈ 0.19 V and the tube is probed at a length larger than the corresponding mean free path. The model can be improved upon by including the entire phonon spectrum, incorporating the spatial current densities within matrix NEGF, and coupling with a rate of phonon equilibration with an added escape rate $\beta[N_\omega - N_\omega^{eq}]/(N_\omega^{eq} + 1)$ into a bath (Fig. 24.1). Including self-heating effects, we get an NDR (red dashed line), also seen experimentally in long suspended tubes where optical phonons fail to equilibrate with contacts and create enhanced high-voltage backscattering.*

phonons and the increased backscattering forces the electron current to go down (Fig. 24.9). In our model, this can be captured with a voltage-dependent λ, for instance, $\lambda[nm] = 10/(\alpha V^2 + 1)$ with α a measure of the phonon self-heating. A coupled electron-phonon-bath model in matrix NEGF should capture these physical effects naturally.

24.7   Comparison with photon assisted tunneling

It is worth emphasizing that the presence of sidebands and the scaling of their heights arises immediately when we add any oscillating potential to an electron field, for instance, an AC field

$$\hat{H} = \epsilon_0 c^\dagger c + q V_{ac} \cos \omega_0 t \tag{24.61}$$

This process is much simpler in many ways from the phonon scattering. We have *fully coherent photons* here not requiring any thermal averaging over an equilibrium Bose Einstein distribution. Furthermore, the coupling is *separable* instead of a linear cross coupling between the dynamically varying electron number $n(t)$ and the oscillating vibrational coordinate $x(t)$ for

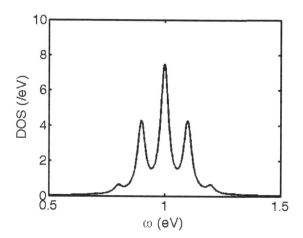

Fig. 24.10 *Symmetric sidebands for coherent, photon assisted tunneling. Stimulated emission and absorption processes for a single coherent mode are symmetric. The asymmetry in Fig. 24.8 arises because the phonon population is set by Boltzmann statistics that includes spontaneous emission. The other difference is in the nature of the coupling. Photonic bands arise by a simple phase transformation of the electron waves (Eq. (24.62)) while phonons have a strong linear coupling between the electron wave and the oscillating phonon variable (Eq. (24.49)) that gives more complexity even for a single mode.*

phonons (which needed a completion of squares). The evolution term would then have an extra trivial factor

$$X(t) = e^{-iqV_{ac}\cos\omega_0 t/\hbar} = \sum_{n=-\infty}^{n=\infty} J_n\left(\frac{qV_{ac}}{\hbar\omega_0}\right) e^{-in\omega_0 t} \qquad (24.62)$$

(we use here the expansion of the cosine of a cosine as the generating function for Bessel equations) so that

$$G^R(t) = -i\Theta(t)\left\langle\{c(t), c^\dagger(0)\}\right\rangle_{\text{el}}\left\langle\{X(t), X^\dagger(0)\}\right\rangle_{\text{ph}}$$

$$= G_0^R\left|e^{-i\epsilon_0 t}\sum_{n=-\infty}^{n=\infty} J_n\left(\frac{qV_{ac}}{\hbar\omega_0}\right) e^{-in\omega_0 t}\right|^2 \qquad (24.63)$$

so that we get the Tien-Gordon formula

$$A_{PAT}(\omega) = \sum_{n=-\infty}^{n=\infty}\left|J_n\left(\frac{qV_{ac}}{\hbar\omega_0}\right)\right|^2 \delta(\omega - \epsilon_0/\hbar + n\omega_0) \qquad (24.64)$$

The coherence manifests itself as a well ordered Bessel function determining the weights, rather than the low temperature stochastic Poisson process

for coherent phonon sidebands in Eq. (24.52), and the associated Bessel
times an overall exponential when we bring in thermal averaging of the
equilibrated (incoherent) phonon distribution in Eq. (24.59)

24.8   Can phonons be coherent?

Note that there is an underlying assumption that we are evaluating the
averages $\langle\ldots\rangle$ between states with fixed phonon numbers, i.e., using the
eigenstates as basis, which is why the number operator $a_k^\dagger a_l$ is diagonal
and gives us N_k when we set $k = l$ (Section 11.4). N_k is given at equilibrium by the Bose-Einstein distribution above, and describes thermalized,
incoherent phonons. In principle, we could also look at *coherent* phonons,
which require very specialized mechanisms to create and last for very short
times before they thermalize through a 'Q factor' with the surrounding
thermal bath. For such coherent phonons, even averages like $\langle a\rangle$ can be
non-zero, and the phonon correlation Green's functions (Eq. (11.28)) will
need to be reevaluated accordingly.

Coherent phonons look like the classical harmonic oscillator solutions. Recall that the quantum harmonic oscillator eigenstates each have a single
well defined eigenvalue, and this makes its wavefunction evolve trivially,
$\Psi_n(x,t) = \Psi_n(x)\exp[-i\omega_n t]$, so that its probability density $|\Psi_n(x,t)|^2$ is
time-independent. Eigenstates are therefore often called 'stationary states',
and their probabilities *do not evolve in time*. A superposition of states however, can evolve with time,

$$\Phi_C^\alpha(x,t) = \sum_{n=0}^{\infty} C_n^\alpha \Psi_n(x,t) \tag{24.65}$$

The corresponding probability distribution

$$P_C^\alpha(x,t) = |\Phi_C^\alpha(x,t)|^2 = \sum_{n,m}\left[(C_n^\alpha)^* C_m^\alpha\right]\Psi_n^*(x,t)\Psi_m(x,t)$$

$$= \sum_{n,m}\left[(C_n^\alpha)^* C_m^\alpha\right]\Psi_n^*(x)\Psi_m(x)e^{i(\omega_n-\omega_m)t} \tag{24.66}$$

is explicitly time-dependent. For incoherent, thermalized modes, we assume
that the thermally averaged density matrix

$$\rho_{nm} = e^{-\mathcal{H}/k_B T}(C_n^\alpha)^* C_m^\alpha \tag{24.67}$$

has no off diagonal terms (hence 'incoherent') due to random stochastic
fluctuations of the phases $\propto \omega_n - \omega_m$, while its diagonal terms are time-

independent and pick up a simple Boltzmann factor. For coherent states, however, the coefficients are carefully chosen so that

(a) we get a Gaussian wavepacket whose envelope itself does not spread, (in fact, is a minimum uncertainty solution). The packet holds together because the expansion coefficients C_n^α above are time-independent and stay in phase (hence the name 'coherent'),

(b) however the 'center' of the packet, specifically, its expectation value oscillates with time and resembles a classical harmonic oscillator solution.

The non-zero time average coordinate means that $\langle a \rangle \neq 0$ (since $\langle x \rangle$ is proportional to $\langle a + a^\dagger \rangle$). It is, as stated above, cosinusoidal with the oscillator frequency. The above property of Φ_C^α arises from the fact that it is an eigenstate of the annihilation operator a (so a reproduces rather than annihilates the state, and thus gives a non-zero $\langle a \rangle$). The corresponding eigenvalue α is the single adjustable parameter labeling each coherent state,

$$a|\alpha\rangle = \alpha|\alpha\rangle \qquad (24.68)$$

where $|\alpha\rangle$ is the coherent state wavefunction Φ_C^α in Dirac notation (i.e., independent of any representation or basis set).

The eigenvalue α ends up having a transparent interpretation. The probability distribution arising from the expansion coefficients, $|C_n^\alpha|^2$, can be readily evaluated, since

$$a|\alpha\rangle = \sum_{n=0}^\infty C_n^\alpha \underbrace{a|n\rangle}_{\sqrt{n}|n-1\rangle} \qquad \text{(From Eq. (24.65) and then Eq. (7.13))}$$

$$= \sum_{n=0}^\infty C_{n+1}^\alpha \sqrt{n+1}|n\rangle \qquad \text{(Removing trivial n} = 0 \text{ term and relabeling n} \to n+1\text{)}$$

$$= \alpha \sum_{n=0}^\infty C_n^\alpha |n\rangle \qquad \text{From Eq. (24.68)} \qquad (24.69)$$

This means the ratio of the probabilities $|C_{n+1}^\alpha|^2$ and $|C_n^\alpha|^2$ is $|\alpha|^2/(n+1)$, which is the signature of a Poisson process (as one expects for n independent phonons equally likely to occur in a time interval),

$$P_n^\alpha = |C_n^\alpha|^2 = \frac{\langle n \rangle^n e^{-\langle n \rangle}}{n!} \qquad (24.70)$$

with the mean number of phonons $\langle n \rangle = |\alpha|^2$ and an equal variance.

It is straightforward to show that $|\alpha|$ also determines the amplitude for the coherent oscillations that distinguish one coherent state from another.

$$\langle x(t) \rangle = \frac{x_0}{\sqrt{2}} \langle a(t) + a^\dagger(t) \rangle$$

$$= \frac{x_0}{\sqrt{2}} \langle ae^{-i\omega t} + a^\dagger e^{i\omega t} \rangle$$

$$= \sqrt{2} x_0 \alpha \cos \omega t \qquad (24.71)$$

for real α, i.e., an oscillation of the average of the wavepacket. However, the variance

$$\langle \delta x^2(t) \rangle = \langle x^2(t) \rangle - \langle x(t) \rangle^2 = x_0^2 \qquad (24.72)$$

is independent of time, hence the packet moves coherently. This is easy to establish, once one realizes that $\langle a^2 \rangle = \langle a \rangle^2 = \alpha^2$ (i.e., there is zero correlation between the a operators), and similarly no correlation for $(a^\dagger)^2$ and $a^\dagger a$, but there *IS* a correlation for $aa^\dagger$ (since a can only act on $|\alpha\rangle$ to its right, and $a^\dagger$ can act on its adjoint to its left, so the way to evaluate the $aa^\dagger$ average is to use the commutation rule to write it as $a^\dagger a + 1$, and the one stays behind to give the above result).

Finally, we can show the overall probability distribution is a Gaussian wavepacket. Rather than do this brute force by evaluating $|\Phi^\alpha(x,t)|^2$ and go through multiple algebraic complexities, it is easier to simply follow the above steps to establish that

$$\langle \delta x^4 \rangle = 3 \langle \delta x^2 \rangle^2 \qquad (24.73)$$

which ends up being a unique signature of a Gaussian. We can thus establish that the coherent state has a probability

$$P_\alpha(x,t) = |\Phi_C^\alpha(x,t)|^2 = \frac{1}{\sigma\sqrt{2\pi}} e^{-\left[x^2 - \langle x(t) \rangle^2\right]/2\sigma^2} \qquad (24.74)$$

i.e., a Gaussian wavepacket that keeps rocking without falling apart.

PART 7

When Electrons Tango!

Dealing with Nonequilibrium Correlations

Chapter 25

What are correlations and why should we care?

In most studies on electronic devices, we spend little time on interactions among charges, beyond just an average potential that each electron swims in and contributes to at the same time. Complexities due to spin and Pauli exclusion are usually included 'after the fact'. We take care to avoid double occupancy but do not worry about their impact on the bandstructures themselves. We worry about spins when they can be separately manipulated with magnetic fields, strain, spin-orbit coupling or exchange. What is less prevalent in device models is any distinction between spins in a purely electrostatic environment. But Pauli exclusion forces such a distinction and creates internal magnetic fields that correlate the charges and spins, especially when they need to avoid each other in the presence of Coulomb repulsion. In 3d and 4f transition metals with localized electrons ($l = n - 1$, so no radial nodes), quantum dot arrays and molecular chains with large unit cells, the overlap of electrons setting their kinetic energies is small compared to their Coulomb cost. A rich tapestry of physical phenomena arises from such strong correlation effects, effects that dominate whenever the energy cost of adding an electron to a level exceeds its broadening by the environment. They range from implicit effects such as many body corrections to band-gaps or optical spectra, to more explicit effects such as superconductivity, superfluidity, polarons, ferromagnetism, excitons and heavy fermions, where particles devise ingenious ways to coordinate their dynamics in order to cut down their overall energy cost.

Correlations certainly make the physics richer and considerably more complex. We can no longer keep track of a single electron wavefunction and treat the rest of the electrons through their average potential. At best, we can use such one electron wavefunctions as 'basis sets' to, in turn, construct various many electron basis sets, and then calculate all those states self consistently. Such a calculation must anticipate all possible configurations of all the electrons, sampling not just their most likely ground state configurations, but even their excited states. In practice, we can limit ourselves to the most relevant excitations or variational many body wavefunctions (say Cooper pairs for superconductors or magnons in a ferromagnet), if we have an inspired guess on what excitations are relevant and what are not. However, to solve the problem in the absence of such a guess, we need to realize that every one electron state expands to two configurational states (filled or empty), so that an $N \times N$ one electron Hamiltonian morphs into a $2^N \times 2^N$ many body Hamiltonian constructed out of it. Clearly, there is no classical computer on earth that can diagonalize such large matrices for a modest sized molecule or device geometry!

There is, however, some device level incentive to study correlated systems, even gazed through the narrow prism of switching applications. As we have pointed out earlier in this book, the energy dissipated in a binary operation is proportional to the number of active information carrying charges per bit, times the energy cost of switching each charge. The latter is set by thermal fluctuations while the former is determined by error thresholds (electrons crave solidarity in numbers). A way to achieve both without making their product large is if we could somehow 'staple' together many charges, spins or other variables, so that toggling just one variable automatically switches the others. This however requires the variables to be strongly correlated, so that the bulk of the work, the heavy lifting, is done by their internal fields, and the dissipative external sources simply initiate the process, hopefully at modest energy costs. Examples include nanomagnetic logic, where many ($\sim 10^4$) spins can be rotated roughly for the energy cost of a few, ferroelectrics that can perhaps accomplish the same with charge, mechanical cantilevers such as the correlated opening of voltage gated ion channels in axonal conduction, and metal insulator transition switches ("MITs") where many charges pool together to form a large condensate.

Whether these ideas pan out depends usually on the overhead cost of creating the switching fields in the first place (an existing proof-of-concept is voltage gated ion channels that are known to beat the Boltzmann limit by a factor of ~ 3 because they tend to correlate three charges). We do however need to devise a formal way to study them quantitatively. Our purpose in this chapter is to discuss how to include correlation in a small molecule (we will use hydrogen as our example), and then a solid, and the nontriviality when we seek to include them in nonequilibrium properties such as transport calculations.

25.1 A bird's eye view: electrons in H_2

A particularly simple system to study correlation is a two-atom two electron molecule such as H_2. We studied bonding in the uncorrelated, one-electron H_2^+ ion earlier (Section 4.6). Our aim here will be to show how Coulomb interactions correlate the two electrons in H_2 beyond just exchange interactions/Pauli exclusion. Let us label the two atoms as A and B. We can take the simplistic (and ultimately incorrect) view that the wavefunction for the two-electron molecule is just the product of the atomic wavefunctions of each electron. This is called the *Linear-Combination-of-Atomic-Orbitals-Molecular-Orbital* (LCAO-MO) approximation. Since the simple product of the spatial orbitals is not antisymmetric with respect to exchange, we will need to tag on an antisymmetric spin component (this antisymmetrization alone enforces a correction to the bandstructure which we know as the 'exchange' correction). But as we shall shortly see, the antisymmetrized product is not enough! The antisymmetrized product reads

$$\Psi_{12}^{\text{LCAO-MO}} = \left(\Psi_\uparrow(1)\Psi_\downarrow(2) - \Psi_\downarrow(1)\Psi_\uparrow(2) \right)/\sqrt{2} \qquad (25.1)$$

To avoid having to spell out the Slater determinants each time, we use the new notation,

$$||\mathcal{AB}|| = [\mathcal{A}(1)\mathcal{B}(2) - \mathcal{B}(1)\mathcal{A}(2)]/\sqrt{2} \qquad (25.2)$$

Aside from antisymmetrization, what would our simplest guess be for a two electron wavefunction in a covalently bonded molecule? When we write down the electronic structure of a material, say, a Silicon atom, we denote it as a many electron configuration that looks like $1s^2 2s^2 2p^6 3s^2 3p^2$, where the $1s^2$ simply means we have two copies of a silicon 1s orbital carrying two spins, $||\Psi_{1s^2}|| = ||\Psi_{1s\uparrow}\Psi_{1s\downarrow}||$. We can make a similar *ansatz* for a H_2 molecule, keeping in mind however that the ground state single electron wavefunction is a *bonding* state (Eq. (4.30)), i.e., a symmetric spatial combination $\sim \Psi_A + \Psi_B$ which minimizes kinetic energy by spreading the electron wavefunction around. Accordingly, our simplest guess for the two electron wavefunction would be a duplication, with two opposite spins cohabiting the same ground state molecular wavefunction (much like the $1s^2$ in Silicon)

$$\Psi_{12}^{LCAO-MO} = \left\|\left\|\Psi_\uparrow \Psi_\downarrow\right\|\right\|$$

$$= \left\|\left\|\left(\Psi_{A\uparrow} + \Psi_{B\uparrow}\right)\left(\Psi_{A\downarrow} + \Psi_{B\downarrow}\right)\right\|\right\|$$

$$= \left[\underbrace{\left\|\left\|\Psi_{A\uparrow}\Psi_{A\downarrow}\right\|\right\|}_{A^-B^+} + \underbrace{\left\|\left\|\Psi_{B\uparrow}\Psi_{B\downarrow}\right\|\right\|}_{A^+B^-}\right] + \left[\underbrace{\left\|\left\|\Psi_{A\uparrow}\Psi_{B\downarrow}\right\|\right\| + \left\|\left\|\Psi_{A\downarrow}\Psi_{B\uparrow}\right\|\right\|}_{AB}\right]$$

$$(25.3)$$

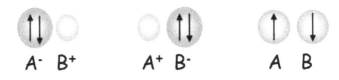

In the equation above, the first term corresponds to the electrons occupying a fully ionized species A^-B^+, with both electrons on A and none on B, endowing the former atom with a net negative charge and thus rendering the latter species positive. The second term shows both electrons on B while A is devoid of electrons. Finally, the last two terms show the electrons equally distributed between the two atoms.

The equation above cannot be strictly true. If we take the H atoms far apart, intuition suggests that the electrons should linger with their individual parent atoms, and not find a compelling reason to jump onto a distant atom, ionizing both in the process. The limiting case for atoms far apart, known as the Heitler-London approximation, must avoid the doubly occupied nuclei and give us a different two electron wave function.

$$\Psi_{12}^{Heitler\text{-}London} = \left[\underbrace{\left\|\left\|\Psi_{A\uparrow}\Psi_{B\downarrow}\right\|\right\| + \left\|\left\|\Psi_{A\downarrow}\Psi_{B\uparrow}\right\|\right\|}_{AB}\right] \qquad (25.4)$$

Note that at this stage the two electrons are completely correlated, in that the action of one electron is inextricably related to that of the other electron. While we may expect two distant atoms to act uncorrelated and independent, the Heitler-London state is still energetically cheaper (if only marginally compared to the simple product when atoms are far away). The act of projecting out the double occupancy states entangled the electrons.

In real life, we expect the result to be somewhere in between, i.e.,

$$\Psi_{12} = \eta \left(\underbrace{||\Psi_{A\uparrow}\Psi_{B\downarrow}|| + ||\Psi_{A\downarrow}\Psi_{B\uparrow}||}_{AB} \right) + \left(1-\eta\right) \left(\underbrace{||\Psi_{A\uparrow}\Psi_{A\downarrow}||}_{A^- B^+} + \underbrace{||\Psi_{B\uparrow}\Psi_{B\downarrow}||}_{A^+ B^-} \right)$$

$$(25.5)$$

where η is a measure of the degree of *correlation*, and of our ability to 'project' out the doubly occupied atomic states. An explicit expression for Ψ will be offered within a simple one orbital model for H_2 in Eq. (25.18). For $\eta = 0.5$, we recover LCAO-MO, while for $\eta = 1$ we recover Heitler-London. Note that in such a correlated state, we can no longer factorize the two-electron wavefunction into single-electron wavefunctions, i.e.,

$$\Psi_{12} \neq \Psi_1 \Psi_2 \tag{25.6}$$

and this difference is called *correlation*. To make matters complicated, once the two hydrogen atoms are far apart, there exists no pursuasive reason for the two electrons to have different spins in the first place or be entangled in any way, and states like $||\Psi_{A\uparrow}\Psi_{B\uparrow}||$ must also start to make an appearance!

Our aim is to see (a) where such correlation effects come from, and (b) how they influence transport far from the independent electron approximation. We will build various models of correlation. The exact problem is intractable in practice; for N electrons, we have 2^N possible combinations (2^{2N} if each electron can have arbitrary spins). However, certain combinations are clearly not allowed, the ones which violate exchange (i.e., Pauli exclusion), so the number of actual combinations is more complicated.

25.2 Solving the H_2 molecule by exact diagonalization

In the previous section, we outlined possible two-electron states for H_2. Each electron had a choice of two atomic centers and two spins. We will now solve the problem using exact diagonalization of the many-body Hamiltonian. This is in general impractical, but we will limit ourselves to a limited subspace of only one ground state s orbital per atom, whereupon the problem becomes solvable. For a chain with N_g atoms and p orbitals per atom, we will deal with 2^{2pN_g} sized matrices. In what follows, we have two atoms, 1 orbital per atom, so that $N_g = 2$ and $p = 1$.

For 2 orbitals and 2 spins we have $2^{2\times2} = 16$ possible arrangements of electrons. This number can be divided into $^N C_0$ zero electron states, $^N C_1$

one electron states, $^N C_2$ two electron states and so on, the subspace sizes of the various N electron states adding up to the combinatorial identity

$$2^N =^N C_0 +^N C_1 + \ldots ^N C_N \tag{25.7}$$

with $N = 4$ here. Let us work out the energetics of each of the N-electron sectors. The hydrogen molecule can be described with a simplified Hamiltonian, written in many body notation as described earlier (Eq. (6.26))

$$\hat{\mathcal{H}} = \sum_{i=A,B} \epsilon_0 n_i - t_0(c_A^\dagger c_B + c_B^\dagger c_A) + \sum_{i=A,B} U n_{i\uparrow} n_{i\downarrow} \tag{25.8}$$

What makes this Hamiltonian 'interacting' is the last Coulomb term, in the absence of which we could have easily expressed the Hamiltonian as a square matrix in the basis set of creation operators $\{c_A^\dagger, c_B^\dagger\}$. That matrix is easy to diagonalize with a suitable basis transformation that reexpresses it in terms of independent non-interacting particles. The extra $n_{i\uparrow} n_{i\downarrow}$ term amounts then to interactions among those particles. In most cases, we attempt an approximate solution using a perturbative expansion in the interaction parameter U normalized to the kinetic energy parameter t_0. For our minimal model for a hydrogen molecule however, we can do this exactly without the need for perturbation theory.

25.2.0.1 *The various N electron sectors for* H_2

(i) N = 0. Start with zero electrons. We have $4C_0 = 1$ possible state, the empty one, with energy $E_0 = 0$. We describe this state as simply $||0||$.

(ii) N = 1. The one-electron state has $4C_1 = 4$ possible states. We will use the following 1-electron basis sets $\{\Psi_{A\uparrow}, \Psi_{A\downarrow}, \Psi_{B\uparrow}, \Psi_{B\downarrow}\}$. In this basis set, the one-electron Hamiltonian is given by

$$H = \begin{array}{c} \\ \Psi_{A\uparrow} \\ \Psi_{B\uparrow} \\ \Psi_{A\downarrow} \\ \Psi_{B\downarrow} \end{array} \begin{array}{c} \Psi_{A\uparrow} \quad \Psi_{B\uparrow} \quad \Psi_{A\downarrow} \quad \Psi_{B\downarrow} \\ \left(\begin{array}{cccc} \epsilon_0 & -t_0 & 0 & 0 \\ -t_0 & \epsilon_0 & 0 & 0 \\ 0 & 0 & \epsilon_0 & -t_0 \\ 0 & 0 & -t_0 & \epsilon_0 \end{array} \right) \end{array} \tag{25.9}$$

with solutions $E_1 = \{\epsilon_0 - t_0, \epsilon_0 - t_0, \epsilon_0 + t_0, \epsilon_0 + t_0\}$, and eigen-states $(\Psi_{A\uparrow} + \Psi_{B\uparrow})/\sqrt{2}$, $(\Psi_{A\downarrow} + \Psi_{B\downarrow})/\sqrt{2}$, $(\Psi_{A\uparrow} - \Psi_{B\uparrow})/\sqrt{2}$ and $(\Psi_{A\downarrow} - \Psi_{B\downarrow})/\sqrt{2}$. The bonding states, the ones with plus signs in their eigenfunctions, have lower energy than the antibonding states, because in the absence of any

Coulomb cost in the single-electron sector, the system can lower its kinetic energy by spreading up the electrons and thus reducing the overall curvature of the wave-function (recall kinetic energy $\propto -\nabla^2$, i.e., curvature).

(iii) $N = 2$. This is the most-interesting sector, the one with the largest number (six) of possible states. The two-electron basis set reads

$$\left\{ \; \left\|A_\uparrow(1)A_\downarrow(2)\right\|, \;\; \left\|B_\uparrow(1)B_\downarrow(2)\right\|, \;\; \left\|A_\uparrow(1)B_\uparrow(2)\right\|, \right.$$
$$\left. \left\|A_\downarrow(1)B_\downarrow(2)\right\|, \;\; \left\|A_\uparrow(1)B_\downarrow(2)\right\|, \;\; \left\|A_\downarrow(1)B_\uparrow(2)\right\| \right\} \qquad (25.10)$$

$$
\begin{aligned}
\Psi_{A\uparrow A\downarrow} &= \left\|A_\uparrow(1)A_\downarrow(2)\right\|, & \Psi_{B\uparrow B\downarrow} &= \left\|B_\uparrow(1)B_\downarrow(2)\right\| \\
\Psi_{A\uparrow B\uparrow} &= \left\|A_\uparrow(1)B_\uparrow(2)\right\|, & \Psi_{A\downarrow B\downarrow} &= \left\|A_\downarrow(1)B_\downarrow(2)\right\| \\
\Psi_{A\uparrow B\downarrow} &= \left\|A_\uparrow(1)B_\downarrow(2)\right\|, & \Psi_{A\downarrow B\uparrow} &= \left\|A_\downarrow(1)B_\uparrow(2)\right\|
\end{aligned}
$$
$$(25.11)$$

The onsite energies are given by the single particle energies on the single atoms, $\epsilon_{A,B}$ while double occupancy on an atom has an on-site energy U. We are assuming the Coulomb term is very localized, so that only onsite electrons repel each other. Also, we assume that hopping between two atoms while conserving spins has a Hamiltonian contribution $-t_0$. Thus, the *many-body Hubbard Hamiltonian* in the two-electron sector is

$$
H = \begin{array}{c|cccccc}
 & \|A_\uparrow A_\downarrow\| & \|B_\uparrow B_\downarrow\| & \|A_\uparrow B_\downarrow\| & \|A_\downarrow B_\uparrow\| & \|A_\uparrow B_\uparrow\| & \|A_\downarrow B_\downarrow\| \\
\hline
\|A_\uparrow A_\downarrow\| & 2\epsilon_0 + U & 0 & -t_0 & t_0 & 0 & 0 \\
\|B_\uparrow B_\downarrow\| & 0 & 2\epsilon_0 + U & -t_0 & t_0 & 0 & 0 \\
\|A_\uparrow B_\downarrow\| & -t_0 & -t_0 & 2\epsilon_0 & 0 & 0 & 0 \\
\|A_\downarrow B_\uparrow\| & t_0 & t_0 & 0 & 2\epsilon_0 & 0 & 0 \\
\|A_\uparrow B_\uparrow\| & 0 & 0 & 0 & 0 & 2\epsilon_0 & 0 \\
\|A_\downarrow B_\downarrow\| & 0 & 0 & 0 & 0 & 0 & 2\epsilon_0
\end{array}
$$
$$(25.12)$$

We can solve this 6×6 Hamiltonian numerically with Matlab (in fact, by rearranging the basis sets a bit, using bonding-antibonding combinations that can only hybridize with likewise combinations, we can diagonalize this matrix easily on paper). But let's get to the results.

Defining $\Delta = \sqrt{U^2 + 16t_0^2}$, the two highest eigenvalues are

$$E_1 = 2\epsilon_0 + U$$
$$E_2 = 2\epsilon_0 + (U + \Delta)/2 \approx 2\epsilon_0 + U \qquad (25.13)$$

where we have assumed $U \gg t_0$. The corresponding eigenvectors are

$$\Psi_1 = \big|\big|A_\uparrow A_\downarrow - B_\uparrow B_\downarrow\big|\big|, \qquad \Psi_2 \approx \big|\big|A_\uparrow A_\downarrow + B_\uparrow B_\downarrow\big|\big| \qquad (25.14)$$

As expected, we have doubly occupied atoms with large Coulomb energies. Following these two high energy states, there is a *triplet* of levels with energy

$$E_3 = E_4 = E_5 = 2\epsilon_0 \qquad (25.15)$$

with eigenvectors

$$\Psi_3 = \big|\big|A_\uparrow B_\uparrow\big|\big|, \qquad \Psi_4 = \big|\big|A_\downarrow B_\downarrow\big|\big|, \qquad \Psi_5 = \big|\big|A_\uparrow B_\downarrow + A_\downarrow B_\uparrow\big|\big| \qquad (25.16)$$

The energy cost is given by the onsite energy alone, which could in principle include longer ranged Coulomb repulsion between the two atoms that we chose to ignore here.

Finally, the lowest energy eigenvalue is a *singlet*

$$E_6 = 2\epsilon_0 + (U - \Delta)/2 \qquad (25.17)$$

with eigenvector (unnormalized here)

$$\Psi_6 = \left|\left|\left(A_\uparrow A_\downarrow + B_\uparrow B_\downarrow\right) + \left[\frac{U + \Delta}{4t_0}\right]\left(A_\uparrow B_\downarrow - A_\downarrow B_\uparrow\right)\right|\right|$$

$$\approx \big|\big|A_\uparrow B_\downarrow - A_\downarrow B_\uparrow\big|\big| \quad (\text{for } U \gg t_0) \qquad (25.18)$$

Because the spins are opposite the spatial part is symmetric, and the result is a *bonding* wavefunction which minimizes the kinetic energy.

The lowest energy singlet is separated from the triplet by an energy gap of $(\Delta - U)/2 \approx -4t_0^2/U$ for large U. It's also important to notice that
• The ground state (Eq. (25.18)) is a strongly correlated state that cannot be factorized into single electron wavefunctions, and has the form Eq. (25.5).

• Also, notice that for finite U/t_0 (Eq. (25.18)), there is a mixture between the singlet state and a double occupied state. For large values of U/t_0, we reach a pure singlet reminiscent of Heitler-London in that there is a separate electron on each atom, with spins paired opposite to each other.

• It is clear then that when the Coulomb energy is costlier than the kinetic energy $(U \gg t_0)$, the singlet state splits away as the lowest energy configuration. The state however is driven by the hybridization t_0 between the two dots. If we expand the geometry with a single dot coupled to a metal, then the singlet state will be seen to grow in spectral weight due to hybridization with the large number (formally an infinite sum) of electrons in the metallic reservoir, giving rise ultimately to a sharp Kondo resonance in the molecular density of states, located at the energy of the metal electrons, i.e., at the Fermi energy (Section 27.6).

(iv) N = 3. There are $4C_3 = 4$ 3-electron states. Within basis set $||A_\uparrow A_\downarrow B_\uparrow||, ||A_\uparrow A_\downarrow B_\downarrow||, ||A_\uparrow B_\uparrow B_\downarrow||, ||A_\downarrow B_\uparrow B_\downarrow||$, the 3-electron Hamiltonian

$$
H = \begin{array}{c} ||A_\uparrow A_\downarrow B_\uparrow|| \\ ||A_\uparrow A_\downarrow B_\downarrow|| \\ ||A_\uparrow B_\uparrow B_\downarrow|| \\ ||A_\downarrow B_\uparrow B_\downarrow|| \end{array}
\begin{array}{cccc}
||A_\uparrow A_\downarrow B_\uparrow|| & ||A_\uparrow A_\downarrow B_\downarrow|| & ||A_\uparrow B_\uparrow B_\downarrow|| & ||A_\downarrow B_\uparrow B_\downarrow|| \\
\end{array}
\left(
\begin{array}{cccc}
3\epsilon_0 + U & 0 & -t_0 & 0 \\
0 & 3\epsilon_0 + U & 0 & -t_0 \\
-t_0 & 0 & 3\epsilon_0 + U & 0 \\
0 & -t_0 & 0 & 3\epsilon_0 + U
\end{array}
\right)
$$

(25.19)

with energies $3\epsilon_0 + U \pm t_0$, each arising in a doubly degenerate pair. In fact, the system is isomorphic to the 1-electron problem, except now instead of 1 electrons, we can think of the system as a filled inactive singleton state with all 4 electrons minus one electron, i.e., a single hole system. Once again, the lowest energy states are bonding while the higher ones are antibonding.

(v) N = 4. Only $4C_4 = 1$ possibility here, with double occupancy on both atoms $||A_\uparrow A_\downarrow B_\uparrow B_\downarrow||$, energy $4\epsilon_0 + 2U$.

Figure 25.1 summarizes the *equilibrium* progression of electron count as we ramp up the Fermi energy. The black line shows the exact energies, while the red dashed lines show the mean-field approach, which we discuss again in a couple of sections.

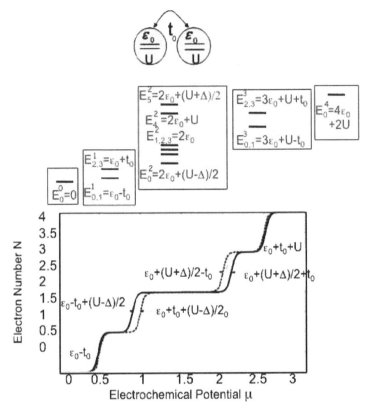

Fig. 25.1 *The various N electron sectors for H_2, and the equilibrium electron number N showing transitions among them, driven by an overall gate driven electrochemical potential. The many body approach (black line) captures the singlet formation energy Δ that the mean field theory (red dashed line) misses*

25.3 Labeling excitations: Configuration Interaction (CI)

Note that every N electron basis set can be described in terms of an enumerable set of excitations out of the N electron ground state, labeled by the orbitals that were swapped in the process. For instance, the two electron ground state is $|\Psi_0\rangle \approx ||A_\downarrow B_\uparrow - B_\downarrow A_\uparrow||$, so that all the other 2 electron states can be described with the swap indices,

$$||A_\uparrow B_\uparrow|| \rightarrow |\Psi_{B_\downarrow}^{B_\uparrow}\rangle, |\Psi_{A_\downarrow}^{A_\uparrow}\rangle, \quad \text{(replace } B_\downarrow \text{ in } |\Psi_0\rangle \text{ with } B_\uparrow, \text{ and } A_\downarrow \text{ with } A_\uparrow)$$

$$||A_\uparrow A_\downarrow|| \rightarrow |\Psi_{B_\downarrow}^{A_\downarrow}\rangle, |\Psi_{B_\uparrow}^{A_\uparrow}\rangle, \quad \text{(replace } B_\downarrow \text{ in } |\Psi_0\rangle \text{ with } A_\downarrow, \text{ and } B_\uparrow \text{ with } A_\uparrow)$$

$$\ldots \tag{25.20}$$

so that we can write the generic N electron wave function as

$$|\Psi\rangle = c_0|\Psi_0\rangle + \sum_{ra} c_a^r|\Psi_a^r\rangle + \sum_{a<b,r<s} c_{ab}^{rs}|\Psi_{ab}^{rs}\rangle + \sum_{a<b<c,r<s<t} c_{abc}^{rst}|\Psi_{abc}^{rst}\rangle + \dots$$

$$(25.21)$$

In other words, we go beyond a single Slater determinant of ground state orbitals (the Hartree Fock approximation), to include all possible Slater determinants involving all possible excitations out of the ground state as well, i.e., mixing in virtual unoccupied orbitals. Mixing those higher orbitals actually changes the energy as they determine the shapes of the orbitals as well. We call each of these excitation states a *configuration*, and the idea of going beyond Hartree Fock to include such configurations in our many body basis set the *configuration interaction* (CI) approach. In Eq. (25.12) we wrote down the entire 2 electron configuration space within the 1s atomic orbitals, but we could limit this by looking at only a subset of possible excitations, or augment this description by including higher unfilled orbitals such as 2s or 2p and write down the many body Hamiltonian including those expanded basis set elements.

Figure 25.2 summarizes the Hartree Fock ground state (a bonding state $||\Phi_{b\uparrow}(1)\Phi_{b\downarrow}(2)||$ labeled '0') and the various excited states that one can derive out of it, specifically, four single excitations $S_1 - S_4 = ||\Phi_{a\sigma}(1)\Phi_{b\sigma'}(2)||$ with four possible σ, σ' combinations, and one double excitation $||\Phi_{a\uparrow}(1)\Phi_{a\downarrow}(2)||$. In this new basis, one can actually show that the only off diagonal non zero term connecting two states is the one between the Hartree Fock ground state 0 and the doubly excited state D. There is no term connecting the ground state with any of the four singly excited states, a manifestation of a general principle known as **Brillouin's theorem**. The theorem can be proved easily by realizing that the sum of two determinants differing by a single row or column (i.e., single excitation) is another determinant in the same basis space, but since the Hartree Fock determinant is already determined to be the lowest energy solution, this new determinant cannot have any overlaps with it or perturb the Hartree Fock ground state in any way. For the specific case of H_2, one can establish this easily by comparing the spins and the spatial symmetry about the bond axis for the various states.

The complete CI approach enumerates all possible excitations out of the ground state. In a full basis set, it provides a *complete* description of the

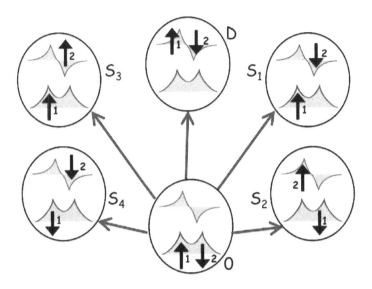

Fig. 25.2 *0: Hartree Fock Ground State for* H_2. $S_1 - S_4$: *Single spin excitations.* D: *Double spin excitation. Each oval is a Slater determinant swapping indices 1 and 2. The full CI matrix block diagonalizes by spin, so states* S_3 *and* S_4 *are already diagonal and the other four form a coupled matrix. Compared with our exact diagonalized states for* $U \to \infty$, *we can make the associations* $0 \to \Psi_6$, $S_1 \to (\Psi_1 - \Psi_6)$, $S_2 \to (\Psi_2 - \Psi_6)$, $S_3 \to \Psi_3$, $S_4 \to \Psi_4$, $D \to \Psi_5$. *The two left transitions do not conserve spin, while the two right ones do not conserve spatial symmetry/parity. Only the* $0 \to D$ *transition can be proved to be symmetry allowed, keeping with the so-called 'Brillouin's theorem'.*

many-body space. However, it spans an impractically large space and is not amenable to brute force computation for realistic structures. In practice, the CI Hamiltonian can be block diagonalized into smaller chunks labeled by spin and structure symmetry, indices that often do not mix (e.g., if we have no spin scattering), occasionally making the problem more tractable. The problem can be further reduced in size by dealing with just the lowest few excitations from the ground state, a method called *partial CI*.

25.4   Where are the single particle energy levels?

We can now extract the one-electron energy levels for a H_2 molecule in its ground state. The ground state corresponds to $N = 2$ in a singlet state given by Eq. (25.18). The one-particle levels are given by transitions between various many electron levels (we see this spelt out shortly in Eq. (26.6)). The lowest unoccupied molecular orbital (LUMO)

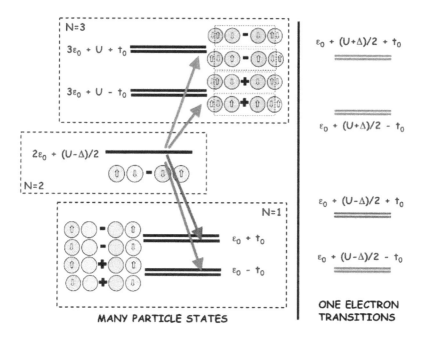

Fig. 25.3 *(Left) Many particle states showing just the $N = 2$ ground state and all the $N = 1$ and $N = 3$ states to map out their one electron excitations. The addition levels $(2 \rightarrow 3)$ give us the various LUMO levels that would form the conduction band if we made an extended network of molecules. The removal levels $(2 \rightarrow 1)$ give us the HOMO levels that stand for the valence band. The one electron transitions (right) are what we usually refer to as one electron levels or simply 'levels' in the previous chapters.*

levels correspond to an addition of either spin to either atomic site, which will naturally complement one of the components on the 2 electron singlet state. The highest occupied molecular orbital (HOMO) levels correspond to removal of one of the spins from the singlet state. The bandgap $E_G = \sqrt{U^2 + 16t_0^2} - 2t_0$. For small charging energies U, the results are what one would expect from simple mean-field theory — low lying energies at $\epsilon_0 \pm t_0$, and higher energies at $\epsilon_0 \pm t_0 + U$ with a bandgap of $2t_0$. But for appreciable charging, the exact bandgap $2t_0 + \Delta \approx 2t_0 + U$ is larger than the noninteracting bandgap $2t_0$ (Fig. 25.1) because of the correlated interaction between the electrons, specifically, self interaction correction (i.e., *the electron does not feel a potential due to itself*). The Coulomb cost of adding an electron to an existing N electron pool is clearly different from the Coulomb cost of removing an electron from that pool (N electrons repel for addition, but only $N - 1$ electrons attract for removal), and this difference is what shows up as the difference between exact vs noninteract-ing bandgap for our model H_2 molecule.

So far, we were looking at just the ground state of the $N = 2$ H_2 molecule and its transitions to various $N = 1$ states interpreted as HOMO, HOMO-1, HOMO-2, .. and $N = 3$ states interpreted as LUMO, LUMO+1, LUMO+2, ... However, remember that the $N = 2$ sector itself has several total energy states, of which the singlet is the ground state. The two elec-trons can indeed be in various excited states of the $N = 2$ sector, and then the transitions from those excited states to the other 1 or 3 electron states give us the other single particle energy levels which go beyond conventional HOMO-LUMO states. Since the electrons can correlate themselves, we can treat *the exact many-body ground state as a mixture of the* <u>*uncorrelated*</u> *ground and excited states.*

25.5   An approximate solution for the H_2 molecule: the Gutzwiller projector

There are approximate ways of extracting the lowest energy solution and the many body configuration, in the spirit of the variational principle we used for helium. In other words, we postulate a form for the many-body wavefunction, introduce a variational parameter to describe the extent of Coulomb interaction/screening, and then minimize the total energy with respect to that parameter.

But how would we construct a variational *many-body wave function* for H_2? A peek at the exact answer (Eq. (25.18) line 1) gives us a clue. Notice that as the Coulomb cost escalates, the doubly occupied states automatically get filtered out of the many body eigenstate. We can thus postulate that the variational wavefunction $|\psi_G\rangle$ projects out the doubly occupied states from the uncorrelated LCAO-MO ground state $|\Phi_0\rangle$ with an efficiency factor g, that is now our variational parameter

$$|\psi_G\rangle = g^D|\Phi_0\rangle \qquad (25.22)$$

where $D = \sum_i n_{i\uparrow}n_{ii\downarrow}$ simply counts the number of such doubly occupied states (note that each term in the sum is non-zero if and only if both the up and down state is occupied). A moment's reflection will show that the exponentiation actually eliminates the doubly occupied states out of $|\Phi_0\rangle$. As the Coulomb cost increases, the optimal g will be seen to reduce, whereby the only way to get a non-zero wavefunction is by making $D = 0$, i.e., removing the doubly occupied states. For the math inclined, we write

$$g^D = e^{\sum_i n_{i\uparrow} n_{i\downarrow} \ln g}$$

$$= \Pi_i (1 - \eta n_{i\uparrow} n_{i\downarrow}) \tag{25.23}$$

where we replaced the exponential of a sum as a product of exponentials, expanded the exponent, set $\eta = 1 - g$, and used the fact that $(n_{i\uparrow} n_{i\downarrow})^2 = n_{i\uparrow} n_{i\downarrow}$ because each n is either zero or one.

We can now calculate the wavefunctions

$$|\Phi_0\rangle = \left(|\Psi_{A\uparrow} \Psi_{A\downarrow}\rangle + |\Psi_{B\uparrow} \Psi_{B\downarrow}\rangle + |\Psi_{A\uparrow} \Psi_{B\downarrow}\rangle - |\Psi_{A\downarrow} \Psi_{B\uparrow}\rangle \right)$$

$$|\psi_G\rangle = g^D |\Phi_0\rangle = \left(g|\Psi_{A\uparrow} \Psi_{A\downarrow}\rangle + g|\Psi_{B\uparrow} \Psi_{B\downarrow}\rangle + |\Psi_{A\uparrow} \Psi_{B\downarrow}\rangle - |\Psi_{A\downarrow} \Psi_{B\uparrow}\rangle \right)$$

$$\tag{25.24}$$

where the Gutzwiller projector g^D gives us a coefficient of g for the doubly occupied states (for H_2, these states have $D = 1$) and unity for all other states $(D = 0)$. With this projected wavefunction, we can calculate the expectation value of the energy from the Hubbard Hamiltonian (Eq. (25.8)), assuming $\epsilon_0 = 0$, as

$$\langle \psi_G | \psi_G \rangle = (1 + g^2)$$

$$\langle \psi_G | H | \psi_G \rangle = \langle \psi_G \Big| -t_0 \sum_{ij\sigma} c_{i\sigma}^\dagger c_{j\sigma} + U \sum_i n_{i\uparrow} n_{i\downarrow} \Big| \psi_G \rangle$$

$$= -4t_0 g + g^2 U$$

$$E(g) = \frac{\langle \psi_G | H | \psi_G \rangle}{\langle \psi_G | \psi_G \rangle} = \frac{-4t_0 g + g^2 U}{1 + g^2} \tag{25.25}$$

In the spirit of the variational principle we can now minimize the energy with respect to the efficiency of projection, $\partial E(g)/\partial g = 0$, to get

$$g^* = \frac{-U + \Delta}{4t_0} = \sqrt{\frac{\Delta - U}{\Delta + U}}, \quad \text{where } \Delta = \sqrt{U^2 + 16t_0^2} \tag{25.26}$$

The ratio of the singly and doubly occupied states is $1/g^* = (U + \Delta)/4t_0$, which is actually the correct answer (Eq. (25.18) line 1). The corresponding ground state energy is

$$E(g^*) = (U - \Delta)/2 \tag{25.27}$$

which again is the correct answer (Eq. (25.17)).

25.6   Gutzwiller for a sequence of atoms: the Brinkman-Rice transition

Generalizing this process to many levels gets challenging for calculating $E(g)$. The process traditionally involves approximating the sum with combinatorial factors that can occasionally miss important physical details. The description of the uncorrelated ground state $|\Phi_0\rangle$ as an equal superposition of all configurations (LCAO-MO) instead of say, a filled Fermi sea, is also a source of error. But even with such simplifications, the Gutzwiller approach predicts important experimentally observable features, such as the so-called Brinkman-Rice transition. That transition, actually turns out to be correct only in the limit of infinite dimensions.

For the H_2 molecule, we saw that the main role of the projector was to renormalize the hopping terms in a nontrivial way. The Coulomb term simply boils down to UD, where D is the number of doublets (we see this if we just consider the singlet terms with coefficient $1/g$ instead of the double terms with coefficient g). But how much is this renormalization? A reasonable guess is to assume that this scaling is proportional to the density matrices, which in turn are proportional to the square root of the product of the probabilities of the initial and final states (similar to a chemical reaction rate).

$$t_0 \rightarrow t_0 \frac{\langle c_{i\uparrow}^\dagger c_{j\uparrow}\rangle}{\langle c_{i\uparrow}^\dagger c_{j\uparrow}\rangle_0} = t_0 \sum_{\alpha\beta} \sqrt{\frac{p^\alpha(ij)p^\beta(ij)}{p_0^\alpha(ij)p_0^\beta(ij)}} \tag{25.28}$$

for various initial configurations α and final configurations β of down spins, each corresponding to an up spin hopping between sites i and j. Let us look at all possible down spin configurations corresponding to the hopping considered here of an up spin from site j to site i. If we designate the probability of empty, single and doubly occupied sites as e, s_σ (for spin $\sigma =\uparrow$ or $\downarrow$) and d respectively, then the probabilities are (site i sitting to the left, j to the right)

No $\downarrow$: $\alpha = 0 \ \uparrow \rightarrow \beta =\uparrow \ 0$

$$p^\alpha(ij) = s_{j\uparrow}e_i, \quad p^\beta(ij) = s_{i\uparrow}e_j$$

Left $\downarrow$: $\alpha =\downarrow \ \uparrow \rightarrow \beta =\uparrow\downarrow \ 0$

$$p^\alpha(ij) = s_{j\uparrow}s_{i\downarrow}, \quad p^\beta(ij) = d_{i\uparrow\downarrow}e_j$$

Right $\downarrow$: $\alpha = 0 \ \uparrow\downarrow \rightarrow \beta = \uparrow \ \downarrow$

$$p^\alpha(ij) = e_i d_{j\uparrow\downarrow}, \quad p^\beta(ij) = s_{i\uparrow} s_{j\downarrow}$$

Both $\downarrow$: $\alpha = \downarrow \ \uparrow\downarrow \rightarrow \beta = \uparrow\downarrow \ \downarrow$

$$p^\alpha(ij) = s_{i\downarrow} d_{j\uparrow\downarrow}, \quad p^\beta(ij) = d_{i\uparrow\downarrow} s_{j\downarrow}$$

$$(25.29)$$

We need to re-express these for a given set of up and down spins $n_\uparrow, n_\downarrow$ (i.e., a given total charge $\propto n = n_\uparrow + n_\downarrow$ and a given total magnetization $\propto (n_\uparrow - n_\downarrow)$). Since an up spin can appear in isolation (s) or in a doublet (d), we need to distinguish them for a correlated system. We can use the identities

$$d_{\uparrow\downarrow} = d$$

$$s_\sigma = n_\sigma - d \quad \text{(exclude doublets)}$$

$$e + s_\uparrow + s_\downarrow + d = 1 \quad \text{(probability normalization)} \qquad (25.30)$$

which in the above probability descriptions replaces

$$d_{\uparrow\downarrow} \rightarrow d, \quad s_\sigma = n_\sigma - d, \quad e = 1 - n + d \qquad (25.31)$$

The uncorrelated denominator for the hopping is more straightforward, its probability equal to $n_\uparrow(1 - n_\uparrow)$ for both p_0^α and p_0^β as we do not need to worry about double occupancy vs single (the spins are uncorrelated!).

We now use Eq. (25.31) to rewrite (25.29) and substitute into Eq. (25.28). The end result then is the renormalized Hubbard energy per site

$$E(d)/N = -t_0 \sum_{ij\sigma} \gamma_\sigma(d) \langle c_{i\sigma}^\dagger c_{j\sigma} \rangle_0 + U d$$

$$\gamma_\uparrow(d) = \frac{(n_\uparrow - d)(1 - n + d) + (n_\downarrow - d)d + 2\sqrt{(n_\downarrow - d)(n_\uparrow - d)(1 - n + d)d}}{n_\uparrow(1 - n_\uparrow)}$$

$$= \frac{\left[\sqrt{(n_\uparrow - d)(1 - n + d)} + \sqrt{(n_\downarrow - d)d} \right]^2}{n_\uparrow(1 - n_\uparrow)} \qquad (25.32)$$

and analogously for $\gamma_\downarrow(d)$ by swapping $\uparrow \leftrightarrow \downarrow$. This expression has the correct behavior. For an uncorrelated system, $d = n_\uparrow n_\downarrow$ (not true in general!), we get back $\gamma = 1$.

To understand the implication of this equation, let's resort to a half-filled, unpolarized structure, where $n_\uparrow = n_\downarrow = n/2$. The equation then simplifies

$$E_{\text{half-filled}}(d) = -8d(1-2d)\underbrace{t_0 \sum_{ij} \langle c_i^\dagger c_j \rangle_0}_{\epsilon_0} + Ud$$

$$= -8d(1-2d)\epsilon_0 + Ud \qquad (25.33)$$

Going back to the variational approach, we find the optimal number of doublets by setting $\partial E(d)/\partial d = 0$, to get $d^* = (8\epsilon_0 - U)/32\epsilon_0$. Substituting,

$$E^* = -\epsilon_0 \left(1 - U/U_c\right)^2, \quad U_c = 8\epsilon_0$$

$$\gamma^* = \left(1 - U^2/U_c^2\right) \qquad (25.34)$$

We can see that for large $U > U_c$, γ^* is negative and we get an insulating phase, while for $U < U_c$, we get a metallic phase. This is called the Brinkman-Rice transition, and is a discontinuous first order phase transition. At the transition, the number of doublets d^* vanishes, as expected and we recover the Heitler-London limit (Eq. (25.4)). Beyond that, the level stays half-filled, but its effective mass $m^*/m = \gamma^{-1}$ keeps increasing.

The opposite limit is easy to see. If $U = 0$, then $d^* = 1/4 = n_\uparrow n_\downarrow$, $\gamma = 1$ and we are in an uncorrelated ground state.

Chapter 26

Transition rates between states

26.1 Equilibrium transition rates between states

In contrast to the rest of the book, for strongly correlated systems, we want to work directly with a set of many body states, instead of dealing with an effective potential projected onto a one-particle subspace. To do so, we first exact diagonalize the many body Hamiltonian to find its many body eigenenergies E_i^N and wavefunctions $\{|\Phi_i^N\rangle\}$. Current flow requires addition and removal of electrons that take the entire system to a different many body state differing by one electron (in principle, one can also have double excitation or cotunneling processes). The one electron spectrum is generated by the differences between the various many body energies.

As we discussed in Chapter 3, nonequilibrium problems require two Green's functions G and G^n, the retarded and the correlation Green's functions, one to establish their static properties (bandstructure) and the other to establish their occupancies (state filling). *For interacting systems, these two are not proportional or simply related.* Even for a many body problem, we deal with the same two Green's functions (Eq. (11.27)) The primary difference is that we will need to extract the time dependence of the c, $c^\dagger$ operators using the entire many body interacting Hamiltonian $\mathcal{H}$. What makes this Hamiltonian nontrivial is that it is not simply decomposed into linear or bilinear operator products like $n \propto c^\dagger c$ for electrons or $x^2 \propto a^\dagger a$ for vibrations that are expressible as a simple square matrix. Instead, the Hamiltonian includes higher order terms (e.g., Coulomb terms $\propto n_i n_j$, anharmonic terms $\propto x^3$), which cannot be diagonalized through a simple basis transformation and needs the full machinery of many body physics while paying attention at every stage to the wavefunction symmetrization rules for Fermions and Bosons.

At steady state, where the Green's functions depend only on the difference between the two times, $|t-t'|$, we can Fourier transform and get the Green's functions in energy representation. Let us start with a complete set of many body N particle basis states $\{|\Phi_i^N\rangle\}$ and corresponding N electron energy eigenvalues $\{E_i^N\}$. We then expand each term in the anticommutator in the definitions (Eqs. (11.27))

$$\langle c_\beta^\dagger(t')c_\alpha(t)\rangle$$

$$= \sum_{iN}\langle\Phi_i^N|\underbrace{\hat\rho}_{\text{density matrix}}\underbrace{e^{iHt'}c_\beta^\dagger e^{-iHt'}}_{c_\beta^\dagger(t')}\underbrace{e^{iHt}c_\alpha e^{-iHt}}_{c_\alpha(t)}|\Phi_i^N\rangle$$

(Introduce completeness relation before and after $c_\beta^\dagger(t')$)

$$= \sum_{iN}\langle\Phi_i^N|\hat\rho\underbrace{\sum_{kL}|\Phi_k^L\rangle\langle\Phi_k^L|}_{=1}e^{iHt'}c_\beta^\dagger e^{-iHt'}\underbrace{\sum_{jM}|\Phi_j^M\rangle\langle\Phi_j^M|}_{=1}e^{iHt}c_\alpha e^{-iHt}|\Phi_i^N\rangle$$

(Each exponent with many body H acts on $|\Phi\rangle$ adjoining it)

$$= \sum_{ijkNML}\langle\Phi_i^N|\hat\rho|\Phi_k^L\rangle\langle\Phi_k^L|e^{iE_k^L t'}c_\beta^\dagger e^{-iE_j^M t'}|\Phi_j^M\rangle\langle\Phi_j^M|e^{iE_j^M t}c_\alpha e^{-iE_i^N t}|\Phi_i^N\rangle$$

$$= \sum_{ijkNML}\rho_{ik}^{NL}e^{i(E_k^L-E_j^M)t'}e^{i(E_j^M-E_i^N)t}\langle\Phi_k^L|c_\beta^\dagger|\Phi_j^M\rangle\langle\Phi_j^M|c_\alpha|\Phi_i^N\rangle \qquad (26.1)$$

Let us ignore all off-diagonal correlations in the density matrix ρ_{ik}^{NL}, so that $N = L$ and $i = k$ in the sum and ρ gets replaced by probability P. Such correlations are important when states are degenerate in energy, or even when they are close enough to be under the same broadening manifold (i.e., the contact state dispersion couples them), but we will ignore them for now. Note that electrons enter or leave one at a time and multielectron processes are much less likely. Accordingly, the creation operators will only take an M electron state to an L electron state if $L = M + 1$, while the destruction operators can take an N electron state to an M electron state if $M = N - 1$. We then get

$$\langle c_\beta^\dagger(t')c_\alpha(t)\rangle = \sum_{ijN}P_i^N e^{i(E_j^{N-1}-E_i^N)(t-t')}\langle\Phi_i^N|c_\beta^\dagger|\Phi_j^{N-1}\rangle\langle\Phi_j^{N-1}|c_\alpha|\Phi_i^N\rangle$$

$$(26.2)$$

and similarly,

$$\langle c_\alpha(t)c_\beta^\dagger(t')\rangle = \sum_{ijN}P_i^N e^{i(E_i^N-E_j^{N+1})(t-t')}\langle\Phi_i^N|c_\alpha|\Phi_j^{N+1}\rangle\langle\Phi_j^{N+1}|c_\beta^\dagger|\Phi_i^N\rangle$$

$$(26.3)$$

so that on Fourier transforming, the retarded Green's function (Eq. (11.27)) satisfies the **Lehmann equation**

$$G_{\alpha\beta}(E) = \sum_{ijN} P_i^N \left[\underbrace{\frac{\langle \Phi_i^N | c_\beta^\dagger | \Phi_j^{N-1} \rangle \langle \Phi_j^{N-1} | c_\alpha | \Phi_i^N \rangle}{E - (E_i^N - E_j^{N-1}) + i0^+}}_{\text{removal}} + \underbrace{\frac{\langle \Phi_i^N | c_\alpha | \Phi_j^{N+1} \rangle \langle \Phi_j^{N+1} | c_\beta^\dagger | \Phi_i^N \rangle}{E - (E_j^{N+1} - E_i^N) + i0^+}}_{\text{addition}} \right]$$

(26.4)

This equation has the same structure as a one electron Green's function expanded in a complete set of one electron basis sets $\{|\phi_n\rangle\}$

$$G_{\alpha\beta}(E) = \langle \alpha | (EI + i0^+ - H)^{-1} | \beta \rangle = \sum_n \frac{\langle \alpha|\phi_n\rangle\langle\phi_n|\beta\rangle}{E - \epsilon_n + i0^+} \qquad (26.5)$$

which prompts us to interpret the molecular HOMO and LUMO levels as the equivalent of ionization potentials and electron affinities taking us from the N electron ground state (labeled '0') to one of the numerous $N \pm 1$ electron excited states.

$$\epsilon_i^{LUMO} \to E_i^{N+1} - E_0^N$$
$$\epsilon_i^{HOMO} \to E_0^N - E_i^{N-1}$$

(26.6)

Standard electronic structure theories often do not satisfy these relations, the ionization potential not matching the HOMO level or the electron affinity the LUMO. This is, however, a reflection of our inability to capture correlation effects properly in practical electronic structure theories, which try to weave these many body interactions into model potentials and small basis sets to keep the matrix sizes reasonable ($N \times N$ rather than $2^N \times 2^N$). It is also worth realizing that instead of just ground state to excited state transitions (Eq. (26.6)), the Green's function in (26.4) also includes excited state to excited state transitions in the denominator (i.e., from an arbitrary i for the N electron state to an arbitrary j in the $N \pm 1$ electron state).

The numerators in Eq. (26.4) describe the partial probabilities analogous to Eq. (26.5), with $|\beta\rangle$ replaced by $c_\beta^\dagger|\Phi_j^{N-1}\rangle$. In addition, we have a thermal average described by P_i^N. For an equilibrium noninteracting system, the addition terms after summing over state index n are proportional to the Fermi function f, while the removal terms are proportional to $1 - f$, and their sum G ends up being independent of occupancy, giving us the spectral distribution and density of states. For an interacting nonequilibrium system however, the Ps deviate from the equilibrium value and the result is more complicated, with the multiple f dependent terms not canceling out in general.

The nonequilibrium dynamics of the interacting many body system is captured by the evolving probability distribution P (more generally, the density matrix ρ_{ik}^{NL} with off diagonal terms). In contrast to the usual NEGF approach, where the electron potential U (and thus the one-electron eigenspectrum itself) evolves self consistently under bias, the many-body spectrum is fixed. Rather, it is P, the probability of occupying the spectral states, that evolves. At equilibrium, the probability distribution is given by the Boltzmann equation,

$$P_i^N|_{\text{eq}} = e^{-(E_i^N - NE_F)/k_B T}/\mathcal{Z} \tag{26.7}$$

where $\mathcal{Z} = \sum_{iN} e^{-(E_i^N - NE_F)/k_B T}$ is the grand partition function that normalizes P. The use of a single temperature T and single electrochemical potential E_F implies that the system is in thermal and chemical equilibrium.

Our aim is to extract the interacting probability distribution under nonequilibrium conditions. This proceeds in two steps. The first step, the static *electronic structure problem* boils down to (a) computing the excitation energies, i.e., the total energy differences in Eq. (26.6), and then (b) extracting the matrix elements in the numerator setting the weights of the various excitations. Instead of the equilibrium Green's function $P_i^N|_{eq}$ entering Eq. (26.4) with the equilibrium probability (Eq. (26.7)), we will solve a *transport problem* at our second step. As we will see in the next section, the transport problem will require us to (c) use the matrix elements to get the excitation-resolved broadenings (Eq. (26.14)), (d) use the broadenings to get the transition rates (Eq. (26.13)) and then finally (e) solving a master equation for the nonequilibrium P_i^N and thence the terminal current I_1 (Eq. (26.12)).

26.2 Nonequilibrium transitions between many body states

26.2.1 *Role of excitations*

In the previous sections, we saw that the correct approach to establishing electronic spectra is to calculate transitions among various many particle energy levels. While this may not be a practical route to modeling large devices, let us focus for the moment on how would we use these levels and their transitions to calculate their nonequilibrium properties such as current.

We start with a small isolated quantum dot whose many-body spectrum is extracted by exactly diagonalizing its Hamiltonian within the restricted subspace of its frontier orbitals, for instance the 1s hydrogen orbitals for H_2 or a single pz orbital per carbon atom in benzene. Let us add a set of leads that are weakly coupled so that we do not need to worry immediately about the level broadening and the associated redistribution of spectral weight. In other words, we assume $U \gg \Gamma$, valid for a small weakly coupled quantum dot. Setting $U = q^2/C_E$ where C_E is the net capacitance of the dot to the leads, $\Gamma = \hbar/\tau_{sc}$, and replacing τ_{sc} by the R-C time constant $\tau_{sc} = RC_E$ at the junctions, we find that this inequality amounts to $R \gg h/q^2$. In this limit, charge addition and removal occurs one at a time. We can then visualize electron transport in terms of electron addition and removal processes that move the entire system among its various many-electron configurations (Fig. 26.2). In the corresponding many body 'Fock' space, we can then set up a master equation for transport. The resulting IVs will have a lot more features than usual, ranging from blockading the current over significant voltage ranges to higher order features arising from the simple

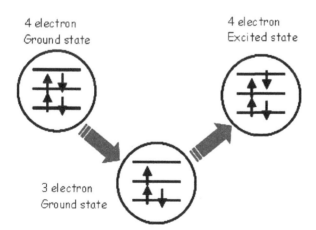

Fig. 26.1 *Charge reorganization under bias takes electrons out of a ground state and promotes them to an excited state. Such low energy excitations are easily accessible for molecular dots and must be accounted for in a transport formalism.*

reorganization of electronic configuration between available filled and empty states, creating excited states in addition to the usual ground state configuration (Fig. 26.1). The parameter denoting the strength of the interaction is $U/\Gamma \propto D(E_F)U$, the product of the low bias density of states and the single electron charging energy. Indeed, many exciting many body physical concepts arise in the limit $UD \gg 1$, which quantifies the effect of Coulomb screening and Fermi level pinning at a contact, the *Stoner criterion* for the onset of ferromagnetism, and the critical temperature $\propto (\hbar\omega_D/k_B)e^{-1/UD}$ for the onset of singlet superconductivity.

We will discuss the formal derivation of the master equation shortly (that derivation is instructive in as much as it allows us to recover some of the broadening physics that we will momentarily jettison for now). The master equation builds on the fact that for weakly coupled electrodes current flows

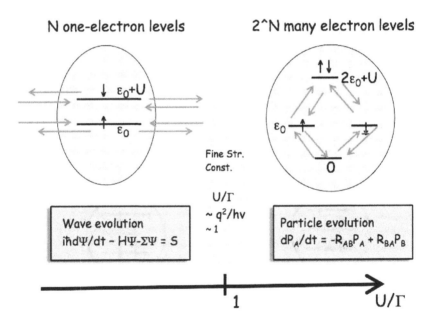

Fig. 26.2 *As the transport 'fine structure constant' U/Γ increases, we move from a continuum wave like evolution in one electron Hilbert space to a correlated particle like evolution in many body Fock space. In the former, we have a single electron entering or leaving a set of dynamic levels that dance around because of interactions with the self consistent field that they settle in. In the many electron approach, we get a set of static many particle energy eigenstates, and the entry or exit of electrons takes the entire system vertically between various many body states.*

via the addition and removal of charges one at a time, driving transitions between many-body states differing by one electron. The balance between entry and exit processes at each contact sets that terminal current.

26.2.2 The 'master equation'

Let us first examine the balance of interfacial charge flow within NEGF. In addition to the electron correlation function G^n, we define a hole correlation function G^p, so that all states, whether empty or full, are occupied either by electrons or by holes,

$$G^n + G^p = A, \tag{26.8}$$

where the diagonal elements of the spectral function A give us the local densities of states. We can see this relation because G^n involves terms like $\langle c_\beta^\dagger(t')c_\alpha(t)\rangle$ while the hole term involves terms like $\langle c_\alpha(t)c_\beta^\dagger(t')\rangle$, so that their sum gives us an anticommutator which relates to G (Eq. (11.27)), whose imaginary part is A.

Much like the inscattering function $\Sigma^{in} \propto \langle SS^\dagger\rangle$ (Eq. (11.16)), we can define an outscattering function $\Sigma^{out} \propto \langle S^\dagger S\rangle$, satisfying the relation

$$\Sigma^{in} + \Sigma^{out} = \Gamma \tag{26.9}$$

We can then rewrite the general NEGF transport equation (11.17) and the corresponding terminal current as

$$
\boxed{
\begin{array}{c}
\underline{\text{NEGF equation (11.17) rephrased}} \\[2mm]
\dfrac{\partial G^n}{\partial t} = - \left[\underbrace{\left(\dfrac{\Sigma^{out}}{h}\right) G^n}_{\text{``outflow''}} - \underbrace{\left(\dfrac{\Sigma^{in}}{h}\right) G^p}_{\text{``inflow''}} \right] \\[6mm]
I_1 = \dfrac{q}{h} \int dE \, Tr\left[\Sigma_1^{in} G^p - \Sigma_1^{out} G^n \right]
\end{array}
}
\tag{26.10}
$$

clearly showing the current as the difference between entrance and exit processes. For a system driven by two contacts with no other scattering mechanisms, it is easy to show that $\Sigma_{1,2}^{in} = \Gamma_{1,2} f_{1,2}$ and $\Sigma_{1,2}^{out} = \Gamma_{1,2}(1-f_{1,2})$, so that

$$\Sigma^{in}(E) = \sum_{\alpha=1,2} \Gamma_\alpha f(E - \mu_\alpha)$$

$$\Sigma^{\text{out}}(E) = \sum_{\alpha=1,2} \Gamma_\alpha [1 - f(E - \mu_\alpha)] \tag{26.11}$$

The many body (Fock space) transport process is quite analogous (Section 26.2.3), except it involves 'vertical' transitions between many electron states rather than horizontal transitions across one electron states (Fig. 26.2). The role of G^n is played by the occupation probability P_i^N of the ith N electron excited state $|N, i\rangle$, with total energy E_i^N. The master equation involves transition rates $R_{(N,i)\to(N\pm1,j)}/h$ (playing the role of $\Sigma^{\text{in,out}}/h$ between states differing by a single electron, leading to a set of independent equations defined by the size of the Fock space. The steady-state solution to this equation for the probabilities then gives us the left terminal current. The equations look analogous to Eq. (26.10), albeit generalized to many-body configurations

---

Multielectron Master Equation (Beenakker)

$$\frac{dP_i^N}{dt} = -\sum_{N,j} \left[\underbrace{\left(\frac{R_{N,i\to N\pm1,j}}{h} \right) P_i^N}_{\text{"outflow"}} - \underbrace{\left(\frac{R_{N\pm1,j\to N,i}}{h} \right) P_j^{N\pm1}}_{\text{"inflow"}} \right] \tag{26.12}$$

$$I_1 = \pm\frac{q}{h} \sum_{N,i,j} \left[R^1_{N,i\to N\pm1,j} P_i^N - R^1_{N\pm1,j\to N,i} P_j^{N\pm1} \right]$$

---

where states corresponding to a removal of electrons by the left electrode involve an additional negative sign in I_1. The first equation needs to be solved at steady state (setting $dP_i^N/dt = 0$), along with the normalization condition, $\sum_{i,N} P_i^N = 1$, and then the solutions are used to solve for I_1.

Analogous to the rate constants for NEGF (Eq. (26.11)) we can write down

$$R_{(N-1,j)\to(N,i)} = \sum_{\alpha=1,2} \Gamma^{Nr}_{ij\alpha} f(\epsilon^{Nr}_{ij} - \mu_\alpha).$$

$$R_{(N,i)\to(N-1,j)} = \sum_{\alpha=1,2} \Gamma^{Nr}_{ij\alpha} \left[1 - f(\epsilon^{Nr}_{ij} - \mu_\alpha) \right] \tag{26.13}$$

for the removal levels $(N, i \to N - 1, j)$, and replacing $(r \to a, f \to 1 - f)$ for the addition levels $(N, i \to N + 1, j)$.

We can use the Lehmann equation (Eq. (26.4)) to help us interpret the various terms and coefficients. The single particle removal and addition transport energies are given by the difference in the many-body energies, $\epsilon_{ij}^{Nr} = E_i^N - E_j^{N-1}$, and $\epsilon_{ij}^{Na} = E_j^{N+1} - E_i^N$. The rate constants $\Gamma_{ij\alpha}^{Nr}$ play the role of $\Gamma_{1,2}$ in Eq. (26.11), *except we are now resolving the partial transitions between the various excited states in addition to the usual ground state to ground state transition.* For weakly coupled dispersionless contacts parametrized using single-electron tunneling rates γ_α, (α: left/right contact), the Fock space rate constants are written down readily using the numerators in Eq. (26.4), setting $\beta = \alpha$ and using the continuity relation $\sum_j |\Phi_j^{N-1}\rangle\langle\Phi_j^{N-1}| = 1$

$$\Gamma_{ij\alpha}^{Nr} = \gamma_\alpha |\langle N, i|c_\alpha^\dagger|N-1, j\rangle|^2$$
$$\Gamma_{ij\alpha}^{Na} = \gamma_\alpha |\langle N, i|c_\alpha|N+1, j\rangle|^2, \qquad (26.14)$$

$c_\alpha^\dagger, c_\alpha$ are the creation/annihilation operators for an electron on the molecular end atom coupled with the corresponding electrode. Note that the terminal current depends on only the component of R arising from processes at that terminal, so that I_1 depends only on R^1, i.e., the terms from R in Eq. (26.13) involving $\alpha = 1$.

Contrast the above approach with standard perturbative NEGF where we deal with a single electron potential or self energy that must include the interactions *self consistently*, as each electron contributes to an overall dynamic field and feels its influence at the same time. The many body approach, however, is a one shot calculation since interactions are already built into its exact diagonalized many body eigenstates. What it does not include is a simple way to broadening of the states, which needs us to formally partition the contact subspace out of the composite channel lead system directly in many body (Fock) space, like we did in Eqs. (11.7)–(11.9). The broadening however is important because we should be able to transfer charges even when we are not precisely resonant but *near resonant* with the single particle transitions. In a couple of sections, we will approach this same problem formally and include some amount of cotunneling that gives us some level of broadening beyond just a simple Golden Rule that will miss partial coherences between energy degenerate levels.

26.2.3 *Deriving the master equation*

The master equation can be formally derived starting with Eq. (3.27) for the evolution of the many body density matrix, which introduced an integral

$\sim \int_0^t dt' \left[\hat{\mathcal{H}}_{TI}(t), \left[\hat{\mathcal{H}}_{TI}(t'), \hat{\rho}_I(t') \right] \right]$. As discussed there, we are interested in tracking the reduced density matrix of the dot, by selectively tracing out the contact degrees of freedom $\rho_{dI}(t) = \text{Tr}_C \rho_I(t)$, a patently non-unitary process. In order to make this tractable, we need to invoke four main simplifying assumptions that will allow us to decouple the leads.

• *Weak coupling ('Born') approximation:* We approximate $\rho_I(t) \approx \rho_{dI}(t) \otimes \rho_C(t)$, which disentangles the channel and contact states (throwing out Kondo correlations for instance).

• *Equilibrium approximation* for each contact state that is now separated out of the channel and integrated above. We assume

$$\left\langle c^\dagger_{\alpha k \sigma} c_{\beta k' \sigma'} \right\rangle = \text{Tr}_C \left(\hat{\rho}_C c^\dagger_{\alpha k \sigma} c_{\beta k' \sigma'} \right) = \delta_{\alpha\beta} \delta_{kk'} \delta_{\sigma\sigma'} \underbrace{\text{Tr}_C \left(\hat{\rho}_C \hat{n}_{\alpha k \sigma} \right)}_{f(\epsilon_{\alpha k \sigma} - \mu_\alpha)}$$
(26.15)

as we saw earlier for equilibrium Coulomb Blockade. Similarly, products of contact states in the form $c_{\alpha k \sigma} c^\dagger_{\alpha k \sigma}$ should give us the hole occupancy $1 - f(\epsilon_{\alpha k \sigma} - \mu_\alpha)$, while $\langle c_{\alpha k \sigma} c_{\alpha k \sigma} \rangle = \langle c^\dagger_{\alpha k \sigma} c^\dagger_{\alpha k \sigma} \rangle = 0$.

• *Markov ('Bloch-Redfield') approximation* for the dot, stipulating that the dot states have no 'memory', which gives us $\rho_{dI}(t') \approx \rho_{dI}(t)$. This can be justified if the contact states, upon which the dot states imprint their detailed phase information, irretrievably lose that memory rapidly in the process of equilibration, robbing the dots of their memory as well.

• *Steady-state approximation* which means all time integrals such as the one above and in Eq. (3.27) will be extended to ∞. This too arises from the Markov approximation.

When doing the integrals to $t = \infty$, the diagonal components of the reduced density matrix (i.e., probabilities) give us the master equations in Eq. (26.12) through the Pleimjl formula (Eq. (8.10)). We get terms like $\int d\epsilon_k \delta(\epsilon_{\alpha k \sigma} - \epsilon_p + \epsilon_q) f(\epsilon_{\alpha k \sigma} - \mu_\alpha)$ which morph into the transition rates 26.13. Off diagonal terms involve Hilbert transforms of the Fermi-Dirac distribution that describe virtual transitions and therefore level broadening $\sim (i/\pi) \int d\epsilon_k \ f(\epsilon_{\alpha k \sigma} - \mu_\alpha)/(\epsilon_{k \sigma \alpha} - \epsilon_p + \epsilon_q)$.

26.2.4 *Lumping excitations together: the orthodox model*

In experiments on quantum dots, one often measures secondary lines in the Coulomb diamond arising from many body excitations (Fig. 1.3(a)). To capture those detailed transitions, many of which are relevant for organic molecules as well (after all a small molecule has a large charging energy and can be a good quantum dot), we will need to use the full machinery described earlier to get those excitations. A major simplification occurs when the levels are closely spaced and there is sufficient inelastic scattering to mix the states and wash out their individual identities, dropping the i, j level indices and simplifying the master equations greatly. The rate constants are obtained by summing incoherently over a continuum of excited states, the incoherence introducing an additional exclusion term $\propto f_1(E_1)[1 - f_2(E_2)]$, as we saw in the context of IETS spectra in Eq. (24.32). Thereupon, we get

$$R^1_{N \to N+1} = \int_{-\infty}^{\infty} 2\pi \Gamma_1^{Na} D_1(E - \mu_1) f(E - \mu_1)$$

$$\times D_m(E - \mu_m) \left[1 - f(E - \mu_m)\right] dE, \qquad (26.16)$$

where $D_1(E)$ and $D_m(E)$ are the densities of states of the right electrode and the middle channel, and μ_m is the local electrochemical potential of the molecular system. If the level separation between one-electron levels is very small, like in a metallic quantum dot, the densities of states are approximately constant for the calculation of the removal and addition rates. By additionally assuming Γ_α^{Na} is energy-independent, the master equations simplify considerably. We can use the relation $\int_{-\infty}^{\infty} f(E -$

$\mu_1)[1 - f(E - \mu_m)] dE = \Delta\mu / \left[e^{\Delta\mu/k_B T} - 1\right]$ with $\Delta\mu = \mu_m - \mu_1$, so that

$$R^1_{N \to N+1} \approx 2\pi \underbrace{\Gamma_1^{Na}}_{\hbar/RC} D_1 D_m \frac{\Delta\mu}{e^{\Delta\mu/k_B T} - 1} \qquad (26.17)$$

Under this condition, the Beenakker equations simplify into an RC network (Fig. 26.3) described as the 'orthodox' model, with the terminal current

'Orthodox' Equation

$$\frac{dP^N}{dt} = -\frac{1}{\hbar} \sum_N [R_{N \to N \pm 1} - R_{N \pm 1 \to N}] \, P^N$$

$$I_1 = \pm \frac{q}{\hbar} \sum_N [R^1_{N \to N \pm 1} - R^1_{N \to N \mp 1}] \, P^N$$

$$R^j_{N \to N \pm 1} = \frac{\hbar}{R_j q^2} \left(\frac{\Delta E^\pm_j}{\exp(\Delta E^\pm_j / k_B T) - 1} \right)$$ (26.18)

$$\Delta E^\pm_1 = \Delta U^\pm \mp \frac{q C_2}{C_E} V_D \mp \frac{q C_G}{C_E} V_G,$$

$$\Delta E^\pm_2 = \Delta U^\pm \pm \frac{q C_1}{C_E} V_D \mp \frac{q C_G}{C_E} V_G$$

the expression with R_j being the junction resistance that absorbs the densities of states $D_{1,m}$. $\Delta E^\pm_j$ is the transition energy $\Delta \mu$ corresponding to adding or removing an electron at the jth contact calculated from simple electrostatics, with gate and drain transfer factors $C_{G,2}/C_E$. $\Delta U^\pm$ is the Coulomb offset for the addition or removal of an electron. We will later use this circuit model to describe experimental data on Coulomb Blockade in organic molecular dots (Fig. 27.8).

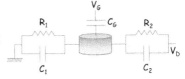

Fig. 26.3   Blockade circuit model

Chapter 27

Why it matters: subtleties from many particle excitation

In Fig. 25.1 we showed the equilibrium ground state occupancy as we ramped up the Fermi energy. The red dashed line came from adding a self consistent mean field obtained by Taylor expanding the Hamiltonian from Eq. (25.8) around the mean and ignoring products of small fluctuations, known as the Hartree Fock or self consistent field approximation

$$\mathcal{H} = \langle\mathcal{H}\rangle + \sum_{i\sigma}\langle\delta\mathcal{H}/\delta n_{i\sigma}\rangle\delta\hat{n}_{i\sigma} + \underbrace{\sum_{i\sigma j\sigma'} O(\delta\hat{n}_{i\sigma}\delta\hat{n}_{j\sigma'})}_{\text{Drop}}$$

$$\approx \text{const} + \sum_{i\sigma}\epsilon_{i\sigma}^{scf}\delta\hat{n}_{i\sigma}$$

$$\epsilon_{i\sigma}^{scf} = \langle\delta\mathcal{H}/\delta n_{i\sigma}\rangle = \epsilon_0 + Un_{i\bar{\sigma}} = \epsilon_0 + U(n_i - n_{i\sigma}) \qquad (27.1)$$

with $\bar{\sigma}$ being the opposite spin as σ. In principle, the Hartree Fock approximation can include longer ranged interactions within a mean field approach, through a nonlocal potential derived variationally from the many body terms in Eq. (6.25)

$$\hat{V}_{HF}\psi_\alpha(\vec{r}_1) = \frac{\delta(U_H + U_F)}{q\delta\psi_\alpha^*(\vec{r}_1)} = \underbrace{\left(\int d\Omega_2 \frac{q\sum_\beta n_\beta(\vec{r}_2)}{4\pi\epsilon_0 r_{12}}\right)}_{V_{\text{Hartree}}(\vec{r}_1)}\psi_\alpha(\vec{r}_1)$$

$$- \int d\Omega_2 \sum_\beta \underbrace{\frac{qn_{\alpha\beta}(\vec{r}_1,\vec{r}_2)}{4\pi\epsilon_0 r_{12}}}_{V_{\text{Fock}}(\vec{r}_1,\vec{r}_2)}\psi_\beta(\vec{r}_2) \qquad (27.2)$$

For localized orbitals, the Fock term only kicks in for equal spins and cancels the equal spin onsite Hartree term, reducing to the equation above (Eq. (27.1)). Either way, the approximation captures exchange, but it ignores correlation (it does so by executing the sums above over *only occupied orbitals*), which boils down to a single Slater determinant $|\Psi_0\rangle$ and ignores the other excited configurations (Eq. (25.21)) that CI captures.

The above mean field equation captures exchange and self interaction correction, by positing that the potential on one species depends on the occupancy of the other spin species, in other words, *an electron does not interact with itself* (this self energy correction is critical to the blockade physics). The equilibrium mean field occupancy is thereafter obtained from Boltzmann physics,

$$\langle n_{i\sigma} \rangle = f(\epsilon_0 + U n_{i\bar{\sigma}} - E_F) \qquad (27.3)$$

where f is the Fermi Dirac distribution, with the ns calculated self consistently. This result agrees fairly well with the exact diagonalized result (black line in Fig. 25.1), except for the singlet formation energy Δ on either side of the stable $N = 2$ plateau. We miss that physics because of our ignorance of correlation, the missing virtual unoccupied orbitals in our configuration space. In principle however, we could make the effective t or U occupation dependent in a complicated way to make those two effects agree. However, as we will see shortly, such an 'adjustment' is much harder to do for the nonequilibrium properties such as current.

Indeed, the nontriviality of many-body physics gets amplified away from equilibrium. Let us go back to a single quantum dot (Hydrogen atom) with a minimal one orbital two spin model, weakly coupled with leads. On applying a drain voltage across it, there is a rise in current at resonance when a single spin is added, followed by a current plateau over which this spin blockades the next incoming charge. Finally the second spin state gets populated, generating a second rise in current after which the current saturates once again. This overall shape of the current voltage curve arises readily out of a mean field calculation, where the energy of one spin depends on the occupancy of the other (Eq. (27.1)). Once a spin state is filled up, the other state gets Coulomb repelled, but since the latter stays empty as a result, the former does not feel a potential due to it and is unaffected. This feature can be captured qualitatively from our mean field model. What is noteworthy however are *the relative heights* of the current plateaus. In the master equation study we find that for equal coupling to the contacts, the first spin carries two third of the current and the second spin carries the balance (Eq. 27.6). However, this ratio changes depending on the contact Fermi functions and the coupling ratios. For instance, making one contact dominant makes the middle plateau half the final plateau (i.e., both spins

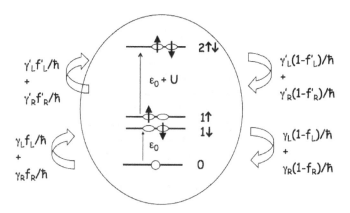

Fig. 27.1   *Fock space transitions for a minimal hydrogen atom (one orbital, two spins).*

are equal) when injected from the stronger contact, but going the other way the middle plateau disappears altogether.

For the two level system, we can easily work out the master equation (Eq. (26.12)). The various electron addition processes follow rates $\gamma_{L,R} f_{L,R}/\hbar$ and $\gamma_{L,R} f'_{L,R}/\hbar$ for injection from the left ('L') and right ('R') contacts (Fig. 27.1), with each Fermi function evaluated at a single energy equal to the addition energy in each case (ϵ_0 for unprimed, $\epsilon_0 + U$ for primed). The removal processes accordingly follow $\gamma_{L,R} \bar{f}_{L,R}/\hbar = \gamma_{L,R}(1 - f_{L,R})/\hbar$ for removal by the left/right contact at the corresponding removal energy. The rate equations are listed below for the four many body states, also describing the processes symbolically at the bottom

$$\frac{dP_0}{dt} = \frac{1}{\hbar}\left[-\underbrace{2(\gamma_L f_L + \gamma_R f_R)P_0}_{R_{0\to 1\uparrow}+R_{0\to 1\downarrow}} + \underbrace{\gamma_L \bar{f}_L + \gamma_R \bar{f}_R (P_\uparrow + P_\downarrow)}_{R_{1\uparrow,1\downarrow\to 0}} \right]$$

$$\frac{dP_\uparrow}{dt} = \frac{1}{\hbar}\left[(\gamma_L f_L + \gamma_R f_R)P_0 + \underbrace{(\gamma'_L \bar{f}'_L + \gamma'_R \bar{f}'_R)P_{\uparrow\downarrow}}_{R_{2\uparrow\downarrow\to 1\uparrow}} - (\gamma_L \bar{f}_L + \gamma_R \bar{f}_R + \gamma'_L f'_L + \gamma'_R f'_R)P_\uparrow \right]$$

$$\frac{dP_\downarrow}{dt} = \frac{1}{\hbar}\left[(\gamma_L f_L + \gamma_R f_R)P_0 + (\gamma'_L \bar{f}'_L + \gamma'_R \bar{f}'_R)P_{\uparrow\downarrow} - (\gamma_L \bar{f}_L + \gamma_R \bar{f}_R + \gamma'_L f'_L + \gamma'_R f'_R)P_\downarrow \right]$$

$$\frac{dP_{\uparrow\downarrow}}{dt} = \frac{1}{\hbar}\left[-2(\gamma'_L \bar{f}'_L + \gamma'_R \bar{f}'_R)P_{\uparrow\downarrow} + \gamma'_L f'_L + \gamma'_R f'_R (P_\uparrow + P_\downarrow) \right]$$

$$(27.4)$$

where we use the notation $\bar{f} = 1 - f$. As expected, the sum of the four equations gives zero on the right hand side, meaning that the total probability $P_0 + P_\uparrow + P_\downarrow + P_{\uparrow\downarrow} = 1$ is conserved. Assuming $U \gg t$ so that $P_{\uparrow\downarrow} \approx 0$, then at steady state, we get $P_\uparrow = P_\downarrow = \dfrac{1 - P_0}{2} = \dfrac{(\gamma_L f_L + \gamma_R f_R)}{\gamma_L(1 + f_L) + \gamma_R(1 + f_R)}$.
The terminal current is then given by

$$I_L = \frac{q}{h}\left[R^L_{0\to1\uparrow} + R^L_{0\to1\downarrow} - R^L_{1\uparrow\to0} - R^L_{1\downarrow\to0} \right]$$

$$= \frac{q}{h}\left[2\gamma_L f_L P_0 - \gamma_L \bar{f}_L(P_\uparrow + P_\downarrow) \right]$$

$$= \frac{q}{h}\left[\frac{2\gamma_L\gamma_R}{\gamma_L(1 + f_L) + \gamma_R(1 + f_R)} \right](f_L - f_R) \qquad (27.5)$$

The maximum current that one can get out of this system corresponds to occupying the shell $1 \uparrow 1 \downarrow$ at energy $\epsilon_0 + U$ and that is given by $2\gamma_L\gamma_R/(\gamma_L + \gamma_R)$. Contrast this with the height of the first plateau we just derived, which equals $2\gamma_L\gamma_R/(2\gamma_L + \gamma_R)$ when we inject from the left and remove from the right ($f_L = 1, f_R = 0$) and $2\gamma_L\gamma_R/(\gamma_L + 2\gamma_R)$ when we inject from the right and remove from the left ($f_L = 0, f_R = 1$). Thus, we see a nontriviality in the *ratio of the secondary plateau to the primary one* (Fig. 27.2)

$$\text{Ratio} = \begin{cases} 2/3 & \text{if equal coupling, } \gamma_L = \gamma_R, \text{ injection from either contact} \\ 1/2 & \text{if strongly asymmetric coupling, inject from stronger} \\ & \text{contact} \\ 1 & \text{(i.e., only one plateau), if strong asymmetry, inject from} \\ & \text{weaker contact} \end{cases}$$

$$(27.6)$$

How can two otherwise identical injected spins behave differently depending on their sequence of injection? The answer lies in the difference between the spin addition and removal channels, which in turn depend on the resistance ratio at the contacts. In the exact diagonalized master equation with equal contact coupling, the first spin can indeed be added in two ways (up or down), but the second spin can only be added one way thereafter, the other option getting promptly eliminated by Pauli exclusion. This accounts

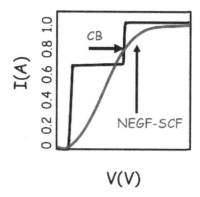

Fig. 27.2 *Current voltage curve in our minimal hydrogen atom shows Coulomb blockade (CB) plateaus, compared with self-consistent field (SCF). The first CB plateau exceeds the second in jump because of nonequilibrium spin correlations.*

for the factor of 2/3. If one contact is much stronger however, then it dominates the transport physics, injecting the first electron into an empty level in two ways (up or down spin) but removing it from a filled level only one way (depending on what spin sat there), giving us an asymmetric current voltage curve with plateau height on one side half that on the other.

What we see therefore is that there is nontrivial physics in the *plateau heights* associated with populating the incomplete shells, arising from the enforcing of Pauli exclusion in the full $2^N \times 2^N$ Fock space. Needless to say these heights are not easy to recover if we simply project onto a one particle subspace. In fact, the heights of the plateau, related to the combinatorial number of excitations, are hard to improve upon simply by improving on the self consistent potential U, for instance, by replacing $n_i n_j \approx (1 - g_{ij})\langle n_i \rangle \langle n_j \rangle$ with a phenomenological pair correlation function g_{ij}. Improving the one electron potential primarily alters the *width* of the plateau, but not its *height* (indeed, for a complicated system we do not even recover the same number of plateaus).

27.1 Is the master equation necessary?

The subtleties introduced by Pauli exclusion on top of Coulomb repulsion are completely ignored in the one-electron viewpoint, which gets incorrect both the plateau heights and the plateau widths. This is borne out by extending our simulation to a double quantum dot (H2 molecule) in Fig. 27.3 (left), solved both using spin unrestricted mean field theory and the rate equation approach. The discrepancy is prominent for strong non-equilibrium transport where the contact couplings are equal ($\gamma_L = \gamma_R$). Making one contact dominate (say $\gamma_L \gg \gamma_R$) brings the system into equilibrium with it, whereby a spin unrestricted approach does match up with the exact I-V (Fig. 27.3 (right)). *It should be clear by now that what works for equilibrium does not often work for nonequilibrium.*

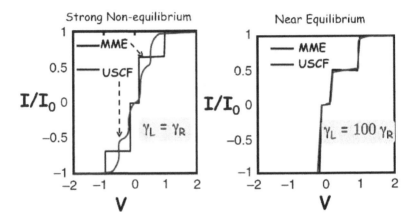

Fig. 27.3   *(Left) The I-V of a minimal H_2 molecule (two coupled quantum dots), computed using the multi-electron master equation (black) and the spin unrestricted mean field theory (red). USCF gets half the charging energy (hence half the plateau widths) and incorrect plateau heights (roughly half instead of 2/3rd) for the open shells. The 2/3 arises because the first spin can be added in two ways, but thereafter the second can be added only in one way because of Pauli exclusion. (Right) A dominant contact brings the system into equilibrium with it, whereupon the two methods merge. While correlation effects can be mimicked by mean-field approaches at equilibrium (right), nonequilibrium is less forgiving (left)*

The master equation may seem like an unnecessarily complex way to compute transport properties. However, it is easy to show that even for the simplest correlated system, it makes a difference! Because of the much bigger Fock space ($2^N \times 2^N$ instead of $N \times N$), correlations generate a profusion of partially occupied excited states that are experimentally seen in quantum dots. How can we possibly get such a large number (e.g., $2^{12} = 4096$ Fock space states for a minimal one C orbital model for benzene) from a $2 \times 6 = 12$ by 12 one electron mean field Hamiltonian? The only way to do this is to add a 12×12 self energy matrix $\Sigma(t, t')$ that is *nonlocal in time*, so that upon Fourier transforming in $|t - t'|$ the resulting $\Sigma(E)$ has a complicated pole structure in complex energy space to recover all those missing many body eigenstates from a 12×12 matrix. It is hard to arrive at such a matrix using model potentials, as one is wont to do in normal band-structure theory. One could of course reverse engineer such a matrix, but that can only be done after one has solved the exact many body problem already. Alternately we need inspired guesses that focus on a few relevant excitations (we will present an example shortly, Eq. (27.19)).

27.2 A case study: Negative Differential Resistance (NDR) in terms of 'dark states'

In Section 19.1 we described NDR using a self-consistent field theory for inter-level electrostatics, coupled with the Landauer theory for current. The operational mechanism in this treatment was the Coulomb electron-electron repulsion between the trap and the channel. A more generic theory can be constructed at this time involving an unspecified 'dark' state, which is simply a configuration of the entire many-body system that conducts less current. Such a dark state could well represent the action of a trap (Fig. 20.3), but it could also describe a coupled electron-phonon system describing a conformational localization of the electrons, or a different kind of NDR that involves two sides of a localized bond sweeping past each other (Fig. 20.1), coming into resonance to give a current peak and then falling out of resonance to reduce the resulting current. The problem is easily

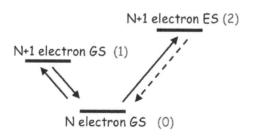

illustrated with an N electron ground state labeled '0', an N+1 electron ground state labeled '1' and an N+1 electron *excited* state labeled '2' (we follow here the treatment by Muralidharan and Datta). We thus have two $N \to N + 1$ single electron transitions. The master equations (Eq. (26.12)) for the first two states as well as the left terminal current

$$\frac{dP_0}{dt} = -(R_{01}P_0 - R_{10}P_1)$$

$$\frac{dP_1}{dt} = -(R_{10}P_1 - R_{01}P_0)$$

$$1 = P_0 + P_1$$

$$I_L = \frac{2q}{h}(R_{01}^L P_0 - R_{10}^L P_1) \tag{27.7}$$

(each R with no superscript is a sum of the corresponding $R^L + R^R$). At steady state, we set the time derivatives to zero and solve for the probabilities, and thence the first current plateau, giving us

$$I_L^{(1)} = \frac{2q}{h}\left(\frac{R_{01}^L - R_{10}^L R_{01}/R_{10}}{R_{01}/R_{10} + 1}\right) \tag{27.8}$$

Once state 2 is accessed, we need to solve for all three probabilities,

$$\frac{dP_0}{dt} = -(R_{01}P_0 - R_{10}P_1) - (R_{02}P_0 - R_{20}P_2)$$

$$\frac{dP_1}{dt} = -(R_{10}P_1 - R_{01}P_0)$$

$$\frac{dP_2}{dt} = -(R_{20}P_2 - R_{02}P_0)$$

$$1 = P_0 + P_1 + P_2$$

$$I_L = \frac{2q}{h}(R_{01}^L P_0 + R_{02}^L P_0 - R_{10}^L P_1 - R_{20}^L P_2) \tag{27.9}$$

At steady state we can again solve for the probabilities and plug back to get the second current plateau height.

$$I_L^{(2)} = \frac{2q}{h}\left(\frac{R_{01}^L + R_{02}^L - R_{10}^L R_{01}/R_{10} - R_{20}^L R_{02}/R_{20}}{R_{01}/R_{10} + R_{02}/R_{20} + 1}\right) \tag{27.10}$$

To achieve an NDR, we now require that the second plateau be *smaller* than the first. The algebra simplifies considerably if we assume positive bias on the right contact, so that all processes going out from 0 involve only the left contact, whereas all processes going back to 0 involve only the right contact. The reverse is true for realizing an NDR for negative bias on the right contact. The conditions for NDR work out to

$$\frac{1}{R_{20}^R} > \frac{1}{R_{01}^L} + \frac{1}{R_{10}^R} \quad (positive\ right\ contact)$$

$$\frac{1}{R_{20}^L} < \frac{1}{R_{01}^R} + \frac{1}{R_{10}^L} \quad (positive\ left\ contact) \tag{27.11}$$

Assuming we start with an empty state initially, we can rewrite the condition for the NDR in terms of couplings $\gamma_{L,R}$ for the ground-state to ground-state (LUMO) transition, and couplings $\gamma'_{L,R}$ for the ground-state to excited-state (LUMO+1) transition. This gives us (compare with Eq. (20.3))

$$\left(\frac{1}{\gamma'_L}, \frac{1}{\gamma'_R}\right) > \frac{1}{\gamma_L} + \frac{1}{\gamma_R} \qquad (right\ sided\ NDR)$$

$$\left(\frac{1}{\gamma'_L}, \frac{1}{\gamma'_R}\right) < \frac{1}{\gamma_L} + \frac{1}{\gamma_R} \qquad (left\ sided\ NDR)$$

$$\frac{1}{\gamma'_R} > \frac{1}{\gamma_L} + \frac{1}{\gamma_R} > \frac{1}{\gamma'_L} \qquad (both\ sided\ NDR)$$

$$\frac{1}{\gamma'_L} > \frac{1}{\gamma_L} + \frac{1}{\gamma_R} > \frac{1}{\gamma'_R} \qquad (no\ NDR) \qquad (27.12)$$

Simply put, this means that *the NDR is realized by the right contact striving to keep the LUMO+1 state occupied over the time scales that the LUMO can react to.* For positive bias on the right contact, an NDR requires that the right contact empty the LUMO+1 level *slower* than the time to fill the LUMO from the left and empty from the right. Conversely, for negative bias on the right contact, NDR requires that the right contact fill the LUMO+1 level faster than the emptying and filling of the LUMO level.

For HOMO levels, the story runs in reverse, so that *it is the job of the left contact to keep the HOMO-1 level empty on the time scales that the HOMO can react to* in order to see an NDR.

A minimum charging energy is needed to effectively expel the level and reduce the current adequately, which in our one-electron picture was set by $U = \epsilon_t - \epsilon_0$. This condition is also implicit in the many-body picture, as one needs the Fermi energy to sit closer to the ground-state to ground-state than the ground-state to excited state transitions. The advantage of the many-body picture is that this can be readily adapted to any kind of dark state, such as processes involving polaron formation or Coulomb Blockade due to charge trapping, or conformational reorganization — processes that cut off the transmission not just at that energy, but throughout the entire conducting bias window.

27.3 Can we capture nonequilibrium correlations with a one-electron theory?

In the previous sections, we saw that an average potential, even computed self-consistently like we do in device theory with Poisson's equation, does not capture many features of correlated systems, especially the current plateau heights. What we will show now is that one can in fact capture these nonequilibrium correlations in a 'NEGF-like' approach by directly *averaging the Green's functions* instead, evaluating them using the full manybody Hamiltonian. While the equations below are not easy to matricize to multiple level systems, the approach can capture broadening and Coulomb Blockade on a somewhat equal footing. It is noteworthy that in contrast to both the SCF NEGF approach and the master equation approach, the equation of motion technique discussed here is inherently non-perturbative in the interaction parameter U_0/Γ, and amounts to ignoring certain diagrams '*ad-hoc*'.

We begin by outlining how the Green's functions get progressively more complex and as a result richer, as we keep evolving longer and longer before truncating the process. The maths behind these approximations will be spelt out in the next section.

Let us begin with potential averaging first. We take the simplest incarnation of *self-interaction correction* (SIC), whereby adding an electron to a level keeps it unperturbed but repels all other levels through Coulomb interaction. The fact that an electron does not feel a potential due to itself is often ignored in band-structure theory – the proper inclusion of it would otherwise lead to a split in the electron density of states. If the single-electron charging energy U_0 is much larger than the level broadening Γ, this split creates a gap that we refer to as Coulomb Blockade (for an extended solid, this gap expands to give a Mott metal-insulator transition). For a two-spin system, the easiest way to include this effect is to make the potentials spin-dependent (Eq. (27.1))

$$U_\uparrow = U_L + U_0(n_\downarrow - n_0)$$
$$U_\downarrow = U_L + U_0(n_\uparrow - n_0) \tag{27.13}$$

This equation, was derived rigorously by starting with a Hubbard Hamiltonian with a Coulomb term and then performing a mean-field decomposition

(Eq. (27.1)) to give us an unrestricted self-consistent field (USCF) potential

$$V_{i\sigma} = U_i \langle n_{i\bar{\sigma}} \rangle + \sum_{j \neq i} U_{ij} \langle n_j \rangle, \qquad \sigma = \uparrow, \downarrow, \quad \bar{\sigma} : \text{Opposite Spin} \qquad (27.14)$$

describing this self-interaction correction explicitly, and reducing to Eq. (27.13) in the case of a doubly spin-degenerate level derived from a single orbital. The correct treatment of correlation must therefore have an *orbital dependent potential*. Indeed, this is the logic behind the LDA + U approach to electronic structure, in contrast with regular LDA which has no such orbital dependences and thus underestimates the bandgap.

For conduction through a single spin-degenerate chemical state with electron charging energy U_0, the USCF mean-field Green's functions are then written down by using this mean field potential (Eqs. (27.1), (27.14))

$$\boxed{\begin{array}{c} \textbf{Mean Field} \\ G_\sigma^{MF}(E) = \dfrac{1}{E - \epsilon_\sigma - U_0 \langle \Delta n_{\bar{\sigma}} \rangle - \Sigma_\sigma} \end{array}} \qquad (27.15)$$

with ϵ_σ including the Laplace potential U_L, and Δn denoting its deviation from the equilibrium value (the equilibrium charging energy having been subsumed into ϵ). Note how the up-spin Green's function G_σ depends on the down-spin occupancy $\Delta n_{\bar{\sigma}}$, extending the USCF approach of Eq. (27.13) directly from the potentials to the response functions. This equation suffices to qualitatively describe the splitting of the density of states due to Coulomb Blockade.

As we crank up the strength of the interaction, the use of an average potential becomes questionable. Instead of averaging the potential, we will need to transition to *averaging the Green's functions themselves*, as these are the thermodynamically averaged dynamical variables by definition (Eq. (11.27)). One can derive the responses by solving the equation of motion for these formally defined Green's functions using the many-body Hamiltonian and making various approximations to truncate the hierarchy. We will do this explicitly in the next section. The next order interaction correction, worked out shortly, will amount to averaging the Green's functions directly (instead of the potentials), weighted by the occupation probabilities, giving us Anderson's expression for the local moment

$$
\boxed{
\begin{array}{c}
\textbf{Local Moment} \\[2mm]
G_\sigma^{LM}(E) = \underbrace{\frac{1 - \langle \Delta n_{\bar\sigma} \rangle}{E - \epsilon_\sigma - \Sigma_\sigma}}_{[0 \to \sigma]} + \underbrace{\frac{\langle \Delta n_{\bar\sigma} \rangle}{E - \epsilon_\sigma - U_0 - \Sigma_\sigma}}_{[\bar\sigma \to \sigma\bar\sigma]}
\end{array}
} \tag{27.16}
$$

The above equation, derived in the next section, says that the total Green's function is a *weighted average* of
(i) The Green's function for the process $[0 \to \sigma]$ (Fig. 27.4, left) filling an empty level with energy ϵ_σ, with probability $1 - \langle \Delta n_{\bar\sigma} \rangle$ that it is empty, and
(ii) The Green's function for the process $[\bar\sigma \to \sigma\bar\sigma]$ (Fig. 27.4, right) filling an occupied level with opposite spin with energy $\epsilon_\sigma + U_0$, with probability $\langle \Delta n_{\bar\sigma} \rangle$ that it is filled.
For isolated systems with $\langle n_\sigma \rangle$ taking values of zero and one only, G^{MF} and G^{LM} agree. It is in the presence of coupling with contacts under increasing U_0 that the difference emerges.

It is also worth noting that at this stage, we can treat the problem as a one electron problem with an effective self energy Σ_{eff} using Dyson's equation (Eq. (7.31))

$$
[G_\sigma^{LM}]^{-1} = \underbrace{[G_{0\sigma}]^{-1}}_{E-\epsilon_\sigma-\Sigma_\sigma} - \Sigma_{\text{eff}} \tag{27.17}
$$

to give us

$$
\Sigma_\sigma^{LM}(E) = U n_{\bar\sigma} + \frac{U^2 n_{\bar\sigma}(1 - n_{\bar\sigma})}{E - U(1 - n_{\bar\sigma})} \tag{27.18}
$$

Finally, at the next level of interaction, worked out in the next section, one gets a similar expression, albeit with additional self-energy terms $\Sigma_{\sigma E, \sigma F}$ *which depend on the contact Fermi-functions* ('E': Empty complementary spin state, 'F': Filled complementary spin state). Such an occupation dependent self energy carries additional poles that can capture the excitations not covered by the $N \times N$ mean field Hamiltonian. They also describe the beginnings of Kondo-interactions where the channel electron spins create singlets by hybridizing with an infinite sea of conduction electrons in the contact, activated primarily near the Fermi energy. The Green's function

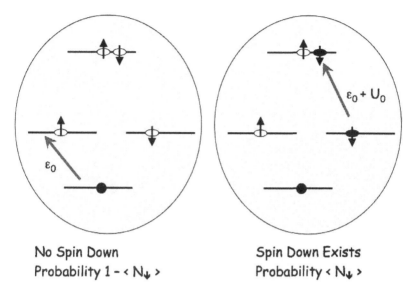

**No Spin Down
Probability 1 - < N↓ >**

**Spin Down Exists
Probability < N↓ >**

Fig. 27.4 *Emergence of local moments from two possible paths for filling an up spin, namely, in the absence vs presence of the complementary down spin state and the associated Coulomb cost. The local moments arise when we calculate the probability weighted averages of the propagators/Green's functions for these two processes, instead of including them in a single Green's function with a weighted potential that is the essence of mean field theory.*

now looks like

$$
\begin{array}{|l|}
\hline
\textbf{Interacting Second Order} \\[4pt]
G_\sigma^K(E) = \underbrace{\frac{1 - \langle \Delta n_{\bar\sigma} \rangle}{(E - \epsilon_\sigma - \Sigma_\sigma) + \Sigma_{\bar\sigma E}}}_{[0 \to \sigma]} + \underbrace{\frac{\langle \Delta n_{\bar\sigma} \rangle}{(E - \epsilon_\sigma - U_0 - \Sigma_\sigma) - \Sigma_{\bar\sigma F}}}_{[\bar\sigma \to \sigma \bar\sigma]} \\[12pt]
\Sigma_{\bar\sigma E}(E) = \sum_{\beta = L,R} \sum_k \left[f_\beta(\epsilon_{k\bar\sigma}) \Sigma'_\beta(E, \epsilon_{k\bar\sigma}) + \bar f_\beta(\epsilon_{k\bar\sigma}) \Sigma''_\beta(E, \epsilon_{k\bar\sigma}) \right] \\[8pt]
\Sigma_{\bar\sigma F}(E) = \sum_{\beta = L,R} \sum_k \left[\bar f_\beta(\epsilon_{k\bar\sigma}) \Sigma'_\beta(E, \epsilon_{k\bar\sigma}) + f_\beta(\epsilon_{k\bar\sigma}) \Sigma''_\beta(E, \epsilon_{k\bar\sigma}) \right] \\
\hline
\end{array}
\tag{27.19}
$$

where we used the notational simplification, $\bar f = 1 - f$. The processes are described in Fig. 27.5 and the equations for the processes are spelt out in Eq. (27.29). The blue dashed lines describe the Coulomb driven cotunneling processes with a direct transition to the doubly occupied state through

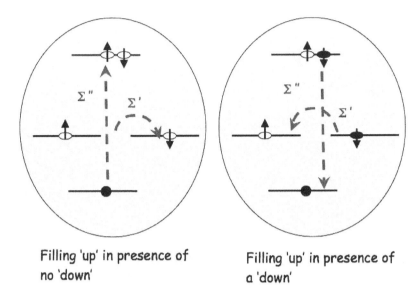

Filling 'up' in presence of no 'down'   **Filling 'up' in presence of a 'down'**

Fig. 27.5 *Cotunneling processes with a direct transition to the doubly occupied state while simultaneously involving spin flip with a second complementary spin state. The first involves the joint processes* $\uparrow\to\downarrow$, $0\to\uparrow\downarrow$, *while the second involves* $\downarrow\to\uparrow$, $\uparrow\downarrow\to 0$.

a simultaneous spin flip involving the complementary spin state. For an empty $\downarrow$ state, we add an $\uparrow$ electron simultaneously with a second $\uparrow$ converted through spin flip to a $\downarrow$ state. For a filled $\downarrow$ state, we convert a second $\downarrow$ electron through spin flip to an $\uparrow$ electron and then remove the first downarrow electron simultaneously.

Note also that there is a clear dependence on the contact electron momenta k and lead index β, which means this process as it stands includes broadening and non-equilibrium transport. We will discuss the impact of these nonequilibrium correlations on the current in a few sections.

27.4 Deriving the diagrams: the equation of motion technique

How did we derive the averaging procedure over the Green's functions? To understand this, let us work out patiently the evolution of the retarded Green's functions using the full many body Hamiltonian (we need the language of creation and annihilation operators, cumbersome though this maybe, to make sure we satisfy wavefunction antisymmetry and Pauli exclusion at every stage). We start with the many body Hamiltonian for the molecular dot, leads and their coupling (Eq. (6.26)). In our notations, c_σ refers to destroying an electron with spin σ sitting on the dot and $c_{\alpha k \sigma}$ refers to destruction of spin σ and momentum k on the αth lead (α: L,R)

$$\hat{\mathcal{H}} = \underbrace{\sum_\sigma \epsilon_\sigma \hat{n}_\sigma + U \hat{n}_\uparrow \hat{n}_\downarrow}_{\hat{\mathcal{H}}_d} \quad \textbf{(Dot)}$$

$$+ \underbrace{\sum_{\substack{k\sigma \\ \alpha=L,R}} \epsilon_{\alpha k \sigma} \hat{n}_{\alpha k \sigma}}_{\hat{\mathcal{H}}_C} \quad \textbf{(Leads)} \quad (27.20)$$

$$+ \underbrace{\sum_{k\sigma\alpha} V_{\alpha k \sigma} c^\dagger_{\alpha k \sigma} c_\sigma + hc}_{\hat{\mathcal{H}}_T} \quad \textbf{(Hybridization)}$$

where $\sigma = \uparrow, \downarrow$ are the two spin states and $\alpha = L, R$ are the two leads. As before, 'hc' refers to Hermitian conjugate, i.e., the term proportional to $V^*_{\alpha k \sigma} c^\dagger_\sigma c_{\alpha k \sigma}$.

The retarded Green's function is already defined in Eq. (11.27),
$$G_\sigma(t) = -\frac{i\Theta(t)}{\hbar} \langle \{c_\sigma(t), c^\dagger_\sigma(0)\} \rangle.$$
The equation of motion of each operator follows the continued application of

(i) the Heisenberg equation of motion $i\hbar d\langle O\rangle/dt = i\hbar \langle \partial O/\partial t\rangle + \langle [O, H]\rangle$, and

(ii) the equal time anticommutation relations $\{c_\alpha, c^\dagger_\beta\} = \delta_{\alpha\beta}, \{c_\alpha, c_\beta\} = 0$ (Eq. (24.4), (24.5)) that enforce antisymmetry and Pauli exclusion.

The coupled interacting nature of the system leads to a proliferation of higher order Green's functions. We will need an approximation to truncate the series in order to close the coupled Green's function loop. Figure 27.6 shows the sequence of approximations, laid out shortly, that gives us the hierarchy of interacting Green's functions, starting with mean-field (Hartree-Fock), then the emergence of local moments and spin polarizations, and then finally through the emergence of occupancy dependent self-energies, Kondo like correlations.

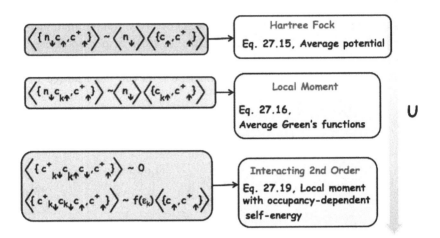

Fig. 27.6   *Sequence of approximations describing elimination of correlations (replacing averages of products with products of averages) and the emergence of corresponding Green's functions of increasing complexity.*

The equation of motion for the operators can be obtained with the use of commutator and anticommutator algebra. The following rules, easily proved using Eqs. (24.4), (24.5), come in handy

$$[c_\alpha, n_\beta] = \delta_{\alpha\beta} c_\alpha$$

$$[c_\alpha^\dagger, n_\beta] = -\delta_{\alpha\beta} c_\alpha^\dagger$$

$$[n_\alpha, n_\beta] = 0$$

$$[AB, C] = A[B, C] + [A, C]B$$

$$[A, BC] = [A, B]C + B[A, C]$$

$$[AB, CD] = [A, C]BD + AC[B, D] + A[B, C]D + C[A, D]B \quad (27.21)$$

giving us (dot represents a time derivative)

$$i\hbar\dot{c}_\sigma(t) = \epsilon_\sigma c_\sigma + \sum_{k\alpha} V_{\sigma k\alpha}^* c_{\sigma k\alpha} + U \underbrace{[n_\sigma n_{\bar\sigma}, c_\sigma]}_{=c_\sigma n_{\bar\sigma}}$$

$$i\hbar\dot{c}_{\sigma k\alpha}(t) = \epsilon_{\sigma k\alpha} c_{\sigma k\alpha} + V_{\sigma k\alpha} c_\sigma \quad (27.22)$$

27.4.0.1 *Noninteracting (Eq. (11.16))*

For a *non interacting* Hamiltonian (no Coulomb term above, $U = 0$), the equation of motion for the Green's function

$$\textbf{(Noninteracting)} \quad \begin{cases} i\hbar\dot{G}_\sigma(t) = \delta(t) + \epsilon_\sigma G_\sigma + \sum_{k\alpha} V^*_{\sigma k\alpha} G_{\sigma k\alpha} \\ i\hbar\dot{G}_{\sigma k\alpha}(t) = \epsilon_{\sigma k\alpha} G_{\sigma k\alpha} + V_{\sigma k\alpha} G_\sigma \end{cases}$$

$$(27.23)$$

involves a new Green's function that we defined as $G_{\sigma k\alpha}(t) = -\dfrac{i\Theta(t)}{\hbar} \langle \{ c_{\sigma k\alpha}(t), c_\sigma^\dagger(0) \} \rangle$. Because the system is non interacting, the two equations above form a complete set and can be solved exactly. From the second equation we can Fourier transform and solve for $G_{\sigma k\alpha}(\omega)$ in terms of G_σ and then plug back into the first equation, yielding the familiar Green's function (Eq. (11.16))

$$G_\sigma(\omega) = \left(\hbar\omega - \epsilon_\sigma - \underbrace{\sum_{k\alpha} |V_{\sigma k\alpha}|^2 / (\hbar\omega - \epsilon_{\sigma k,\alpha})}_{\Sigma_\sigma(\omega)} \right)^{-1}$$

$$= \left(EI - H - \Sigma(E) \right)^{-1} \qquad (27.24)$$

27.4.0.2 *Interacting zeroth Order (Mean Field, Eq. (27.15))*

If we now turn on the interactions, we get an extra Green's function to the right of Eq. (27.23)

$$\textbf{(Interacting)} \quad \begin{cases} i\hbar\dot{G}_\sigma(t) = \delta(t) + \epsilon_\sigma G_\sigma + \sum_{k\alpha} V^*_{\sigma k\alpha} G_{\sigma k\alpha} + \underbrace{U G_{\bar{\sigma}\sigma}}_{\text{extra term}} \\ i\hbar\dot{G}_{\sigma k\alpha}(t) = \epsilon_{\sigma k\alpha} G_{\sigma k\alpha} + V_{\sigma k\alpha} G_\sigma \end{cases}$$

$$(27.25)$$

Notice that the interaction brought in an extra Green's function

$$G_{\bar{\sigma}\sigma}(t) = -\frac{i\Theta(t)}{\hbar} \langle \{ n_{\bar{\sigma}} c_\sigma(t), c_\sigma^\dagger(0) \} \rangle \qquad (27.26)$$

which we now need to solve for. We could approximate it by replacing $-i\Theta(t)\langle \{ n_{\bar{\sigma}} c_\sigma(t), c_\sigma^\dagger(0) \} \rangle / \hbar \approx -i\Theta(t) \langle n_{\bar{\sigma}} \rangle \langle \{ c_\sigma(t), c_\sigma^\dagger(0) \} \rangle / \hbar = \langle n_{\bar{\sigma}} \rangle G_\sigma$, which leads to the mean field (Hartree Fock) Eq. (27.15).

27.4.0.3 *Interacting First Order (Local Moment, Eq. (27.16))*

If we want to proceed beyond this approximation, we can get to the next level of sophistication by writing out the equation for $G_{\bar{\sigma}\sigma}$. This in turn brings in three additional higher order Green's functions

$$i\hbar \dot{G}_{\bar{\sigma}\sigma} = \left\langle n_{\bar{\sigma}} \right\rangle \delta(t) + \left(\epsilon_{\sigma} + U \right) G_{\bar{\sigma}\sigma}$$

$$+ \sum_{k\alpha} \left(V_{k\bar{\sigma}\alpha}^* G_{\bar{\sigma}\bar{\sigma}k\alpha} - V_{k\bar{\sigma}\alpha} G_{\bar{k}\alpha\bar{\sigma}\sigma} - V_{k\bar{\sigma}\alpha}^* G_{\bar{\sigma}\sigma\bar{k}\alpha} \right)$$

$$G_{\bar{\sigma}\bar{\sigma}k\alpha}(t) = -\frac{i\Theta(t)}{\hbar} \left\langle \left\{ c_{\bar{\sigma}}^{\dagger}(t) c_{\bar{\sigma}}(t) c_{k\sigma\alpha}(t), c_{\sigma}^{\dagger}(0) \right\} \right\rangle$$

$$G_{\bar{k}\alpha\bar{\sigma}\sigma}(t) = -\frac{i\Theta(t)}{\hbar} \left\langle \left\{ c_{k\bar{\sigma}\alpha}^{\dagger}(t) c_{\bar{\sigma}}(t) c_{\sigma}(t), c_{\sigma}^{\dagger}(0) \right\} \right\rangle$$

$$G_{\bar{\sigma}\sigma\bar{k}\alpha}(t) = -\frac{i\Theta(t)}{\hbar} \left\langle \left\{ c_{\bar{\sigma}}^{\dagger}(t) c_{\sigma}(t) c_{k\bar{\sigma}\alpha}(t), c_{\sigma}^{\dagger}(0) \right\} \right\rangle \qquad (27.27)$$

We could do a similar decomposition by replacing

$$-i\Theta(t) \left\langle \left\{ c_{\bar{\sigma}}^{\dagger}(t) c_{\bar{\sigma}}(t) c_{k\sigma\alpha}(t), c_{\sigma}^{\dagger}(0) \right\} \right\rangle / \hbar$$

$$\approx -i\Theta(t) \left\langle n_{\bar{\sigma}} \right\rangle \left\langle \left\{ c_{k\sigma\alpha}(t), c_{\sigma}^{\dagger}(0) \right\} \right\rangle / \hbar$$

$$= \left\langle n_{\bar{\sigma}} \right\rangle G_{k\sigma\alpha}$$

for correlations between like spins in the dot, and then set the other two Green's functions to zero as they explore correlations between unlike spins in the dot. With this approximation, we close the loop and recover the local moment Eq. (27.16).

27.4.0.4 *Interacting Second Order, Eq. (27.19)*

We can continue to forge ahead by patiently writing down the equations for each of the higher order Green's functions without simply factorizing, instead evaluating their commutators with H and using the anticommutation relations (a good exercise for students!) As before, this introduces several additional Green's functions and the hierarchy continues

$$i\hbar \dot{G}_{\bar{\sigma}\bar{\sigma}k\alpha} = \epsilon_{k\sigma\alpha} G_{\bar{\sigma}\bar{\sigma}k\alpha} + V_{k\sigma\alpha} G_{\bar{\sigma}\sigma}$$

$$i\hbar \dot{G}_{\bar{k}\alpha\bar{\sigma}\sigma} = \left(\epsilon_\sigma + \epsilon_{\bar{\sigma}} + U - \epsilon_{k\bar{\sigma}\alpha} \right) G_{\bar{k}\alpha\bar{\sigma}\sigma} - V^*_{k\bar{\sigma}\alpha} G_{\bar{\sigma}\sigma}$$

$$- \sum_{k'} \left(V^*_{k'\sigma\alpha} G_{\bar{k}k'\alpha\bar{\sigma}\sigma} - V^*_{k'\bar{\sigma}\alpha} G_{\bar{k}\bar{k}'\alpha\sigma\sigma} \right)$$

$$i\hbar \dot{G}_{\bar{\sigma}\sigma\bar{k}\alpha} = \left(\epsilon_\sigma - \epsilon_{\bar{\sigma}} + \epsilon_{k\bar{\sigma}\alpha} \right) G_{\bar{\sigma}\sigma\bar{k}\alpha} - V_{k\bar{\sigma}\alpha} G_{\bar{\sigma}\sigma}$$

$$+ \sum_{k'} \left(V_{k'\bar{\sigma}\alpha} G_{\bar{k}'\bar{k}\alpha\sigma\sigma} - V^*_{k'\sigma\alpha} G_{\bar{k}k'\alpha\bar{\sigma}\sigma} \right)$$

$$G_{\bar{k}k'\alpha\bar{\sigma}\sigma} = -\frac{i\Theta(t)}{\hbar} \left\langle \left\{ c^\dagger_{k\bar{\sigma}\alpha} c_{k'\sigma\alpha} c_{\bar{\sigma}}(t), c^\dagger_\sigma(0) \right\} \right\rangle$$

$$G_{\bar{k}\bar{k}'\alpha\sigma\sigma} = -\frac{i\Theta(t)}{\hbar} \left\langle \left\{ c^\dagger_{k\bar{\sigma}\alpha} c_{k'\bar{\sigma}\alpha} c_\sigma(t), c^\dagger_\sigma(0) \right\} \right\rangle$$

$$G_{\bar{k}'\bar{k}\alpha\sigma\sigma} = \frac{i\Theta(t)}{\hbar} \left\langle \left\{ c^\dagger_{k'\bar{\sigma}\alpha} c_{k\bar{\sigma}\alpha} c_\sigma(t), c^\dagger_\sigma(0) \right\} \right\rangle$$

$$G_{\bar{k}\alpha k'\bar{\sigma}\sigma} = \frac{i\Theta(t)}{\hbar} \left\langle \left\{ c_{k\bar{\sigma}\alpha} c_{k'\sigma\alpha} c^\dagger_{\bar{\sigma}}(t), c^\dagger_\sigma(0) \right\} \right\rangle \tag{27.28}$$

We could continue this exercise further, but seeing as we are now evaluating correlations between electrons in the contact (multiple c_k terms), it is prudent to do one more decomposition. To that end, we will mandate that opposite spins in the contact are uncorrelated, while like spins are correlated by the local Fermi Dirac distribution

$$\left\langle \left\{ c^\dagger_{k\bar{\sigma}\alpha} c_{k'\sigma\alpha} c_{\bar{\sigma}}(t), c^\dagger_\sigma(0) \right\} \right\rangle \approx \left\langle c^\dagger_{k\bar{\sigma}\alpha} c_{k'\sigma\alpha} \right\rangle \left\langle \left\{ c_{\bar{\sigma}}(t), c^\dagger_\sigma(0) \right\} \right\rangle,$$

$$\left\langle \left\{ c^\dagger_{k\bar{\sigma}\alpha} c_{k'\bar{\sigma}\alpha} c_\sigma(t), c^\dagger_\sigma(0) \right\} \right\rangle \approx \left\langle c^\dagger_{k\bar{\sigma}\alpha} c_{k'\bar{\sigma}\alpha} \right\rangle \left\langle \left\{ c_\sigma(t), c^\dagger_\sigma(0) \right\} \right\rangle,$$

and analogous ones for the other two Green's functions, meaning that the contact states and the dot states are not correlated. We also assume each contact is in a spin uncorrelated local equilibrium state, so that $\left\langle c^\dagger_{k\bar{\sigma}\alpha} c_{k'\sigma\alpha} \right\rangle = 0$ for opposite spins, and $\left\langle c^\dagger_{k\bar{\sigma}\alpha} c_{k'\bar{\sigma}\alpha} \right\rangle = f_\alpha(\epsilon_{k\bar{\sigma}})\delta_{kk'}$, the Fermi Dirac distribution for the lead $\alpha \in L, R$. If we now plug this into the previous equation and work all the way through, we again close the loop by terminating the sequence of higher Green's functions, connecting them with each other. If we now Fourier transform the six equations for the six Green's functions and solve them, we end up with the one particle Green's function G_σ^K described in the last section (Eq. (27.19)) that captures the so-called *cotunneling* processes. We also get an explicit expression for the second order cotunneling transition rates/broadenings

$$\Sigma'_\beta(E, \epsilon_{k\bar\sigma}) = \frac{|V_{k\beta\bar\sigma}|^2}{E - \Delta\epsilon - \epsilon_{k\beta\bar\sigma}} \times \frac{U_0}{E - \epsilon_\sigma - U_0 - \Sigma_{tot}}$$

$$\Sigma''_\beta(E, \epsilon_{k\bar\sigma}) = \frac{|V_{k\beta\bar\sigma}|^2}{E - E_{\uparrow\downarrow} - \epsilon_{k\beta\bar\sigma}} \times \frac{U_0}{E - \epsilon_\sigma - U_0 - \Sigma_{tot}}$$

$$\Sigma_{tot} = \Sigma_\sigma + \sum_{\beta k} \left[\Sigma'_\beta(E, \epsilon_{k\bar\sigma}) + \Sigma''_\beta(E, \epsilon_{k\bar\sigma}) \right] \qquad (27.29)$$

where $\Delta\epsilon = \epsilon_\sigma - \epsilon_{\bar\sigma}$ and $E_{\uparrow\downarrow} = \epsilon_\sigma + \epsilon_{\bar\sigma} + U_0$. If we wish to make this process a little more rigorous, we should calculate the energies ϵ_σ self consistently by adding in the same self energy to the bare value they started off with.

27.5 Capturing correlations in NEGF with a non-perturbative Σ

To calculate the current, we will need to do the same laborious exercise for the nonequilibrium Green's function $G_\sigma^n(t) = \langle c_\sigma^\dagger(0) c_\sigma(t) \rangle$. At this stage however, we could make the simplifying (and reasonable) assumption that in the Keldysh equation, the in-scattering function Σ^{in} stays unaffected by the interactions, since the leads are spatially separated and noninteracting. In other words, the only change with interaction is in the overall broadening that is occupation dependent, so that

$$G_\sigma^n = G_\sigma^K \underbrace{\left(\Gamma_L f_L + \Gamma_R f_R \right)}_{\Sigma^{in}} \left(G_\sigma^K \right)^\dagger \qquad (27.30)$$

with G_σ^K worked out in Eq. (27.19). The current is given by the Meir Wingreen expression (Eq. (11.17)), but since the only place where the interactions enter is in the broadening in the retarded Green's function G_σ^K but not in the rest of the Keldysh equation, the current simplifies to the Landauer equation with the Fisher Lee formula (Eq. (11.20)), with conductance given by the transmission $\bar T = \mathrm{Tr}\left[\Gamma_L G_\sigma^K \Gamma_R \left(G_\sigma^K \right)^\dagger \right]$. It is noteworthy that what we are doing at this stage is to include a *nonpertur-bative self energy inside the conventional NEGF equations.*

Let us explore the intervening current plateau for large Coulomb coupling, $U_0 \gg \Gamma$. In Eq. (27.19), the second term with filled complementary spin (process $[\bar{\sigma} \to \sigma\bar{\sigma}]$) gets eliminated because of the large Coulomb cost in the denominator, while the cotunneling self energy $\Sigma_{\bar{\sigma}E}(E)$ for the first term simplifies. From Eq. (27.29) we see that $\Sigma''_{\beta}(E, \epsilon_{k\bar{\sigma}})$ drops out, while in $\Sigma'_{\beta}(E, \epsilon_{k\bar{\sigma}})$ the second factor becomes unity while the first approaches $-\Sigma_{\beta\bar{\sigma}}$, assuming the two spin states have the same energy $\epsilon_{\sigma} = \epsilon_{\bar{\sigma}} = \epsilon_0$. This means from Eq. (27.19), we approach for large U_0

$$\langle \Delta N_{\bar{\sigma}} \rangle \to 0$$

$$G_{\sigma}^K \to \frac{1}{E - \epsilon_0 - \Sigma_L(1 + f_L) - \Sigma_R(1 + f_R)}$$

$$T \to \frac{2\gamma_L\gamma_R}{\gamma_L(1 + f_L) + \gamma_R(1 + f_R)} \tag{27.31}$$

in contrast with the noninteracting transmission $T = \gamma_L\gamma_R/(\gamma_L + \gamma_R)$, and in agreement with the multielectron master equation (Eq. (27.5)). In fact, we recover the 2/3 plateau for equal coupling that we spent some time emphasizing, *but with the bonus that we also can now include cotunneling processes that capture some amount of broadening.*

27.6 Occupancy dependent broadening: Kondo effect

When the Coulomb force is large, it promotes a singlet formation (Eq. (25.18)) between a localized quantum dot electron and the sea of electrons at the Fermi energy in the contacts. The hybridization with a staggeringly large number of contact electrons creates a sharp resonance at the Fermi energy (Figs. 1.3(b) and 27.7). The resistivity of normal metals decreases with temperature, so the Kondo resonance from dilute magnetic impurities creates enhanced backscattering at lower temperature. The added $\delta\rho \sim \ln T/T_K$ correction creates a minimum in the resistivity vs temperature plot. In quantum dots, the resonance generates a conductance peak in the gap (Fig. 1.3(b)) and promotes forward scattering, creating the opposite effect as metals — as argued in Section 23.2. The effect is size-

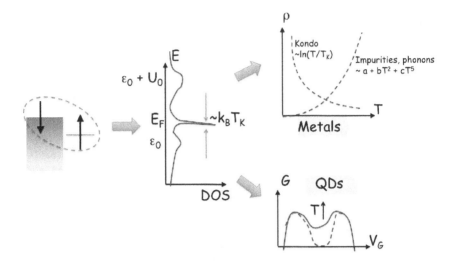

Fig. 27.7   *Evolution of Kondo singlet between a localized level and the contact Fermi sea.*
Metals see an increase in resistivity while quantum dots see an increase in conductance.

able at low temperature compared to the Kondo binding energy. In order
to see this effect, let us assume large U_0, whereupon we can simplify the
self-energies in Eqs. (27.19), (27.29). Assuming a weakly energy-dependent
V and DOS D, and spin degenerate levels $\Delta\epsilon = 0$, and lower bandwidth
Δ, we get

$$G_\sigma^K(E) \approx \frac{1 - \langle \Delta n_{\bar{\sigma}} \rangle}{E - \epsilon_\sigma - \Sigma_\sigma + \Sigma_{\bar{\sigma}E}},$$

$$\Sigma_{\bar{\sigma}E}(E) = -\sum_{\beta,k} \frac{f_\beta(\epsilon_{k\beta\bar{\sigma}})|V_{k\beta\bar{\sigma}}|^2}{E - \Delta\epsilon - \epsilon_{k\beta\bar{\sigma}}},$$

$$Re[\Sigma_{\bar{\sigma}E}(E)] \approx -\sum_\beta \int_{-\Delta}^{\infty} |V_{\beta\bar{\sigma}}(\epsilon)|^2 D(\epsilon) \frac{f_\beta(\epsilon)d\epsilon}{E - \epsilon} \qquad (27.32)$$

$$\approx \sum_\beta D|V_{\beta\bar{\sigma}}|^2 \ln\left|\frac{\mu_\beta - E}{-E - \Delta}\right|$$

The Green's function now has a pole at ϵ_0, and another of width $\sim k_B T_K$
near $\mu_\beta = E_F$ arising from the logarithm (from the term $f(\epsilon)/[E-\epsilon]$, which
we also encountered at the end of Section 26.2.3. For normal electrons, the

terms $\tau^2 f$ in Σ^{in} and $\tau^2(1 - f)$ in Σ^{out} remove the f in Γ. Here the factors replacing τ^2 are $(S_\mu \sigma_\mu)(S_\nu \sigma_\nu)$ and $(S_\mu \sigma_\nu)(S_\nu \sigma_\mu)$ which do not commute and leave behind an f-dependent Γ which creates the Kondo peak).

We can estimate the peak width by setting $E \approx E_F + k_B T_K$ for the pole, which from Eq. (27.32) amounts to setting $E_F + k_B T_K - \epsilon_\sigma + i\Gamma/2 + DV^2 \ln(k_B T_K/\Delta)$ equal to zero. Assuming wide bandwidth Δ and $E_F \gg k_B T_K$, this gives the Kondo temperature for $U_0 \to \infty$

$$E_F + k_B T_K - \epsilon_0 \approx -D|V|^2 \ln\left|\frac{k_B T_K}{\Delta}\right| \to k_B T_K \approx \Delta e^{-|E_F - \epsilon_0|/DV^2} \quad (27.33)$$

The pre-factor is actually determined by the lowest of the energy scales Δ, U_0 and Γ. For finite U_0, the equation changes a bit,

$$k_B T_K \approx \frac{1}{2\pi} \sqrt{\frac{2\Gamma U_0}{\pi}} e^{(E_F - \epsilon_0)(E_F - \epsilon_0 - U_0)/DV^2 U_0} \quad (27.34)$$

In a quantum dot, the resonance contributes to modes around a localized energy ϵ_0 with decreasing temperature (Fig. 27.7)

$$\Delta G \propto G_0 T(\epsilon_0) \frac{\Gamma}{E_F - \epsilon_0} \ln\left|\frac{\Delta}{k_B T}\right| \quad (27.35)$$

but also backscattering in a metal with dilute magnetic impurities.

27.7   Multiband and multilevel generalization

Extending the formalism to multiple bands isn't easy. A matrix equivalent of NEGF in Fock space would be wonderful, and there have been some valiant efforts towards it. We will focus on the local moment approximation without the cotunneling terms and focus on generalizing it to multiple bands. For a molecule with multiple levels, the use of the equation of motion technique followed by a similar decoupling at the LM level gives

$$G_m^{LM} \approx \sum_{l \neq m} \frac{\langle \Pi_{l \neq m}(1 - n_l) \rangle}{E - \epsilon_m + i0^+} + \frac{\langle n_l \Pi_{s \neq l \neq m}(1 - n_s) \rangle}{E - \epsilon_m - U + i0^+}$$

$$+ \frac{\langle \Pi_{l \neq m} n_l \rangle}{E - \epsilon_m - (M - 1)U + i0^+} \quad (27.36)$$

Clearly extending this to a band of electrons is challenging, but we could make some simplifications at this stage. Within the local moment approximation, we have already extracted an atomic self energy Σ^{LM} in Eq. (27.18).

We can now include this in the Green's function of a band, with the atomic energy replaced with the band energy, to give us

$$G(\epsilon_{k\sigma}) = \frac{1}{E - \epsilon_{k\sigma} - \Sigma_\sigma^{LM}(E)} \qquad (27.37)$$

We can set the denominator to zero to get the poles. Since $\Sigma_\sigma^{LM}(E)$ brings its own complexity, we get a quadratic equation whose solution is two spin split bands (we are using the notation $-\sigma = \bar{\sigma}$ for compactness),

$$E_\sigma^\pm = \frac{(\epsilon_{k\sigma} + U) \pm \sqrt{(\epsilon_{k\sigma} - U)^2 + 4Un_{-\sigma}}}{2} \qquad (27.38)$$

The Green's function above can be reexpressed in terms of its simple poles with residues $Z_\sigma^\pm$

Hubbard I approximation

$$G(\epsilon_{k\sigma}) = \frac{Z_\sigma^+}{E - E_\sigma^+} + \frac{Z_\sigma^-}{E - E_\sigma^-}$$

$$Z_\sigma^\pm = \pm \frac{E_\sigma^\pm - U(1 - n_{-\sigma})}{E_\sigma^+ - E_\sigma^-}$$

$$\qquad (27.39)$$

---

to be calculated self consistently with the charges n by integrating the spin split densities of states

$$n_{\pm\sigma} = \int_{-\infty}^{E_F} dE \, D_\sigma^\pm(E) = \int_{-\infty}^{E_F} dE \, D_{0\sigma}^\pm(E) Z_\sigma^\pm(E) \qquad (27.40)$$

with $D_{0\sigma}^\pm$ the density of states of the unrenormalized Hubbard bands $E_\sigma^\pm$. Equation (27.39) shows how the local moment version of Coulomb Blockade broadens to create the Hubbard gap responsible for the Mott transition. More to the point, the bandgap and the weights of the Hubbard bands evolve with doping n, and provide a crude starting point to study strongly correlated solids such as doped Mott-Hubbard insulators.

We can improve on this approximation by making the Hubbard contributions different for the two channels. We can further invoke the Coherent Potential Approximation (CPA), also called the Alloy Analogy, by mandating that the $-\sigma$ states are fixed at random 'alloy' sites, introducing an overall shift in the local moment self energy $\Sigma^{CPA}(E) = \Sigma^{LM}\left(E - \Delta_\sigma(E)\right)$, self consistently with a concomitant shift in the Green's function

$$G_\sigma(E) = \sum_k \frac{1}{E - \epsilon_{k\sigma} - \Sigma^{CPA}(E)} = \frac{1}{E - \Delta_\sigma(E) - \Sigma_\sigma(E)} \qquad (27.41)$$

Finally, we can correct for the static $-\sigma$ states (**Hubbard III approximation**) by accounting for the events where the $-\sigma$ state leaves and for events where σ and $-\sigma$ jointly occupy a site, i.e., add corrections $\Delta_{-\sigma}(E)$ and $-\Delta_{-\sigma}(U - E)$ to $\Delta_\sigma(E)$.

27.8   Experimental signs of nonequilibrium correlation

Figure 27.8 shows a comparison between experimental and computed I-Vs of a set of molecular conductors. Several features are noticeable, ones that are not easy to explain from a regular Landauer SCF approach. For starters, there is a pronounced blockade region after an initial rise in current, where an added electron repels the next oncoming electron. While a spin dependent mean field theory will capture that effect (Eq. (27.15)), its blockade region is given by the HOMO-LUMO gap, distinct from the split between ionization potential ($E_N \to E_{N-1}$) and electron affinity ($E_N \to E_{N+1}$). Assuming an interaction term $U_0 N(N-1)/2$ term in E_N between N electrons ($^N C_2$ pairs), the ionization potential $E_N - E_{N-1} \propto U_0(N-1)$ while the electron affinity $E_{N+1} - E_N \propto U_0 N$, which means the HOMO and the LUMO levels should be shifted by different amounts, i.e., an orbital-dependent potential. Instead, typical DFT-LDA theories calculate the potential using an orbital-independent potential $\propto \partial E_N/\partial N = U_0(N - 1/2)$ which thereby underestimates the bandgap by U_0.

Figure 27.9 shows the computed current flow through a single orbital benzene ring. The Coulomb energies and hopping parameters in the Hamiltonian were fitted in their mean field limit with LDA, and then employed in a multielectron master equation that was solved by exact diagonalization. The total Fock space of size $2^{12} = 4096$, although the Hamiltonian can be readily block diagonalized into smaller chunks by sorting states by symmetry such as spin and particle number. Two differences spring to attention when compared with mean field Landauer results. The HOMO LUMO gap as extracted from the mean field method in both the density of states and the I-V is smaller than the split between ionization potential and electron affinity (due to the extra charging energy arising from self interaction correction, discussed above). In addition, the exact diagonalized treatment yields a lot more one particle levels through transitions such as shown in the transition diagram, with a lot more spikes in the density of states through such higher order transitions, while maintaining the same sum rule overall.

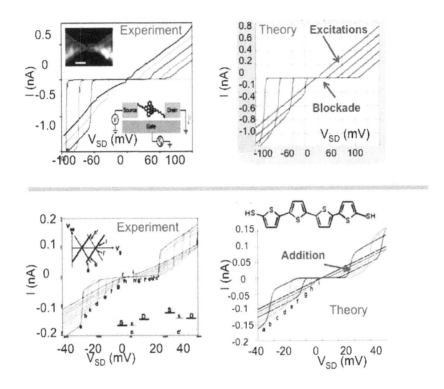

Fig. 27.8 *(Left Column) Experiment (see papers by Park et al. and Zhitenev et al. in bibliography), vs (Right Column) Orthodox Theory show common features such as the presence of a pronounced blockade region of zero conductance, followed by jumps in current followed by ramps created from closely spaced excitations. The appearance of plateaus vs ramps can be rationalized from an energy transition diagram that outlines the sequence in which the transitions electronic, or in this case, vibronic are populated.*

In addition to the suppressed zero bias conductance, sharp current jump and subsequent plateaus, the experimental IVs reveal features that need us to invoke many body excitations explicitly. These excitations aren't necessarily all electronic, and may involve vibronic excitations as well, but the overall physics is similar. In particular, there is a clear cross-over between plateaus and ramps depending on the relative energies of (i) the ground state to ground state (**G to G**) vs (ii) ground to excited (**G to E**) state transitions. When G to G occurs first, we see a jump followed by a plateau and then a ramp of closely spaced excitations when the G to E is reached. If however G to E occurs first, we see no features as there are as yet no electrons to excite. Once the subsequent G to G occurs, the G to E is

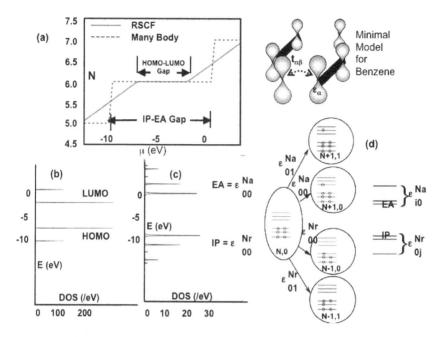

Fig. 27.9 *(a) Calculated IV for benzene in a minimal pz orbital Hubbard model shows a blockade over a HOMO LUMO gap extracted from single particle theory in a Landauer approach, vs the true ionization potential to electron affinity gap in a multielectron master equation approach. (b) LDA transitions vs (c) many body transitions show the difference in gap as well as the presence in the latter case of several satellite peaks arising from transitions between ground states and excited states (d).*

enabled rightaway, meaning we see a jump followed instantly by a ramp without any intervening plateau.

Let us dwell on one final intricacy arising from nonequilibrium correlations, and that is in the asymmetry in the molecular IVs in response to an asymmetry in the contacts. Depending on whether we are in the weak correlation ($U_0 \ll \Gamma$) versus strong correlation ($U_0 \gg \Gamma$) regime, the output can be shown to switch between conductance asymmetry (unequal IV slopes) vs current asymmetry (unequal IV plateau heights). Both asymmetries arise due to *shell filling*, where we inject rapidly from the stronger contact and remove slowly by the weaker contact leading to a buildup of charge and its associated Coulomb cost. For the opposite polarity, the charges do not linger but are pulled out immediately, suffering no Coulomb repulsion, the levels thereby filled as predicted from one particle physics (*shell tunneling*).

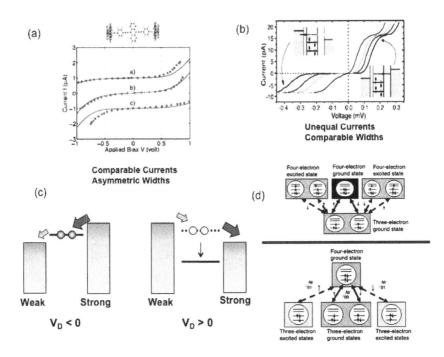

Fig. 27.10 *(a) Measured and modeled IVs for anthracene show a conductance asymmetry (data taken from J. Reichert et al., Phys. Rev. Lett. 88, 176804 (2002)). In contrast, molecular dots show (b) a current plateau height asymmetry (data taken from M. M. Deshmukh et al., Phys. Rev. B 65, 073301 (2002)). (c) Conductance asymmetry arises from asymmetry in polarization, whereby shell filling by the stronger contact charges a level and postpones the onset of conduction, while the opposite polarity leads to a noninteracting shell tunneling regime. (d) Current asymmetry happens because the weaker, rate determining contact encounters different numbers of addition vs removal levels for forward vs reverse bias (spelt out here for transitions from 3 particle ground state to 4 particle states in one polarity, and the reverse for the other polarity)*

The *conductance asymmetry* is easier to see. If we apply a positive bias on the stronger contact, the molecule gets reduced (electron pulled out) by the stronger contact faster than it is oxidized by the weaker. For a p-type (HOMO level) molecule, the net charge removal costs a Coulomb energy that pulls the HOMO level down further from the bias window, postponing the onset of conduction and dragging out the IV. For negative bias, the HOMO stays filled and the current stays unperturbed. In other words, we see a *polarization asymmetry* or *conductance asymmetry* whose polarity tells us if we have a *p*-type or an *n*-type molecule.

In the Coulomb Blockade limit, we see instead a *current asymmetry* arising from the dynamics of the partially filled levels. For positive bias on the stronger contact we go from the 4-electron to the 3-electron ground state in 2 ways. The weaker contact brings us back to the 4-electron ground state in 2 ways, but can also bring us back to the 4 possible 4-electron excited states. For negative bias on the stronger contact, we go to the 5-electron ground state, and return to the 4-electron ground and excited states. This creates a *current asymmetry*. Furthermore, this asymmetry switches polarity around the charge neutrality point in the $G - V_d - V_g$ diamond diagram.

As we vary the ratio U_0/Γ, we expect there to be a smooth transition between the current and the conductance asymmetries. While this is hard to capture formally (since we do not yet have a proper theory for the intermediate coupling regime $U_0/\Gamma \approx 1$), we can accomplish this by raising the temperature artificially, creating thereby a thermal broadening in lieu of a contact-induced broadening. The plateaus corresponding to shell tunneling get absorbed in the broadened manifold, qualitatively leading to the expected conductance asymmetry (Fig. 27.8). The quantitative mismatch, however, underscores the difficulty of achieving this transition smoothly due to our inability to capture non-equilibrium processes, especially, broadening and coherence, within the many-body configuration space.

PART 8

Looking Back and Looking Ahead:

Novel Transport vs Novel Devices

Chapter 28

Looking back and looking ahead

28.1 Looking back: a microscopic view of transport physics

The idea behind the book was to provide a flavor for the complex and varied dynamics of an electron in its exciting journey along a channel. This dynamics covers a wide swath of behavioral patterns, from classical particulate to quantum wave propagation, from ballistic to diffusive, non-interacting to strongly correlated. The equations presented in this book provide a framework to understand and model these dynamical systems. In the process, we spelt out a hierarchy of transport formalisms from Newtonian drift-diffusion to quantum Landauer, along with a hierarchy of electronic structure methods from empirical tight-binding to 'first principles' Density Functional Theory (DFT) and configurational interaction (CI) methods.

One of the important messages we seek to promote in this book is the fundamental novelty of transport properties. By this, I mean properties measured *far from equilibrium* — far from 'linear response' where we could have still used equilibrium relations based on the Kubo formula (Eq. (12.18)). Transport isn't just a trivial extension of equilibrium, nor a matter of simply identifying the correct Hamiltonian. To extract transport properties we need to go beyond Kubo, even beyond Landauer in general, and add two intricacies: (a) a complex self-energy matrix with components $\Sigma_{ij}(E)$ (more generally, $\Sigma(x_i, x_j, t, t')$ with interconnected real and imaginary parts), to describe the *non-unitary* evolution of the electron and the shifting and broadening of device levels by the outside world not captured by our device Hamiltonian. (b) In addition, we need an in-scattering function $[\Sigma^{\text{in}}(E)]$ or $[\Sigma^{\text{in}}(t, t')]$ that describes the filling of these levels under *non-equilibrium* conditions by the multitude of reservoirs — real and virtual contacts maintained at separate equilibria (set at different μs) by energy sources such as batteries. We saw how to write down Σ and Σ^{in} within reasonable approximations starting with microscopic Hamiltonians, not just for complicated materials and geometries, but also for interactions from phonons, polarons, Coulomb screening and impurities that can all be looked upon as part of the *outside*.

What makes the bottom-up, 'molecular' view of nanoelectronics relevant today are two parallel developments. First of all, experiments have become sophisticated enough to directly measure the complex dynamics in many systems down to molecular dimensions. We witnessed the emergence of an entire class of single atom thick 2-D materials, with considerable nontriviality not just in their E-ks but often in their molecular Bloch states (recall pseudospins in graphene and spins in topological insulators). We also encountered several developments in nano-magnetism including spin torque and spin Hall effects. Researchers have directly measured the current and current noise through a single hydrogen atom, through benzene and other small groups of aromatic molecules, nanotubes of precise structures, single atom nanowires and point contacts. Looking back at the terrain we covered, the purpose of the bottom-up viewpoint is to provide a natural framework to understand the diverse transport physics in these nanosystems, along with the modeling platform to simulate their behavior.

More excitingly, a number of emergent devices have appeared on the horizon that seek to exploit novelties at the nanoscale, from tunnelFETs to STTRAMs. It is conceivable that in these novelties lie the seeds of unconventional switching that could complement, and perhaps even outperform silicon CMOS in some way. Looking ahead, we can view these emergent candidate devices as *case studies*, to employ our tools and understand their behavior. This is what we will focus on in these parting chapters — specifically, low power *subthermal* switching, deconstructed in the language developed in the book. We explore how all the different transport modes of an electron, from tunneling to scattering, spin torque to Mott transition could conceivably act as an unconventional, low power switch.

28.2   Looking ahead: the high cost of switching

One way to quantify the cost of switching is to minimize its 'action', the product of its free energy and delay, $F\Delta t$, where $F = E - TS$ depends on the operating temperature T and the entropy S. Statistical mechanics teaches us that the entropy $S = k_B \ln \Omega$ where Ω is the number of allowed microstates with a given constraint (e.g., total energy and particle number). The error probability $\Pi_{\text{err}} \propto 1/\Omega$ is inversely related to the number of microstates. The least action boils down to minimizing the product

$$\underbrace{E_{\text{diss}}}_{\text{Energy}} \times \underbrace{\Delta t}_{\text{Delay}} \times \underbrace{\ln(\Pi_{\text{err}})}_{\text{Error}} \qquad (28.1)$$

for a fixed volume. In other words, an ideal switch must be fast, waste little energy/volume and be highly accurate. An equivalent criterion is maximizing bits/joule/second, with bits proportional to the information entropy.

It is perhaps intuitively obvious that speed costs energy. The analysis above suggests that accuracy costs energy as well. Therein lies the trade-off, and also the demands of a particular application. Not all applications require extreme speed or accuracy. An electronic sensor does not need to operate constantly at a high frequency as a data server does. An image processor acts as a comparator and does not need nearly the same level of accuracy as the accounting process in a banking software. In general, we can often offset one metric against another, for instance lower the power budget by running a circuit slowly. An STTRAM could run at low dissipation if we just lowered its drive current and reduced Joule loss. A low current however reduces the probability of torque induced switching at finite temperature. A small fraction of spins tend to stagnate at a configuration precisely parallel to the hard magnet and see zero torque. The stagnation increases the write-error rate, unless we wait long enough for random thermal kicks to dislodge the spins and then for the torque to take over (Fig. 21.7) — underscoring the crucial point that maintaining a low value for the overall product is the real challenge in ultrascaled devices.

Experts believe we are rapidly nearing the end of the roadmap for silicon CMOS. To be fair, negative predictions have been lurking for ever — VLSI textbooks once predicted a fundamental scaling limit of 250 nm from oxide tunneling and impurity related fluctuations. Today's challenges however venture well beyond material and design constraints up against physical and more importantly financial limits. The lithographic challenge of patterning small feature sizes is considerable, not to mention that state of the art lithography tools cost nearly $100M today while a leading edge fab costs upwards of $6B. Many of these issues go well beyond the scope of this book, so let us focus instead on physics — the high cost of switching that spelt the demise of Dennard scaling. At its most basic, a transistor is a channel turned on and off by a gate capacitor connected to a power supply voltage. As we saw in the introduction, the power dissipated in the interconnects during the process of charging such a capacitor to a voltage V_D is $\alpha C V_D^2 f + I_{\text{off}} V_D$ where f is the clock frequency, I_{off} is the off state (leakage) current, and $\alpha < 1$ is the activity factor that describes the fraction of gates active at a given switching cycle. Including typical numbers from the semiconductor roadmap, we find that an end of the roadmap device operating under conventional scaling would dissipate > 100 MW/cm^2 beyond 2020, well in excess of known limits on solid state heat removal.

The semiconductor industry has responded to the rapidly burgeoning thermal budget by abandoning conventional 'Dennard' scaling, in particular the scaling of the voltage and the clock frequency. In addition, we shifted to multicore and hyperthreading strategies. We also kept scaling the wafer size with concomitant growth in the vertical dimension for upcoming technology nodes (not to mention that 'nodes' are defined more indirectly instead of simply gate length or half-pitch, while the time between nodes keeps getting pushed out). The end of planar technology commenced a few years ago with the arrival of FINFETs and their cousins such as trigates, Omega-gates and dual gate MOSFETs. The scaling of the in-plane fin dimension, even at the expense of its vertical out-of-plane growth, reduces the overall interconnect footprint and restores computational power, even though the individual device performance no longer scales aggressively. Overall, Moore's law, interpreted generally as the *scaling of the number of functions per dollar*, plods along (although by some analyses, it may already have peaked some years ago). There still is effort at improving channel mobility, generating interest in III-V, Ge and 2-D materials. However there is clearly a growing urgency for disruptive architectural and algorithmic solutions, including perhaps an overhaul of the way computation is done.

We can reduce the dissipated energy by running a circuit slower than the RC time constant, allowing its charges to stay in phase with the voltage source (Eq. 2.17). We can also do certain kinds of computation, for instance pattern recognition, which brings down the dissipation (Eq. 2.20). Not withstanding adiabatic computing, approximate or probabilitic computing, and reversible computing, which requires precise timing circuitry, we must reduce either the number of charges Q or the voltage V_D. We saw earlier in Chapter 2 that the former is limited by drivability, the latter by Boltzmann physics plus desired error rates. To reduce Q, more generally the number of bits, we need to 'staple' them together as in magnets or correlated solids. To reduce ΔV_D we need to bypass the Boltzmann limit. Either approach usually requires us to change the nature of the carriers and use non-charge based logic.

28.3 Limit to dissipation: the Boltzmann 'tyranny'

The voltage needed to move a charge reliably from source to drain is determined by (i) the ON/OFF ratio, i.e., the number of decades that the current must swing through below threshold, and (ii) the steepness of the gate transfer characteristic $I - V_G$ that converts the ON/OFF ratio into a voltage swing. Between the two, the ON/OFF ratio is set by imposed standards for circuit level error rates Π_{err}, and independently by the need for a high ON current to reduce delay, and a low OFF current to reduce standby dissipation. The steepness of the transfer curve (known as subthreshold swing — lower is steeper) is set by fundamental physics, specifically the rate of spontaneous thermal excitation over an energy barrier. We can invoke the equations developed throughout this book to explore this intrinsic bound, known as the Boltzmann limit. We emphasize the word intrinsic because the performance of an actual device may well be limited by poor contact resistances and parasitic capacitances, completely masking an otherwise superlative switching curve.

Conventional switches operate by electrostatically raising and depressing the gate potential, allowing the charges to thermally jump over the barrier. The rate of thermal emission is given by the Boltzmann tail of the Fermi-Dirac distribution of electrons in the contacts. The Landauer equation can be written as a convolution of the total transmission over all modes, and the thermal broadening function $F_T = -\partial f / \partial E$ (Eq. (12.8))

$$I = \frac{q}{h} \int_{\mu_1}^{\mu_2} dET \left[M \otimes F_T \right] \tag{28.2}$$

Since the convolution of two functions is limited by the spread of the slower one, for the best case of an abrupt change in transmission over an energy $\sim k_B T$, the swing is set by the gate dependent shift of the broadening F_T relative to the mode spectrum. In subthreshold, we use the Boltzmann approximation, $f \approx e^{-E/k_B T}$, so that $F_T \approx f/k_B T$. Assuming the modes simply shift rigidly with a capacitive gate transfer factor α_G, $M(E, V_G) \approx M_0\Theta(E - q\alpha_G V_G)$, we get the normalized transconductance $\bar{g}_m$ that gives us the steepness of the gate transfer curve, inverse to the subthreshold swing

$$S^{-1} = \bar{g}_m = \frac{\partial \log_{10} I}{\partial V_G} = q\alpha_G \beta / \ln 10, \quad \beta = 1/k_B T \qquad (28.3)$$

leading to the well known expression for subthreshold swing, $S_0 = k_B T \ln 10 / q\alpha_G \sim 60$ mV/decade for perfect gate control with $\alpha_G = 1$. This steepness limit in turn limits the operational voltage and energetics of conventional CMOS.

28.4   Bypassing the 'Boltzmann tyranny' — quantum physics to the rescue?

In the above derivation, we have ignored a possible gate voltage dependence of the transmission function $\bar{T}(E, V_G)$ itself, which can add another term

$$S^{-1} = (q\alpha_G \beta + \overline{\partial \ln T/\partial V_G}) / \ln 10 \qquad (28.4)$$

where the bar denotes that the second term is mode averaged over the energy bias window. We see that the subthreshold swing is reduced if the quantum mechanical transmission function increases in area within the bias window under the action of a gate voltage. This would happen for example if the system were to shrink its bandgap or open additional conducting channels under a gate bias, in addition to the band realignment. We will now see how the various modes of electronic conduction can each lead to a way to accomplish such a radical shape shift in the transmission function and sharpen the transfer characteristic beyond the Boltzmann limit.

The discussions from here on are speculative, with no clear winners despite intense efforts under way. However, they are illustrative, so let us focus purely on the intellectual merit of switching a transistor in a way that beats the Boltzmann limit. As it turns out, all the discussed approaches rely somewhat explicitly on quantum physics. Tunnel FETs and NEMFETs both rely on the abrupt onset of quantum tunneling, intrinsically a temperature independent process. Graphene p–n junctions rely on band to band transfer limited by their orbital (pseudospin) degree of freedom, involving chiral and Klein tunneling. Finally, Mott switches rely on many body quantum effects and strong correlation. Due to their heavy dependence on quantum effects, we have to venture beyond classical drift-diffusion analyses, relying instead on the Landauer equation and beyond.

Chapter 29

Case study: a new kind of switch?

Let us put our theoretical machinery to the test — by exploring the subthermal switching of transistors beyond the Boltzmann limit. As discussed in the previous chapter, we require a gate dependent change in shape of the transmission function, in addition to the shift in Fermi energy relative to the band-edge, to draw more current than usual. We will discuss a few ways to do this. Many of these devices may prove inadequate in retrospect; — they may have other trade-offs, or end up too expensive, unreliable or difficult to integrate into existing CMOS process flows. The primary purpose of this exercise is to showcase how we can take a practical problem and walk through an analysis using the Landauer-NEGF approach.

29.1 Gating transmission band-width: Tunnel FETs

The difficulty with continuous downscaling of transistors is mainly the onset of tunneling across the oxides, which until recently were very thin. Quantum tunneling and associated current leakage became an undesirable but unavoidable trait, an inconvenient bit of physics that device engineers needed to account for. More recently however, there is growing effort to actually exploit this tunneling effect to help with abrupt switching, the operating idea behind a tunnel field effect transistor (tunnel FET).

Let us discuss switching in a bipolar junction transistor (BJT), which has a pnp (or an npn) structure — two back to back pn junctions. The first (emitter-base) junction is forward biased into injecting a large number of majority carriers into the base, fabricated to be thinner than the diffusion length so they reach the collector. The second (base-collector) junction is reverse biased so these carriers are whisked away by the large collector fields. The few minority carriers that are injected back from the base into emitter are then used as the 'control' — increasing that base current a little bit unleashes a large proportional increase in collector current, leading to a high current gain in the common emitter configuration (input at base, output at collector, emitter grounded), the gain mainly set by the doping ratio across the emitter-base region.

A MOSFET starts with the same pnp or npn structure, but operates differently using majority carrier (unipolar) thermionic emission followed by drift. Turning the transistor on requires making the central region homologous with the contact polarity, with the application of a negative or positive gate bias V_G for pnp or npn respectively (Fig. 18.2). This however amounts to pulling down the channel barrier till the electrons from the contacts can thermionically excite over it (for npn), meaning that the rate of change of current is determined by the Boltzmann tail $\sim \exp\left(-qV_G/k_BT\right)$, and is thus limited to a rate of q/k_BT, i.e., a 40X increase in current per volt in gate voltage. Conversely, every order of magnitude increase in current requires at least $S_0 = (k_BT/q)\ln 10 \approx 60$ mV of applied gate voltage. It is this Boltzmann rate that ultimately limits voltage scaling, and accounts for the large power dissipation in today's CMOS. The tunnelFET attempts to realize a rate of increase in current larger than the Boltzmann limit stipulates.

The TFET starts with a similar geometry but has both pn junctions forward biased, i.e., a *p-i-n* junction (Fig. 29.1). To turn it on we apply a voltage to the 'i' region to bring it into the band to band tunneling regime with the source, so that the conduction band of one sweeps past the valence band of the other. The abrupt turn on of temperature-independent, band-to-band tunneling is steeper than the Boltzmann limit, driven by a gate modulation of the transmission band-width. To quantify the actual turn on mechanism, we go back to the Landauer equation, and use the Zener tunneling expression for a positive barrier between the two crossing bands (Eq. (14.9)). Here, the barrier height $U_0 = E_G + \Delta\Phi$, the bandgap plus the added energy $\Delta\Phi$ above crossing, while the separation is Λ, so that the electric field $\mathcal{E}_0 = (E_G + \Delta\Phi)/q\Lambda$ in Eq. (14.9). Finally, a positive $\Delta\Phi$ requires that the electron has an energy E lying in a window, $E_V^p < E < E_C^i$

$$T_{\text{Zener}} \sim e^{-4q\Lambda\sqrt{2m}E_G^{3/2}/3\hbar(E_G+\Delta\Phi)} \left[\Theta(E - E_C^i) - \Theta(E - E_V^p)\right] \quad (29.1)$$

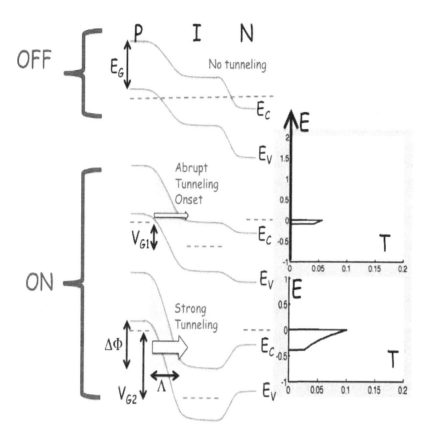

Fig. 29.1 *Top to bottom: Band diagram of p–i–n junction under OFF, nearing ON threshold at V_{G1} and well above ON at $V_{G2} > V_{G1}$. Quasi-Fermi levels are shown as red dashed lines. As the conduction band edge E_C^I of the I region sweeps past the valence band edge E_V^P of the P region, the transmission function increases in bandwidth and band amplitude. The expanding band-pass filter arises from progressively increasing band to band tunneling through a rapidly narrowing triangular barrier, shown in grey. The subthreshold swing $S \approx S_0 \Delta\Phi/k_B T$ where S_0 is the Boltzmann limit of ~ 60 mV/decade limiting today's thermally activated switches. Simulations are shown for a gate split 10 nm, bandgap 1.1 eV, dopings 10^{16} and 10^{12} cm^{-3} across the PI region, $m^* = 0.19$ m_0.*

with $E_C^i = E_V^p - (E_G + \Delta\Phi)$. The current is then given by Landauer theory

$$I = \frac{2q}{h} \int_{E_C^i}^{E_C^i + E_G + \Delta\Phi} dE\, e^{-4q\Lambda\sqrt{2m}E_G^{3/2}/3\hbar(E_G+\Delta\Phi)} M(E)$$

$$\times \left[f(E - E_p) - f(E - E_i) \right] \qquad (29.2)$$

Note that we now have two window functions, one set by the gate and drain voltage dependent Fermi functions, and the other by the relative locations of the conduction and valence bands. The actual window is thus decided by the overlap of those two window functions, and this overlap changes abruptly since the band edges themselves do not come with any thermal smearing (in practice, they would smear due to the presence of phonons, but that is a higher order process). From this current we can then calculate the subthreshold swing. There will be two contributions to this, one that involves controlling the transmission window, and the other involves modulating the amplitude of the mode averaged transmission

$$S^{-1} = \frac{\partial \log_{10} I}{\partial \Delta \Phi} \approx \left(\underbrace{\left[f(E - E_p) - f(E - E_i) \right]}_{\text{window modulation}} + \underbrace{\frac{4q\Lambda\sqrt{2m}E_G^{3/2}}{3\hbar(E_G + \Delta\Phi)^2}}_{\text{amplitude modulation}} \right) \frac{1}{\ln 10}$$

(29.3)

Assuming the Fermi energies lie near the bandedges, i.e., $E_p \approx E_V^p$ and $E_i \approx E_C^i + E_G$, the first term in the bracket gives us $S \sim \Delta\Phi \ln 10/q$. As the two band edges sweep past each other, $\Delta\Phi$ and thus the subthreshold swing rises from zero and reaches a finite value.

The switch operates unconventionally as it relies on tunneling — intrinsically a temperature independent process. Recall in conventional transistors the rate of change of current is given by the convolution between the transmission bands and the fermi tails, f for the conduction band and $1 - f$ for valence. In a tunnelFET however, because of band to band (Zener) tunneling across the source channel junction, the distribution function is flipped, with the valence band of the central region populated by the conduction electron f, not $1 - f$, from the source. The bandgap chops off the tail of the f function and effectively cools the electrons, which accounts for the low subthreshold swing. Put another way, the overall transmission shows an additional window effect above and beyond those imposed by the contact Fermi functions, and this window is gate voltage dependent (Fig. 29.1), creating an aggressive cut off to the Fermi distribution and allowing the system to switch sharper than the Boltzmann limit.

At the time of writing this book, the main challenge with TFETs was maintaining a low subthreshold swing over a few decades of voltage while simultaneously maintaining adequate overall current drive for appreciable switching speed. Since the ON current is limited by tunneling, there is a lot of effort on reducing the effective tunnel barrier width Λ by using a staggered bandgap semiconductor pair to form a suitable heterojunction. Such heterojunctions however come with interface traps that create a leakage current that raises the current floor. In other words, raising the ON current compromises the subthreshold swing.

29.2 Gating transmission gap: chiral tunneling across graphene $p–n$ junctions

To get a change in current, what we need ultimately is a transmission gap and not an outright bandgap. One way to do this is to allow momentum rather than energy filtering. For TFETs, the problem is that the ON state is still limited by tunneling. In contrast, graphene and other Dirac cone systems (such as Bi_2Se_3 surfaces) have zero bandgap, which is great for the ON current but not for the OFF. A zero gap however has the advantage that opening a gap somehow (e.g., electrostatically) leads to a large fractional change in conductance. Opening a transmission gap electrostatically not only reduces the OFF current, but in addition the gate tunability of this transmission gap would invariably generate a low subthreshold swing. As the Fermi energy moves towards a bandedge, that gap closes in addition, pulling out more current for the same voltage.

A transmission gap based on momentum filtering can be engineered using electrostatically gated $p–n$ junctions in 2D chiral systems such as graphene. Recall that the bandstructure of graphene resembles photons and follows a Dirac cone. Furthermore, there is a strong locking between electron momentum and the chiral degree of freedom, which for graphene is a pseudospin (the phase angle describing the mixing of the pz dimer orbitals). In fact we worked out the 2 component spinor eigenstate (Eq. (5.14)) that shows this explicit dependence on wavevector orientation θ. The angle dependence makes the wavefunctions for forward ($\theta = 0$) and backward ($\theta = \pi$) orthogonal, and are comprised respectively of the bonding (pseudospin $+1$) and antibonding (pseudospin -1) combinations of the dimer p_z basis sets.

The orthogonality of the normal modes implies that a normally incident electron at a junction can only go forward, at least at low energies (at higher energies they can Umklapp scatter to another valley, or break the symmetry selection rules by involving optical phonons). But even at low energy, modes corresponding to higher angles of incidence can mix their pseudospins, so that *the transmitted modes lie within a cone whose angle depends on the voltage barrier across the junction.* Since the low angle transmission stays pinned to unity, the overall transmission lobe changes width with gate voltage across the p–n junction.

We have already worked out the transmission $T(\theta_1)$ (Eq. (15.14)) across a p–n junction and showed only a modest gate dependence (Fig. 15.6),

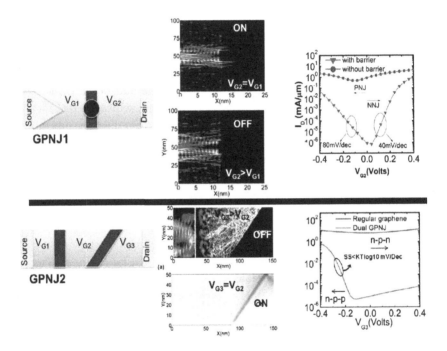

Fig. 29.2 *(Top) GPNJ filter consisting of a collimating point contact and a hole at the split-gated pn junction. Atomistic NEGF simulations show that in the ON state where the critical angle for transmission exceeds the angle subtended by the hole at the point contact ($\theta_C > \theta_B$), we get a strong electron transmission. Conversely at a higher gate bias collimation becomes aggressive and the critical angle reduces till $\theta_C < \theta_B$ and all electrons are reflected. The resulting gate tunable transmission gap creates a large ON-OFF ratio with a transfer curve steeper than the Boltzmann limit (shown here ignoring edge reflections by directly employing Eq. (15.14)). (Bottom) An alternate geometry consists of two GPNJ junctions tilted relative to each other, where again NEGF simulations show that at high gate bias the critical angle of collimated modes drops below the tilt angle, $\theta_C < \delta$, and once again we get a steep transfer curve. Simulations are referenced at the end.*

because of the preponderence of Klein tunneling at normal incidence. If the maximal Klein mode can now be suppressed geometrically, we will get a complete quenching of the OFF current. One way to do this is to put a cut inside the graphene sheet at the *p–n* junction (Fig. 29.2, top), which subtends an angle θ_B at the source, assumed to be a collimating point contact. If the critical angle θ_C corresponding to the maximum transmitted mode is smaller than θ_B, i.e., $\theta_C < \theta_B$, then all modes are rejected back to the source and T has an abrupt cut off over a range of angles and energies

$$T(\theta_1, V_{G1}, V_{G2}) = T(\theta_1)\Theta(\theta_1 - \theta_B) \tag{29.4}$$

where Θ is the Heaviside step function. We can also accomplish this quenching with an extended rather than a point contact using a second *p–n* junction tilted at angle δ relative to the first (Fig. 29.2 bottom). If $\theta_C < \delta$, once again the charges are reflected back to the source, with the role of θ_B now played by δ. In both cases, we have a range of energies corresponding to $\theta_C < \theta_B, \delta$ where there is no transmission. This provides a low OFF current while only marginally degrading the ON current. But in addition, the transmission gap is directly controlled by the gate dependent voltage barrier across the junction, vanishing when we reach the homogeneous limit. The conduction and valence bands are obtained by setting $\delta = \theta_C$, with the critical angle obtained from Snell's Law (Eq. (15.3)) as

$$\sin\theta_C = -\left(\frac{E_V + qV_{G2}}{E_V + qV_{G1}}\right) = \left(\frac{E_C + qV_{G2}}{E_C + qV_{G1}}\right) \tag{29.5}$$

which gives us the gate tunable band-edges of the transmission gap, and the corresponding gate tunable transmission gap

$$E_{C,V} = -qV_{G2} \mp q\frac{(V_{G1} - V_{G2})\sin\delta}{1 \pm \sin\delta}$$

$$E_G = E_C - E_V = q\Delta V_G\frac{2\sin\delta}{\cos^2\delta} \tag{29.6}$$

The gate tunability of the transmission gap leads to a voltage upconversion of the applied gate bias. From the shift of band edges over and above the applied gate bias $-qV_{G2}$ (assuming perfect gate control here), we can extract the final subthreshold swing

$$S = k_B T \ln 10/q \times (1 \pm \sin\delta) \tag{29.7}$$

with a higher degraded threshold as we move from intrinsic PI to the heterogeneous *p–n* region and a lower improved threshold when we move towards the homogeneous PP region where we expect no transmission gap.

Ultimately, it is the gate dependent collapse of the transport gap near homogeneous doping that makes the switching significantly subthermal and beats the Boltzmann bound.

A sequence of non-parallel graphene p–n junctions can thus eliminate the undesirable electrons by redirecting them towards the source and accomplish the same steep transfer curve beating the Boltzmann limit (Fig. 29.2). The challenge however is edge scattering, which can mess up momentum filtering by bouncing around the rejected electrons, increasing their chance for being normally incident at the second interface upon repeat attempts. Innovative engineering design is needed to make sure the rejected electrons return through the first junction instead of finding their way through the second upon multiple attempts.

29.3   Gating transmission amplitude: ion channels and nanomechanical relays

A nanomechanical cantilever provides yet another example of a gate tunable transmission function, in this case, the tunneling probability of an electron from the cantilever to a drain contact. In the chapter on gating, we already developed the transport equations for a nanomechanical relay. The gist was a conformation dependent Landauer equation, where the most likely conformational coordinate θ^* was set by the gate voltage by minimizing the net potential $dU/d\theta^* = 0$. Let us now work out the phase transition physics and the associated reduction in subthreshold swing. Differentiating the Landauer current (Eq. (19.1)) with the gate voltage, we get

$$\frac{\partial I}{\partial V_G} = \frac{q}{h} \int_{\mu_1}^{\mu_2} dE M(E) \left[\frac{\partial \bar{T}}{\partial V_G} \otimes F_T + \bar{T} \otimes \frac{\partial F_T}{\partial V_G} \right] \qquad (29.8)$$

The tail of the thermal broadening function approaches the bandedge with an applied gate bias, making $\partial F_T/\partial V_G \approx \alpha_G \beta = \alpha_G q/k_B T$ for an enhancement mode transistor, while the derivative of the first term in the bracket can be written as $(\partial \bar{T}/\partial V_G) = (\partial \bar{T}/\partial \theta^*)(\partial \theta^*/\partial V_G)$. The transmission is limited by electron tunneling into the drain from the edge of the cantilever through the vacuum barrier of width $t_{air}(\theta) = h - L \sin \theta$, h being the bare height above the drain before bending and L being the length of the cantilever. Tunneling can be captured with a WKB approximation, $\bar{T}(E, \theta) \propto \exp[-2\kappa t_{air}(\theta)]$. We then get $\partial \bar{T}/\partial \theta^* = -2\kappa \bar{T} dt_{air}(\theta^*)/d\theta^*$.

From geometry, $t_{air}(\theta^*) = h - L\sin\theta^* \approx L(\theta_0 - \theta^*)$ for a long cantilever where $0 < \theta^* < \theta_0 \approx h/L \ll 1$. Then $dt_{air}(\theta^*)/d\theta^* \approx -L$, and we get the normalized transconductance as the separable sum of electronic and conformational contributions

$$\bar{g}_m = \frac{\partial \ln I}{\partial V_G} = \underbrace{\beta}_{\bar{g}_m^{el}} + \underbrace{2\kappa L\left(\frac{d\theta^*}{dV_G}\right)}_{\bar{g}_m^{conf}} \qquad (29.9)$$

assuming perfect gate control ($\alpha_G = 1$). The normalized transconductance is inversely proportional to the subthreshold swing to within a constant factor of $\ln 10/q$, so that the corresponding subthreshold swings add in parallel, reducing the overall value

$$\boxed{S^{-1} = S_{el}^{-1} + S_{conf}^{-1}} \qquad (29.10)$$

It is easy to see from Eq. (29.9) that the transconductance is already larger (and thus the subthreshold swing smaller) than the purely electrostatic limit β, and that this extra transconductance can diverge if either the cantilever is very long ($\kappa L \gg 1$), or if the cantilever can be made to collapse abruptly with gate voltage V_G so that $d\theta^*/dV_G \gg 1$.

In an earlier chapter, we worked out the cantilever angle $\theta^*(V_G)$ (Eq. (19.8)) by setting the slope $\partial U/\partial\theta$ to zero, and the destabilization point by also setting the curvature $\partial^2 U/\partial\theta^2$ of the cantilever energy-configuration relation to zero. This process identifies the critical point where the two wells merge and drive the transition of the cantilever from a metastable to a stable well. From Eq. (19.8) we can now extract the slope of the $\theta^*(V_G)$ curve and plug into the normalized transconductance $\bar{g}_m$

$$\bar{g}_m = \beta + 2\kappa\frac{\sqrt{\mu^2\theta_0^2 + U_{eff}\epsilon_{air}AL}}{3U_{eff}(\theta^* - \theta_{destab}^*)(\theta_0 - \theta^*)} \qquad (29.11)$$

where $U_{eff} = U_0^{Bend} + k_{VDW}$, with symbols defined in Eq. (19.6). Note that we can make the transconductance really high by either ramping up the total dipole moment μ in the numerator, or by collapsing the air gap in the denominator at the destabilization point $\theta^* \approx \theta_{destab}^*$. In the first case, we switch with a low voltage because several charges that go on to constitute the dipole are being rotated for the 'price' of one. This is in fact the operational mechanism behind the ultrasteep gating of sodium channels in axonal systems (Fig. 29.3). In the second case, we switch efficiently near the phase transition point where a small voltage input creates

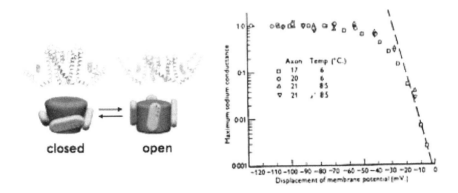

Fig. 29.3   *(Left) Movement of a cantilever protein in an ion channel, upon accumulation of a critical number of charges [X-ray structure of a voltage-dependent K^+ channel, Y. Jiang et al., Nature 423, 33 (2003)]. (Right) The resulting transfer characteristic [Fig from Chapter 20, Ref. 3] shows a subthreshold swing that beats the Boltzmann limit, usually attributed to the correlated movement of three charge units.*

a large output (effective 'gain') through internal amplification. This is the operational mechanism behind Nano Electro Mechanical Field Effect Transistors (NEMFETs), and to some degree, negative capacitance devices that also operate on a similar flipping in curvature $C = \partial^2 U/\partial V_G^2$ of the energy landscape. For NEMFETs, our normalized transconductance reaches infinity, an artifact of the fact that we only included quadratic terms in our potential, and ignored thermal fluctuations around the expected minimum θ^* assuming an infinitely stiff cantilever. We can readily calculate a proper thermally averaged treatment beyond the delta function approximation directly from Eq. (19.1). Straightforward derivation of the Boltzmann averaged transmission gives us

$$(\bar{g}_m)_{\text{conf}} = \beta \underbrace{\left[\frac{\langle \bar{T}_0 \rangle \left\langle \dfrac{\partial U}{\partial V_G} \right\rangle - \left\langle \bar{T}_0 \dfrac{\partial U}{\partial V_G} \right\rangle}{\langle \bar{T}_0 \rangle} \right]}_{C(\theta, V_G)} = \beta C \qquad (29.12)$$

where the term in square brackets is the correlation term $C(\theta, V_G)$ which quantifies the fluctuations around the mean. The correlation term is nonzero because the transmission $\bar{T}$ and the charge $Q = \partial U/\partial V_G$ must be entangled, so that the angle that minimizes U (the stable configuration) must also be the one that maximizes the transmission. Such fluctuations will average out the sharp transition over an angular width $\Delta\theta \approx \sqrt{2k_B T/U''}$

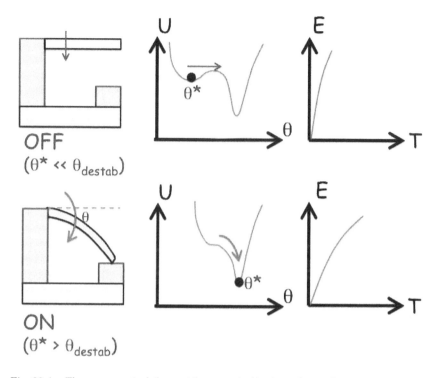

Fig. 29.4 *The movement of the cantilever angle θ^* along the conformational potential landscape leads to a gate tunable transmission amplitude when tunneling is activated at the edge. The gate voltage V_G shifts the metastable local minimum of the first well, until it merges into the global minimum of the electrostatic (capacitive plus Van Der Waals) well and the cantilever shows a conformational phase transition.*

that we had ignored earlier before Eq. (19.1), when we assumed a delta function distribution at low temperature. Near the inflexion point, that approximation breaks down and higher order non-quadratic terms become important. For a long cantilever, $\partial U/\partial V_G$ computes to

$$\frac{\partial U}{\partial V_G} \approx -\frac{\mu\theta + \epsilon_{\text{air}}AV_G}{L(\theta_0 - \theta)} \tag{29.13}$$

but instead of being evaluated at a single θ^* that depends on V_G, we have an independent variable θ that we will ultimately integrate over. The finite integration domain will keep the overall transconductance large but finite. The subthreshold swing has been measured to go down until as little as 7 mV/decade in biological systems and 0.1 mV/decade in mechanical relays.

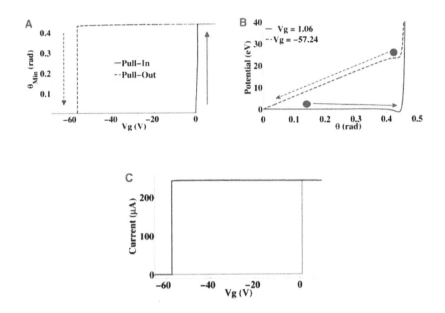

Fig. 29.5   *A. Hysteretic evolution of the cantilever coordinate, B. the conformational potential profile and C. the cantilever tunnel current. While the transitions are steeper than Boltzmann along the walls of the hysteresis curve, the hysteresis compromises the average subthreshold swing.*

After executing the voltage-gated phase transition, the cantilever sticks to the drain. It is the pull in from this adhesion that makes the cantilever angle change abruptly and causes its subthreshold swing to plummet. However, this means that for the reverse cycle, we will need to tear the cantilever away from the drain by catapulting it with a high negative gate bias out of the deep VDW potential back to the weaker parabolic potential. This adhesion creates a strong hysteresis loop during the return from one dominant potential well to two. As a consequence the *average slope* over the loop becomes quite large even though the walls of the loop remain steep (Fig. 29.5). We can capture this hysteresis in our Landauer equation by including a modified distribution $P(\theta, V_G(t))$ to replace the Boltzmann factor $e^{-\beta U}$. The distribution function P describes the nonequilibrium occupancy under a finite scan rate $V_G(t)$ that keeps the cantilever stuck to its metastable state long enough. The probability distribution can be obtained by numerically solving the Fokker Planck equation (Eq. (9.14)), discussed earlier, for a time dependent gate voltage measured until a finite time

$$\gamma \frac{\partial P}{\partial t} = \frac{\partial}{\partial \theta}\left(P\frac{\partial U(\theta, V_G)}{\partial \theta}\right) + k_B T \frac{\partial^2 P}{\partial \theta^2} \qquad (29.14)$$

where γ is the angular damping constant with units J-s. For slow scanning rates we recover the Boltzmann distribution, and the hysteresis — fundamentally a nonequilibrium process, disappears from the I-V.

29.4 Gating bandgap and effective mass: Mott switches and metal-insulator transitions

In the chapter on correlated transport, we saw how the onset of Coulomb charging can open a bandgap in a molecular density of states through blockade, besides precipitating a cascade of excitations. For a solid, the Hubbard approximation showed how the split morphs into a Mott-Hubbard gap. We will discuss how the abrupt onset of such a gap and the associated metal insulator transition (MIT) could possibly help with switching and subthreshold swing. In experiments the observation of a purely electronically driven MIT is complicated by the gateability and the presence of co-existing mechanisms such as thermally driven polarons. Without going into details, let us consider it purely an intellectual exercise to see how an electronic Mott-Hubbard transition, such as in a VO_2 film, can reduce the subthreshold swing. At its simplest, there is an increase in Coulomb screening relative to bandwidth with increased doping or decreasing pressure, leading to an abrupt opening of a gap in a metallic band and creation of a paramagnetic insulator. This correlation dominates for narrow d bands. Depending on whether the repelled lower Hubbard d-band stays highest occupied, vs is shoved below an oxygen 2p-band, we label it a Mott (e.g., VO2) vs a charge transfer insulator (e.g., cuprates). In addition, a Peierls transition is also observed, when the formation of a charge density wave deforms the underlying lattice, and the dimerization opens a gap at the Fermi surface when the latter coincides with the Brillouin zone (such nesting happens only for specific geometries, facilitated in lower dimensions).

Opening a Coulomb gap purportedly kills conductivity in two ways, namely, shutting off the mobility or abruptly depleting the free carrier density. The conventional Mott transition reduces the carrier density exponentially, creating an insulating state. The steepness of the corresponding transfer curve is expected in either case to show a reduced subthreshold swing.

Nanoelectronics: A Molecular View

A 'toy' model for an MIT can be set up readily, qualitatively explaining the reduction in subthreshold swing near the phase transition. For a metal the bandgap is influenced by the Coulomb screening from conduction electrons. The applied gate bias shifts the Fermi energy $E_F(V_G) = E_F + qV_G$, which increases the charge density and screens the Coulomb gap

$$n = n_0 e^{[qV_G - U(n)]/k_B T} \qquad (29.15)$$

Since the gap reduces with gate voltage in addition to the shift in Fermi energy, the net current drawn is higher than the Boltzmann limit.

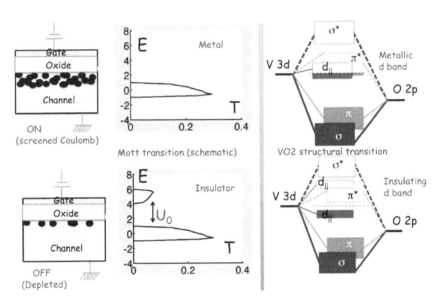

Fig. 29.6 *Gate induced depletion of electrons in the metallic state of a complex oxide decreases Coulomb screening until its bare Coulomb energy exceeds its the band-width, whereupon a Mott transition occurs to an insulating state. Model calculations are performed with a Lorentzian DOS and a self-consistent Coulomb term (see text). The right column shows the proposed energy level diagram in VO_2 with a Mott-Hubbard gap opening in the $d_\parallel$ state (Nagaphani et al., Nature Physics 9, 661666 (2013))*

Let us start with a single atom, modeled as a strongly correlated quantum dot. Such a dot shows pronounced Coulomb Blockade when we add a single charge of a given spin to it. The complementary spin state gets pushed up by Coulomb repulsion from the existing spin, while the latter remains unchanged as it cannot feel a force due to itself. The self interaction correction can be included in a simple Hartree Fock model (Eq. (27.15)), interpolating between the filled charge and empty charge cases with and without repulsion for the one electron Green's function $G = (EI - H)^{-1}$

$$G_\sigma(E) \approx \frac{1}{E - \epsilon_\sigma - U(n)n_{\bar{\sigma}} - \Sigma_\sigma} \quad (29.16)$$

where σ is a particular spin (up or down) and $\bar{\sigma}$ is its compliment. The electron density is obtained self consistently with the density of states

$$n_\sigma = \int D_\sigma(E)f(E - E_F)dE$$

$$D_\sigma(E) = -Im\Big(G_\sigma(E)\Big)/\pi \quad (29.17)$$

As we deplete the electron density with a gate, the Coulomb potential gets underscreened. The exponential Thomas-Fermi screening for a solid (Eq. (4.16)) becomes a Lorentzian in k-space upon Fourier transforming

$$U(n) = \frac{U_0}{1 + a_0^2\kappa^2(n)}, \quad \kappa(n) = (3\pi^2 n)^{1/3} \quad (29.18)$$

where a_0 is the Bohr radius, and $\kappa(n)$ is the screening vector that reduces with reduced electron charge density n (Eq. (13.9)). When the electron overlap becomes smaller than the Bohr radius, $n^{1/3}a_0 \ll 1.19$ and $\kappa a_0 \ll 1$, the unscreened Coulomb potential U_0 creates a gap that exceeds the bandwidth given by $Im\Big(\Sigma_\sigma\Big)$ and a metal insulator transition occurs.

The physics can be extended from a single quantum dot to a solid by treating the latter as an array of coupled quantum dots and replacing $\epsilon_\sigma \to \epsilon_{k\sigma}$ in Eq. (29.16). The Coulomb Blockade now morphs into a Mott insulator with two bands. The sudden removal of spectral weight at the contact Fermi energy leads to a sudden drop in conductivity. The kinetic energy for a free electron gas $3E_F/5 \sim n^{2/3}$ matches the Coulomb repulsion $q^2/a_0 \sim n^{1/3}$ at that critical electron density, leading to the Mott criterion above. The modulation of bandgap increases the mobile charge density sharply and can be directly mapped onto the creation of a correlated condensate, which reduces the subthreshold swing analogous to the mechanism behind an ion channel relay, i.e., many charges moving for the price of one.

Photoemission experiments on V_2O_3 show an added feature beyond the Hubbard gap, namely, a secondary quasiparticle peak near the Fermi energy that keeps narrowing and losing spectral weight to the main Hubbard bands. We alluded to such a Brinkman-Rice mechanism for a half-filled insulator using Gutzwiller projectors (Eq. (25.34)), especially in the limit of infinite connectivity/dimension. The corresponding increase in effective mass m^* from spectral narrowing, $m^*/m = (1 - U/U_C)^2$, slows down the electrons and provides an added mechanism for charge localization (Fig. 29.7). Note that unlike the Hubbard gap where there are no states at the Fermi energy, in the Brinkman Rice transition the state at the Fermi energy stays half filled but infinitely narrow and thus ultimately insulating.

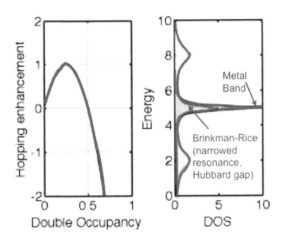

Fig. 29.7 *(Left) Renormalization of hopping with occupancy (Eq. (25.34)). (Right) Change in spectral weight with doping.*

The aim of this chapter was to provide simplified 'toy' models that explain how a gate dependent transmission function leads to an added contribution to current flow that lowers the voltage requirement and ultimately the subthreshold swing. Such gate dependences take varied forms, from bandwidth modulation (Brinkman Rice) to bandgap modulation (Hubbard) to transmission gap modulation (chiral tunnel FETs) to amplitude modulation (NEMFETs) to both bandwidth and amplitude modulation (tunnel FETs). Analogous arguments can be made for devices not covered in this writeup, like polaron mediated switches where a bandgap opens at the Brillouin zone due to dimerization. We can model it using a polaronic self energy in the Born approximation proportional to the density of states shifted by the polaron formation energy (Fig. 24.6, Eq. (24.11)).

Chapter 30

A molecular view of electronics — lessons learned

30.1 Biology's secrets

Much of this book and corresponding wisdom in the semiconductor industry focuses on precise, high throughput Boolean information processing of small chunks of data. This may well run its course with the current slow-down and inevitable ultimate demise of Moore's Law. Industry may still continue to eke out a profit margin through the scaling of size and footprint. But it is time to look at other ways to process information, such as fast, non Boolean techniques to harness Big Data, involving quantum or neuromorphic computing for instance. Compelling proofs-of-concept exist for both. Seth Lloyd argues that our known universe provides a functioning example of a quantum computer (it seems to solve incredibly complicated strong correlation problems in real time), while processes in our bodies provide working templates for neuromorphic computing.

We continue to glean useful lessons from biology, which embodies a very different kind of information processing that has evolved through natural selection over the millenia. Our overall goals — optimizing energy-delay-reliability products under spatial constraints, are analogous, though our needs are different. Our survival as a species relied less on our ability to add large numbers and more on our evolutionary skills for pattern recognition, motion sensing and edge detection — adaptive, contextual pre-processing often done directly at the detection site, in order to quickly distinguish predator from prey. This is enabled by the unique architectural layout, material embodiment, resource allocation and specialized algorithms for biological detection, communication, processing and energy management.

Almost all biological information processing relies on the shuffling around of ions, heavy entities that are not limited by quantum tunneling. As a result, individual molecular sub-units can stack at dimensions a lot closer than possible in solid-state electronics (DNA base pairs are separated by mere Angstroms, compared to oxides in FETs that are bound by a 1 nm effective thickness). Close packing allows efficient space utilization as well as highly branched, fractal like structures. In contrast to a fanout of 3 or 4 in solid state electronics, the brain consists of 100 billion nerve cells, each neuron connecting with another 10,000 (i.e., a fanout of $\sim 10^4$), amounting to about 10-100 trillion synaptic connections firing 5 to 50 times per second. The slower speeds compared to CMOS processors running at GHz speeds are thus offset by the massively parallel architecture of the neural processes. In addition, the packing extends into three dimensions, allowing for efficient heat removal, its power consumed estimated at ~ 10 Watts.

Biological information is often stored in analog form, as the temporal separation between voltage spikes, making that information flow rather inexpensive. Voltage spikes encoded in our action potentials are no more than a couple of $k_B T s$ tall at best. Analog information is readily corruptible with noise, but is compensated by redundancy accorded by the massively parallel architecture. In addition, biological systems make good use of their stochastic environment. Noise allows a system to quickly fall into a basin of attraction, such as identifying a pattern that a set of neurons have trained for. It is believed that non-equilibrium noise is actually utilized in driving ions against chemical gradient in ion pumps — in effect, an ATP powered Brownian ratchet. On the other hand, ion channels open efficiently using subthermal switching mechanisms reminiscent of nanoelectromechanical relay based field effect transistors (NEMFETs).

Scrutiny of cortical processes continues to make deep impacts in the areas of information processing and machine learning. It is however not trivial to translate these bio paradigms easily into their solid-state analogs. Nanorelays create unacceptable amounts of hysteresis in solid-state environments, while ratchets relevant for ion pumps lose their energy advantage once we convert the low power currents into capacitive voltages in order to drive other devices downstream. Significant creativity is needed to figure out how to make good use of the lessons learned from biology for CMOS compatible high performance low power applications.

30.2 Electronics with molecules?

In this book I tried to promote a molecular, 'bottom-up' view of nanoelectronics, one that spans transport physics, quantum chemistry, materials science and device applications. In doing so, I invoked several molecular concepts, from correlation energies to orbital structures. It is worth distinguishing here between two definitions of molecular electronics. One that interests me is how electrons flow in 'molecular' structures - including nanocrystals, dots, tubes and sheets. The other definition, the one that gets technologists and futurists excited, is the possibility of assembling organic molecular components like miniature lego blocks to do designer electronics.

On the issue of actually doing conventional electronics with molecules, technology remains distant at the moment. I personally consider this a secondary issue. The field of 'molecular electronics' has suffered through complications associated with poor yields, thermal floppiness, contact dominated states and overall reproducibility issues. We may not realize our ultimate dream of designer molecular circuits — due to considerable fabrication challenges, but more so because an ultimately scaled organic molecular channel suffers from poor electrostatics and contact dominated states and seemingly offers no obvious advantages vis-a-vis current saturation, power dissipation or speed for digital logic. Not unless we somehow exploit attributes that molecules are actually good at, - utilizing their strong correlation energies, conformational flexibilities, or dynamical stereochemistry to perhaps build entirely new kinds of unconventional logic such as non-Boolean means of Big Data processing or pattern recognition.

On the issue of a molecular view of electronics — how electrons flow through molecules, we have indeed learned a lot. I consider this bottom-up view the most exciting legacy of our scientific explorations over the years. That is, in fact, what this entire book is all about! A molecule serves as a suitable laboratory to explore quantum transport at the nanoscale. Such a 'molecule' can stand in for an organic solar cell, a 2-D transition metal dichalcogenide, a quantum dot, or even a piece of bulk silicon. Transport invariably requires opening up each system to external contacts, ultimately turning these diverse materials into a monolithic continuum, where metallic dots exhibit molecular signatures even as molecules like hydrogen turn metallic. Along this continuum, quantum chemistry, surface physics, solid state electronics and many-body interactions, circuit theory and Feynman diagrams, reaction kinetics and Green's functions all conspire together towards a common goal — watching the resolute electrons jiggle and jostle, pirouette and tunnel as they march forth on their incredible journey.

References

Chapter 2

For the energetics of switching, see

(1) R. E. Landauer, 'Irreversibility and heat generation in the computing process,' IBM Journal of Research and Development 40, pp. 183191 (1961).

(2) C. H. Bennett, 'The thermodynamics of computation — a review', Int. J. Theoretical Physics 21, 905-940 (1982).

(3) C. H. Bennett, 'Logical reversibility of computation,' IBM Journal of Research and Development, vol. 17, no. 6, pp. 525-532 (1973).

(4) R. P. Feynman, *Feynman lectures on computing*, Westview Press; 1 edition (2000).

(5) L. B. Kish, 'End of Moore's Law: thermal (noise) death of integration in micro and nano electronics', Physics Letters A 305, 144-149 (2002).

(6) S. Mukhopadhyay 'Switching Energy in CMOS Logic: How far are we from physical limit?', NanoHUB (https://nanohub.org/resources/1250)

(7) V. V. Zhirnov, R. K. Cavin, J. A. Hutchby and G. I. Bourianoff, 'Limits to binary logic switch scaling — a gedanken model', IEEE 91, 1934 (2003).

(8) V. V. Zhirnov and R. K. Cavin II, *Microsystems for Bioelectronics, 1st Edition: the Nanomorphic Cell,* Elsevier 2010.

(9) G. L. Snider, A. O. Orlov and C. S. Lent, 'Quantum Dots in Nanoelectronic Devices', in *Nano and Molecular Electronics Handbook*, Ed. S. E. Lyshevski, CRC Press, Taylor and Francis Group, 2007.

For the industrial solutions to technology scaling, instead of highlighting specific reviews, I would defer to numerous online resources, which is perhaps ideal given that technology is changing so rapidly (Intel and Micron announced their 3D XPoint memory mere hours before I type this). The economic aspects of Moore's law discussed at the start of the chapter came from a lecture given by Ralph K. Cavin of the Semiconductor Research Corporation. Discussions on ballisticity of transistors follow a lecture given by Daniel Connelly of Synopsis at the Device Research Conference in Santa Barbara in summer 2014. A good book that summarizes the state of technology today, marking the 35th anniversary of the IEEE Electron Devices Society, is *Guide to State-of-the-Art Electron Devices*, Ed. J. N. Burghartz, John Wiley and Sons Ltd. 2013. The book won the 2013 PROSE award.

For the physics of ion channels, see 'Dynamical systems in neuroscience: the geometry of excitability and bursting', EM Izhikevich MIT press (2007).

For logic with dipole coupled magnets, see

(1) J. Atulasimha and S. Bandyopadhyay, 'Bennett clocking of nanomagnetic logic using multiferroic single-domain nanomagnets', Applied Physics Letters 97, 173105 (2011).
(2) S. Fashami, K. Munira, S. Bandyopadhyay, A W Ghosh and J Atulasimha, 'Switching of dipole coupled multiferroic nanomagnets in the presence of thermal noise: reliability analysis of hybrid spintronic-straintronic nanomagnetic logic', IEEE TNano, 12, 1206 (2013).

For perspective on the 'nano' buzz by two Nobel winning physicists, see

(1) H. L. Stormer, 'Silicon forever! Really?' Special Issue: Papers Selected from the 35th European Solid-State Device Research Conference - ESSDERC05, Solid-State Electronics 50, 518 (2006).
(2) H. Kroemer, 'Nano-whatever: Do we really know where we are heading?', Physica Status Solidi (a) 202, 957 (2005).

Both articles are reproduced in *Into The Nano Era: Moore's Law Beyond Planar Silicon CMOS*, edited by Howard Huff, Springer Series in Materials Science, Springer-Verlag, Berlin Heidelberg 2009.

Chapter 3

For an introduction to first quantization, see 'standard' quantum mechanics text books, such as

(1) Claude Cohen-Tannoudji, Bernard Diu and Frank Laloe, *Quantum Mechanics, Vol. 1, Vol. 2*, Wiley-Interscience, 1977.
(2) J. J. Sakurai and J. J. Napolitano, *Modern Quantum Mechanics (2nd Edition)*, Pearson Education Limited 2014.
(3) D. J. Griffiths, *Introduction to Quantum Mechanics (2nd Edition)*, Pearson Education Limited 2014.

Chapter 4

For discussions on quantum chemistry, see

(1) A. Szabo and N. S. Ostlund, *Modern Quantum Chemistry: Introduction to Advanced Electronic Structure Theory* (Dover Books on Chemistry), 1996.
(2) R. P. Feynman, R. B. Leighton and M. Sands, *The Feynman Lectures on Physics, Vol. 3*, Addison Wesley (1971).
(3) E. Madelung. 'Die Mathematischen Hilfsmittel des Physikers', 6th edn., Berlin: Springer, 611 (1950).
(4) D. Pan Wong, 'Theoretical justification of Madelung's rule', J. Chem. Educ., 56, 714 (1979).
(5) L. Pauling, *The Nature of the Chemical Bond*, Cornell University Press 1960.
(6) D. A. McQuarrie, *Statistical Mechanics*, University Science Books 1st edition (May 2000).

Chapter 5

For tight-binding bandstructures of 1-D polymers, organic molecules, 2-D graphene, graphene nanoribbons, bilayer graphene, topological insulators and bulk silicon, see

(1) A. J. Heeger, S. Kivelson, J. R. Schrieffer, W. P. Su, 'Solitons in conducting polymers', Rev. Mod. Phys. 60, 781 (1998).
(2) J.N. Murrell, A.J. Harger, *Semi-Empirical SCF MO Theory of Molecules* (Wiley, New York, 1972.
(3) G-C. Liang, A. W. Ghosh, M. Paulsson and S. Datta, 'Electrostatic potential profiles of molecular conductors', Phys. Rev. B 69, 115302 (2004).

(4) A.H. Castro Neto, F. Guinea, N.M.R. Peres, K.S. Novoselov, A.K. Geim, 'The electronic properties of graphene', Rev. Mod. Phys. 81, 109 (2009).

(5) C. T. White, J. Li, D. Gunlycke and J. W. Mintmire, 'Hidden One-Electron Interactions in Carbon Nanotubes Revealed in Graphene Nanostrips', Nano Lett. 7, 825 (2007).

(6) F. Tseng, D. Unluer, M. R. Stan and A. W. Ghosh, 'Graphene Nanoribbons: From Chemistry to Circuits', in *Graphene Nanoelectronics: Metrology, Synthesis, Properties and Applications*, H. Raza (ed.), Springer Berlin Heidelberg 2012.

(7) E. McCann, D. S. L. Abergel and V. Falko, 'The low energy electronic band structure of bilayer graphene', European Physical Journal - Special Topics. 148, 91 (2007).

(8) X-L Qi and S-C Zhang, 'Topological insulators and superconductors'm Rev. Mod. Phys. 83, 1057 (2011).

(9) M. Z. Hasan and C. L. Kane, 'Colloquium: Topological insulators', Rev. Mod. Phys. 82, 3045 (2010).

(10) N. W. Ashcroft and N. D. Mermin, *Solid State Physics*, Cengage Learning; 1 edition (1976).

(11) P. Vogl, H.P. Hjalmarson, and J. Dow, 'A semi-empirical tight-binding theory of the electronic structure of semiconducors', J. Phys. Chem. Solids 44, 365 (1983).

(12) T. B. Boykin, G. Klimeck and F. Oyafuso, 'Valence band effective-mass expressions in the sp3d5s* empirical tight-binding model applied to a Si and Ge parametrization', Phys. Rev. B 69, 115201 (2004).

(13) Y. M. Niquet, C. Delerue, G. Allan, and M. Lannoo, 'Method for tight-binding parametrization: Application to silicon nanostructures', Phys. Rev. B 62, 5109 (2000).

(14) J. von Neumann and E. Wigner, 'Uber merkwurdige diskrete Eigenwerte', Z. Phys. 30, 467 (1929).

For spin-transistor, the most relevant reference is S. Datta and B. Das, 'Electronic analog of the electro-optic modulator', Appl. Phys. Lett. 56, 665 (1990).

Chapter 6

For discussions on second quantization and the second quantized Hamiltonian, see

(1) G. D. Mahan, *Many-Particle Physics* (Physics of Solids and Liquids), Springer; 3rd ed. (2000).
(2) A. A. Abrikosov, L. P. Gorkov and I. E. Dzyaloazhinski, *Methods of Quantum Field Theory in Statistical Physics* (Dover Books on Physics), 1975.
(3) A. L. Fetter and J. D. Walecka, *Quantum Theory of Many-Particle Systems* (Dover Books on Physics), 2003.
(4) R. D. Mattuck, *A Guide to Feynman Diagrams in the Many-Body Problem* (Dover Books on Physics), 1992.
(5) R. G. Parr and W. Yang, *Density-Functional Theory of Atoms and Molecules* (International Series of Monographs on Chemistry), 1994.

and Ref (1) of Chapter 4 (Szabo and Ostlund). For applicability of Extended H'ueckel Theory to bandstructure and transport, see

(1) D. Kienle, J-I. Cerda and A. W. Ghosh, 'Extended Huckel theory for bandstructure, chemistry and transport. Part I: Carbon Nanotube', J. Appl. Phys. Vol. 100, 043714 (2006).
(2) D. Kienle, K. Bevan, G-C. Liang, L. Siddiqui, J-I. Cerda and A. W. Ghosh, 'Extended Huckel theory for bandstructure, chemistry and transport. Part II: Silicon', J. Appl. Phys. 100, 043715 (2006).
(3) F. Zahid, M. Paulsson, E. Polizzi, A. W. Ghosh and S. Datta, 'A self-consistent transport model for molecular conductors with applications to some real systems', J. Chem. Phys. Vol. 123 , 064707 (2005).

Chapter 7

For discussions on Green's functions

(1) S. Doniach and E. H. Sondheimer, *Green's Functions for Solid State Physicists*, World Scientific Publishing Company (1998).
(2) G. Rickayzen, *Green's Functions and Condensed Matter* (Dover Books on Physics), Dover Publications; Reprint edition (2013).
(3) E. N. Economou, *Green's Functions in Quantum Physics* (Springer Series in Solid-State Sciences), 2006.

Chapter 8

For the role of environmental coupling on dissipation, see

(1) A. Caldeira and A. J. Leggett, 'Influence of dissipation on quantum tunneling in macroscopic systems, Phys. Rev. Lett., 46, 211 (1981).
(2) R. P. Feynman and F. L. Vernon, 'The Theory of a General Quantum System Interacting with a Linear Dissipative System', Ann. Phys. (N. Y.) 24, 118 (1963).

For an overview of transport in the semiclassical and quantum limits, see

(1) M. Lundstrom and C. Jeong, *Near-equilibrium Transport: Fundamentals and Applications, in Lessons from Nanoscience: A Lecture Notes Series: Volume 2*, World Scientific, 2013.
(2) S. Datta, *Lessons from Nanoelectronics: A New Perspective on Transport*, World Scientific, 2012.

For the distinction between contacts and channel, see Liang *et al.*, reference from Chapter 5, PRB 2004.

Chapter 9

For classical transport, thermodynamics, the Langevin and Fokker-Planck equations, and colored noise

(1) H. B. Callen, *Thermodynamics and an Introduction to Thermostatistics*, Wiley; 2 edition (1985).
(2) C.W Gardiner, *Handbook of Stochastic Methods for Physics, Chemistry and the Natural Sciences*, Second Edition, Springer-Verlag 2004.
(3) H. Risken, *The Fokker-Planck Equation: Methods of Solution and Applications* (Springer Series in Synergetics), 1996.
(4) W. Horsthemke and R. Lefever, *Noise-Induced Transitions*, Springer Series in Synergetics (1984).
(5) F. Moss and P. V. E. McClintock (eds.), *Noise in Nonlinear Dynamical Systems. Vol. 1: Theory of Continuous Fokker-Planck Systems. Vol. 2: Theory of Noise Induced Processes in Special Applications. Vol. 3: Experiments and Stimulations.* Cambridge etc., Cambridge University Press 1989.
(6) A. W. Ghosh and S. V. Khare, 'Rotation in an Asymmetric Multidimensional Periodic Potential due to Colored Noise', Phys. Rev. Lett. 84, 5243 (2000).

Chapter 10

For the transition from classical to semi-classical transport, see

(1) R. F. Pierret, *Semiconductor Device Fundamentals*, Addison Wesley; 2nd edition (1996).
(2) M. Lundstrom, *Fundamentals of Carrier Transport*, Cambridge University Press; 2 edition (2009).
(3) C. Cercignani, *The Boltzmann Equation and Its Applications*, Applied Mathematical Sciences 67 (1988).
(4) C. Cercignani and E. Gabetta, *Transport Phenomena and Kinetic Theory: Applications to Gases, Semiconductors, Photons, and Biological Systems* , Modeling and Simulation in Science, Engineering and Technology (2007).
(5) For Gibbs-Duhem, see H. B. Callen, reference from Chapter 9.

Chapter 11

For references on quantum transport, see

(1) S. Datta, *Electronic Transport in Mesoscopic Systems* (Cambridge Studies in Semiconductor Physics and Microelectronic Engineering), Cambridge University Press (1997).
(2) S. Datta, *Quantum Transport: Atom to Transistor*, Cambridge University Press; 2nd edition (2005).
(3) S. Datta, 'Steady-state quantum kinetic equation', Phys. Rev. B 40, 5830(R) (1989).
(4) D. K. Ferry, S. M. Goodnick and J. Bird, *Transport in Nanostructures*, Cambridge University Press; 2 edition (2009).
(5) J. Rammer and H. Smith, 'Quantum field-theoretical models in transport theory of metals', Rev. Mod. Phys. 58, 323 (1986).
(6) G. Stefanucci and R. van Leeuwen, *Nonequilibrium Many-Body Theory of Quantum Systems: A Modern Introduction*, Cambridge University Press; 1 edition (2013).
(7) H. Haug and A.-P. Jauho, *Quantum Kinetics in Transport and Optics of Semiconductors*, Springer Series in Solid-State-Sciences, 1996.
(8) H. Haug and S. W. Koch, *Quantum Theory of the Optical and Electronic Properties of Semiconductors*, World Scientific 1990.
(9) L. V. Keldysh, 'Diagram Technique for Nonequilibrium Processes', Zh.

Eksp. Teor. Fiz. 91, 1815 (1986) [Sov. Phys. JETP 64 1075 (1986)].

(10) L. P. Kadanoff and G. Baym, *Quantum Statistical Mechanicss: Greens Function Methods in Equilibrium and Nonequilibrium Problems* (Benjamin, New York, 1962).

(11) P. Danielewicz, ' Quantum Theory of Nonequilibrium Processes,' Ann. Phys. (N. Y.) 152, 239 (1984).

(12) Y. Meir and N. S. Wingreen, 'Landauer formula for the current through an interacting electron region', Phys. Rev. Lett. 68, 2612 (1992).

(13) D. S. Fisher and P. A. Lee, 'Relation between conductivity and transmission matrix', Phys. Rev. B 23, 6951 (1981).

(14) F. Sols, 'Scattering, dissipation, and transport in mesoscopic systems', Ann. Phys. (N.Y.) 214, 386 (1992).

(15) Y. Imry, *Introduction to Mesoscopic Physics* (Mesoscopic Physics and Nanotechnology), Oxford University Press; 2 edition (2008).

(16) T. T. Heikkilä, *The Physics of Nanoelectronics: Transport and Fluctuation Phenomena at Low Temperatures*, Oxford Master Series in Condensed Matter Physics 2013.

(17) For time-dependent NEGF, see S. Vasudevan, K. Walczak and A. W. Ghosh, 'Coupling optical and electrical gating for electronic readout of quantum dot dynamics', Phys. Rev. B 82, 085324 (2010).

Chapter 12

For the Landauer approach we suggest the references from Chapter 11 (e.g. books by Datta, Ferry, Haug and Jauho). For finite temperature/voltage effects and thermal conductance quantum we suggest

(1) P. F. Bagwell and T. P. Orlando, 'Landauers conductance formula and its generalization to finite voltages', Phys. Rev. B 40, 1456 (1989).

(2) L. G. C. Rego and G. Kirczenow, 'Quantized Thermal Conductance of Dielectric Quantum Wires', Phys. Rev. Lett. 81, 232 (1998).

(3) K. Schwab, E. A. Henriksen, J. M. Worlock and M. L. Roukes, 'Measurement of the quantum of thermal conductance', Nature 404, 974 (2000).

(4) J. Callaway, 'Model for Lattice Thermal Conductivity at Low Temperatures', Phys. Rev. 113, 1046 (1959).

Chapter 13

For the Drude limit, see the book by Datta (Last reference in Chapter 8). For transport with hopping, see

(1) N. Mott, *Conduction in Non-Crystalline Materials*, Oxford Science Publications 1987.
(2) P. Stallinga, 'Electronic Transport in Organic Materials: Comparison of Band Theory with Percolation/(Variable Range) Hopping Theory', Adv. Mater. 23, 3356 (2011).

Chapter 14

For treatments of tunneling, see

(1) C. B. Duke, *Tunneling in Solids*, Academic, New York (1969).
(2) M. V. Berry and K. E. Mount, 'Semiclassical approximations in wave mechanics', Rep. Prof. Phys. 35, 315 (1972).
(3) J. G. Simmons. 'Generalized Formula for the Electric Tunnel Effect between Similar Electrodes Separated by a Thin Insulating Film', J. Appl. Phys. 34, 1793 (1963).
(4) K. Gloos, P. J. Koppinen and J P Pekola, 'Properties of native ultrathin aluminium oxide tunnel barriers', J. Phys.: Condens. Matter 15, 1733 (2003), Fig. 9.
(5) L. D. Landau and I. M. Lifshitz, *Quantum Mechanics, Third Edition: Non-Relativistic Theory (Volume 3)*, Butterworth-Heinemann; 3 edition (1981).
(6) H. B Akkerman and B. de Boer, 'Electrical conduction through single molecules and self-assembled monolayers', J. Phys.: Condens. Matter 20, 013001 (2008).

Chapter 15

For references on tunnel magnetoresistance, symmetry filtering, and chiral/Klein tunneling, see

(1) Y. Xie, K. Munira, I. Rungger, M. T. Stemanova, S. Sanvito and A.W. Ghosh, 'Spin Transfer Torque: A Multiscale Picture', invited book chapter, Nanomagnetic and Spintronic Devices for Energy Efficient

Memory and Computing, ed. S. Bandyopadhyay and J. Atulasimha, Wiley 2016.

(2) W. H. Butler, X.-G. Zhang, T. C. Schulthess, and J. M. MacLaren, 'Spin-dependent tunneling conductance of Fe—MgO—Fe sandwiches', Phys. Rev. B 63, 054416 (2001).

(3) W. H. Butler, 'Tunneling magnetoresistance from a symmetry filtering effect', Sci. Technol. Adv. Mater. 9, 014106 (2008).

(4) R. Sajjad and A. W. Ghosh, 'Manipulating chiral transmission with gate geometry: switching with graphene with transmission gaps', ACS Nano 7, 9808 (2013).

(5) R. N. Sajjad, S. Sutar, J. Lee and A. W. Ghosh, 'Manifestation of Chiral tunneling in tilted graphene pn junction', Phys. Rev. B 86, 155412 (2012).

(6) R. Sajjad and A. W. Ghosh, 'High efficiency switching using graphene based electron optics', Appl. Phys. Lett. Vol. 99, 123101 (2011).

(7) S. Chen, Z. Han, M. M. Elahi, K. M. M. Habib, L. Wang, B. Wen, Y. gao, T. Taniguchi, K. Watanabe, J. Hone, A. W. Ghosh and C. R. Dean,'Electron optics with ballistic graphene junctions', arXiv:1602.08182.

(8) U. Sivan, M. Heiblum, C. P. Umbach and H. Shtrikman, 'Electrostatic electron lens in the ballistic regime', Phys. Rev. B 41, 7937 (1990).

(9) V. G. Veselago, 'The electrodynamics of substances with simultaneously negative values of ϵ and μ', Sov. Phys. Usp. 10, 509 (1968).

(10) V. V. Cheianov, V. Fal'ko and B. L. Altshuler, 'The Focusing of Electron Flow and a Veselago Lens in Graphene p-n Junctions', Science 315, 1252 (2007).

Chapter 16

For I-V of graphene as well as papers on conductivity quantization and tunneling across graphene, see

(1) P. N. First, W. A. DeHeer, T. Seyller, C. Berger, J. A. Stroscio, J. Moon, 'Epitaxial Graphenes on Silicon Carbide', MRS Bulletin 35, 296 (2010).

(2) S. Adam, E. H. Hwang, V. M. Galitski and S. Das Sarma, 'A self-consistent theory for graphene transport', PNAS 104, 18392 (2007).

(3) M. I. Katsnelson, K. S. Novoselov and A. K. Geim, 'Chiral tunnelling and the Klein paradox in graphene', Nat. Phys. 2, 620 (2006).

Chapter 17

For references on the derivative discontinuity and Density Functional Theory, see Parr-Yang reference from Ch. 6.

For I-Vs of molecules, see

(1) A. Aviram and M. A. Ratner, 'Molecular rectifiers', Chem. Phys. Lett. 29, 277 (1974).
(2) X. Xiao, B. Xu and N. J. Tao, 'Measurement of Single Molecule Conductance: Benzenedithiol and Benzenedimethanethiol', Nano Lett. 4, 267 (2004).
(3) A. W. Ghosh, P. S. Damle, S. Datta and A. Nitzan, 'Molecular Electronic Devices', MRS Bull. 29, 391 (2004) (Special issue on Molecular Transport Junctions).
(4) P. S. Damle, A. W. Ghosh and S. Datta, 'First-Principles Analysis of Molecular Conduction Using Quantum Chemistry Software', Chem. Phys. 281, 171 (2002) (Special issue on Molecular Nanoelectronics, Ed. Mark Ratner).
(5) P. S. Damle, A. W. Ghosh and S. Datta, 'Unified description of molecular conduction: from molecules to metallic wires', Phys. Rev. B 64 Rapid Comms., 201403 (R) (2001).

For interface properties, see

(1) A. M. Cowley and S. M. Sze, 'Surface States and Barrier Height of Metal-Semiconductor Systems', J. Appl. Phys. 36, 3212 (1966).
(2) E. Abad, J. Ortega and F. Flores, 'Metal/organic barrier formation for a C_{60}/Au interface: from the molecular to the monolayer limit', Phys. Stat. Solid A 209, 636 (2012).
(3) N. Koch, N. Ueno, A. T. S. Wee, *The Molecule-Metal Interface*, Wiley, 2013.
(4) R. T. Tung, 'Recent advances in Schottky barrier concepts', Mat. Sci. and Eng. R 35, 1 (2001).
(5) F Garcia-Moliner and F Flores, 'Theory of electronic surface states in semiconductors', J. Phys. C 9, 1609 (1976).
(6) Prashant S. Damle PhD. thesis, Purdue University.

Chapter 18

For references on electrical gating, see the book by Pierret (Ref. 1 of Chapter 10). In addition,

(1) S. M. Sze and K. K. Ng, *Physics of Semiconductor Devices*, 3rd Edition, Wiley-Interscience, 2007.

(2) M. V. Fischetti, D. A. Neumayer, E. A. Cartier, 'Effective electron mobility in Si inversion layers in metal-oxide-semiconductor systems with a high-kappa insulator: The role of remote phonon scattering', J. Appl. Phys. 90, 4587 (2001).

(3) D.M. Caughey and R.E. Thomas. 'Carrier Mobilities in Silicon Empirically Related to Doping and Field', Proc. IEEE 55, 2192 (1967).

(4) S. Takagi, 'On the Universality of Inversion Layer Mobility in Si MOSFET's: Part I-Effects of Substrate Impurity Concentration', IEEE Trans. El. Dev. 41, 2357 (1994).

(5) S. Takagi, A. Toriumi, M. Iwase and H. Tango, 'On the Universality of Inversion Layer Mobility in Si MOSFETs: Part 11-Effects of Surface Orientation', IEEE Trans. El. Dev. 41, 2363 (1994).

(6) A. Rahman, A. Ghosh and M. Lundstrom, 'Generalized effective mass approach for n-type metal-oxide-semiconductor field-effect transistors on arbitrarily oriented wafers', J. Appl. Phys. 97, 053702 (2005).

(7) A. Rahman, M. Lundstrom and A. W. Ghosh, 'Effective Mass Approach for n-MOSFETs on arbitrarily oriented wafers', J. Comp. El. 3, 281 (2004).

(8) A. W. Ghosh, T. Rakshit and S. Datta, 'Gating of molecular transistors: electrostatic and conformational', Nano Lett. 4, 565 (2004).

(9) J.-O. Lee, G. Lientschnig, F. Wiertz, M. Struikj, R. A. J. Jansen, R. Egberink, D. N. Reinhoudt, P. Hadley, C. Dekker, 'Absence of strong gating effect in electrical measurements on phenylene-based conjugated molecules', Nano Lett. 3, 113 (2003).

(10) C. R. Kagan, A. Afzali, R. Martel, L. M. Gignac, P. M. Solomon, A. Schrott, B. Ek, 'Evaluations and considerations for self-assembled monolayer field-effect transistors', Nano Lett. 3, 119 (2003).

(11) M Lundstrom, 'Elementary scattering theory of the Si MOSFET', IEEE El. Dev. Lett. 18, 361 (1997).

(12) *Nanoscale Transistors: Device Physics, Modeling and Simulation*, M. S. Lundstrom and J. Guo, Springer Science and Business Media, Jun 18, 2006, and references therein.

(13) A Rahman, J Guo, S Datta, MS Lundstrom, 'Theory of ballistic nan-

otransistors', IEEE Transactions on Electron Devices 50, 1853 (2003).
(14) K. Natori, 'Ballistic metal-oxide-semiconductor field effect transistor', J. Appl. Phys. 76, 4879 (1994).

Chapter 19

For solid-state relays for low-power electronics, see

(1) H. Kam, D. Lee, R. Howe, and T.-J. King, 'A new nano-electromechanical field effect transistor (nemfet) design for low-power electronics', in IEDM Technical Digest 463, Electron Devices Meeting, 2005.
(2) N. Abele, R. Fritschi, K. Boucart, F. Casset, P. Ancey, and A. Ionescu, 'Suspended-gate mosfet: bringing new mems functionality into solid-state mos transistor', in IEDM Technical Digest 479, Electron Devices Meeting, 2005.

For relays in action potentials, see

(1) D. A. Doyle, J. M. Cabral, R. A. Pfuetzner, A. Kuo, J. M. Gulbis, S. L. Cohen, B. T. Chait, and R. MacKinnon, 'The structure of the potassium channel: Molecular basis of k+ conduction and selectivity', Science 280, 69 (1998).
(2) Y. Jiang, A. Lee, J. Chen, V. Ruta, M. Cadene, B. T. Chait, and R. MacKinnon, 'X-ray structure of a voltage-dependent k+ channel', Nature 423, 33 (2003).

Chapter 20

For references on GOTO pair, ionic gating and action potentials, and random telegraph noise, see

(1) E. Goto, K. Murata, K. Nakazawa, K. Nakagawa, t. Moto-Oka, Y. Matsuoka, Y. Ishibashi, T. Soma, E. Wada, 'Esaki diode high-speed logical circuits', IRE Trans. Electr. Comp. EC-9, 25 (1960).
(2) E. I. Izhikevich, *Dynamical Systems in Neuroscience: The Geometry of Excitability and Bursting* (Computational Neuroscience), The MIT Press (2010).
(3) A. L. Hodgkin, A. F. Huxley, 'Currents carried by sodium and potassium ions through the membrance of the giant axon of loligo', J. Physiol. 116, 449 (1952).

(4) H. Haken, *Brain Dynamics: Synchronization and Activity Patterns in Pulse-Coupled Neural Nets with Delays and Noise*, Springer Verlag 2007.

(5) D. B. Strukov, G. Snider, D. R. Stewart, and R. S. Williams, 'The missing memristor found', Nature 453, 80 (2008).

(6) D. B. Strukov, J. L. Borghetti, and R. S. Williams, 'Coupled ionic and electronic transport model of thin-film semiconductor memristive behavior', Small 5, 1058 (2009).

(7) Y. N. Joglekar and S. J. Wolf, 'The elusive memristor: properties of basic electrical circuits', Eur. J. Phys. 30, 661 (2009).

(8) *Advances in Neuromorphic Memristor Science and Applications*, Ed. R. Kozma, R. E. Pino and G. E. Pazienza, Springer Series in Cognitive and Neural Systems 2012.

(9) *Memristors and Memristive Systems*, Ed. R. Tetzlaff, Springer 2014.

(10) J. J. Yang, D. B. Strukov and D. R. Stewart, 'Memristive devices for computing', Nat. Nano 8, 13 (2013).

(11) M. Xiao, I. Martin, E. Yablonovitch and H. W. Jiang, 'Electrical detection of the spin resonance of a single electron in a silicon field-effect transistor', Nature 430, 435 (2004).

(12) G. Dearnaley, A. M. Stoneham and D. V. Morgan, 'Electrical phenomena in amorphous oxide films', Rep. Prog. Phys. 33, 1129 (1970).

(13) J. G. Simmons and R. R. Verderber, 'New conduction and reversible memory phenomena in thin insulating films', Proc. Roy. Soc. A 301, 77 (1967).

(14) J. Chan, B. Burke, K. Evans, K. A. Williams, S. Vasudevan, M. Liu, J. Campbell and A. W. Ghosh, 'Reversal of current blockade through multiple trap correlations', Phys. Rev. B 80, 033402, (2009).

(15) S. Vasudevan and A. W. Ghosh, 'Using room temperature current noise to characterize single molecular spectra', ACS Nano 8, 2111 (2014).

Chapter 21

There are numerous monographs on magnetoconductance and Quantum Hall effect, which is now a textbook material. See for instance, D. Yoshioka, *Quantum Hall Effect*, Springer 2002; R. E. Prange and S. M. Girvin, *The Quantum Hall Effect*, Springer 2012.

For spintronic, magnetoelectronic devices and torque, look at the reference by Xie *et al.* in Chapter 15, the references on straintronics from the Bandyopadhyay group in Chapter 1, and also

(1) J.C. Slonczewski, 'Current-driven excitation of magnetic multilayers', Journal of Magnetism and Magnetic Materials 159, L1L7 (1996).
(2) L. Berger, 'Emission of spin waves by a magnetic multilayer traversed by a current', Phys. Rev. B 54, 9353 (1996).
(3) K. Munira, W.H. Butler and A. W. Ghosh, 'A Quasi-Analytical Model for Energy-Delay-Reliability Tradeoff Studies During Write Operations in a Perpendicular STT-RAM Cell', IEEE Tran. El. Dev. 59, 2221 (2012).
(4) K. M. Habib, R. N. Sajjad and A. W. Ghosh, 'Chiral Tunneling of Topological States: Towards the Efficient Generation of Spin Current Using Spin-Momentum Locking', Phys. Rev. Lett. 114, 176801 (2015).
(5) B. Behin-Aein, D. Datta, S. Salahuddin, S. Datta, 'Proposal for an all-spin logic device with built-in memory', Nature Nano 5, 266 (2010).
(6) E. I. Rashba, 'Theory of electrical spin injection: Tunnel contacts as a solution of the conductivity mismatch problem', Phys. Rev. B 62, R16267(R) (2000).
(7) B. E. Kane, 'A silicon-based nuclear spin quantum computer', Nature 393, 133 (1998).
(8) *Spintronics: From Materials to Devices*, edited by Claudia Felser, Gerhard H Fecher, Springer Science and Business Media, March 2013.
(9) L. Liu, C-F. Pai, Y. Li, H. W. Tseng, D. C. Ralph and R. A. Buhrman, 'Spin-torque switching with the giant Spin Hall effect of Tantalum', Science 336, 555 (2012).
(10) A. Pushp, T. Phung, C. Rettner, B. P. Hughes, S-H. Yang and S. S. P Parkin, 'Giant thermal spin-torque assisted magnetic tunnel junction switching', Proc. Nat. Acad. Sci. 112, 6585 (2015).

Chapter 22

For references on scattering, see

(1) E. T. Swartz and R. O. Pohl, 'Thermal boundary resistance', Rev. Mod. Phys. 61, 605 (1989).
(2) C. W. J. Beenakker and H. van Houten, 'Quantum transport in semiconductor nanostructures', Solid State Physics 44, 1 (1991).

(3) M. Büttiker, 'Role of quantum coherence in series resistors', Phys. Rev. B 33, 3020 (1986).

(4) J. L. DAmato and H. M. Pastawski, 'Conductance of a disordered linear chain including inelastic scattering events', Phys. Rev. B 41, 7411 (1990). (1990).

(5) P. G. Klemens, D. F. Pedraza, 'Thermal conductivity of graphite in the basal plane', 547, 735 (1994).

(6) J. L. D'Amato and H. M. Pastawski, 'Conductance of a disordered linear chain including inelastic scattering events', Phys. Rev. B 41, 741 (1990).

(7) P. W. Anderson, D. J. Thouless, E. Abrahams and D. S. Fisher, 'New method for a scaling theory of localization', Phys. Rev. B 22, 3519 (1980).

(8) V. Mujica, M. Kemp, M.A. Ratner, 'Electron conduction in molecular wires. I. A scattering formalism', J. Chem. Phys. 101, 68496855 (1994).

Chapters 23 and 24

For concepts on scattering, see Maximilian Schollhauer, *Decoherence and the Quantum-to-Classical Transition*, Springer 2008.

For graphene conductivity, see references in Chapter 16.

For references on vibrational scattering, see

(1) R. Golizadeh-Mojarad and S. Datta, 'Nonequilibrium Greens function based models for dephasing in quantum transport', Phys. Rev. B 75, 081301(R) (2007).

(2) L. Siddiqui, A. W. Ghosh and S. Datta, 'Phonon runaway in nanotube quantum dots', Phys. Rev. B 75, 085433 (2007).

(3) N. A. Zimbovskaya, *Transport Properties of Molecular Junctions*, Springer Tracts in Modern Physics (Book 254), Springer; 2013 edition (2013).

(4) A. C. Johnson, C. M. Marcus, M. P. Hanson and A. C. Gossard, 'Coulomb Modified Fano-Resonance in a One-Lead Quantum Dot', Phys. Rev. Lett. 93, 106803 (2004).

(5) A. Troisi and M. A. Ratner, 'Modeling the inelastic electron tunneling

spectra of molecular wire junctions', Phys. Rev. B. 72, 03340 (2005).

(6) M. Paulsson, T. Frederiksen, H. Ueba, N. Lorente and M. Brandbyge, 'Unified Description of Inelastic Propensity Rules for Electron Transport through Nanoscale Junctions', Phys. Rev. Lett. 100, 226604 (2008).

(7) Y-C. Chen, M. Zwolak and M. Di Ventra, 'Inelastic current-voltage characteristics of atomic and molecular junctions', Nano Lett., 4, 1709 (2004).

(8) R. N. Sajjad, F. Tseng, K. M. M. Habib, A. W. Ghosh, 'Quantum transport at the Dirac point: Mapping out the minimum conductivity from pristine to disordered graphene', Phys. Rev. B 92, 205408 (2015).

(9) W. Wang, T. Lee, I. Kretzschmar and M. A. Reed, 'Inelastic Electron Tunneling Spectroscopy of an Alkanedithiol Self-Assembled Monolayer', Nano Lett., 4, 643 (2004).

(10) P. K. Tien, J. P. Gordon, 'Multiphoton process observed in the interaction of microwave fields with the tunneling between superconductor films', Phys. Rev. 129, 647 (1963).

(11) J. Tucker, 'Quantum limited detection in tunnel junction mixers', IEEE Journal of Quantum Electronics 15, 1234 (1979).

(12) A. W. Ghosh, L. Jönsson and J. W. Wilkins, 'Bloch oscillations in the presence of plasmons and phonons', Phys. Rev. Lett. 85, 1084 (2000).

For spin scattering, see F. Meier and B. P. Zakharchenya, *Optical Orientation*, North-Holland 1984.

Chapters 25, 26

For correlation effects at equilibrium, see

(1) P. Fulde, *Electron Correlations in Molecular and Solids*, (Springer Series on Solid-State Sciences), 1995.

(2) A. C. Hewson, *The Kondo problem to Heavy Fermions*, Cambridge University Press 1993.

For transport with correlation, see

(1) B. Muralidharan, L. Siddiqui and A. W. Ghosh, 'The role of Many Particle Excitations in Coulomb Blockaded Transport', J. Phys. Cond. Mat. 20, 374109 (2008).

(2) B Muralidharan, A. W. Ghosh, S. K. Pati and S. Datta, 'Theory of

high bias Coulomb Blockade in ultrashort molecules', IEEE-TNT 6, 536 (2007).

(3) B. Muralidharan, A. W. Ghosh and S. Datta, Molecular Simulation (Special Issue ed. D. Beratan), 'Conductance in molecular quantum dots – fingerprints of wave-particle duality?', 32, 751 (2006).

(4) B. Muralidharan, A. W. Ghosh and S. Datta, 'Probing electronic excitations in molecular conduction', Phys. Rev. B 73, 155410 (2006).

(5) *Single Charge Tunneling*, Ed. H. Grabert adn M. H. Devoret, NATO ASI series 294, Plenum Press, New York.

(6) O. D. Miller, B. Muralidharan, N. Kapur and A. W. Ghosh, 'Rectification by charging: Contact-induced current asymmetry in molecular conductors', Phys. Rev. B. 77, 125427 (2008).

(7) C. W. J Beenakker, 'Theory of Coulomb-blockade oscillations in the conductance of a quantum dot' Phys. Rev. B. 44, 1646 (1991).

(8) Bakkers FPAM, Hans Z, Zunger A, et al. 'Shell tunneling spectroscopy of the single-particle energy levels of insulating quantum dots' Nano Letters 1: 551.

Chapter 27

For ways to deal with correlation in transport, see

(1) J. K. Freericks, *Transport in Multilayered Nanostructures: The Dynamical Mean-field Theory Approach*, Imperial College Press, 2006.

(2) A. Georges, G. Kotliar, W. Krauth, and M. J. Rozenberg, 'Dynamical mean-field theory of strongly correlated fermion systems and the limit of infinite dimensions', Rev. Mod. Phys. 68 (1996).

(3) F. Gebhard, *The Mott Metal-Insulator Transition: Models and Methods* (Springer Tracts in Modern Physics), Springer; 1997 edition (2013).

(4) Y. Meir, N. S. Wingreen and P. A. Lee, 'Transport through a strongly interacting electron system: Theory of periodic conductance oscillations', Phys. Rev. Lett. 66, 3048 (1991).

(5) M. Imada, A. Fujimori and Y. Tokura, 'Metal-insulator transitions', Rev. Mod. Phys. 70, 1039 (1998).

(6) N. H. March, *Electron Correlations in Molecules and Condensed Phases*, Physics of Solids and Liquids, Plenum Press, New York 1996.

(7) A. Levy Yeyati, F. Flores, and A. Martn-Rodero, 'Transport in Multilevel Quantum Dots: From the Kondo Effect to the Coulomb Blockade Regime', Phys. Rev. Lett. 83, 600 (1999).

(8) N. Tsuda, K. Nasu, A. Yanase, K. Siratori, *Electronic Conduction in Oxides*, Springer Series in Solid-State Sciences, Springer-Verlag Berlin Heidelberg 1990.

(9) N. A. Zimbovskaya, 'Electron transport through a quantum dot in the Coulomb blockade regime: Nonequilibrium Green's function based model', Phys. Rev. B 78, 035331 (2008).

(10) B. Muralidharan and S. Datta, 'Generic model for current collapse in spin-blockaded transport', Phys. Rev. B 76, 035432 (2007).

(11) W. F. Brinkman and T. M. Rice, 'Application of Gutzwiller's Variational Method to the Metal-Insulator Transition', Phys. Rev. B 2, 4302 (1970).

(12) F. Zahid, A. W. Ghosh, M. Paulsson, E. Polizzi, and S. Datta, 'Charging-induced asymmetry in molecular conductors', Phys. Rev. B 70, 245317 (2004).

Experimental data are from

(1) J. Park, A. N. Pasupathy, J. I. Goldsmith, C. Chang, Y. Yaish, J.R. Petta, M. Rinkovski, J. P. Sethna, H. D. Abruna, P. L. McEuen, and D. C. Ralph, 'Coulomb blockade and the Kondo effect in single-atom transistors', Nature 417, 722 (2002).

(2) N. B. Zhitenev, H. Meng, and Z. Bao, 'Conductance of Small Molecular Junctions', Phys. Rev. Lett. 88, 226801 (2002).

Chapters 29, 30

For references on ultra low-power switching, see

(1) A. Seabaugh, 'The tunneling transistor,' IEEE Spectrum, 50, 35 (2013)

(2) 'Emerging Devices', S. Bandyopadhyay, M. Cahay and A. W. Ghosh, Chapter 5, pps 59-69 in *Electron Devices: An Overview by the Technical Area Committee of the IEEE Electron Devices Society*, Ed. Joachim Burghartz, John Wiley and Sons Ltd, 2013.

(3) A. W. Ghosh, 'Transmission engineering as a route to subthermal switching', IEEE JEDS 3, 135 (2015).

(4) O. M. Nayfeh, C. N. Chleirigh, J. Hennessy, L. Gomez, 'Design of Tunneling Field-Effect Transistors Using Strained-Silicon/Strained-Germanium Type-II Staggered Heterojunctions', IEEE EDL 29, 1074 (2008).

(5) U. Zaghloul, G. Piazza, 'Sub-1-volt Piezoelectric Nano- electromechanical Relays With Millivolt Switching Capability', IEEE Electron Dev. Lett., 35, 669 (2014).

(6) A. M. Ionescu, and H. Riel, 'Tunnel field-effect transistors as energy-efficient electronic switches', Nature 479, 329 (2011).

(7) A. Declan Doyle, J. M. Cabral, R. A. Pfuetner, A. Kuo, J. M. Gulbis, L. Steven, B. T. Chait and R. Mackinnon, 'The Structure of the Potassium Channel: Molecular Basis of K+ Conduction and Selectivity', Science 280, 69 (1998).

(8) N. Shukla, A. V. Thathachary, A. Agrawal, H. Paik, A. Aziz, D. G. Schlom, S. K. Gupta, R. Engel-Herbert and S. Datta, 'A steep-slope transistor based on abrupt electronic phase transition', Nat. Comm. 6, 7812 (2015).

(9) S. Salahuddin and S. Datta, 'Interacting systems for self correcting low power switching', Appl. Phys. Lett. 90, 093503 (2007).

(10) Y. Zhou, and S. Ramanathan, 'Correlated electron materials and field effect transistors for logic: a review', Crit. Rev. Solid State Mater. Sci. 38, 286 (2013).

(11) W. Haensch, E. J Nowak, R. H Dennard, P. M Solomon, A. Bryant, O. H Dokumaci, A. Kumar, X. Wang, J. B Johnson, M. V Fischetti, 'Silicon CMOS devices beyond scaling' IBM Journal of Research and Development 50, 339 (2006).

(12) R. K. Cavin, P. Lugli, and V. V. Zhirnov, 'Science and engineering beyond Moores law', Proc. IEEE 100, 1720 (2012).

(13) I. L. Markov, 'Limits on fundamental limits to computation', Nature 512, 147 (2014).

(14) http://www.human-memory.net/brain_neurons.html

(15) *The Universe Within* by Morton Hunt (Simon and Schuster, 1982).

(16) *Noise*, B. Kosko, Viking Penguin 2006.

(17) *Vision: Coding and Efficiency*, Ed. C. Blakemore, Cambridge University Press 1990.

(18) *Information Theory and the Brain*, Ed. R. Baddeley, P. Hancock, P. Földiák, Cambridge University Press 2008.

(19) 'Heterogeneity and Efficiency in the Brain', V. Balasubramanian, Prof. IEEE 103, 1346 (2015).

(20) 'Energy Efficient Neural Codes', W. B. Levy and R. A. Baxter, Neural Computation 8, 531 (1996).

(21) 'Energy Limits to the Computational Power of the Human Brain', R. C. Merkle, Foresight Update No. 6, Aug 1989.

(22) 'Computing with Nonequilibrium Ratchets', M. Kabir, D. Unluer, L. Li, A. W. Ghosh and M. R. Stan, IEEE TNano 12, 330 (2013).

(23) 'ATP synthase: two motors, two fuels', G. Oster and H. Wang, Structure 7, R67, 1999.

Index

Printed in the United States
By Bookmasters